TOWN AND INFRASTRUCTURE PLANNING FOR SAFETY AND URBAN QUALITY

PROCEEDINGS OF THE XXIIIRD INTERNATIONAL CONFERENCE 'LIVING AND WALKING IN CITIES' (LWC 2017), 15–16 JUNE 2017, BRESCIA, ITALY

Town and Infrastructure Planning for Safety and Urban Quality

Editors

Michèle Pezzagno & Maurizio Tira

Department of Civil, Environmental, Architectural Engineering and Mathematics, University of Brescia, Brescia, Italy

CRC Press is an imprint of the
Taylor & Francis Group, an **informa** business
A BALKEMA BOOK

CRC Press/Balkema is an imprint of the Taylor & Francis Group, an informa business

Typeset by V Publishing Solutions Pvt Ltd., Chennai, India

Published by: CRC Press/Balkema
Schipholweg 107C, 2316 XC Leiden, The Netherlands
e-mail: Pub.NL@taylorandfrancis.com
www.crcpress.com – www.taylorandfrancis.com

ISBN: 978-0-8153-8731-2 (Hbk)
ISBN: 978-1-351-17336-0 (eBook)

Town and Infrastructure Planning for Safety and Urban Quality – Pezzagno & Tira (Eds)
© 2018 Taylor & Francis Group, London, ISBN 978-0-8153-8731-2

Table of contents

Town and Infrastructure Planning for Safety and Urban Quality – Pezzagno & Tira (Eds)
© 2018 Taylor & Francis Group, London, ISBN 978-0-8153-8731-2

Preface

The International Conference "Living and Walking in Cities" was organized under the auspices of the University of Brescia. The conference was organized under the patronage of: the Region of Lombardy, Italian National Institute of Town Planning (INU), Italian Society of Urban planners (SIU), National Centre for Urban Studies (CeNSU), Municipality of Brescia, European Transport Safety Council (ETSC); and with grants from: Fondazione ASM, Brescia Mobilità and Fondazione della Comunità Bresciana.

The conference was held June 15 – 16, 2017 in Brescia, Italy.

The purpose of the conference was to gather researchers, road users, administrators, technicians, city representatives and experts to discuss problems that affect the safety of pedestrians in cities, especially of children and persons with reduced mobility. The conference, which attracted both practitioners and researchers; offered detailed presentations on policy issues, best practices and research findings across the broad spectrum of urban and transport planning. The conference covered international issues, national and local policies and the implementation of projects at the local level. Furthermore, it presented a great opportunity for networking and forming career-spanning professional relationships.

The conference focused on a wide spectrum of topics and areas of urban planning, transport planning and road safety as listed below:

1. Town planning
   - Transport policies and mobility challenges towards cleaner cities
   - Reflecting climate change in transportation solutions
   - Current and Future Challenges in Mobility: From policy to implementation
   - SUMPs: Sustainable Urban Mobility Plans
   - Mobility as a service
   - The relationship between "clean mobility" and shaping public space
   - Transportation solutions in (historical) urban centres
2. Road safety
   - Measures and strategies to enhance the safety of vulnerable road users through urban planning, vehicle design, behavioral change, innovative road safety campaigns
   - Pedestrian injury prevention
   - Cyclist injury prevention
   - Pedestrian infrastructure
   - Cycling infrastructure
   - Traffic calming/30 km /hr
   - Elderly and disabled mobility
   - Shared mobility

Forty-five papers from 11 countries were accepted for publication in the proceedings of the Conference, in addition to eight lectures of international experts. Each of the accepted papers was reviewed by selected members of the Scientific Committee in accordance with the scientific area and orientation of the papers.

The editors would like to express their many thanks to the members of the Organizing and Scientific Committees. The editors would also like to express a special thanks to all reviewers, sponsors and conference participants for their intensive cooperation to make this conference successful.

Maurizio Tira Michèle Pezzagno

Town and Infrastructure Planning for Safety and Urban Quality – Pezzagno & Tira (Eds)
© 2018 Taylor & Francis Group, London, ISBN 978-0-8153-8731-2

Scientific committee

Published with the contribution of
Fondazione ASM, Brescia Mobilità,
Fondazione della comunità bresciana

Introduction

Town and Infrastructure Planning for Safety and Urban Quality – Pezzagno & Tira (Eds)
© 2018 Taylor & Francis Group, London, ISBN 978-0-8153-8731-2

A safer mobility for a better town: The need of new concepts to promote walking and cycling

M. Tira
Università degli Studi di Brescia, Italy

Several different ideas are included today in the current concept of sustainability: each could influence the management of urban public space of European towns in the XXI Century.

More than half of the world's population now live in urban areas. By 2050, that figure will have risen to 6.5 billion people—two-thirds of all humanity. *Sustainable development cannot be achieved without significantly transforming the way urban areas are designed, built and managed. Making cities safe and sustainable (…) also involves investment in public transport, creating green public spaces, and improving urban planning and management in a way that is both participatory and inclusive* (UNDP, 2015).

The target specifically set by the United Nations' Sustainable Development Goals for transport by 2030 is: "provide access to safe, affordable, accessible and sustainable transport systems for all, improving road safety, notably by expanding public transport, with special attention to the needs of those in vulnerable situations, women, children, persons with disabilities and older persons and provide universal access to safe, inclusive and accessible, green and public spaces." (UNDP, 2015).

The main goal of the Living and Walking in City Conferences is exactly that: improving urban quality while promoting safer mobility, especially for the most vulnerable road users. The way vulnerability is tackled is one of the most important indicators of civilisation and the "design for all" approach appears to be the most promising one.

Cities and towns differ immensely from each other, but once housing demand is no longer the main focus and priority, citizens call for greater environmental preservation and better services; these include urban public spaces, which have increasingly been used for mobility, and furthermore have been disproportionately allocated to cars. In such spaces, quality often suffers due to congestion, lack of maintenance, noise and pollution. The realisation of public transport networks, cycling lanes and pedestrian facilities is not only meant to develop modes of transport alternative to cars, but it is often a chance for improving common spaces and overall liveability.

Just as housing demand has been the lever of urban development, involving a great portion of actual urban public space, mobility management and road safety can play the role of a lever for improving town management and townscape as well. For this reason this topic is relevant for urban planners, not just transportation and traffic engineers. *It is obvious that the action of the town planner influences safety, and that the analyses of this influence should be done according to a system approach* (as Fleury states in this book).

Even the most traditional approach to road safety, i.e. that of treating 'clustered' accidents, often hides the inner relation between town structure and road (un)safety. The so-called 'most dangerous spots' represent just the tip of the iceberg that can be depicted as a generalised problem of friendly and safe mobility. On the contrary, the proportion of 'scattered' accidents (accounting sometimes for 50% of the total) increases the importance of area-wide techniques.

The 'area-wide safety approach' is a systematic approach of particular interest and forces the consideration of 'urban planning and design'. The Urban Safety Management scheme (TRL et al., 2003), set in UK and developed through several EU funded projects (see among them the DUMAS project; Tira, 2003) is maybe the most advanced and comprehensive frame for developing a safer urban form.

'Urban areas form a complex, dynamic system in which various factors inter-relate in many different ways. What is the place of road and transport safety within this system and how does it link with other system components?' (Dumas, 1999). Several key issues referring both to mobility and town planning can be listed such as, transport network, urban design, traffic management practices, planning policies, transport policies, parking strategies, public transport, local public opinion, employment, social policies, town image, and environmental concerns. These issues correspond to political processes that are critical to urban development and hence the safety of the road network (Fleury, 1998).

There is now a considerable effort to combine the planning of land use and transport in many European countries, although this is mainly because environmental problems caused by private motorized transport in the form of car-emitted gases, noise, water and land pollution, have become crucial issues. However, this situation can often be used as a catalyst for change and it may lead to a reduction in car trips that can benefit road safety and vulnerable road user comfort. It is worth reflecting on what could be a strategic lever to start those different processes that would culminate in a safer town.

The other illuminating challenge is that of car dependence; the topic has been addressed by, among others, Dupuy (2002). A deep reflection is also merited on how cities and towns designed before cars have to cope with those "private cells moving in a public space" (Dupuy, 2002).

Focusing on pedestrians and cyclists is not just a matter of lobbying for their own benefit, rather a better mix of modes can be an important factor in improving safety for all. However, modal shift is a difficult challenge, and the lack of safety imposed by cars is paradoxically one of the reasons for using cars even more. Furthermore, the perception of space from within a car is totally different. Drivers and passengers have the feeling of being in a private, safe and comfortable cell moving on a dedicated track. There is a strong sense of separation from the outside environment, growing in citizens a particular attitude towards their living space. The space itself, its maintenance and quality then influence the perception and the relationship between public and private interests. It could be then said that good public space design promotes walking and cycling and, at the same time, one of the main consequences of promoting walking and cycling is that of returning urban space to citizens and shaping towns for people (not cars) *(Busi and Ventura, 1995; Tolley and Thomas, 2001; Bonanomi, 2002)*.

The same results of the international debate about town and infrastructure planning for safety and urban quality were discussed and developed during the International Conference "Living and Walking in Cities" (LWC, 2017). It took place in Brescia on 15–16 June 2017 and its highlights are reported in this book. The conference's Scientific committee was involved in a challenging work of revision and balance of all contributions received during and after the conference.

I would like to thank the more than 100 conference participants and especially the members of Scientific committee: *Richard Allsop*—Dept of Civil, Environ & Geomatic Eng, Faculty of Engineering Science. UCL University (UK); *Antonio Avenoso*—European Transport Safety Council (BE); *Roberto Busi*—Dipartimento DICATAM. Università degli Studi di Brescia (IT); *Enrique J. Calderon*—ETS de Ingenieros de Caminos, Canales y Puertos. Universidad Politécnica de Madrid (ES); *Agostino Cappelli*—Dipartimento di culture del progetto. Università IUAV di Venezia (IT); *Dominique Fleury*—IFSTTAR LMA, Laboratoire Mécanismes d'Accidents (FR); *Peter Hollò*—Department for Road Safety and Traffic Engineering KTI (HU); Matteo Ignaccolo—Dipartimento Ingegneria Civile e Ambientale. Università degli Studi di Catania (IT); *Chris Lines*—Independent transport consultant. London (UK); *Giulio Maternini*—Dipartimento DICATAM. Università degli studi di Brescia (IT); *Rocco Papa*- Dipartimento di Ingegneria civile, edile e ambientale. Università degli Studi di Napoli Federico II (IT); *Graham Parkhurst*—FET Geography and Environmental Management. University of the West of England (UK); *Michèle Pezzagno*—Dipartimento DICATAM. Università degli Studi di Brescia (IT); *Tim Pharoah*—Transport and urban planning consultant (UK); *Paola Pucci*—Dipartimento di Architettura e Studi Urbani. Politecnico di Milano

(IT); *Michela Tiboni*—Dipartimento DICATAM. Università degli Studi di Brescia (IT); *Rodney Tolley*—Sustainable Transport Consultant (UK); *Paolo Ventura*—Dipartimento DICATeA. Università degli Studi di Parma (IT); *Fred Wegman*—SWOV Institute for Road Safety Research. (NL); *George Yannis*—Department of Transportation Planning and Engineering. NTUA University of Athens (EL).

In conclusion, I believe that the LWC-2017 Conference has increased the knowledge of the topic and the conscience of the urgency for a comprehensive approach to urban mobility and planning as the most successful one. Success can only be built on a multi-disciplinary approach and broad consultation, as shown in the conference, within a clear strategic framework.

REFERENCES

Bonanomi L., 'Urban development: pedestrians as a new departure in town planning', in Fleury, D (ed), A city for pedestrians: policy making and implementation. Final Report of COST Action C6, EU Directorate for Research, (Brussels: 2002).

Busi R. and Ventura V., Living and walking in cities: town planning and infrastructure projects for urban safety, Proceedings of the 1st International Brescia-Cremona Conference (1994), (Brussels: 1995).

Dept. of Transport, TRL and The Institution of Highway and Transportation, Urban safety management guidelines, London, 2003.

DUMAS—Developing Urban Management and Safety Project, Linking framework with other initiatives. Management of urban areas and urban structure, transport, malfunctions (WP8), (INRETS, Paris: 1999).

Dupuy G., 'Cities and automobile dependence' revisité: les contrariétés de la densité, Revue d'Economie Régionale et Urbaine, I: (2002), pp. 141–156.

Fleury D., Sécurité et urbanisme: la prise en compte de la sécurité routiére dans l'aménagement, (Presses de I'ENPC, Paris : 1998).

Tira M., "Safety of pedestrians and cyclists in Europe: the DUMAS approach", in Tolley R. (ed.) Sustainable transport, Woodhead Publishing, Cambridge (UK), 2003; pp. 339–350.

Tolley R. and Thomas C., 'Walking research and communication: new initiatives', Proceedings of the 8th International Brescia-Cremona Conference (2001) (Brussels: 2001).

United Nations Development Programme (UNDP), Sustainable development goals, 2015.

Best practices and case studies to improve urban quality through mobility management, shaping public spaces and road safety policies

Town and Infrastructure Planning for Safety and Urban Quality – Pezzagno & Tira (Eds)
© 2018 Taylor & Francis Group, London, ISBN 978-0-8153-8731-2

Improving urban quality through mobility management: What is the right balance of encouragements, incentives and restraints?

G. Parkhurst
Centre for Transport and Society, University of the West of England, Bristol, UK

ABSTRACT: Despite two decades of mobility behaviour-change initiatives, progress in Europe remains limited. The presentation addresses why there has not been greater success by reviewing the role that structural constraints play in limiting mobility choices. These include not only urban form, which can change only slowly, but also economic influences and cost trends which incentivise behaviour contrary to policy objectives, and finally social and cultural factors, given that some cities and countries have different mobility profiles than others, despite otherwise similar conditions. Findings from the EC-funded EVIDENCE project, which reviewed 22 categories of Sustainable Urban Mobility Plan measures, are considered.

The presentation concludes that promoting urban forms which are accessible by a range of modes will result in a resilient economy and society that will not be highly disrupted by energy supply difficulties. Also it is essential that transport system costs are aligned in a way that promotes policy objectives.

1 INTRODUCTION

There is growing evidence that rapidly rising greenhouse gas concentrations are resulting in climate change: urban areas contribute 23% of European transport-related climate change emissions. They also exhibit persistent and widespread poor air quality. Despite European societies showing high mobility overall, there are considerable inequalities in accessibility. Although 'technical fix' solutions are emerging to address some of these problems, Greene and Parkhurst (2017)[1] recently reviewed the evidence for the US Transportation Research Board and the European Commission, concluding that technology substitution on its own would not be a sufficient or reliable strategy, to achieve 95% greenhouse gas mitigation. Behaviour change must therefore be a key part of our transport strategies, including our Sustainable Urban Mobility Plans (SUMPs).

However, after two decades of behaviour change initiatives, progress remains limited. Indeed, the European Union transport policy foresees a doubling in passenger and freight traffic by 2050.[2] It is therefore very relevant and urgent to demand why there has not been greater success. Three key questions are:

- How far is behaviour genuinely constrained by circumstances?
- Or does observed behaviour result from inappropriate incentives?
- What is a fair balance of incentives and restraints in order to achieve important policy objectives?

---

1. Greene, D., and Parkhurst, G., (2017). Decarbonizing Transport for a Sustainable Future: Mitigating Impacts of the Changing Climate. White Paper to the 5th EU-US Transportation Research Symposium. TRB, Washington DC and EC, Brussels.
2. European Environment Agency (EEA), 2016c. Transitions towards a More Sustainable Mobility System. TERM 2016: Transport Indicators Tracking Progress Towards Environmental Targets in Europe. EEA: Copenhagen.

The presentation will review the role that structural constraints play in limiting mobility choices, recognising that some aspects, such as urban form, can change only slowly. However, in the case of economic influences, which can in principle be changed by policy, it is observed that the cost trends are incentivising behaviour which is in clear contrast to policy objectives. Social and cultural factors are also identified as influential, with some cities and countries showing a different mobility profile to others, despite otherwise similar conditions. Here, the work of the EC-funded EVIDENCE project,[3] which reviewed 22 categories of Sustainable Urban Mobility Plan measure, is considered.

Finally, turning to consider communication on mobility behaviour change, it is observed that successful behaviour change campaigns on smoking and road safety have used clear messaging, whereas in the case of mobility behaviour choices the messaging is often weak and vague. On the one hand it does not achieve the shock messaging of health campaigns, on the other it rarely achieves the sophisticated sociocultural associative messaging employed by advertising in the automotive sector.

It is concluded that, whilst structural barriers may be hard to pull down, we must at least avoid making them higher by increasing the provision of car-dependent environments. Promoting urban forms which are accessible by a range of modes is also a matter of constructing a resilient economy and society that will not be highly disrupted by energy supply difficulties.

Second, it is essential that transport system costs are aligned in a way that promotes policy objectives. It is not a politically easy measure to 'manage' up the cost of private car travel, but essential if more efficient, shared, mobility solutions are to be successful.

Finally, it is essential that the meanings of behaviour change and quality of urban life are better understood and communicated more effectively in order to support the structural and economic changes in an integrated way.

3. http://www.evidence-project.eu/.

Town and Infrastructure Planning for Safety and Urban Quality – Pezzagno & Tira (Eds)
© 2018 Taylor & Francis Group, London, ISBN 978-0-8153-8731-2

Different viewpoints on improving safety: Enforcement, planning, modal shifts, integrated policies

D. Fleury
Research Director Emeritus IFSTTAR, France

ABSTRACT: Improving road safety has many perspectives including enforcement, planning, modal shifts, and integrated policies. This presentation first discusses the work carried out at IFSTTAR for over 30 years. Based on the conclusions drawn from the analysis of blackspots versus citywide accidents, questions are raised about how to take safety into account in local action, particularly through new design of urban road networks.

The presentation then describes the two styles of thought on road safety: 1) "pragmatic", based on beliefs and the immediate interests of organizations and 2) "rational", the reduction in the number of accidents and their severity, based on the ability to forecast the consequences of decisions and actions. The reasons for the shift from the first style "pragmatic" to the second "rational" are described. The presentation finishes by addressing the need for integrating safety with local management including precise regulation, training of road safety specialists, and further research.

1 INTRODUCTION

Safety is a field that is crisscrossed by detectable theories that are widely quoted and classified in the literature (Häkkinen, 1979; OECD, 1984; Gunnarsson, 1985, cité par Rumar 1988; Salusjärvi, 1989; Fleury, 1998; 2012) and which have produced points of view that have evolved and may converge and/or diverge in the way they perceive facts.

Roughly, these theories that will support prevention policies can be classified into:

- Fatality
- Human factors
- monofactorial
- multifactorial
- System approach

What can be seen today is the superposition of all these theories through action policies. For example, in France, policies focused on the human factor (enforcement) become predominant. Road Safety has gone from the Ministry of Transport to the Ministry of the Interior in charge of the police.

Nevertheless, for those concerned with safety in relation to road design, urban planning and mobility, it is necessary to adopt a systemic vision integrating the complexity of the system. By complexity I will hold to a definition insisting on the feedbacks and on the non-predictability of the result that will be achieved.

Thus in our fields it has become more than obvious that we must study the effects of an action undertaken on space. No one will be able to say "I am modifying a design, and the situation will be better". There will always be someone—be it an elected official or a citizen's group via social networks—to ask for an analysis of the induced effects.

2 ABOUT MY INVOLVEMENT IN ROAD SAFETY

I started in the field of road safety in the early 1970s, when public policy was being started in many OECD countries. But this is not the subject here. I want to talk about my experience on subjects in relation with this conference.

2.1 *What are accidents in urban areas?*

When I started working in the field of road safety in urban areas, I studied the reports written by the police of the accidents that occurred in a city to try to understand, beyond the statistics what this phenomenon could be. We could conclude globally that certain spatial configurations will correspond to accident patterns. That means, behaviors are not spatially located randomly, but are closely related to contexts of occurrence. It was possible to conceive a typology of these phenomenon (for example, rear impact accident when there is an upstream traffic light, crossing pedestrians in a secondary commercial zone, turning left in front of a motorcycle on fast road…). The consequence is that accident can be avoid with a modification of the design.

2.2 *Blackspots analyses*

One of the express demands of this period were on blackspots analysis. It should not be forgotten that the number of victims was 4 to 5 times higher than today and that the notion of blackspots had a real meaning. Spatial concentrations appeared on maps and we developed analysis procedures, disseminated through the technical network. A blackspot comes from a localized dysfunction, resulting from a design error or a local singularity. This dysfunction succeeds in «trapping» a user. It results from this that the treatment is often light to be carried out and therefore inexpensive, and moreover, the more light it is, the more the behaviors are little modified and will therefore produce little adverse feedback. These treatments have very high cost-effectiveness.

Blackspots investigation also emphasizes the essential importance of driver expectations and understanding.

- Importance of driver understanding/visible and understandable situations
- Importance of expectations/road legibility
- The more the intervention is light, the better the future is controlled

2.3 *Accident analyses of a street*

At another level, we have undertaken analyses of axes to relate the nature of design to the levels/types of accidents (Tira & Brenac, 1999). The streets differ according to their function in the structure of the circulation network of the city and according to the distance to the center. Of course this is an extreme simplification, but results from the analysis of accidents. What is questioned then is the level of equipment, the type of design corresponding to types of traffic, modes and activities in these places. The accidents then make it possible to point out the problematic elements that should then be taken into account.

But at this stage, safety is obviously not the only objective of a development and numerous other objectives—aesthetics, economy, accessibility, cycle layout… - are at stake. The development will therefore take into account these different aspects. A modification of the existing situations in greater depth means a higher risk of producing new dysfunctions. The challenge is therefore to take safety into account among other objectives that are essential to the design of urban public spaces.

2.4 *Accident analyses of a whole city*

Work on a whole city or a whole neighborhood raises the same questions to which are added the modifications of hierarchization of networks (using traffic plan) as well as modal shift strategies.

The planning must take into account the actual speeds, the modal splits, the potential conflicts, the priorities to be given to certain modes, certain functions, activities in the space, the built environment, local activities, etc.

The analyses of the dynamics of movement and of the accidents require that priority be given to:

- The possibilities of encounters/conflicts that must be integrated into the design
- Among them, those that are more difficult to predict
- Encounter scenarios that are more likely to result into an injury accident
- The following of different sequences during a travel
- To a lesser degree, the dynamic difficulties of trajectories

And this according to speeds.

The layout is based on a set of tools known for a very long time (apart from so-called intelligent devices and automatisation). Finally, the only strategies available are separation (segregation in space or time) and integration by making meetings/conflicts between users visible, predictable and acceptable.

2.5 *About the action of the town planner*

It is obvious for each one in this conference that the action of the town planner influences safety, and that the analyses of this influence should be done according to a system approach. This means that the impact on behavior—in the very broad sense—is not completely predictable, that unplanned, even deviant, offending behaviors still will appear.

Safety is a game with two players between the planer and the road user, so adaptation is one of the difficulties encountered. The more extensive the treatment, the more difficult it becomes to predict the induced effects. That is why the evaluation is necessary. This evaluation enables the improvement of knowledge and thus the practices.

Is the analysis of accidents the solution? It is necessary, but the explanation of accident factors does not necessarily indicate what action to take. We must recognize this distance between the factor and the action. If speed is an important factor, it is possible to bridle the engines, put humps, change the design, put radars or a police officer at all times, or make the risks of pedestrian crossings more visible, put a localized speed limit in the vicinity of playgrounds where children are playing or of bars from which alcoholic people will emerge… For each accident factor, many types of action are possible.

2.6 *How is safety integrated into planning?*

I shall speak here only of my experience on the French situation.

We have observed a number of processes, and in particular urban mobility plans, which aim at improving mobility at the level of a whole urban area. The guidebooks insist on a prior analysis of accidents, and specifically their mapping. In practice, this is realize but the information is quickly forgotten in the discussion.

The conception of a urban mobility plan is organized around an "action scene" made up of people with special interests—responsible for the tramway network, shopkeepers, cycling associations … - Discussions build alliances that produce an urban model often based on a structure rejecting the automobile, thanks to increasingly distant ring roads, and a structuring of the network by public transport (tramway) and cycling lanes (Hernandez, 2003).

In this context, safety is not an objective, but an argument for promoting action. Some call for the development of bicycle lanes to increase the use of bicycles, but in the discussion, they emphasize safety. Although evaluations do not necessarily conclude with an increase in safety due to cycling networks.

Some conclusion we had met:

- Safety, an exhibited goal, an argument used
- During discussions the safety consequences are based on "common sense" and not based on scientific results

- Safety is a problem that engineering must resolve at the end of the decision-making process.

2.7 *Safety in local organizations*

There were some years ago discussions about the place of a safety cell in local organizations. Nowadays in France, the organizations are structured by objective rather than by profession. Some were pushing to integrate such a cell into the mobility department, which takes the questions upstream. But for what I saw, safety was always integrated into the traffic department, in the road design sub-department. It remains therefore a technical problem, linked to the details of the layout rather than as a consequence of the organization of the displacements.

Here we must mention the interesting experience of London during the term of Ken Livingstone. A Road Safety Unit was created, led by Chris Lines. In Britain, it is traditional to only undertake policies with a favorable cost-benefit. This involves performing a priori assessments before taking any action and then afterwards, measuring its consequences. This is not in the French tradition, basing policies more on a priori positions, based on ideological bases—despite some attempts to rationalize choices.

The road safety unit in London had a budget, like other units. Any project must therefore seek its financing by highlighting the objectives sought and by tapping the doors of the various money providers available. This has resulted in a real monitoring of the state of safety and the achievement of most of the reduction targets set by the mayor (Mayor of London, 2006).

Thus it may be considered that it is necessary to

- Integrate safety upstream in the development process
- Consider safety at the same level as other strategic objectives, mobility, modal shift, aesthetics, accessibility, attractiveness, economy, etc.
- Be able to consider the consequences of different scenarios on levels and types of accidents
- rely on safety specialists (not layout specialists as it is generally today)
- develop a specific training

2.8 *Safety from the point of view of populations*

In France, accident reports, which are legal documents belonging to the courts, are digitized. Parliament has given our organization—IFSTTAR—access to this information base. It is then possible, thanks to character recognition software, to access text that is more interesting than statistical databases.

It is then possible to link the inhabitants of certain sectors whose socio-demographic characteristics are known to their risk of accidents in France. Several studies have been carried out showing:

- Over-risk of populations living in depraved neighborhoods, especially young men. Many implications of young children because of their high number in these neighborhoods, even if there is no over-risk compare to children in more favored neighborhoods. (Fleury et al, 2010).
- The higher number of accidents to which the inhabitants of the city centers are exposed, but with much lower gravity than that of the inhabitants of the periphery (Fancelli, 2012).
- The higher risk of people living on streets in intermediate road hierarchies, between main roads and feeder roads. An explanation of this result may be the greatest difficulty in designing these paths between strict spatio-temporal segregation on the main axes and an almost total integration on the feeder roads (Haidar, 2014).

3 POINT OF VIEW ON IMPROVING SAFETY

3.1 *Enforcement*

Ezra Hauer (2007) describes two styles of thought on road safety. The first is "pragmatic", based on beliefs and the immediate interests of organizations. It does not require knowledge or assessments of actions. The second style is "rational", based on the desire to reduce the number of accidents and their seriousness. It is based on the ability to foresee the consequences of decisions and actions. When organizations think in this way, they make use of empirical knowledge and undertake systematic assessments.

Safety actions are first based on a pragmatic style, i.e. emphasizing what is felt to be obvious and shared by elected officials, technicians and the entire population. If mobility can be modeled in the form of a man-vehicle-environment system, this would mean that safety can be obtained by improving each of these components: if each component works as well as possible, the entire system will work better.

This way of thinking about safety can be effective, like for instance, enforcement policies. Thus, the speed cameras set up in France since 2002 (Carnis, 2010) have had an impact that goes far beyond isolated actions in the areas where they have been implemented. Attitudes have changed, with effects even being observed on departmental roads or communal road systems where there was no speed cameras. The objective of this operation, or at least the interest of it, therefore is not only to change behavior through a Pavlovian strategy—there is a speed camera, it is announced and the user slows down—but also a more general modification in attitudes towards speed and compliance with the law that produces a measurable effect on the entire network.

3.2 *Managing road design*

Changing the status of components often makes it possible to make progress toward safety. But it is nonetheless true that the designer is very often confronted with the adaptability of users, drivers, pedestrians—users of public spaces. Like any sociotechnical system, the road system—or more generally the travel system—is a complex system and is therefore unpredictable. The relationship between the decision maker and the population using traffic spaces is a two-player game. Depending on what can be modified, the user will adapt his behavior in a way that is more effective for him, and not necessarily in a way that is better for collective safety. There are many examples of adaptation that is rather negative compared with intentions that were thought to be good at first: improvements to visibility that reduce attention, road resurfacing that increases speed, creating cycling lanes used by powered two-wheelers (Evans, 1991 et 2004). That is why we must monitor changes in terms of safety for possible adaptations, in other words, "to steer the system".

Research then turns toward a more "rational" type of action, as the term is used by Ezra Hauer. As there is feedback from users, the results of the action vary widely: all decisions are taken in a context of uncertainty. It is then necessary to measure the results of actions, not only to evaluate what has been undertaken, but more importantly to accumulate knowledge in order to be able to predict the effects of future actions, and therefore choices and possibility for designing policies (Elvik, 1981; Evans, 1991; Elvik et Vaa, 2004; FHWA, 2008). Such an evaluation policy is necessary to carry out planning for a better adaptation of existing infrastructures.

3.3 *Research and action*

In some countries such as France, where the culture of evaluation is not widespread, this approach can lead to conflicts between researchers and policy decision-makers. Reports that are in line with the action are accepted much better than those demonstrating negative results.

A policy of modal shift in favor of bicycles undertaken for many years in our country aims to increase the use of the bicycle, very low compared to the situation in other countries, such as the Netherlands for instance.

International comparisons show that travel habits affect the number of accidents and the accident rate. Thus, the higher the level of bicycle use, the higher the number of accidents involving cyclists, the higher the number of cyclists killed or injured and the higher the proportion of cyclists among the different categories of accident victims (killed or injured). On the other hand, the higher the level of bicycle use, the lower the proportion of cyclists involved in accidents (killed or injured). The countries where cycling is the most developed are also those where the rate of fatalities per kilometer ridden on bicycles is the lowest.

If we suppose that bicycle use is the same in France and the Netherlands. And then, due to the massification of this use, suppose that the rate of risk for bicycles were to drop to 1.1E-08 (current rate of people killed per km ridden in the Netherlands), vs. 3.3E-08 (current rate of people killed per km ridden in France). It is nonetheless true that the rate of fatalities per inhabitant would rise from 2.45E-06 (current rate in France) to 8.99E-06 (current rate in the Netherlands), which would increase the number of fatalities from 156 to 573 (figures for 2007).

The argument by policy makers for a modal shift will then ignore these latter figures to focus on the former, stressing a reduction in the individual risk of cyclists that can be estimated to go from 3.3E-08 to 1.1E-08.

3.4 *Comprehensive approaches*

These statistical approaches can be used to evaluate the effects of actions directly and to study the impact of accident factors, making it possible to set up actions to try and change them. While they are necessary, they are not enough and a more analytical approach to the system is needed to understand how it works and the origin of dysfunctional situations (Fleury & Tira, 2010).

In 1998, Gabriel Dupuy already denounced the current trend towards the internationalization of design, based in part on the implementation of tools universally recognized as useful for safety but producing standardized layout. Developing knowledge and methods to effectively adapt action to the diverse environments is a necessity to safeguard both the genius of the places and the safety of those who practice them. The integration of safety into all actions on the city (eg modal shifts, mobility planning), requires such a comprehensive approach.

In a way, it is like opening the black box and analyzing the processes at work in driving, piloting or walking. Many disciplines are used to observe, analyze and experiment with the functions at work, which are interactions between the components. This entails gathering information, processing it, choices and decisions, activating commands, vehicle dynamics on the carriageway and in its environment. All these aspects are the subject of many research projects analyzing the processes themselves as well as the impact of possible modifications.

This integration of knowledge in the design processes has been conceived in other socio-technical fields such as aviation, oil or nuclear industry,...

Some authors model the evolution of the safety culture of an organization in different stages. Initially, management believes that the accident is the result of fault or inattention of employees. At these stages (1 and 2), management does not recognize a role in safety. At the envolvment stage, management is changing its viewpoint and designing procedures to take safety into account. In step 4, procedures are in place and management appropriates safety objectives, in particular by dedicating sufficient resources. In step 5, management effectively monitors and manages safety, continuously improving this (Flemming, 1999).

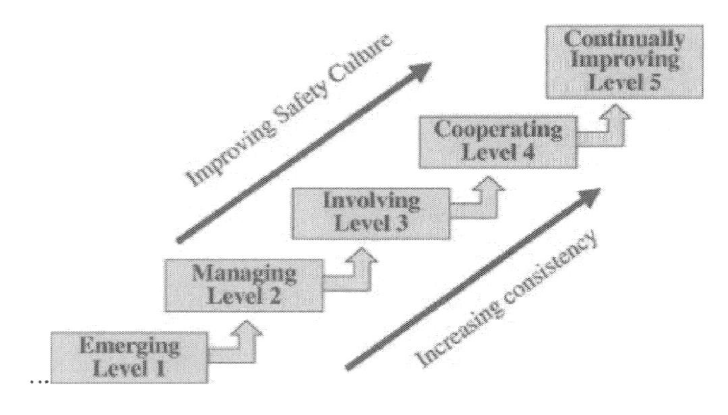

Figure 1. Safety culture maturation model (Fleming, 1999).

4 CONDITIONS FOR INTEGRATION

The second style defined by Ezra Hauer (2007) is "rational", based on the ability to foresee the consequences of decisions and actions. When organizations think in this way, they make use of empirical knowledge and undertake systematic assessments.

Four reasons explain the shift from the first style "pragmatic" to the second "rational". The first is that humanity has always evolved from actions based on intuition toward actions based on scientific knowledge. It would be surprising for road safety not to follow such a general law. Then, while intuition can be used in implementing initial actions, what remains to be done requires more serious knowledge. The third reason is that—in the United States and Canada—legislation requires safety to be taken into account when planning national and local travel networks. Lastly, the author shows—once again in North America—that many initiatives go in the right direction, whether through technical recommendations or the development of skills.

Ezra Hauer, however, observes that there is little demand for people with road safety training, which pleads in favor of the third condition, i.e. resorting to legal obligation. Starting with what training requires both in terms of teachers and the knowledge needed to teach, i.e. the need for research, he draws up a highly critical assessment of the practices in this field. The research community all too often focuses on the past and much of the research is based on strong biases that are not very scientific.

Ezra Hauer describes the North American situation. But his text, while incisive, does situate the place of public decision-making and research in most countries.

However, in a more optimistic view (Fleury, 2005), we can think of a favorable evolution in the way road safety is taken into account in urban planning. We must start from the idea that winning the first deaths is always easier than saving the next ones, because they require much more knowledge.

It is possible to borrow from Ezra Hauer the conditions he puts forward in his text and thus allow the opening of a discussion:

- Greater involvement of local authorities and therefore development of local associations
- More precise regulation in favor of road safety
- Necessary training of high-level specialists in road safety
- The need for further research in the field.

REFERENCES

Carnis, L. (2010). A Neo-institutional Economic Approach to Automated Speed Enforcement Systems. European Transport Research Review, Vol. 2, no. 1, pp. 1–12.

Department For Transport, 1999, Tomorrow's roads: safer for everyone, London, Department for Transport.

Department For Transport, 2008, Road Safety Web Publication No. 7, Neighbourhood Road Safety Initiative, London, Central Team: Final, Report NRSI Central Team.

Dupuy, G. (1998). Introduction à Fleury, D. Sécurité et Urbanisme. La prise en compte de la sécurité routière dans l'aménagement urbain. Presses de l'école nationales des Ponts et Chaussées. pp. 13–15.

Elvik R. (1981). Effects on Road Safety of Converting Intersections to Roundabouts: Review of Evidence from Non-U.S. Studies. Transportation Research Record. Vol. 1847. pp. 1–10.

Elvik, R., Vaa, T (2004). The handbook of road safety measures. Elsevier Science, Oxford.

Evans, L. (1991). Traffic Safety and the driver. Van Nostrand Reinhold. New York. 405 p.

Evans, L. (2004). Traffic Safety. Science Serving Society, Bloomfield Hills, Michigan, 2004.

Fancelli B. (2012). Les personnes vivant en périphérie ou en centre-ville urbains sont-elles exposées aux mêmes risques d'avoir un accident de la route. Rapport de stage Master 2 Professionnel Université de Versailles Saint Quentin-en-Yvelines. Saint Quentin-en-Yvelines 114p.

FHWA (2008) Guidance Memorandum on Consideration and Implementation of Proven Safety Countermeasures. http://safety.fhwa.dot.gov/policy/memo071008/

Fleming, M., 1999. Safety Culture Maturity Model. UK HSE Offshore Technology Report OTO 2000/049. HSE Books, Norwich.

Fleury D. & Tira M. (2010), « Etat des recherches sur l'appréhension de la sécurité routière par le territoire », *Communication au colloque de l'ASRDLF*, Aoste septembre 2010.

Fleury, D. (2012). Sicurezza e urbanistica L'integrazione della sicurezza stradale nel governourbano. Gangemi Editore. Piaificare per reti l'ambiente e il territorio, collana diretta da Maurizio Tira. Roma 254p.
Fleury D., (2005), Villes et réseaux de déplacements, vers un métier de la sécurité routière? Synthèse n°49, Collection de l'INRETS.
Fleury, D. (1998). Sécurité et Urbanisme. La prise en compte de la sécurité routière dans l'aménagement urbain. Presses de l'école nationales des Ponts et Chaussées. 299 p.
Fleury, D., Peytavin, J.F., Alam, T., Brenac, T. (2010). Excess accident risk among residents of deprived areas, Accident Analysis and Prevention, Elsevier, vol. 42, no. 6, pp. 1653–1660.
Fleury, D., Peytavin, J.F., Alam,T., Brenac,T. (2010). Excess accident risk among residents of deprived areas. *Accident Analysis & Prevention, Volume 42, Issue 6, November 2010, Pages 1653–1660.*
Gunnarsson, S.O. (1985). The paradigm of accident research (in Swedish) Department of traffic Planning, Technical University of Gothenburg, Sweden - Report 1986: 2.
Haidar M. (2014) L'intégration de la sécurité routière dans l'action locale: l'inflence de la hiérarchisation du réseau sur le risque routier. Thèse pour obtenir le grade de docteur de l'université Paris Est. Discipline: Transport. 18 Septembre 2014. Marne la Vallée 266p.
Häkkinen S. (1979). Accident Theories. Acta Psychologic Fennica VI, pp. 19–28, biblio. Helsinki.
Hauer, E. (2007). A Case for Evidence-Based Road-Safety Delivery, AAA Foundation for Traffic Safety.
Hernandez F. (2003). Le processus de planification des déplacements urbains entre projets techniques et modèles de ville. Mémoire de thèse "Aménagement de l'espace et urbanisme", ss la dir. D. Pinson, D. Fleury, Institut d'Aménagement Régional, Université d'Aix-Marseille III, Décembre 2003, 351 p + annexes.
Mayor of London (2006) Towards the year 2010: monitoring casualties in Greater London. 130p.
OECD, (1984). Integrated road safety programs. Paris.
Reigner H., (2004), « La territorialisation de l'enjeu « sécurité routière »: vers un basculement de référentiel ? » in *Espaces et sociétés* 2004/3 - no 118, pp.23–41.
Rumar, K. (1988). Collective risk but individual safety. Ergonomics 1988, vol. 31, no. 4, 507–518.
Saint Gérand T., Medjkane, (2009) de l'analyse du risque routier à la problématique territoriale de la sécurité routière. Journée de valorisation GO2. 3–4 décembre 2009. La Grande Arche. Paris.
Salusjärvi, M. (1989). Why and how the direction of traffic safety work is changing? in "Second European workshops on recent developments in road safety research". Paris 26–27 January. INRETS, Acte No. 17, 289 p.
Tira, M., Brenac, T. (1999). Scenari d'incidente stradale e cartografia per la gestione urbana. Transporti Europei. V12. pp. 7–11.

Town and Infrastructure Planning for Safety and Urban Quality – Pezzagno & Tira (Eds)
 ISBN 978-0-8153-8731-2

Against traffic engineering axioms. A change in the hierarchy of the priorities of the city transport planning: Slow mobility and public transport to change user behavior for safety purposes

A. Cappelli
Università IUAV di Venezia, Italy

ABSTRACT: The aim of this presentation is to overcome the idea that traffic flow management and regulation can pursue urban quality, sustainability and security. An organic framework of strategies and operational tools is needed, which means adopting medium-term Integrated Transport Plans (10+ years) including Sustainable Urban Mobility Plans.

Sustainability goals require urban mobility focused on high quality public transport integrated with bicycle routes and pedestrian paths. Motorized private mobility should provide local accessibility (in low density areas) and short-haul routes. Such a redesigned transport system would have considerable savings particularly for families, given the cost of auto ownership.

Innovations in passenger cars (fuels, driverless, etc.) do not change the rules of mobility; they are "product innovation", not a positive evolution of the system. Technological innovation in collective transport today make it possible to redesign cities for people not cars. This is the true challenge of this century for the whole world!

1 INTRODUCTION

The following notes use some paradoxes to highlight how road transport presents intrinsic elements of insecurity that controlled density systems have ruled out since the mid-nineteenth century.

I am aware that a car of today has rules of circulation that are the same as a horse carriage two centuries ago can be a source of perplexity. But we must not let ourselves be deceived by the different technologies. In both cases, the safety is entirely entrusted to the driver's behavior.

As a first consequence, I can only say that road traffic management alone cannot solve the safety problems (both passenger and pedestrian). An integrated approach as defined in the contents of the Sustainable Mobility Plans set out by the European Union is needed [1].

The notes add some economic considerations. Road transport requires considerable resources paid by families. An integrated, collective, sustainable system would free up huge resources for other needs of the community.

2 THE ROAD SYSTEM IS UNSAFE BY DEFINITION

The birth of traffic engineering in the 1960s was an important innovation in managing the increasing traffic flows in a system based on free driver behavior. But the road system is an unsafe system by definition; It's based on free density; circulation safety is completely delegated to the behavior of the driver.

The road codes sanctions this axiom. The Italian Road Code: TITLE V—RULES OF CONDUCT Legislative Decree 30 April 1992 n. 285 and subsequent modifications in the "Article 141. Speed" [3] says this:

THE SAME RULES ?!?

Figure 1. An apparent paradox (Source: Free picture in the web-site).

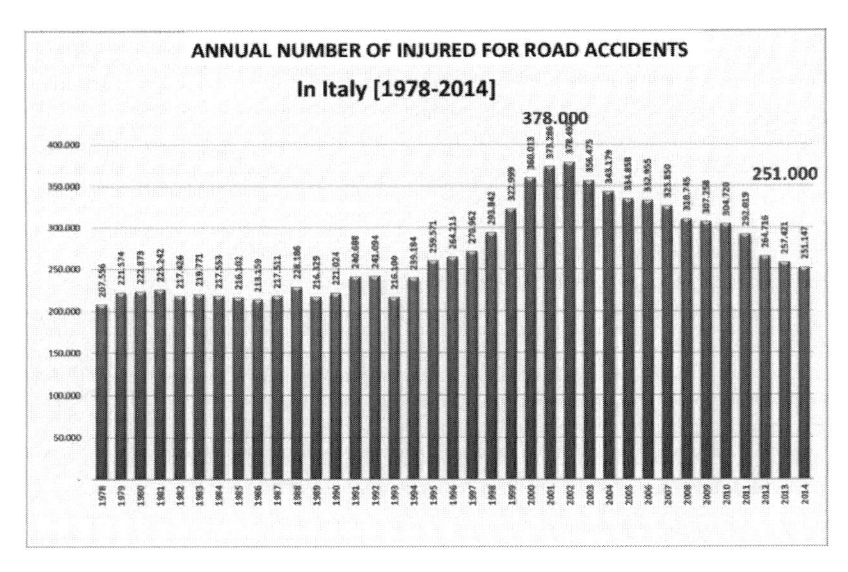

Figure 2. Injured for road accidents in Italy (Conto Nazionale dei trasporti 2016 – Ministero delle Infrastrutture e dei trasporti, IT).

- *It is the obligation of the driver to regulate the speed of the vehicle so that, having regard to the characteristics, state and load of the vehicle itself, the characteristics and conditions of the road and traffic and any other circumstance of any nature, Any danger to the safety of persons and things and any other cause of disorder for circulation is avoided.*
- *The driver must always maintain control of his vehicle and be able to carry out all necessary safety-related maneuvers, in particular the timely shutdown of the vehicle within the limits of his field of vision and any foreseeable obstacle.*

Vehicle technological innovations (ABS, safety belts, shock-absorbing capacity) and sanctions on incorrect behavior (video surveillance systems for speed limits, withdrawal of a driving license for serious infringements, the Laws on Road Crime) confirm that the system is in itself "insecure" [4]. The innovation only seeks to limit damage (to people transported on the vehicle) or to put down behaviors that may cause damage to other people.

The new information technologies for driving a car (warnings for obstacles or driving attenuation, etc.) help the free driver behavior (they are therefore useful for the road safety) but are also an optional.

All this derives from the very nature of the system (evolved as a natural consequence of the horse carriage) while all the densities controlled systems from their origins, in the early eighties, introduced the principle of controlling circulation by external systems and not dependent on the driver's choice (due to the high speeds from the start and the long stopping distances for the wheel-rail coupling), [8].

It may seem an exaggeration but the two vehicles in Figure 1 follow the same rules in their circulation.

Technological innovation and the improvement of the mechanical characteristics of vehicles have positively contributed to a reduction in mortal accidents, road deaths have decreased but accidents and the injured are still high [7] (Fig. 2).

3 CHANGE THE RULES OF THE GAME

The rules of circulation and the Transportation Plans are consequently outdated, and in Italy in particular the only legal obligations concern municipalities with more than 30,000 inhabitants (real or virtual due to tourism) who are required to draw up a Urban Traffic Plan (PUT), [6].

The PUT is designed to manage short-term (2 years) circulation and try to get fluidized traffic.

This is to ensure good levels of road mobility, recalling the containment of air pollution, the integration with public transport and the improvement of safety. They are shared goals but are only partially realized, at least in Italy.

In fact, in many cases PUTs work to define the vehicle circulation plan (circulation directions), regulate the intersections (nowadays more and more often with appropriate roundabouts where space is available), establish restricted traffic areas (to improve urban quality), improve integration with public transport stops (sometimes) and program pedestrian end cycle pathways (often incomplete and mixed with motor vehicles).

Road traffic management is still important but should be included in a strategic plan (as indicated by the EU with the Guidelines for Sustainable Urban Mobility Plans—EU 2014) [5] which identifies:

- the general objectives,
- the actions to be implemented
- the projects to be implemented,
- funding resources and the sources of funding,
- an organic framework of interventions and their timing priorities.

For twenty-three years, the *Living and Walking in Cities Conferences* focus on urban quality, non-motorized mobility, which is defined as weak or slow, and forms for reorganizing the urban road system to control the speed of motor vehicles (the Traffic Calming procedures) and to reorganize the same road infrastructure in a "sustainable way" for all different users.

4 SOME BEST PRACTICES THAT WE ALL KNOW

There are several best-practice references at European level; Only some of the most famous as an example:

- The Copenhagen Finger Plan (built in 1947 but upgraded over time and still confirmed in 2007) centered on the city's five rail connection lines,

- The Emerald project, the winner of the Helsinky 2050 international competition, based on the structure of the rail network,
- The Vauban District of Freiburg, where the community of residents has decided to minimize the use of private cars, recovering urban spaces for collective life.

For this reason, the title of this note is "*Against traffic engineering axioms*": the aim is to overcome the idea that traffic flow management and regulation can pursue urban quality, sustainability and security perceived by citizens.

An organic framework of strategies and operational tools is needed, which in international literature and practice means drafting and managing medium-term Integrated Transport Plans (at least 10 years).

5 SUSTAINABILITY (ENVIRONMENTAL AND ECONOMIC) GOALS

Sustainability goals and the tools available today for Smart Cities can imagine (and design) urban mobility focused on public transport (efficient and secure and therefore generally on guided link and with controlled density systems). Public transport must be integrated By bicycle routes and safe pedestrian paths.

The motorized private mobility system should, above all, retain local accessibility features (in territorial systems with low density) and short-haul routes in areas of poor public transport (better if with shared cars).

A fully redesigned transport system with these goals would also have considerable savings for families; The cost of car management is estimated in the following values:

The cost of car management [2] is estimated **in Europe** equal to **1,200 billion euro per year** [*without infrastructure investment costs*]!

Part of these large resources could be spared and used for other social purposes (welfare, health, education, leisure). Obviously, this is a long-term plan that requires a profound industrial re-engineering of the Automotive World Industry, which employs a large number of workers (though much less than in the past with the introduction of automation in industrial processes) and still represents a Relevant share of the Gross Domestic Product in many countries of the World.

Estimation of yearly car costs in IT/EU/USA

	NUMBER OF VEHICLES (year 2016)	
ITALY	37 millions	
EUROPE (28)	262 millions	**worldwide 1.3 trillion drive**
USA	256 millions	
	COST OF THE FUEL (2016)	
ITALY AND EUROPE	1,30 €/liter	
USA	0,52 €/liter	
	ANNUAL COST OF A CAR MANAGEMENT	
ITALY	4500,00 €	
EUROPE (28)	4000,00 €	
USA	3800,00 €	
	TOTAL ANNUAL COSTS FOR THE FAMILIES	
ITALY	193 billion €	
EUROPE (28)	1200 billion €	
USA	980 billion €	
GROSS DOMESTIC PRODUCT [GDP] – at current market prices & ref. car costs		
ITALY	1600 billion €	[car costs = 12,0%]
EUROPE (28)	14000 billion €	[car costs = 8,6%]
USA	13700 billion €	[car costs = 7,2%]

Figure 3. Our Estimation of car transport costs in the World in the year 2016 [1].

These elements show that innovation on passenger cars (full electric or hybrid vehicles, new fuels such as hydrogen, automatic driving) is still just an economic and industrial goal. These innovations do not change the rules of mobility and its elements of insecurity, and therefore it is product innovation but not a positive evolution of the system.

Finally, if we are to replace the entire world fleet of cars (1 billion and 300 million vehicles) with new vehicles (zero emissions), over the next 20 years, citizens should invest about 30,000 billion US dollars. Why not decide to use them (at least in part) in real sustainable mobility projects based on collective electrical traction systems?

It is clear that this investment is a cost for citizens and not for states with more and more limited public resources, but substantial savings in private spending on their individual mobility would allow for rethinking public transport rates (raising them for improvements Quality and accessibility levels). Technological innovation in collective transport today would make it possible to redesign a city for people and not for car.

This is the true challenge of this century for the whole world! It is necessary to finalize cultural intervention for a collective sharing of a future scenario and then set up a new industrial policy and a different system of quality of life.

REFERENCES

[1] Automobile Club d'Italia, Tabelle nazionali dei costi chilometrici di esercizio di autovetture e motocicli elaborate dall'ACI - Art. 3, comma 1, del decreto legislativo 2 settembre 1997, n. 314. (GU Serie Generale n.298 del 22-12-2016 - Suppl. Ordinario n. 58) - Vigente al: 1-1-2017.

[2] Cappelli A, Libardo A., "Costi del sistema ferroviario a confronto con i sistemi concorrenziali", in Ingegneria dei Sistemi Ferroviari: tecnologie, metodi ed applicazioni, a cura di Stefano Ricci, EGAF Edizioni, Collana Ingegneria dei Trasporti a cura di Agostino Cappelli, Giovanni Corona, Gabriele Malavasi e Stefano Ricci, Forlì dicembre 2013, pag. 481–494, ISBN: 978-88-8482-545-2.

[3] Codice della strada (Edizione giugno 2017) (Decreto Legislativo 285/92), Testo aggiornato alle modifiche introdotte con il Decreto Legge 24 aprile 2017, n. 50 e Decreto Legislativo 29 maggio 2017, n. 98, Gazzetta Ufficiale della Repubblica Italiana, 24 giugno 2017.

[4] Dalla Chiara, B., ITS nei trasporti stradali: Tecnologie di base della telematica per i trasporti, con approfondimento di metodi ed applicazioni, Egaf Edizioni, Forli, Edizione: 1° - Marzo 2013, ISBN: 978-88-8482-477-6.

[5] European Union: GUIDELINES "Developing and implementing a Sustainable urban mobility plan", Funded by the Intelligent Energy Europe co-funded by the Intelligent Energy Europe Programme of the European Union, 2013.

[6] Ministero dei Lavori Pubblici, Direttive per la redazione adozione ed attuazione dei piani urbani di traffico. Art.36 del decreto legislativo 30 aprile 1992, n.285 Nuovo codice della strada, 12 aprile 1995.

[7] Ministero delle infrastrutture e dei trasporti [IT], dipartimento per le infrastrutture, i sistemi informativi e statistici direzione generale per i sistemi informativi e statistici, ufficio di statistica, sistema statistico nazionale, "Conto nazionale delle infrastrutture e dei trasporti 2015/2016", Roma—Istituto Poligrafico e Zecca dello Stato S.p.A. 2017.

[8] Ricci, S., Ingegneria dei sistemi ferroviari: Tecnologie di base dell'ingegneria dei sistemi ferroviari, con approfondimento di metodi ed applicazioni, Egaf Edizioni, Forlì, Edizione: 1°—Dicembre 2013, Edizione: 1°—Dicembre 2013, ISBN: 978-88-8482-545-2.

Town and Infrastructure Planning for Safety and Urban Quality – Pezzagno & Tira (Eds)
 ISBN 978-0-8153-8731-2

Promoting buses to increase walkability

T. Pharoah
Transport and Urban Planning Consultant, UK

ABSTRACT: This paper advocates a fairly obvious, but generally overlooked aspect of promoting more sustainable travel: The more that urban structure and travel patterns are orientated towards public transport, the more walking people will undertake.

The car now unfortunately dominates people's extended geographical range of daily life beyond walking distance, but local buses can provide a less damaging alternative. Walking and public transport are often in transport planning treated as separate modes of travel. In this paper the case is made for treating them together, with the objective of increasing the use of both and thereby increasing sustainability, safety, and quality of travel experience.

Some assertions are made about the interconnection between walking and bus use, and relevant data are explored using the British National Travel Survey. This is followed by some consideration of the planning methods and principles needed to promote the walk-bus partnership.

1 INTRODUCTION

This paper advocates a fairly obvious, but generally overlooked aspect of promoting more sustainable travel: The more that urban structure and travel patterns are orientated towards public transport, the more walking people will undertake. The public transport—walking partnership serves all of the urban sustainability objectives, including a health and fitness agenda, and therefore should be prominent in local planning and transport policy.

For millennia, mankind has sought to get access to places well beyond reasonable walking distance, whether by the use of animals, or wheeled vehicles. Few people today choose to live without using some form of wheeled transport, and towns are rarely sufficiently compact and diverse so that people could choose to rely solely on walking to meet their various needs. The car now unfortunately dominates people's extended geographical range of daily life beyond walking distance, but local buses can provide a less damaging alternative.

Walking and buses together form a natural partnership. It is not often a partnership that is consciously promoted, or even acknowledged, perhaps because it is so ordinary and everyday. A partnership involves two (or more) entities acting together to produce an outcome that is better than could be achieved by the entities acting alone. In this case, walking and bus use together can produce a sustainable transport outcome and a high quality of urban life. This is the case wherever buses operate effectively to absorb a significant proportion of daily travel. Looking at it the other way round, places that do not have effective bus (or other transit) services will almost certainly have less sustainable transport, with a high proportion of travel by private car, and a small proportion of journeys on foot. Because transport investment usually reflects travel patterns, low levels of walking are also associated with poor walking infrastructure, and unsafe or poor quality walking experiences.

Walking and public transport are often in transport planning treated as separate modes of travel. In this paper the case is made for treating them together, with the objective of increasing the use of both and thereby increasing sustainability, safety, and quality of travel experience.

Some assertions are made about the interconnection between walking and bus use, and relevant data are explored using the British National Travel Survey (Ref 1). This is followed by some consideration of the planning methods and principles needed to promote the walk-bus partnership.

2 THE WALK-BUS RELATIONSHIP

The assumption of much urban planning in Western Europe and beyond is that current travel patterns are environmentally unsustainable and damaging to health and lifestyles. As a generalisation, there is too much use of private cars, and concomitantly not enough use of public transport, walking and cycling. In England, for example, the car accounts for 64% of trips and 77% of distance travelled. Only 16% of people use a bus 3 or more times a week, and 59% of people take the bus less than once a month (2015 NTS data). Policies are aimed at achieving "mode switch" from car to the other modes. For example, the UK Government's national planning policy document (Ref 2) states that planning should "*actively manage patterns of growth to make the fullest possible use of public transport, walking and cycling...*" It has to be said, however, that such policies have so far made little difference to what happens on our roads and streets. The car continues to dominate in mode split charts, and areas of urban growth are adding to car dominance and total car kilometres.

The urban form that suits walkability is also the urban form that suits the use and operation of buses. Activities are grouped together at nodes, so that people walking to them can for a single journey satisfy two or more of their purposes—multi purpose journeys are possible. These nodes also can be served by public transport services, which necessarily are limited in the frequency of stops that they can make. Density of development also plays into this relationship, because the greater the proportion of people and activities at or close to these "nodes", the greater will be the proportion that of people that can avoid car use. In fact the urban structure that suits the walking-public transport partnership is diametrically opposite the urban structure that suits travel by car.

The current relationship between walk and bus may be described as passive, but inevitable: *Passive*, because it is rarely deliberately planned for; *Inevitable*, because without walking, buses would be cut off from their passengers, and would in effect be useless.

Transit Oriented Development (TOD) as a concept for organising urban development originated in the United States, perhaps because in so many cities the lack of transit and related development is so obvious. In Europe, by contrast, most cities developed around historic centres and transit corridors, and maybe that is precisely why the relationship has so often been taken for granted. While TOD is important, it is noticeable in the many documents exploring and advocating the concept, that the emphasis is almost wholly on the organisation of urban growth around track-based systems (e.g. Metro, Tram, BRT) (See for example Ref 3). The same is true of the main European initiatives in Transit Oriented Development, for example the large scale new urban quarters in Amsterdam (Ijburg), Stockholm (Hammarby Strand), and Vienna (Seestadt). The original codification of the concept by Calthorpe in the 1980s did, however, consider two levels of TOD: urban level in relation to rail stations, and neighbourhood level in relation to suburban local bus services (Ref 4).

Development that exploits the bus-walk partnership can and should also be characterised as Transit Oriented Development, but this is very rarely made explicit. The difference between local bus and track-based transit is largely one of scale, in particular passenger carrying capacity and the resulting building densities and mix of activities that can be supported; but the principle of maximising connectivity on foot at transit nodes (whether stations or bus stops) remains the same.

Three assertions are made below which help to build the case for promoting the bus-walk relationship in urban development.

2.1 *Assertion 1*

Buses are more important than trains.

Railways and tramways rely on people walking to and from the stations, but local planning should focus on buses first, because they provide more frequent and widespread connectivity. For example, in England, buses cater for 7% of trips, compared to 3% for trains.

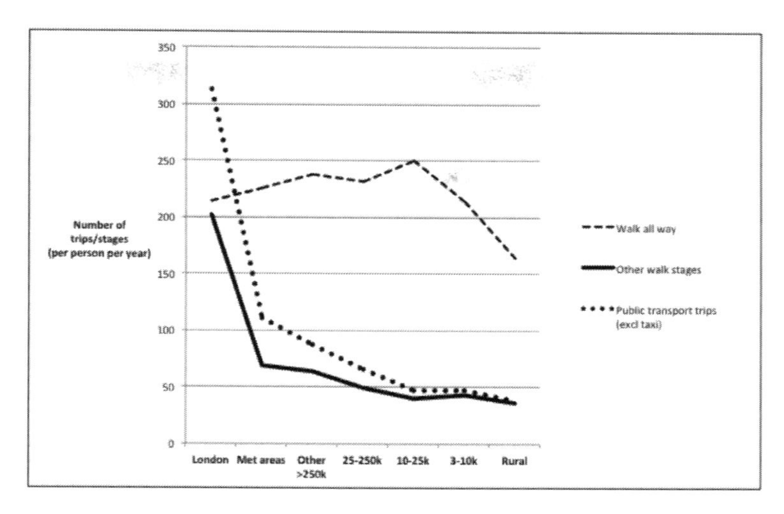

Figure 1. Walk and public transport trip stages by size of settlement, Great Britain 2007–2009 (NTS special tabulations).

2.2 *Assertion 2*

People who own cars make little use of buses.

In England, car users make less than one tenth of the number of bus trips than people with no car. They also make less than half the number of trips entirely on foot. Therefore if car ownership can be suppressed, this will increase both walk trips, and bus journeys together with walk access stages. Can ownership can be suppressed if public transport and access to it is of good quality.

2.3 *Assertion 3*

People who use buses and other public transport make more trip stages than those who don't.

Bus users make more trip stages on foot than car users, because they walk to and from bus stops. Of course, people also walk to and from their cars, but car trips are generally door-to-door, involving very short walks, whereas trips by public transport rarely are. People travelling by public transport will usually need to walk a distance to reach the stop or station, and often a distance at the other end of the journey also. Consequently, the total number of walk stages that people undertake is strongly related to the extent of public transport use, as shown in Figure 1.

The chart shows travel characteristics for different sizes of town/city. It shows that while the number of "walk all the way" trips is more or less constant across the different population sizes (apart from very small settlements), the number of other walk stages is strongly associated with the level of public transport use.

3 PROMOTING THE BUS-WALK RELATIONSHIP

Very often transport and planning policies will advocate "mode switch" from car to walking, cycling and public transport. Achieving mode switch to walking alone can be difficult, since people will already walk to facilities that are close by (perhaps reflected in the stability of walk-all-the-way trips across the spectrum of different settlement sizes, as shown in Figure 1). Getting them to walk instead of making a car trip will be difficult since the car is used mostly to make journeys that are further than a convenient walking distance. In England, only 3% of car trips are under 3 kilometres. It is possible (and desirable) for trips to be switched from car to walk, but this will probably involve people choosing a closer destination, for example using a local shop rather than a more distant supermarket, rather than transferring an existing car trip to walk only.

Consequently, increasing the amount of public transport use can be a more effective way of getting people to walk more. This can be achieved by relatively short-term measures, such as improving the quality of service (e.g. higher frequency, extended operating hours, better reliability), or introducing disincentives for car use (e.g. parking controls and charges). In the medium to longer term, changes in land use structure and density can reinforce the choice of walking and public transport. Given the cost and lead times involved in providing rail-based transport, the conclusion can be narrowed to say that to increase the amount of walking, bus travel should be promoted.

Given the strong negative relationship between bus use and car ownership, suppressing car ownership is a potentially powerful policy aim in support of sustainable transport. The politically motivated view that policy should be directed only towards less car use rather than less car ownership is misplaced. Once a car has been acquired, there is immediately little incentive to use the bus. In order to make it possible for people to avoid car ownership, however, it is necessary to have not only good local connectivity on foot, but also good quality local bus services that operate reliably and at high frequency throughout the day and evening and weekends. If this is not provided, not all trips can be served, thereby compelling people to acquire a car.

4 CYCLING IN RELATION TO THE BUS-WALK PARTNERSHIP

This paper has not so far talked about the role of cycling. It is more often the case, in policy making, that cycling is considered at the same time as walking. But rarely is either considered alongside public transport. Why should this be? Promoting more cycling is surely a good thing if it is achieved at the expense of private motorised transport (less traffic, better health, less pollution, noise and danger). But the benefits are perhaps less clear if more cycling is achieved at the expense of less public transport use, or less walking. Abstracting people from buses to get them onto bikes reduces the passenger base (reduced revenue and hence reduced service quality and/or viability) and also may diminish the benefits of nodal development, and hence degree of walkability. There are additional complications in terms of street design, because cycle tracks around bus stops are tricky, and cyclists and buses do not mix well on general-purpose streets. These consequences and interactions are complex, and difficult to apply in the planning of specific development schemes, but that does not mean that they should be neglected.

My own view is that cycling should be promoted where is can be a realistic alternative to the car, and where mode switch can be achieved without weakening the bus-walk partnership. In practice this means concentrating first on established car-based urban areas where decent public transport is unlikely to be viable, and where the environment is poorly suited to walking. Low density suburbs, out of town retail and business parks, and areas dominated by unattractive roads and parking are where typically cycling is more attractive than walking, or at least can be made so with relatively simple cycle infrastructure. The Netherlands has many very good examples of this.

Achieving mode switch to cycling in car-based areas therefore should be a lot easier than getting mode switch to walking. This conclusion is reinforced by the fact that cycling distances tend to be close to those of local car journeys, as shown in Figure 2.

Whereas walking and public transport (especially local buses) form a perfect partnership, as argued in this paper, the relationship between cycling and public transport is less comfortable and more complex.

The British Government in May 2017 issued its first statutory "Walking and Cycling Investment Strategy" (Ref 4). This emphasises national policy to "*make the fullest possible use of public transport, walking and cycling*", and yet there is no acknowledgement of the interaction between the sustainable modes and their differences, nor any guidance on how to resolve the potential conflicts that can arise. In this document walking and cycling are effectively lumped together as more sustainable ways of catering for short trips, despite the reality that cycling trips are more akin to those of bus or even car, than to walk trips.

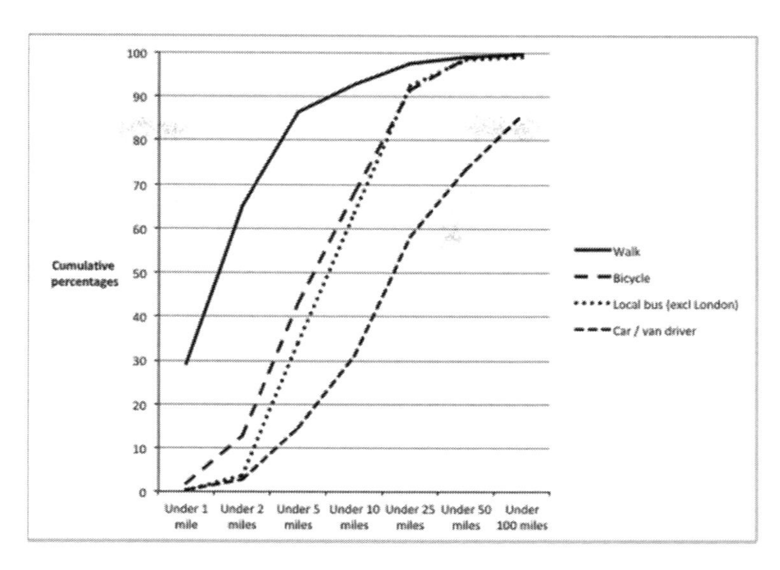

Figure 2. Cumulative percentages of travel by walk, bicycle, bus and car (NTS England 2013).

5 CONCLUSION

Increasing the amount of walking that people undertake as part of their journey making (i.e. excluding purely recreational walks), is desirable for a range of reasons. An effective way of achieving this, especially in the short term and at reasonable cost, is to promote the use of buses. The planning of bus services alongside improved accessibility to bus stops on foot is a neglected planning tool that deserves more attention in local development and transport plans. The development of new areas should be planned from the start to favour access to bus stops, the gathering of activities close to bus stops, and the provision of high frequency, reliable and comfortable bus services.

In summary:

1. Increase public transport accessibility, with better services and better access on foot to stops;
2. Provide homes within walking distance of schools, shops, parks, playgrounds, health facilities;
3. Group non-residential activities together and at bus stop locations (public transport nodes);
4. Build and create facilities that serve a wider population only in centres that are connected by bus (or other public transport) to places where people live, and to local public transport nodes;
5. Stop building at locations that lie outside centres or their connecting radial public transport corridors;
6. Do not provide free and plentiful parking at non-residential facilities.

REFERENCES

[1] Department for Transport (UK) National Travel Survey (series ongoing).
[2] Department for Communities and Local Government, (UK Government), 2012, "National Planning Policy Framework" page 6.
[3] Nordregio (Research Project: CASUAL Urban Europe), 2016, "Transit-oriented development and sustainable urban planning" (available at http://tinyurl.com/ly4xz6d).
[4] Department for Transport (UK), May 2017, "Cycling and Walking Investment Strategy" (available at http://tinyurl.com/mu5jqpw).

Town and Infrastructure Planning for Safety and Urban Quality – Pezzagno & Tira (Eds)
© 2018 Taylor & Francis Group, London, ISBN 978-0-8153-8731-2

Cycling in Amsterdam, if you can do it here, you'll do it anywhere

R. Eenink
Swov Institute of Road Safety Research, The Netherlands

ABSTRACT: Cycling in the Netherlands is healthy, clean and essential for accessibility. It is also a challenge because high volumes of cyclists interact with cars, trams, pedestrians, mopeds and other cyclists. Therefore, the cycling infrastructure was adapted in the last decades and many separate cycle paths were introduced. Cycling fatalities were reduced but not serious injuries. The Netherlands has more serious road traffic injuries than ever (21,300 in 2015), and around 2/3rd are cyclists. Half of all seriously injured are cyclists who were not hit by a motor vehicle. The injury rate has increased as well, so it is less safe now. The main reasons seem to be the quality of the cycling infrastructure: around 50% of all cycling crashes is related to kerbstones, road surface, shoulders etc. Separating cars and bicycles is not enough, attention must be given to making cycling infrastructure safe, including wider cycle paths with no obstacles.

1 INTRODUCTION

Cycling is an essential mode of transport in the Netherlands, on average people in the Netherlands cycle annually 880 kilometres, which amounts to a total of 15 billion bicycle kilometres each year. There are more bicycles (22.5 million) than people (17 million), and soon one in ten cyclists will own an electric bicycle (now 1.3 million). Cycling is cheap, healthy, clean and in our busy cities often the fastest way to travel. In the last decade, 2005–2015, cycle use has increased by 11%. But cyclists are also vulnerable road users, often young and inexperienced or senior and fragile. They are unprotected and seldom wear helmets, using a mode of transport that is inherently unstable, and which balance is easily compromised.

2 ROAD SAFETY

In the Netherlands we had our worst road fatality records in the early Seventies (Fig. 1).

Four modes are dominant: car, moped, bicycle and walking. Measures such as the improvement of safety features in cars, separation of cars and mopeds from cyclists and pedestrians, introduction of traffic calming, pedestrian—and 30-km/h zones, acceptance of helmet and licensing laws for moped riders has improved the safety for car occupants, moped riders and pedestrians, but far less so for cyclists. This may be due to cyclists not wearing helmets, a strong rise in bicycle travel—up 11% in the period 2005–2015, a rise in elderly cyclists because of population ageing, and the popularity of electric bicycles, which may ride at higher speeds.

These unfavourable trend, makes it most unlikely that the Netherlands will meet its National safety target for severely injured, namely less than 10,600 in 2020 (see Figure 2).

Unfortunately, severe injuries are rarely registered by the police, therefore we have limited knowledge about their crash characteristics. What we know is from hospital data linked to the limited police records. Severe injuries in 'crashes with motor vehicles involved' (M-crashes) seem to decrease slightly since 2000, but came to a standstill in 2010. Nowadays, the majority of severely injured are cyclists involved in single accidents, (N-crashes), due to falls from a bicycle, crashing into other cyclists or pedestrian, or bumping into obstacles such as trees or sign posts.

Figure 3 shows the situation in 2015 for various vehicle types.

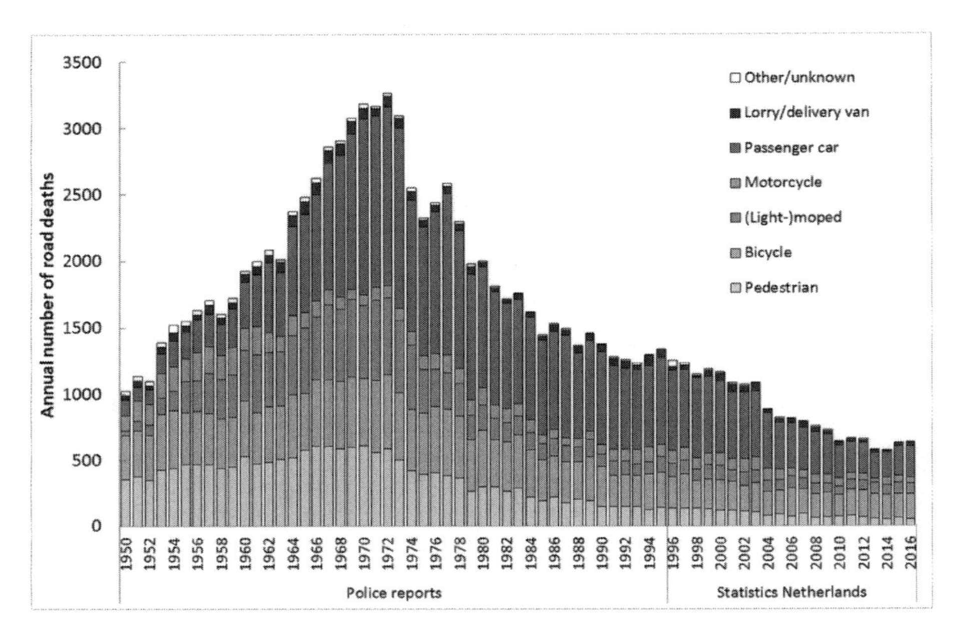

Figure 1. Road fatalities in the Netherlands 1950–2015.

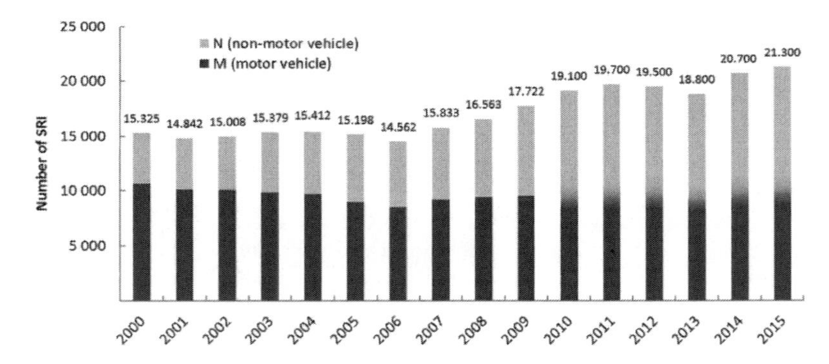

Figure 2. Serious Road Injuries (SRI) in the Netherlands (MAIS2; 2000–2015).

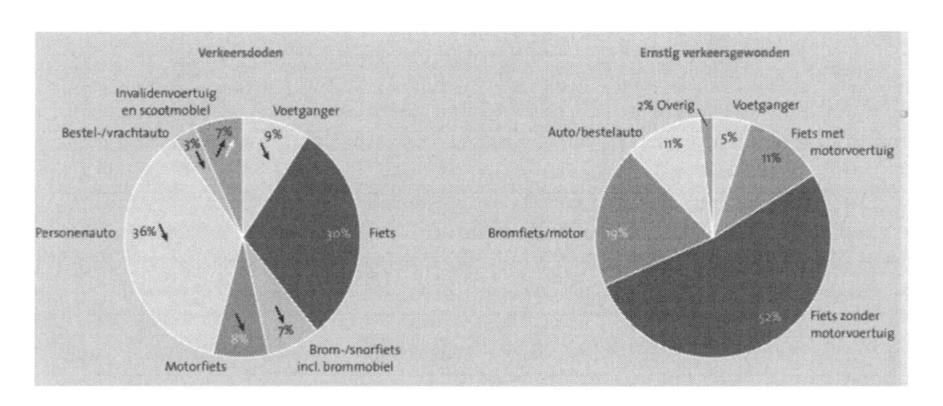

Figure 3. Falatities and serious road injuries in 2015 per vehicle type.

The left pie-chart shows fatalities, the one on the right serious road injuries (SRI). In orange we have the share of fatally injured cyclists (30%). The share of SRI cyclist is divided in 2 categories. For N-crashes the share is 52% and for M-crashes it is 11%, which amount to a total share of 63%. So, almost one third of all road fatalities and two third of seriously injured road users are cyclists. In only a period of three to four decades, road safety in the Netherlands has shifted from a motor-vehicle-centred-problem to a cycling-centred one.

3 RISK FOR CYCLISTS

As mentioned, police reporting of road crashes has deteriorated since 2009, with the unfortunate consequence that since 2009 fatality or SRI-rates are unreliability and no longer usable for analyses.

For this reason of 'poor' data, Figure 4 only presents these figures (fatalities or injuries per distance travelled) for the period 2000–2009. In this period, fatality rate has halved for car occupants Fatality rates for cyclists dropped as well, by about 35%. Given the trend in Figure 1 for cyclists and taking into account that exposure increased by 11% from 2005–2015, one can assume there is still a decrease since 2009, but at a much slower rate. For serious injuries however the trends differ for car occupants and cyclists. For car occupants there was still a 40% decrease, whereas for cyclists the injury rate increased by 35%. Again, looking at the situation since 2009 it is highly probable that this trend had continued since then. Apparently, safety in numbers does not apply here, but is fair to say no corrections were made for contributing factors, such as the ageing society and the increasing popularity of electric bicycles.

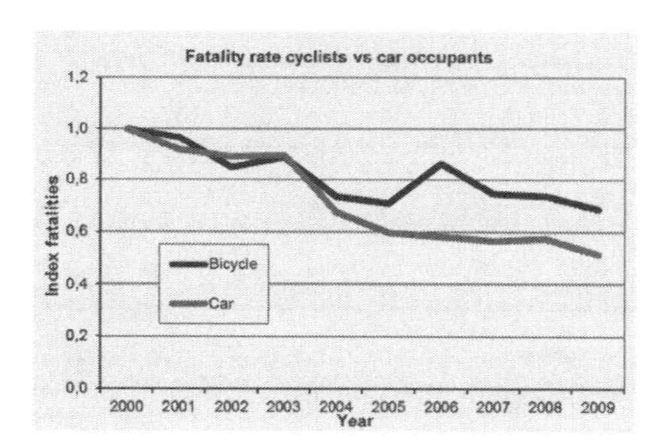

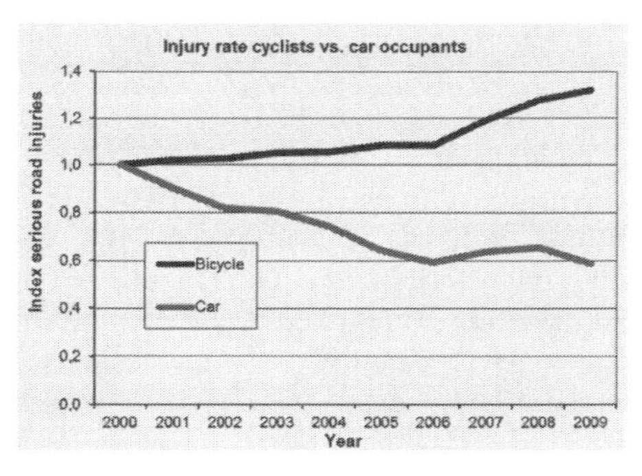

Figure 4. Fatality and injury rate trends (2000–2009) for car occupants and cyclists.

Figure 5. Rijksmuseum, the effect of a traffic light during rush hour (left) and separating pedestrians and cyclists.

Gemeente	Straat	1 of 2 richtingen	Intensiteit (aantal/uur)	Normbreedte (m)	Werkelijke breedte (m)	Druk?
Amsterdam	De Clerqstraat	1	758	4	2	Ja
	Weesperstraat	1	467	3	2,1	Ja
	Geldersekade	2	461	3,5 a 4	3,9	Nee
	Piet Heinkade	2	206	3,5 a 4	3,5	Nee
Den Haag	Prinsegracht	1	720	3	2	Ja
	Laan van Meerdervoort	1	85	2	2,1	Nee
	Laan van Hoornwijck	2	304	4	2,7	Ja
	Waalsdorperweg	2	97	3	3,5	Nee

Figure 6. Busy cycle paths in Amsterdam and The Hague.

4 BUSY CYCLE PATHS

The popularity of cycling also has its downside. That is crowding. This is a picture of the Rijksmuseum in Amsterdam. A top tourist destination, home of the world's most famous paintings, but also located at one of the busiest cycle paths of Amsterdam.

When the museum was renovated recently, the debate was whether the cycle path that goes right through the museum should stay. Of course, the cyclists won: Rembrandt, Vermeer, tourists, pedestrians and cyclists still share the same space. However, during rush hours there is a traffic light for cyclists (left picture), and pedestrians and cyclists are clearly separated (on the right).

This example of the Rijksmuseum is quite unique but illustrates the importance of (mass) cycling in centres of cities like Amsterdam, The Hague, Groningen, Leiden, and Utrecht. Outside city centres, in the residential areas, with a speed limit of 30 km/h, cycling density is usually not a problem. But on distributor roads (50 km/h) with separate cycle path, space is frequently insufficient, resulting in cycle path being too narrow, and not conforming to the National standards. This is illustrated in Figure 6, which holds the result of a recent SWOV study in Amsterdam and The Hague (de Groot-Mesken, 2015):

In this study, we looked at 2 cycle paths in one direction per city and 2 in both directions. According to the National Guidelines (ref) the amount of daily traffic ("intensiteit") is linked to a minimum width ("breedte"). As we see, this minimum standard is not met for half the number of the paths ("Nee").

5 LOW QUALITY CYCLING INFRASTRUCTURE

A high quality cycle infrastructure is of utmost importance for cycling safety. Schepers (2008) for instance estimated that about 50% of all single vehicle bicycle crashes are due to flaws in the infrastructures such as: kerbstones (14%), shoulder (7%), slippery road, grooves (17%), bollard, road narrowing (7%), bumps, holes, garbage on the road (6%). This calls for an assessment tool that can support road authorities in proactively identifying and prioritizing unsafe locations for cyclists. To meet this need, ANWB (The Dutch touring Club), the city of Amsterdam and SWOV collaborated in developing such an assessment tool, called CycleRAP. In terms of objective and approach, CycleRAP is similar to iRAP (EuroRAP) and Road Protection Score (RPS) for rural roads.

CycleRap has three composite indicators of cycling infrastructure

- Quality of surface;
- Obstacles on cycling infrastructure;
- Alignment.

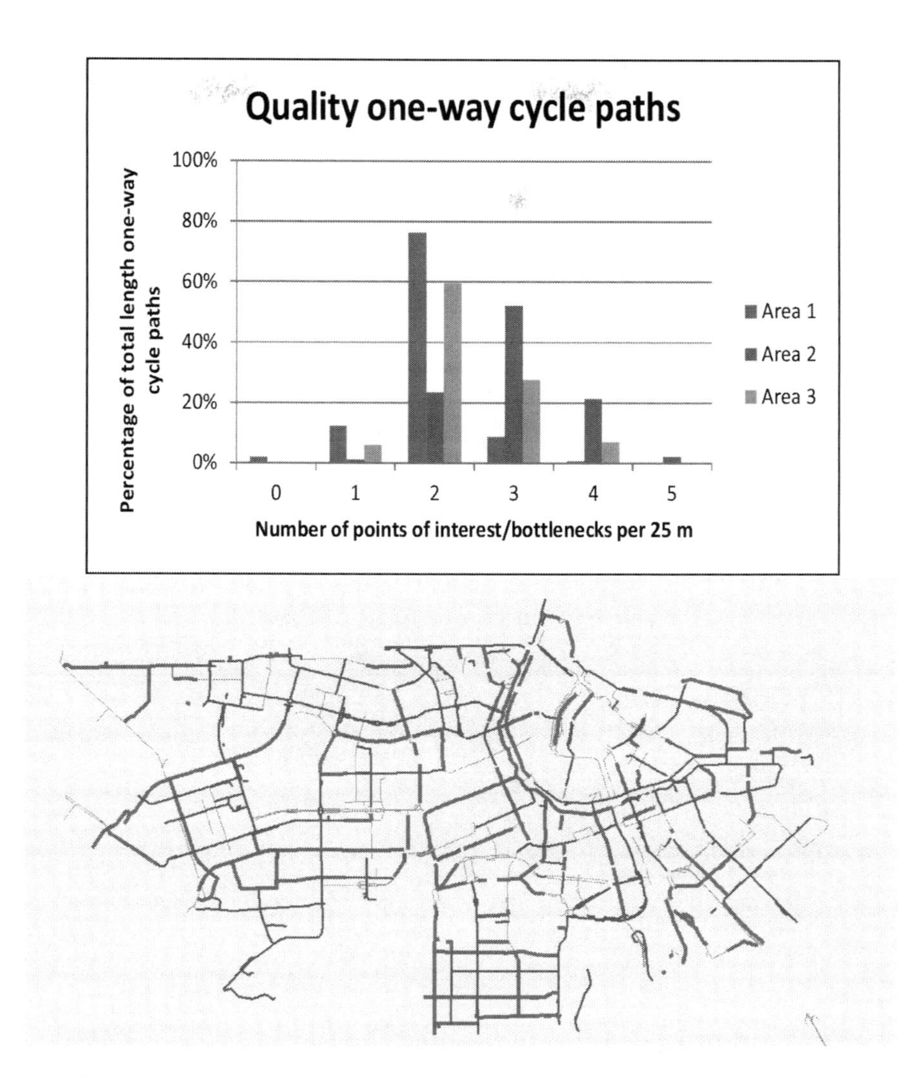

Figure 7. Results CycleRAP per area or on a map.

CycleRap uses the visual information from 360 degrees pictures of roads with a distance of only 25 metres apart, and codes relevant characteristics into a data base. The quality of the infrastructure is scored as the number of points-of-interest or bottlenecks per 25 m. This results into a spatial representation of hazardous locations as illustrated in this map of Amsterdam below.

6 AGEING POPULATION

The Netherlands has an ageing population, with a generation of senior citizens that is more healthy than previous generations. As shown in Figure 8, this results into a rise in cycling mileage in the older age groups.

An increase of about 25% for the 50 to 59 year olds, of 35% for the 60 to 74 year olds, and even 65% for the 75 olds and over. We have no reason to assume that these trends have changed after 2009. This increase in cycle use among the elderly has consequences for safety, because of their higher fragility. For instance, for 75 year olds and older, the risk of a fatal or serious injury is about a factor 10 higher than for middle-aged cyclists, for 50 tot 59 year olds

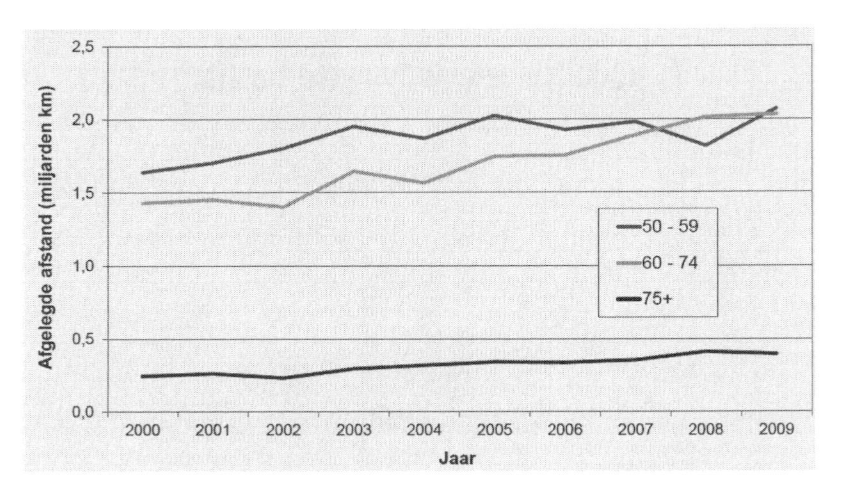

Figure 8. Distance cycled by seniors (2000–2009).

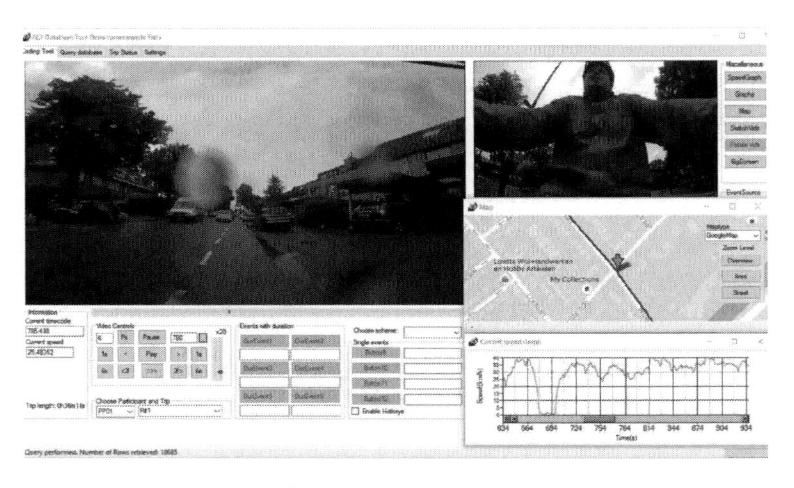

Figure 9. Data analysis tool for naturalistic cycling.

it is a factor 2, and for 60 to 74 it is a factor 4. This higher injury risk for the elderly is one of the reasons why the number of injuries per distance travelled has increased in the Netherlands.

7 ADVANCING RESEARCH METHODS: NATURALISTIC CYCLING

Now that cycling fatalities and injuries have become the core of the road safety problem in the Netherlands many research programmes have aimed to get a better understanding of the problem and to develop effective countermeasures. This has also sparked of research that has employed advanced methods that traditionally were only used for cars. Naturalistic cycling is such a research method. This method—using cameras and sensors fixed on bicycles or in cars, allows us to see how people drive or ride in their everyday life. SWOV has developed naturalistic cycling, using small sensors (GPS, movement etc.), a small on-board computer (data) and two cameras. Results are analysed in a special tool (see Figure 9).

With the tool specific areas of interest in the network, such as situations of harsh braking or high speeds can be analysed. Figure 10 presents some first results and shows the average speed of normal cyclists, eBike riders (support up to 25 km/h) and Speed pedelec riders (45 km/h) in- and outside the built-up area.

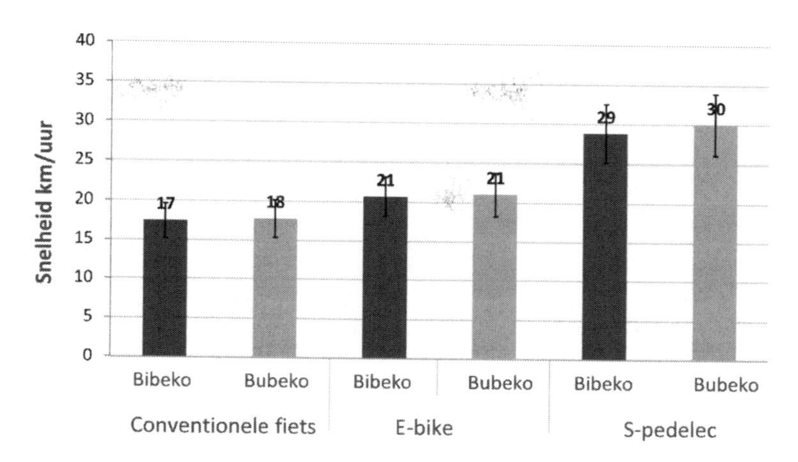

Figure 10. Average speed for normal cycling, eBikes and speed pedelecs.

In contrast to popular belief, there is no significant difference in the speeds of E-bikes and conventional bicycles. Also, there is no difference in speed for both types, whether they are used inside or outside urban areas. In contrast, the Speed pedelec is significantly faster than the other two bicycle types, and these extremely high speeds do not change whether these speed pedelecs ride within the busier urban areas or in the less busy rural areas (light green). Findings such as these can be used to anticipate safety issues or for development of protective measures, such as bicycle helmets that protect at high impacts.

8 CONCLUSIONS

Cycling injuries and fatalities are at the core of the Dutch road safety problem. Tools are now being developed to improve the safety of the current cycling infrastructure. To identify contributing factors to cycling crashes and design countermeasures, advanced research methods are called for.

REFERENCES

De rol van infrastructuur bij enkelvoudige fietsongevallen. Schepers, P. Delft, Directoraat-Generaal Rijkswaterstaat, Dienst Verkeer en Scheepvaart DVS, 2008.

Doelmatigheid handmatige intensiteitsmetingen, betrouwbaarheid beoordelingen infrastructuur en validiteit van het CycleRAP-instrument. Wijlhuizen, G.J.; Petegem, J.W.H. van; Goldenbeld, Ch.; Gent, P. van; Bruin, J. de; Commandeur, J.J.F.; Kars, V. SWOV R-2016-11.

Doorontwikkeling CycleRAP-instrument voor veiligheidsbeoordeling fietsinfrastructuur.

Fietsen in cijfers, Fietsersbond https://www.fietsersbond.nl/ons-werk/mobiliteit/fietsen-cijfers/.

Gebruikers van het fietspad in de stad. Aantallen, kenmerken, gedrag en conflicten. Groot-Mesken, J. de; Vissers, L.; Duivenvoorden, K. SWOV R-2015-21.

FACTSHEETS

https://www.swov.nl/en/facts-figures/factsheet/cyclists.

https://www.swov.nl/en/facts-figures/factsheet/road-deaths-netherlands.

https://www.swov.nl/en/facts-figures/factsheet/serious-road-injuries-netherlands.

Town and Infrastructure Planning for Safety and Urban Quality – Pezzagno & Tira (Eds)
© 2018 Taylor & Francis Group, London, ISBN 978-0-8153-8731-2

Making walking and cycling on Europe's roads safer

A. Avenoso
European Transport Safety Council, Belgium

ABSTRACT: Deaths among pedestrians and cyclists, the most vulnerable road users and whose use of the roads is being encouraged for reasons of health and sustainability, account for 29% of all road deaths across the EU. Pedestrians killed represent 21% and cyclists 8% of all road deaths.

Deaths of unprotected road users have been decreasing at a slower rate than those of vehicle occupants. Since 2010 the reduction in the number of pedestrian and cyclist deaths has slowed down markedly. The safety of unprotected road users should therefore receive special attention from policymakers at the national and European levels. As active travel is being encouraged, the safety of walking and cycling in particular must be addressed urgently.

In 2018 the Commission is going to propose revisions to the Pedestrian Protection Regulation and the General Safety Regulation which provides an opportunity to increase pedestrian and cyclist safety by setting new vehicle standards.

1 INTRODUCTION

Around 138,400 pedestrians and cyclists lost their lives on EU roads between 2001 and 2013. 7,600 were killed in 2013 alone. Deaths among pedestrians and cyclists, who are the most vulnerable road users and whose use of the roads is being encouraged for reasons of health and sustainability, account for 29% of all road deaths across the EU. Pedestrians killed represent 21% and cyclists 8% of all road deaths. But big disparities exist between countries.

Since 2010 the reduction in the number of pedestrian and cyclist deaths has slowed down markedly. The safety of unprotected road users should therefore receive special attention from policymakers at the national and European levels. As active travel is being encouraged, the safety of walking and cycling in particular must be addressed urgently.

There is a high level of underreporting of collisions involving pedestrians and cyclists. Deaths of unprotected road users have been decreasing at a slower rate than those of vehicle occupants. In the last ten years deaths among pedestrians decreased by 41%, those among cyclists by 37% and those among power two wheeler (PTW) users by 34% compared to a 53% decrease for vehicle occupants.

2 HOW DOES THE SITUATION VARY BETWEEN COUNTRIES?

2.1 *Pedestrians*

Pedestrian safety has improved in all EU countries over the last ten years. Yet over 5,500 pedestrians were killed in the EU in 2013 alone, representing 21% of all road deaths. Almost 73,300 have been killed since 2004.

For the EU as a whole, the number of pedestrian deaths has decreased by 5.5% on average each year over the period 2003 to 2013. However, in recent years the reduction in pedestrian deaths has slowed down markedly. Over the three years since the beginning of the decade the annual progress was only around 4%; in 2013 it was 2%. In some countries improvements in pedestrian safety are to a large extent a function of the overall developments in road safety. Countries that have made the biggest improvements in road safety since 2001, namely Latvia, Slovakia, Lithuania, Estonia and Spain, are also rapidly reducing the numbers of pedestrian deaths.

2.2 *Cyclists*

More than 2,000 cyclist deaths were recorded in traffic collisions in the EU in 2013 representing 8% of the total number of road deaths in those countries. Around 25,000 have been killed since 2004.

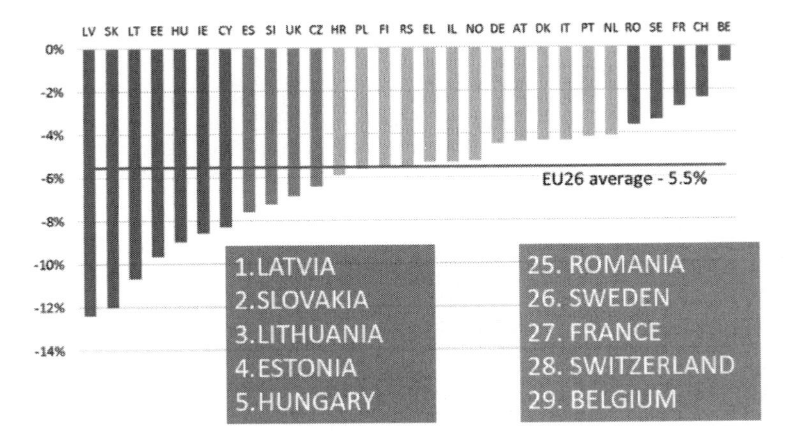

Figure 1. Average annual percentage change in pedestrian deaths over the period 2003–2013.

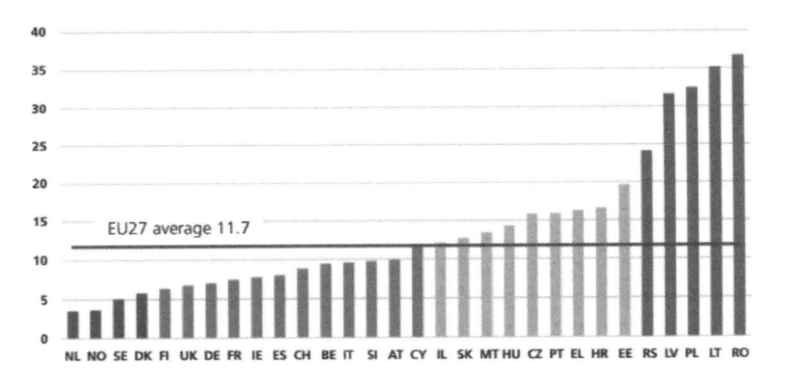

Figure 2. Average annual pedestrian deaths in 2011–2013 per million inhabitants in 2013.

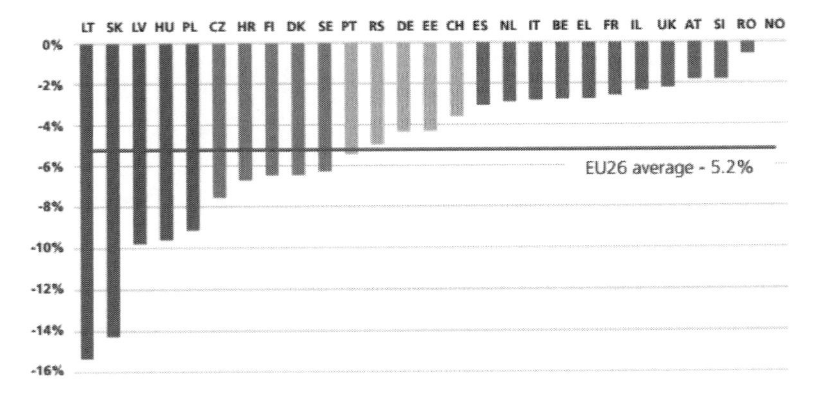

Figure 3. Average annual percentage change in cyclist deaths over the period 2003–2013.

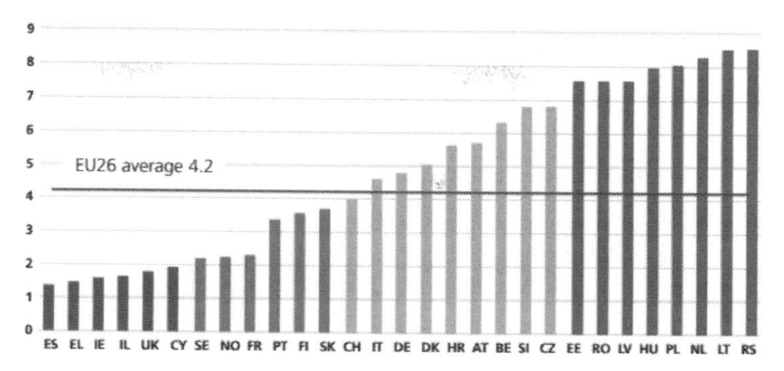

Figure 4. Average annual cyclist deaths in 2011–2013 per million inhabitants in 2013.

In the last ten years all EU countries have seen a reduction in the number of cyclist deaths. However, since 2010 the reduction in the number of cyclist deaths has stagnated with less than a 1% year-to-year reduction in the EU. 2,078 cyclists were killed in the EU in 2010, in 2013 this number dropped only to 2,009. This slowdown in the fall in cyclist deaths may well be partly related to a growing use of bicycles as a form of active travel among EU citizens. An increasing number of EU countries are adopting national strategies to promote Cycling 5, so it is possible that in recent years more people are choosing cycling as a means of transport. However, national cycling strategies should not only encourage cycling, but also promote high safety standards for bicycle users.

A high level of underreporting in the number of collisions involving cyclists exists. This is noticed when police reporting is compared to hospital records.6 Moreover, the rate of reporting is much higher for bicycle collisions with motor vehicles involved than for bicycle only collisions.

3 WHAT FACTORS AFFECT THE RISK OF DEATH FOR PEDESTRIANS AND CYCLISTS?

The risk of death as a pedestrian and cyclists varies, depending on a number of factors, including:

3.1 *Age*

In the EU, the risk of being killed as a pedestrian is consistently lowest for children, with 3.4 deaths per million child population, about double that for adults under 50, with 7.5 million deaths per million adult population. The greatest risks are for people aged 50–64 and especially for those over 65 with 13 and 28 deaths per million population in the age group respectively.

The situation is similar for cyclists. Mortality is least for children under 15, with around 1.1 deaths annually per million child population, more than twice as big for the adult population under 50 with 2.6 deaths per million adult population, and twice as great again for citizens aged 50–64 with 5.3 deaths per million population of this age group. The greatest risk is for people older than 65, with 10 deaths per million elderly population.

3.2 *Environment*

In the EU, over the period 2011 to 2013, 69% of all pedestrian deaths occurred on urban roads. Another 27% occur on rural roads, and 4% on motorways. For the EU as a whole, just over half of cyclist deaths occur in urban areas. These figures vary significantly by Member State.

3.3 *Gender*

There is extensive evidence to show that more males than females are being killed in road collisions in Europe, which is also the case for pedestrians. Figures from 2012 to

2015 show that males represented 64% of all pedestrian deaths and 78% of all cyclist deaths.

3.4 *Interaction with traffic*

Collisions with passenger cars make up 52% of the total number of cyclists deaths in the EU. Collisions with goods vehicles and buses account for 24% of cyclist deaths. 94% of pedestrians and 78% of cyclists are killed in collisions with a motorized vehicle.

4 WHAT CAN CITIES AND COUNTRIES DO TO IMPROVE SAFETY FOR PEDESTRIANS AND CYCLISTS?

Both walking and cycling are encouraged at EU level and also at a national level by a number of Member States. Increasing pedestrian and cyclist safety measures requires a combination of measures. Improved infrastructure in conjunction with developments in other areas related to the traffic system, like in-vehicle technologies and road user behaviour, can deliver high safety standards for all road users, especially the most vulnerable ones.

Areas that countries can focus on include:

- **Pedestrian and cyclist safety in urban areas**
  - Urban road safety characteristics
  - Safety potential of 30 km/h speed limits
- **Pedestrian and cyclist interaction with motorized vehicles**
  - Passenger cars—passive and active safety
  - Goods vehicles—a need for safer trucks design
- **Pedestrian and cyclist behaviour**
- **Passive safety for cyclists**

A major feature of urban road use is frequent and close interaction between unprotected road users and motor vehicles. However, almost half of all car trips in EU urban areas are for distances shorter than 5 km. Making active travel an attractive alternative to motorised transport can result in decreases in traffic noise, pollution and congestion and help unprotected road users. Lower speed limits can help reduce the likelihood of death and serious injury and can help create positive interaction amongst road users. They can also help to encourage people to walk and cycle.

Evidence also shows that the more pedestrians or cyclists there are using the road, the lower the risk to each individual from motor traffic, car drivers and other motorised vehicle users being more used to sharing the road with pedestrians and cyclists when more people walk and cycle.

Although an increase might, at least at first, lead to an increase in the number of people killed and seriously injured, the advantages of walking and cycling (a healthy life through regular exercise, benefit to the environment and higher quality of life) outweigh their disadvantages (in terms of the risk of death or injury).

Moreover, cyclists and pedestrians do not endanger other road users as much as car drivers do because of their lower speed and mass. So shifting a substantial proportion of short-distance car trips to walking, cycling and public transport can, if accompanied by measures to reduce the risks of walking and cycling, increase overall road safety.

Pedestrian and cyclist safety can be improved by setting new technical requirements for motor vehicles sold in the EU market. This is currently underway as part of the revision of the General Safety Regulations, which includes Pedestrian Protection Regulations.

New standards for vehicles' frontal protection systems, active in-vehicle safety technologies and safer goods vehicle cabin designs can all help to improve the safety of pedestrians and cyclists.

Correct user behaviour by pedestrians, cyclists and motor vehicle drivers is required to ensure the effectiveness of improvements to infrastructure and vehicles. It is important that pedestrians and cyclists, who need no licence to travel, have at least a minimum knowledge of road safety education and how to use the roads. Similarly, it is important that cyclists make use of lights, protective equipment, such as helmets, and safe clothing, such as reflective jackets.

Town and Infrastructure Planning for Safety and Urban Quality – Pezzagno & Tira (Eds)
© 2018 Taylor & Francis Group, London, ISBN 978-0-8153-8731-2

Advancing safe system: The need for realistic goals

R. Allsop
Centre for Transport Studies, University College London, UK

ABSTRACT: Safe System is an approach to road safety management that can be advocated as being the current state of the art; it draws comprehensively upon experience of recent decades in road safety management in many countries. Inspired by Vision Zero, the Safe System Sourcebook purports to eliminate death and life-changing injury from use of the roads. This paper questions whether the elimination of death and life changing injuries is at odds with reality. It calls for an adapted vision with realistically ambitious goals: one of zero *preventable* deaths and life-changing injuries.

Realistic goals are important for road safety in the realm of day-to-day political reality. Emphasis on realistic goals rather than remote prospects should not only help to advance Safe System but also help to align progress towards it with the promotion of active travel and the creation of places for living and walking in our cities.

1 INTRODUCTION

Safe System is an approach to road safety management that can be advocated as being the current state of the art; it draws comprehensively upon experience of recent decades in road safety management in many countries, notably in north-west Europe and Australasia. The first sourcebook for the Safe System approach was the ITF report *Towards Zero* (ITF 2008), and this has been supplemented by *Zero Road Deaths and Serious Injuries* (ITF 2016). Both reports were produced by international teams including contributors from many of the countries that have achieved the greatest improvements in road safety over the last four decades.

These two sources contain a wealth of sound advice and together identify as essentials of Safe System that:

- people make mistakes that will continue to lead to collisions;
- the human body can withstand only limited forces in collisions without death or life-changing injury resulting;
- those who design, build, manage or use roads and vehicles or who provide post-collision care all share responsibility for preventing collisions resulting in death or life-changing injury;
- road safety management should be aligned with wider economic, human and environmental goals;
- road safety interventions should be shaped to meet chosen long term road safety goals; and
- different elements of protection for road users should be managed holistically to reinforce one another and minimise the consequence of failure of any one element.

2 VISION ZERO AND THE ELIMINATION OF DEATH AND LIFE-CHANGING INJURY

Safe System owes much of its inspiration to *Vision Zero*, adopted by the Parliament of Sweden in 1997. This envisages road transport from which the risk of death or life-changing injury has been removed on the basis that it cannot be acceptable to trade life or limb for any benefits of road transport. Vision Zero is rightly credited with inspiring fresh ambition for road safety in many countries, and its assertion of the joint responsibility of system providers and road users for safety on the roads has become a cornerstone of Safe System.

Inspired by Vision Zero, the 2008 report conveys a strong presumption that the long term ambition of Safe System should be to eliminate death and serious injury from use of the roads. The report recommends, as a first step towards building a safe system approach, adopting *"the elimination of death and serious injury from use of the road transport system as the level of ambition for long term road safety achievement".*

The chapter describing the safe system approach concludes by stating that Safe System strategies being adopted in various countries are characterised by aiming to "eliminate all fatalities and serious trauma arising from road crashes in the long term".

The 2016 report speaks of several well-performing countries having *"adopted a long-term policy goal that no-one should be killed or seriously injured in a crash on their roads".*

The section entitled 'Description of a Safe System' begins by saying that the term Safe System refers to *"the vision or aspiration that zero fatalities and serious injuries from road crashes are ultimately possible",* but the section on leadership for a Safe System qualifies this by saying that a number of countries, cities and companies adopting Safe System thinking and practice have *"made zero preventable fatalities and serious injuries the ultimate goal of their policy".*

So the ambition to eliminate death and life-changing injury has come to be expressed as the ultimate goal of zero deaths and life-changing injuries, and this has evolved from being a vision to being regarded as ultimately possible—even though it is not known whether, let alone how, it can be achieved, even with autonomous vehicles. This lack of knowledge casts doubt upon the ethics of advocating elimination or zero as a goal of policy.

3 WHY NOT ELIMINATE?—WHY NOT AIM FOR ZERO?

Few would question that it should be the aim of a commercial airline that no-one be killed or suffer life-changing injury while flying in its aircraft—or that it should be the aim of a train operator that no-one using its trains should be killed or suffer life-changing injury through a train colliding, derailing or catching fire.

The fact that many airlines and train operators achieve this aim in most years in respect of deaths is used to support an expectation that this could also be a realistic aim for road transport.

But to do so is to ignore two important differences. First, these near-achievements of elimination or zero relate not to aviation or rail transport as a whole, but only to closely defined elements of these two forms of transport, while there continue to be many deaths elsewhere in aviation and rail transport. Secondly, both train and commercial airline operation are tightly managed activities in which those with responsible roles in the movement of the closely managed fleets of trains and aircraft are all highly trained and closely managed professionals.

In contrast, road transport in a country may well involve tens of millions of people, mostly with no more than general education, of whom more than half are entitled to drive the tens of millions of motor vehicles, most may cycle if they choose to, and almost all use the system on foot, so that those with responsible roles in movement by road comprise almost the whole population.

Vision Zero, in which road transport without risk of death or life-changing injury was first envisaged, was immensely influential in raising ambition greatly to reduce deaths and life-changing injuries on the roads, and can thus be credited with contributing hugely to the substantial reduction achieved across Europe in the last 20 years. But as zero evolves from being a vision to being seen as an achievable goal, there are parts of the basis for Vision Zero that call for examination.

In the 2008 ITF report, the then Swedish Road Administration state that

> "Human life and health are paramount ethical considerations, According to Vision Zero, life and health should not be allowed to be traded off against the benefits of the road transport system, such as mobility".

Yes, human life and health are very important, and should not be lightly traded off, but people and society do not regard them as paramount. In almost every human activity, people as individuals accept risk in return for what they see themselves as gaining from the activity: *safety is for living—living is more than just keeping safe*. Examples of this in use of the roads are cycling in heavy urban motor traffic and leisure motorcycling at high speeds on rural roads. For society as a whole there are many policy areas in which premature death and life-changing injury or illness can be reduced by allocating resources to them, but available resources are finite, so government and organisations have to judge just how many deaths and injuries to try to prevent in each policy area. Even within road transport, active travel, like walking in cities, is encouraged as promoting health and long life, notwithstanding that the risk of death or life-changing injury from falling or from collision with a vehicle is often greater than if most of the journey were made by car or public transport.

So Vision Zero is at odds with realities of the experience of individuals and society—and a vision that fails to adapt to these realities risks fading or rejection. But the whole world would be the loser if Vision Zero were to fade or be rejected. So we all need to reconcile its sound elements with the behavioural reality of choices made by road users and with economic rationality in use of resources.

In looking for an adapted vision it is helpful to distinguish between deaths and injuries that are, under current circumstances, preventable, and others that are, for the time being, not preventable. This distinction is mentioned or hinted at in places, but too rarely, in the ITF reports. It points to an adapted vision of *zero preventable deaths and life-changing injuries*.

A death or injury is preventable when:

- technical or behavioural means of prevention are established;
- society is ready to allocate resources to implement these means; and
- people are ready to accept the changes needed to implement the means of prevention.

The range of deaths and injuries that are preventable can be widened not only by devising new methods of prevention, including ones that require less resources or gain greater public acceptance, but also by increasing the allocation of resources and by gaining greater public acceptance through information and persuasion.

Such an adapted vision is one of *boundless ambition to use all means that are affordable and acceptable to the public to reduce without predetermined limit the number of deaths and life-changing injuries on the roads*.

With this understanding, *zero preventable deaths and life-changing injuries* becomes more than a vision, but instead a realistically ambitious goal.

4 WHY REALISTIC GOALS ARE IMPORTANT?

The danger that unrealistic goals may be counterproductive in terms of reducing deaths and life-changing injury on the roads is twofold: they may

- undermine the case for action in the eyes of decision-makers; and
- distract those working for road safety from upcoming challenges that have to be faced long before even zero preventable deaths and life-changing injuries comes within sight.

This is because high ideals that rightly underpin Safe System have to be applied in the realm of day-to-day political reality.

The principal decision-makers on road safety policy and its implementation are politicians in national, regional and local government. They tend to be both wary of the long term and wary of offering the prospect of outcomes that look too good to be true or are likely to be belied in the eyes of the sceptical public by short-term developments. Elimination and zero are high-risk on all these grounds—the more so because their advocates have no hard knowledge whether they are achievable at all, let alone how they can be achieved, even with the help of autonomous vehicles and related developments. Claims by advocates for road safety that they can be achieved in the long term threaten to undermine sound evidence-based claims that can be made for policies and interventions in the here and now.

Up to now the cases, both for resources for road safety and for interventions that people may at first find unwelcome, have been helped by the fact that the risk of death from using

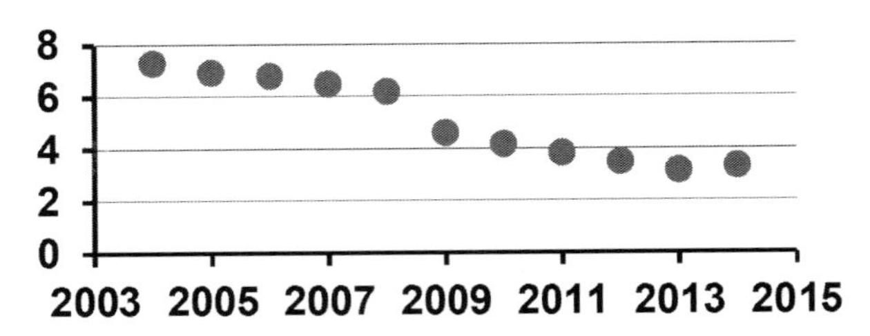

Figure 1. Risk of death per hour spent using the roads for people under 75 in Great Britain between 2004 and 2014.

the roads is substantially higher than in most activities shared by all or large sectors of the population. For any country or region for which mortality statistics by cause of death and an estimate of the time people spend using the roads are available, this disparity can be measured by the ratio shown roughly by way of example in the accompanying chart for those aged under 75 in Great Britain between 2004 and 2014:

The chart shows how the ratio has been falling as using the roads has become less dangerous compared with the rest of everyday life. As this ratio is reduced further, people will look harder at the cost and intrusiveness of further reducing death and injury on the roads compared with those of reducing death and injury elsewhere, for example in healthcare, social care and fire prevention. And as the ratio is brought close to 1, road safety will be competing for resources and public attention on level terms with other areas of preventable death and injury across society.

The challenge that this will present to those working for road safety in the context of increasing pressure on public funding and the administrative capacity of government does not seem to be recognised in the ITF reports—and a ratio of 1 implies numbers of deaths that are still far, far from zero.

5 CONCLUSION

Advancing the objectives of Safe System seems to require different understandings of

- the place of concern for safety in human life, and
- the place of concern about risk on the roads in the wider concerns of society

than those to be found in the ITF reports that are the sourcebooks for Safe System.

Practical progress in reducing deaths and life-changing injuries in road transport through securing resources for the development and implementation of technical interventions and public support for interventions that require changes in the lifestyles and behaviours of road users is likely to be helped by emphasis on the shorter-term goals that are rightly recommended in the ITF reports and reduced emphasis on remote prospects like elimination and zero.

Revised understandings may well also help to align progress towards Safe System with the promotion of active travel and the creation of places for living and walking in our cities.

REFERENCES

ITF (2008) Towards zero: ambitious road safety targets and the safe system approach. Paris: OECD Publishing.

ITF (2016) Zero road deaths and serious injuries: leading a paradigm shift to a safe system. Paris: OECD Publishing.

Town and Infrastructure Planning for Safety and Urban Quality – Pezzagno & Tira (Eds)
© 2018 Taylor & Francis Group, London, ISBN 978-0-8153-8731-2

Some general considerations and examples in the field of road safety

P. Hollò
KTI Institute for Transport Sciences Non Profit Ltd., Hungary

ABSTRACT: Practice shows that countries with quantitative targets in road safety improve more than countries without such targets. In spite of this, targets must be realistic and scientifically based. The paradigm shift of zero fatalities and zero serious injuries i.e. vision zero is an ethical platform and not a quantified target. A political target of a 50% reduction in serious injuries is realistic in theory. The analysis of the Hungarian time series shows clearly that the number of fatalities decreased dramatically and not the number of all injuries. And since people saved from fatalities may then become the ones seriously injured, the target may not be realistic. It is possible to set a quantitative target only after the careful analysis of the collision time series.

The presentation concludes with a short overview about the Hungarian road safety from which is drawn some important conclusions regarding the safety of unprotected road users.

1 TARGETS AND VISIONS

Unfortunately quantitative targets and visions are often mixed up even among professional road safety experts too. Target setting is a scientific activity which must be based on analysis of time series.

The quantitative and ambitious targets of the EU and some international organizations (ETSC, WHO, etc.) are only well-intentioned wishes; they are only pure political objectives and aren't based on any scientific or professional analysis.

Realistic targets can be set only in the following ways:

1.1 *Making a forecast based on the analysis of the previous time series*

Analysing the long time series of the past, the future development of road fatalities can be forecast in a more reliable way than a simple estimation. There are a lot of mathematical models which are applicable for this task.

1.2 *Summarizing the known effects of the planned measures*

There are a lot of information available about the efficiency of different road safety measures [1]. If we know, what kind of measures will be introduced in the future, we are able to estimate the number of saved lives in a more reliable way. Some examples: we can estimate the number of saved lives in case of a 10% decrease in average speed, or in case of a 10% increase in safety belt wearing rates, etc. Summarizing the effects we can get a realistic quantified target.

Practice shows that countries with quantitative target can reach faster development in the field of road safety than countries without such targets. In spite of this, targets must be realistic and scientifically based.

Even Swedish people say that the vision zero is an ethical platform and not a quantified target. There is no doubt, the biggest motivation and the best approach is, if we try to prevent all fatalities. Nowadays people speak about paradigm shift and zero fatalities or zero serious injuries [2]. The fact is that even spaceships have been crashed in spite of the fact that many people tried to guarantee their safety and applied all possible safety measures.

Figure 1. Hungarian time series.

Figure 2. Changes in the number of fatalities and serious injuries in Hungary between 1990 and 2016.

From this point of view, the political target "–50% serious injuries" can also be discussed at least. In theory it could be realistic, but only in case if the active and passive road safety measures were equally successful. I try to illustrate it on the example of the Hungarian time series (Fig. 1).

The analysis of the Hungarian time series shows clearly that mainly the number of fatalities (in other words: the outcome of the injuries) decreased dramatically and not the number of all injury crashes.

It means that the passive safety seems to be more successful nowadays in Hungary than the active one.

The explanation could be the well-known phenomenon of risk compensation.

Other interesting experience, that the trend-line shows a value of 470 fatalities for the year 2020. Taking into account that reaching of the EU target would require a lower value than 319; the EU target seems to be quite unreal at the moment. (In Hungary the number of road fatalities was 638 in 2011. If we multiply this by 0,5, the result is 319). What is more, even the number 470 means a great—in my opinion questionable—challenge.

On the one hand, according to the experience the people saved from the death will be mostly seriously injured. Therefore the target "–50% in serious injuries" does not seem to be entirely realistic at the moment. Not to mention the fact that we do not know the exact number of these injuries according to the MAIS3+ definition on EU level yet, we have only estimation.

Maybe it would be too early yet to set a quantitative target regarding the number of serious injuries. It is possible with responsibility only after the careful analysis of their time series.

On the other hand, the changes in the number of fatalities and seriously injured (according to the Hungarian definition) are very similar (Fig. 2). In spite of this, it can be observed that the two curves are diverging in recent years. In other words, the decrease in fatalities is higher than in serious injuries.

It seems that the same measures are applicable to prevent serious injuries as fatalities.

2 SHORT OVERVIEW ABOUT THE HUNGARIAN ROAD SAFETY SITUATION

In 2016 – after two deteriorating years—the number of road crash fatalities could be decreased. Last year so-called intelligent speed cameras were installed along the Hungarian road network. Yet the so-called "VÉDA" cameras are mainly speed measuring cameras, but later they would be able to detect red light running, the usage of hand-held mobile phone while driving, the non-wearing of safety belts, also, etc. The deterrent effect could be felt already before the installation of cameras. The cameras were installed not only outside but inside built-up areas as well. According to the latest news, now 365 locally installed and 160 mobile cameras are operating. The 365 locally installed cameras "are monitoring" the traffic at 134 different places (Fig. 3).

3 ROAD SAFETY OF DIFFERENT ROAD USER CATEGORIES IN HUNGARY

If we summarize the percentages of vulnerable road users (pedestrians, cyclists, motorcycle and moped riders), we can get 47,7% in 2016. It is almost the same as the value of killed car occupants (47,3%).

It means that almost 50% of killed road users are/were unprotected in Hungary in 2016.

The increasing relative frequency of killed car occupants and the decreasing percentage of killed pedestrians can be explained mostly with the increasing level of motorization. We can say in a very simple form: we travel more and more in car and we walk less and less as pedestrians. In the Figure 4 some interesting changes can be observed.

In 2013 the percentage of killed car occupants was almost 50%. Between 2013 and 2014 this percentage decreased to almost 40%. This favourable development was mainly due to the continuously increasing safety belt wearing rate. The temporary decrease in the safety belt wearing

Figure 3. Newly installed intelligent "VÉDA" cameras.

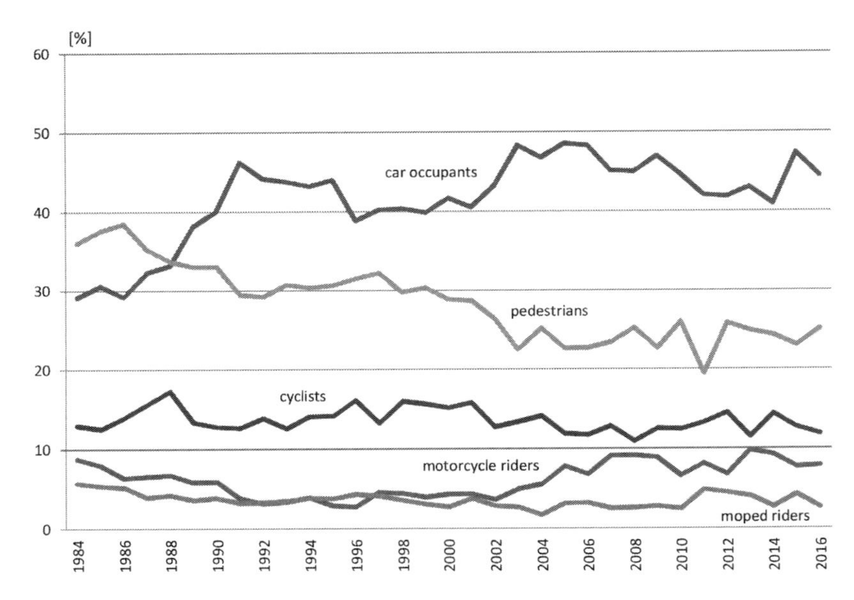

Figure 4. Distribution of road crash fatalities in Hungary.

rate had its immediate negative effect on the percentage of killed car occupants: it was 47,3% in 2015. After the implementation of the mentioned intelligent cameras the safety belt wearing rate increased again and as a result of it the percentage of killed car occupants decreased.

The percentage of the killed cyclists was more or less the same within the investigated period of time. On the one hand, it could mean that the road safety of this group of road users cannot be improved significantly, at least in this form of illustration. On the other hand, it could be the consequence of the increased cycle traffic and/or contradictory measures.

Among the motorized two wheelers the percentage of killed motorbike riders increased in recent years, but according to the Figure, it seems that the situation could be stabilized.

4 CONCLUSIONS

Realistic targets must be based on analysis of crash data and on professional forecast. We have to distinguish between quantified targets and visions.

Fatalities are decreasing much faster in Hungary than injury crashes, in other words: the passive road safety measures seem to be more successful than the active ones.

Speed management is of key importance inside built-up areas too.

In Hungary—after two deteriorating years—the number of fatalities could be decreased mainly due to the installed intelligent camera system.

Almost half of the killed road users are/were unprotected. Vehicle technology made a lot for the safety of vehicle occupants. Same development would be necessary in the field of unprotected road users too.

REFERENCES

[1] Rune Elvik, Alena Hoye, Truls Vaa & Michael Sorensen: The Handbook of Road Safety Measures, Second Edition, Emerald, 2009, ISBN: 978-1-84855-250-0.

[2] ITF (2016), *Zero Road Deaths and Serious Injuries: Leading a Paradigm Shift to a Safe System*, OECD Publishing, Paris.

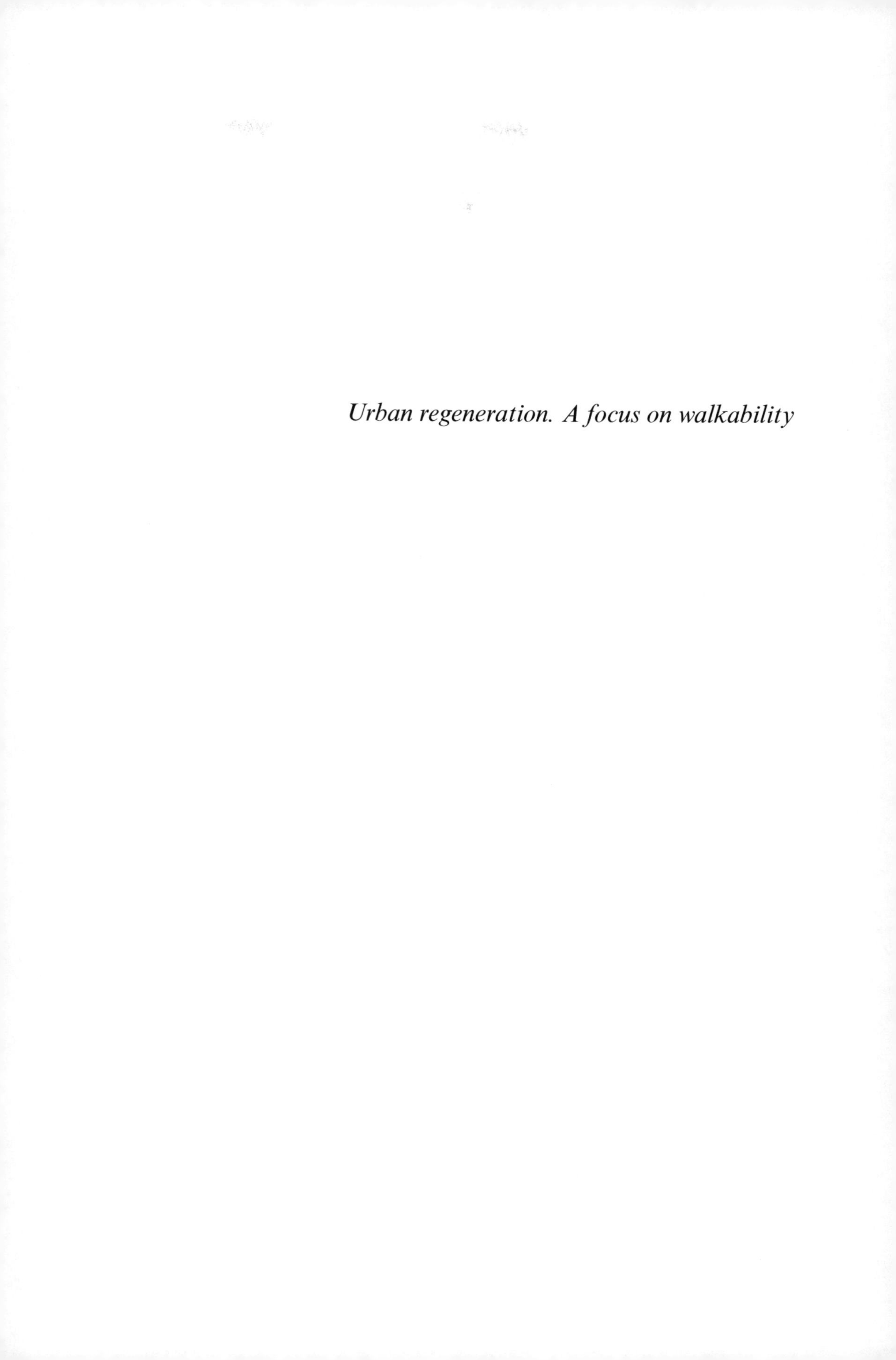

Urban regeneration. A focus on walkability

Town and Infrastructure Planning for Safety and Urban Quality – Pezzagno & Tira (Eds)
© 2018 Taylor & Francis Group, London, ISBN 978-0-8153-8731-2

The Active City perspective

E. Dorato
Università di Ferrara, Italy

A. Borgogni
Università di Cassino e Lazio Meridionale, Italy

ABSTRACT: The emergent Active Cities approach has been firstly promoted by the public health sector (Edward and Tsouros, 2008), and investigated through town planning, socio-educational, and physical activity perspectives (Borgogni, 2012; SUSTRANS, 2015; Dorato, 2015; Borgogni and Farinella, 2017). New urban challenges are emerging and need to be tackled in integrated manners. The world experiences great urbanization trends, such trends are deeply changing not only the social context as well as people's habits and health conditions,. Thus, the promotion of walkable and, more broadly, physical-activity-friendly urban environments and active lifestyles reaches several requirements. The Active City model should be able to guide the practice of urban within the legislation and the urbanism disciplines topics. New instruments and methodologies are being conceived and tested, trying to overcome the paradoxical lack of planning and design-oriented instruments as well as professional contributions within the Active City field: most important, the *Active City Chart*.

1 INTRODUCTION

Especially over the past couple of decades, the volume of scientific literature addressing the relations between the urban environment, public health, and lastly physical activity (PA) has grown tremendously. However, researching in these fields implicates to face and tackle both the complex nature of urban systems and that of human beings: topics of extreme intricacy, operating at multiple levels and through many different factors. Such a multifaceted matter of investigation has brought scholars to the consolidation of a 'modus operandi' conveying both investigated factors and outcomes into categories, in the attempt of simplification.

Among others, and addressing the Active City's complicated and multilayered issues from a socio-ecological perspective, Vlahov and Galea (2002) divided the field of investigation into three sections: the social aspects of urban health; the physical aspects of the urban environment; and access to health and social services. Similarly, Northridge and colleagues (2003) explained that the relationships between the urban environment and health can be understood in terms of the natural environment; macro-social factors; and inequalities. And also, the fundamental WHO healthy cities report by Duhl and Sanchez (1999) identified three different theories connecting urban planning to health: social justice theory, dealing with issues of housing, education, and safety; political-economic theory, including history, age, class, race, sexuality, and gender; and environmental theory, about the physical and social constructs of communities. Another way in which these topics can be categorized is through a discussion on the various scales, a methodology followed by Galea and Vlahov (2005) when comparing rural to urban conditions; cities among them; or examining interurban variations within a single city.

Thus, specific environmental features (e.g. proximity, attractiveness, safety, density, street connectivity); peculiar PA behaviors (e.g. recreational PA, walking or cycling for transportation, walking for recreation); different scales of investigation or spatial unit of interest (e.g. neighborhood, city, metropolitan area); behavioral focus (e.g. overall PA levels, recreational PA, walking, active transport); and methodological approaches (e.g. self-reported measures either later aggregated or not, use of pedometers and accelerometers, direct observations) are only some of the possible categorizations that could be given for supporting the wide range of ways in which the Active City topic can be researched and understood. The level of detail and the scope of a study are generally the ones to determine how specific observed associations can be, how generalizable the findings, and how useful they will be in terms of informing policy or design.

Moreover, the urban planning and design fields have been the ones contributing to a lesser extent to the Active Cities' debate, greatly outscored by epidemiology, public health, sociology, and PA studies, thus causing methodological and disciplinary arguments. Such trend appears clear also in relation to the existing tools and decision-making or planning instruments, which rely very little on urban planners and designers' competences and professionalism, while depending more often on population's surveys, data analysis, or the application of standardized procedures/algorithms. This tendency might also corroborate the idea that architects and planners are not yet doing enough for effectively re-consider and address people's health and levels of daily movement through the planning disciplines; and the fact that the great majority of existing tools do not consider their role is a clear sign.

Urbanism, public health and the PA fields also face, today, the challenge of the lack of a shared vocabulary to completely engage with each other on these issues. In fact, not only are indicators of health and the environment not agreed upon, but one type of measurement might be appropriate for a public health study, while irrelevant for planners and designers trying to respond to the study. Data is collected and understood in different ways as well, and while public health and PA professionals tend to look for statistical significance in actual measurements of behavior and disease, planners mainly collect data through instruments and approaches like plans, drawings, field observations and mapping, and in very few cities—at least around Italy—are public health and planning agencies truly at the same table of discussion.

2 OBJECTIVE OF THE PAPER

Findings from theoretical researches have shown how most of the three macro-families' existing tools for measuring, assessing, and certifying (Dannenberg and Wendel, 2011) the Active City (and therefore most of the evidence that researchers point to when discussing the relationships between the urban environment, public health and PA) are derived from self-reported data and perceived measures of the urban environment, analyzing healthy urban conditions and planning outcomes on population's health[1].

In addition to that, the overall absence of instruments capable of assisting and guide urban planners along the 'healthy and active' planning and design process has been noticed, thus focusing on setting the basis for future urban regeneration rather than only measuring current situations without the aim of actively using those evaluations within the design process. The development of more sophisticated and multilevel ways for measuring the urban environment and tracking these measurements over time will be some crucial steps to better understand how planning and design can actually help to address NCDs and the inactivity pandemic, while minimizing health and PA disparities in cities.

By relying on previously developed works casting a light on the most recent tools and procedures used for assessing, evaluating and certifying the Active City, the overall aim of this contribution is to argue the lack of planning and design-oriented instruments as well

1. Similarly to the status of scientific literature (having longer addressed the relations between the urban environment and urban health, in comparison to PA levels and outcomes which have entered the picture only much more recently), also most existing tools are predominantly oriented towards the assessment of 'healthy' environments, rather than 'active' ones. In such perspective, an additional challenge will be to select and combine together the most effective and PA-related 'healthy tools' or parts of them, in order to create brand new Active City instruments, directly linking urban form and features to population's PA levels and, consequently, health.

as professional contributions within the Active City field, and therefore propose an original hard-copy tool. The 'Active City Chart' (A.C.C.) is conceived for being used mainly by architects, urban planners and designers, municipality's technicians and/or consultants, or other public or private developers and experts, for guiding the planning and design process of urban regeneration, aimed at creating healthier and more walkable, active, and lively neighborhoods and cities.

3 METHODOLOGICAL APPROACH

By recognizing the great potentials of feeding research through professional experimentation and vice versa, we developed an original hard-copy tool (which, possibly, will be tested and experimented in its efficacy in a second and field-base phase) to be used by experts for guiding the planning and design process of urban regeneration[2] aimed at creating healthier and more active and lively neighborhoods and cities. A.C.C. consists of an evaluation grid which also introduces, as additional and yet necessary component, the drafting of site-specific 'intervention guidelines', capable of tracing planning and design priorities for neighborhood transformations, aimed at making the urban environment more active.

A.C.C. is broadly applicable at the neighborhood scale, chosen for several purposes (despite the fact that there is no internationally agreed upon definition or exact dimension of a neighborhood): firstly because, as underlined by many scholars, usually the neighborhood represents a spatial unit around which one could easily move on foot, by bicycle, or other non-motorized vehicles. The much debated 'human scale' can be found in most European neighborhoods' layout as the average, walking distance of maximum 500 meters from one's residence to public, recreational, commercial or educational facilities, transportation system, and so forth.

"What is a neighborhood? (…) that portion of the town you can get around easily in on foot or, to say the same thing in the form of a truism, that part of the town you don't need to go to, precisely because you are already there" (Perec, 1997: p.57). Residents usually develop a feeling of belonging and emotional attachment to the neighborhood they live in, potentially making them more keen on actively participating to the planning process. In more practical terms, neighborhoods are also usually the aggregation units (see, among others, Duany et al, 2010) around which census data and other statistical information are collected. Also, neighborhoods are the city aggregation level—both physical and social—on which the outcomes of urban transformations can be perceived. Interventions for regenerating an ex-industrial site; infrastructural works on the road system or on active mobility infrastructures; new constructions and/or regeneration of existing public spaces such as parks and squares; or the realization of new housing units: these are all urban projects directly and firstly affecting, either in good or bad, neighborhood life.

Differently from the many other existing tools, A.C.C. is thought as a work instrument to be used by professional in the Urbanism field to feedback both the planning process and the participatory one, therefore avoiding the use of questionnaires involving large samples of population. In fact, the use of A.C.C. represents the first, explorative phase of a more articulated planning method also encompassing a participatory process with resident population where urban transformations are about to be implemented, as well as panel discussions with local authorities and involved stakeholders (Fig. 1).

As a possible answer to the observed overall dearth of planners and architects' professionalism when addressing the Active City-related matters (and conscious that including health and PA thinking and interventions in urban planning and design is not yet on every city's agenda) A.C.C. wants to be of practical support when public authorities or also private

2. Given the specific time Italy as well as may other European countries are experiencing in relation to economic possibilities and new constructions, A.C.C. is thought as an operative tool to be used for supporting urban interventions—of whatever kind—on existing neighborhoods, therefore it would not be exploitable as it has been conceived for guiding brand new development processes.

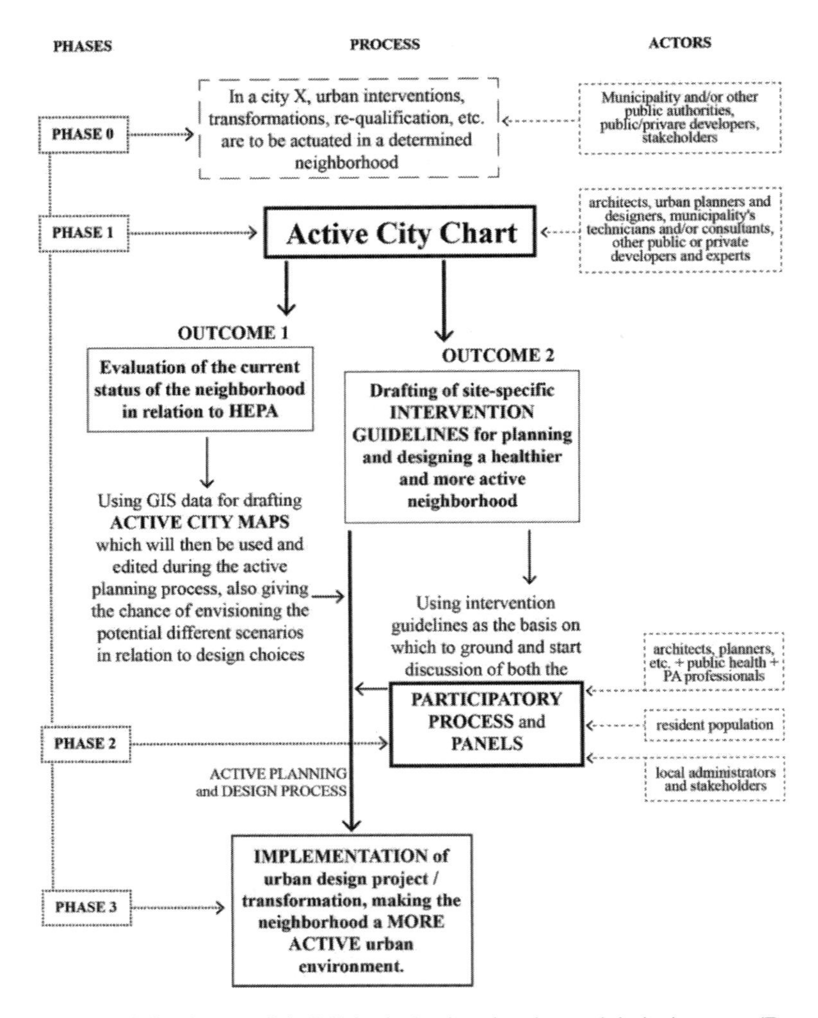

Figure 1. Framework for the use of A.C.C. in the 'active planning and design' process (Dorato, 2017).

developers decide to intervene on a determined neighborhood for operating regeneration interventions. In fact, A.C.C. can be used in a variety of cases, taking advantage of all kinds of needed interventions on the urban realm; in this perspective, the actual pre-condition for the use of A.C.C. for guiding and assisting the planning and design process is political will: urban transformations/interventions of all kinds and for whatever purpose can and should be understood as opportunities for enhancing people's life quality and health, also creating physical activity chances.

Thus, loosely building on the structure of existing audit tools (i.e. SPACES—Systematic Pedestrian and Cycling Environmental Scan; Pikora et al., 2002) and questionnaires (i.e. PREQ & NA—Perceived Residential Environment Quality and Neighborhood Attachment; Bonaiuto et al., 1999; 2006), and integrating with evidence from scientific literature review and other already developed theoretical researches, A.C.C. provides a practical method for collecting street-level data on the principal physical environmental factors potentially influencing PA in local neighborhoods, then integrating such data with census and policies information, to finally transform them into site-specific intervention guidelines. A.C.C. relies both on qualitative and quantitative methods, and is composed by three main 'investigation blocks'.

4 DESCRIPTION

The first investigation block reports the most relevant urban environmental/design factors potentially influencing population's PA levels, which have resulted from previous researches of being: density, land-use mix, street pattern, sidewalks and cycle paths, access to public transportation, access to public spaces/recreational facilities, presence of fringe/neglected areas, safety, and aesthetics (Fig. 2).

It is fundamental to stress that, within both the Active City approach and the A.C.C. framework, 'walkability' is understood as a conducive-feature of the urban structure, as well as a privileged form of HEPA. Previous studies we developed for researching and updating the correlation between specific built environment variables and overall levels of PA showed a quite consistent association between walking as a form of PA and density; mixed land-use; safety; aesthetics; presence and maintenance of walking infrastructures; high street connectivity; and density of shop facilities (see, among others, Saelens et al, 2003; Sallis et al, 2004; Saelens, Handy, 2008), all aspects which have been considered and integrated within the A.C.C.

The second investigation block is made by socio-demographic data as gender and age composition, education, ethnicity, income, and living arrangements. And the third and last block, by the existing municipal (or metropolitan/regional, according to the specific case) policies for enhancing PA and urban health among the urban population.

FACTORS	COMPONENTS	DATA	DESCRIPTION	OBSERVATIONS	GUIDELINES
STREET PATTERN	width of streets; vehicle parking; curb types; traffic volume; traffic speed; traffic control devices; intersection distance; intersection design	GIS / observations	How wide is the street/road? Are there restrictions on vehicle parking on the street/road? Are cyclists able to mount the curb to move away from traffic? How heavy is the traffic volume? What is the posted traffic speed? Are there devices that slow or restrict traffic? Is the distance between intersections short? Are the intersections designed to allow more choice of route?	Excessively wide for the traffic its serves / too narrow for pedestrians and cyclist to be safe / etc. Mountable curbs / non-mountable curbs / no curbs / no ramps / ramps too steep / etc. Traffic control devices present: traffic lights / roundabout / chicanes, /speed humps / autovelox / no devices present / etc.	Reduce road width to only 2 lanes for using the underused street space for pedestrian facilities / new public spaces / the realization of a new cycling path / etc. Implement the existing traffic control devices for slowing down vehicular traffic / allowing safer pedestrian and cycling crossing / etc.
SIDEWALKS and CYCLE PATHS	path type; surface type; path maintenance; path continuity; widths of paths; presence of urban furniture	GIS / observations	Is there a path suitable for walking/cycling? What material is the path made of? Is the path well maintained? Does the path form a useful, continuous and coherent route through the neighborhood? How wide is the cycle path or lane? Are the walking/cycling paths well lighted? Are there benches to rest and chat, handrail along ramps, etc.?	No path / sidewalk / shared path with markings / shared path with no markings / on-road, cycle lane marked / on-road, no markings / etc. Paths have even surfaces / smooth with no holes, cracks, weed or tree root intrusions / many sidewalks present holes and dangerous obstructions / etc.	Re-design road section for obtaining at least a 1.5 meters sidewalk where not present / paint bright markings for better defining cycle lanes / etc. Repair sidewalk discontinuities which could be dangerous for pedestrians' safety / Place benches in the widest parts of sidewalks and in the shadow of trees for allowing all people to sit and rest / etc.
ACCESS TO PUBLIC TRANSIT	transit lines; connections and destinations; transit stops; conditions of transit stops; inter-modality of transit stops	GIS / observations	Are there transit lines serving the neighborhood? What do they connect and which are the destinations? Are transit stops placed in a radius of max 500 meters? Are they safe? Are transit stops adequately furnished for all users fruition? Do transit stops allow for exchanging means of transportation?	Neighborhood is overall well connected / many areas are not served by public transit / few lines badly connecting it to other urban polarities / etc. Transit stops are too sparse / are spaced-out / stops and stations are not safe for people to get in-descend / not well mantained / no siting features / no platform roofs / etc.	Set medium-long term strategies for better serving the neighborhood with more urban transit lines / etc. Implement transit stops in blank areas / refurbish stops with benches, platform roof, ramps for allowing all people to safely use them / equip all buses/trams with mobile ramps / etc.
ACCESS TO RECREATIONAL FACILITIES / PUBLIC SPACES	presence of public spaces / fac; proportion of dedicated land area; average distance; density; maintenance; PA enhancing features; presence of neighborhood sport facilities	GIS / observations	Are there parks/squares/other recreational and sport facilities? Is land area adequate compared to resident population? Is average distance from public spaces and facilities under 500 m? Is density adequate? Are public spaces well maintained? Are there PA and sport enhancing features in public spaces/facilities? Are there neighborhood sport facilities?	Public spaces and facilities ratio is adequate / insufficient for the neighborhood / there are no recreational facilities / etc. Average distance is appropriated / too spaced-out / some areas are poorely provided compared to others / etc. Some parks and gardens have sport and PA equipments / no playgrounds / no game fields / only one bocce field for the elderly / no open-air gyms / etc.	Take advantage of new constructions for providing the neighborhood with new public spaces / intervene on the existing square for making it more accessible and appealing / etc. Install new PA equipment in the local park / design a children playgound by the school / maintenance works to the bocce field / plant new trees for shadowing part of the square / paint bright markings for sport fields / etc.
AESTHETICS	streetscape; public spaces, parks, gardens; cleanliness; architecture; sights	observations	Are there trees along the streets? Are the public spaces in the neighborhood appealing? Is the neighborhood public realm kept clean and neat? Are there diverse and interesting architectural designs in the neighborhood? Are there diverse, interesting and different sights in the neighborhood?	Public spaces are various and appealing / monotonous and unpleasant / badly furbished / no trees / no or almost no rubbish / etc. There is a good range of building designs / all building designs are similar and monotonous / some architectures stand out / etc.	Re-qualify and better characterize unpleasant public spaces / insert benches and plant new trees / take advantage of new constructions for deisigning a new and unique space / etc. Start artistic collaborative processes for painting buildings façades / add value to architectural masterpieces through public spaces' design / etc.
SAFETY	pedestrian/traffic safety; crashes and accidents; crime rates; surveillance	GIS / observations / police records	Are active mobility infrastructures and crossings safe for pedestrians and cyclists? Which is the number of crashes and accidents involving pedestrians or bicyclists in the neighborhood? And which are the worst places? What is the crime rate? Can others observe people on streets and public spaces through passive surveillance?	Sidewalks are well separated from streets / no protective curbs / unsafe street crossings / no ramps / traffic light crossing times too fast / no possibilities for safely crossing roundabouts / etc. Low accident rates in the neighborhood / intersection X displays the highest number of crashes involving pedestrians / etc.	Increase traffic lights' crossing times / implement adequate ramps and curbs / design new pedestrian and cycling crossing systems in roundabouts / etc. Focus on the re-design of intersection X implementing traffic control devices and safer crossings / etc.

Figure 2. Extract of some of the factors composing the Active City Chart first investigation block (Dorato, 2017). 'Observations' and 'Guidelines' (in gray) have been filled with some examples to simplify and show how A.C.C. could potentially be used.

Thence, the chart is composed by three different macro-families of data, each addressing a fundamental variable for the realization of healthy and more active neighborhoods and urban environments: physical and spatial design features; social composition; and political will (expressed through existing policies as well as potentially implementable ones), in the attempt of maintaining a combined and multilevel socio-ecological approach, even though simplified. All the factors are then sub-divided into principal components influencing and further characterizing them; finally, by describing both critical and potential aspects of each component, the current status of the investigated neighborhood emerges, together with a set of site-specific guidelines prioritizing interventions, and capable of guiding the following urban transformation process.

Information about neighbourhood and individual-level characteristics can be collected from multiple sources, including objective and yet easier to manage measures such as GIS, census data, direct observations, epidemiological data, police records, and so forth. There are, of course, also other sources of available or established databases to consider when evaluating neighborhood features with regard to PA (e.g. commercial and residential property information, real-estate data, economic/fiscal data, public health information, community-resources provision, utilization of PA facilities) and also the use of big data could be integrated to the chart in the near future. However, in the attempt of making the A.C.C. easier and quicker to use, GIS, census data, and direct observations could be sufficient for correctly using the chart, therefore starting the 'active planning' process.

5 REFLECTIONS AND RESULT OF THE RESEARCH

The relationships between the urban environment, urban health and PA have been broadly explored during the researches which brought to the definition of A.C.C. As evidence has shown, the physical environment in which people live and grow older plays an important role in determining population's health and also in influencing participation in PA, although which factors of the physical environment have the greatest effect on patterns of activity and to what extent still remain to be better determined. In fact, several are the issues which should be considered when conducting healthy and active urban environments research.

First, no single study could be defined as definitive; this implies that establishing an association between population's health, levels of PA and a determined built environment (its form, physical features, design and characteristics) requires a replication of a study in multiple settings, as well as several testing. Second, measuring exposure is often difficult, and such conceptual and methodological obstacle becomes even stronger when addressing the issue from an Urbanism point of view.

For such reasons, the Active City Chart which we are proposing is a working tool not based on indices or algorithms, also following the belief that it would not be conceptually correct to index a determined and peculiar urban neighborhood, for an urban environment is intrinsically different from another in so many ways. On the contrary, we want to propose a potential active planning and design process which, grounding on the use of A.C.C. (which, however, greatly depends on the capabilities and cognizance of the professionals using it) could support and guide city transformations, with a particular attention to integrating in future urban designs both features and expedients enhancing population levels of PA, and therefore health. Moreover, although local planners and public health professionals in charge of implementing small scale interventions and infrastructure policies may have little capacity for conducting research, they can greatly contribute to the evidence base, by conducting evaluations of policies and projects, either using 'pre-post' or comparison group methods; and foremost to the actual active planning and design process.

Third, outcome analyses should examine full benefits, because most built environment interventions potentially have multiple impacts on health, including of course effects on PA, injuries, disease due to air pollution, and so forth. Fourth, when intervening on the city structure, local communities should help to better frame research questions and to carry out research in collaboration with investigators (an approach known as community-based participatory research), and actively participate to the planning and design process, for they are the final users of the transformed neighborhood, as well as the subjects most acquainted with its critical aspects and potentials.

6 CONCLUSIONS

As already stressed, while scientific evidence has been consolidating on understanding the urban environment and its characteristics as an important determinant, influencing walkability conditions and overall participation in PA, investigation on the relationships between the built environment and PA is a relatively new area of inquiry, and the evidence collected to date has been limited, even though is growing rapidly. Most of the reviewed researches and reports we used as literature in our studies concluded with a specific mention of further research needed on the topic, a symptomatic evidence of a certain degree of 'incertitude' which still wreath this emerging, multidisciplinary field. However, this should be seen as a great potentiality to be exploited for further investigating especially the gaps which research on the Active City topic still displays.

One of these gaps is undoubtedly represented by the extremely scarce use of comprehensive, operative tools capable of guiding and support the planning and design phases for regenerating the urban environment into a more healthy and active one, and therefore a fundamental follow-up of this work will be to test A.C.C.

Thanks to a strong collaboration with the agency for social housing (ACER) of the city of Ferrara, A.C.C. will be used within the public-private transformation project which has recently started for the reuse of Palazzo degli Specchi, and the regeneration of its surroundings. This major intervention will provide for the reuse of the ex-office block into about 250 new housing units, with a mix of public housing, residences for the elderly and for students, rent or property dwellings, tertiary and commercial uses. The challenge launched by ACER, supported by the CITER research lab of the Ferrara University, was to take advantage of such an intervention for a broader reflection on the whole neighborhood in which the establishment stands, exploiting infrastructural works and new designs for making the area more active and healthy, through new PA and recreational facilities; better transit and active mobility connections with the rest of the city; and a more intelligent and active use of the many neglected and fringe spaces which are currently present in this quite stigmatized district. Potentially, the testing of A.C.C. will prove to be an efficient working tool for assisting the active planning and design process—then replicable in other contexts—therefore reversing the "medicalization" trend of the architecture and urbanism fields from cure to care.

REFERENCES

Bonaiuto M., Aiello A., Perugini M., Bonnes M., Ercolani A.P., (1999). Multidimensional Perception of Residential Environment Quality and Neighbourhood Attachment in the Urban Environment. Journal of Environmental Psychology, vol.19(4), pp.331–352.

Bonaiuto M., Fornara F., Bonnes M., (2006). Perceived Residential Environment Quality in middle and low-extension Italian cities. Revue Européenne de Psychologie Appliquée/European Review of Applied Psychology, vol.56(1), pp.23–34.

Borgogni A., (2012). Body, Town Planning, and Participation. The Roles of Young People and Sport. Jyväskylä: Publishing Unit, University Library of Jyväskylä.

Borgogni A., Farinella R., (2017). Le Città Attive. Percorsi pubblici nel corpo urbano. Milan: Franco Angeli.

Dannenberg A.L., Wendel A.M., (2011). Measuring, Assessing, and Certifying Healthy Places. In Dannenberg A.L., Frumkin H., Jackson R.J. (eds.), Making Healthy Places. Designing and Building for Health, Well-being, and Sustainability. Washington-Covelo-London: Island Press, pp.303–318.

Dorato E., (2015). The Active City. New approaches to the design of urban public spaces. Paesaggio Urbano, n.1/2015, pp.52–57.

Duany A.M., Speck J., Lydon M., (2010). The Smart Growth Manual. New York: MacGraw Hill.

Duhl L.J., Sanchez A.K., (1999). Healthy Cities and the City Planning Process: a background document on links between health and urban planning. Copenhagen: WHO Regional Office for Europe.

Edwards P., Tsouros A., (2008). A Healthy City is an Active City: a physical activity planning guide. Copenhagen: WHO Regional Office for Europe.

Galea S., Vlahov D,. (2005). Urban Health: Evidence, Challenges, and Directions. Annual Review of Public Health, vol.26(1), pp.341–365.

Northridge M.E., Sclar E., Biswas P., (2003). Sorting Out the Connections Between the Built Environment and Health: A Conceptual Framework for Navigating Pathways and Planning Healthy Cities. Journal of Urban Health: Bulletin of the New York Academy of Medicine, vol.80, pp.556–568.

Perec G., (1997). Species of Spaces and Other Pieces. London: Penguin Books.

Pikora T., Bull F., Kamrozik K., Knuiman M., Giles-Corti B., Donovan R.J., (2002). Developing a Reliable Audit Instrument to Measure the Physical Environment for Physical Activity. American Journal of Preventive Medicine, vol.23(3), pp.187–194.

Retrieved from: http://www.designedtomove.org/resources (Accessed 20/11/2016).

Saelens B.E., Sallis J.F., Frank L.D., (2003). Environmental correlates of walking and cycling: Findings from the transportation, urban design, and planning literatures. Annals of Behavioral Medicine, vol.25(2), pp.80–91.

Saelens B.E., Handy S.L., (2008). Built environment correlates of walking: a review. Medicine and Science in Sports and Exercise, vol.40(7), pp.S550–S566.

Sallis J.F., Frank L.D., Saelens B.E., Kraft M.K., (2004). Active Transportation and Physical Activity: opportunities for collaboration on transportation and public health research. Transportation Research Part A: Policy and Practice, vol.38(4), pp.249–268.

Sustrans, (2015). Designed To Move: Active Cities. A guide for city leaders.

Vlahov D., Galea S., (2002). Urbanization, Urbanicity, and Health. Journal of Urban Health: Bulletin of the New York Academy of Medicine, vol.79(4), pp.S1–S12.

Town and Infrastructure Planning for Safety and Urban Quality – Pezzagno & Tira (Eds)
 ISBN 978-0-8153-8731-2

Elements towards the protection and promotion of urban spaces in the historical city: The study of the historic center of Brescia

F. Botticini, B. Scala, M. Tiboni & E. Vizzardi
Università degli Studi di Brescia, Italy

ABSTRACT: The Code of cultural heritage and landscape, known as "Codice Urbani" (DLG 42/2004), defines the Archaeological Park as an important archaeological proof. This definition gives life to a project which Brescia Municipality is working on, in order to create the Archaeological Urban Park of the Historic Center of Brescia.

The article retraces the work done to build the system of knowledge that supports the design of the park.

The aim of this project is to promote compatible actions that will aim to promote the site strengths and will create a scenario of sustainable development based on the site peculiarities.

The analyses have been conducted with the help of GIS software that allows the creation of databases. This information establishes the beginning towards the development of design solutions focused on a better use of urban spaces and monuments by pedestrians.

1 INTRODUCTION

The Code of cultural heritage and landscape, known as "Codice Urbani" (DLG 42/2004), defines the Archaeological Park as a portion of territory characterized by important archaeological proofs and by the presence of historical and landscape values, organized like an open-air museum.

Starting from this definition, the Archaeological Superintendence of Brescia and Bergamo has done an observation at the Second General Variation of the Municipality Urban Plan, focusing on the actual Archaeological Urban Park coincident with the UNESCO site, in which is said that it is properly an Archaeological Area so they asked an upgrade with a new perimeter that contain also the Castle and the areas closed to the UNESCO site characterized by the presence of archaeological, landscape and monumental limits (Fig. 1).

Figure 1. The new perimeter asked by the Archaeological Superintendence.

The design of the Brescia Archaeological Urban Park has been the occasion, for the Public Administration, to take in exam some topics that define the relation between the historical and the modern part of the contemporary city. One of the basilar problem of the modern planning is the need of making this two souls of the city living together: technological developments often are incompatible with traditional techniques and this brings at the creation of clashes that requires a careful analysis of problems and accurate responses linked and based on the peculiarities of each part of the 9 site. Example of those problems are the seismic upgrade of ancient buildings and the incompatibility of the actual Public Transport Lines with the historical materials that characterize part of the streets of the city center. This antagonism has been resolved, for example, on the viability level, by paving the streets, this solution creates a lot of arguments because there is the loss of the original cobblestones with a high historical, artistic and esthetic value. Another problem linked to the increase of traffic load is the growing number of vehicles that requires a higher number of car parks inside the city center. Often this require is satisfied by the transformation of squares in car parks with the loss of their original function of places for meeting.

The topic of this article is the ancient part of Brescia: the "Cittadella", or the site shown in the Municipal Urban Plan as the Buffer Zone of the future Archaeological Urban Park, characterized by the presence of various historical, archaeological and artistic limits. This site, settled inside the Main Ancient Part of Brescia (Fig. 2), shows evidences coming from every age in which there were urbanistic transformations of the city, starting from the prehistoric village, going through roman colonization, medieval and renaissance transformations and arriving to the big changes occurred during the industrialization of the XIXth century and the big demolitions of the fascist period. This site is also characterized by the presence of a lot of squares linked by a thick network of streets, most of them show the original features typical of the ages in which they have been built. For this reason, this site is an important source of evidences.

This area is important also for the clashes already mentioned: the streets network (mostly composed by medieval paths) that link the squares has to live near the Public Transport Lines system that creates huge damages to the cobblestones. The site is already characterized by a huge amount of shops and offices and this is linked to a high demand of car parks because of the chronic addiction to vehicle. Another cause of problems is given by the position of "Cittadella" that is built over a dense network of rivers and canals: this creates a lot of problems in buildings, in which, in the most part there are damages linked to humidity. Those buildings also need to be upgraded under the seismic point of view, as demanded by the actual low for civil constructions. The aim of this task is to create a guide line with the goal of regulate the intervention on buildings and open spaces for the enhancement of an area characterized by a huge number of limits.

Figure 2. The Archaeological Urban Park inside the historical city center.

The realization of the new Archaeological Urban Park has the goal of enhance the architectural and cultural potentialities that Brescia oldest portion offers. The appreciation of this site is strictly linked to his fruition. For this reason, it is necessary to study a network of clean mobility paths with the aim to create a relation between the users and the monuments and to fix, with an organic and homogeneous vision, the different areas composing the "Cittadella". This entail the necessity to think about the existing services inside the area and to review the fruition of the roads network with the aim to incentivize pedestrians, cyclists and people with disabilities to come to the Park. To do this, different itineraries have been designed. They join up the peculiarities that the area offers; there are two different kind of paths: the main one that twists and turns between the squares network and links the monuments. The secondary roads, called "thematic" are inserted over the main one and they show the face that Brescia had in the different ages. To promote the fruition of the Park and to guide tourists to his discovery an App has been already developed. To do this a webGIS on line platform has been used and it shows the designed paths and the monuments composing every itinerary. At every object composing the map, data have been associated to and they are useful to the users to have information about the monument they are watching to.

2 METHOD

The enhancement project finds his base on the SWOT analysis of the site; on detail, toward the use of a GIS software it has been possible to create informative basis and their study allows to understand "Cittadella" potentialities and the elements that can damage the monuments or can put in danger the users. To obtain this kind of data it has been necessary to do a huge number of surveys and the results have been flowed into the creation of the reference cartography. From the analysis of this information peculiarities to enhance and protect with the project have been emerged and also the weaknesses and the threats on which is necessary to operate with the aim to create a safety place for the users coming to visit the Archaeological Park. All the observations done in this phase can be a starting point for the Public Administration that has to communicate also with the private sphere with the aim to work on the site potentialities with the goal to create sceneries of sustainable development of the territory.

The surveys have covered two aspects: the open spaces and the buildings. For the first ones, it has been necessary to start from the squares network. The squares have been subdivided in main ones and secondary because of the historical importance they have inside the "Cittadella". This analysis allowed to show how the site has been developed during the time and which are the valuable elements that define the open spaces. To do this it has been necessary to compare historical maps and paintings with contemporary pictures, after this, graphic outlines of the squares have been realized and from them the architectonical valuable elements are emerged (Fig. 3). The analysis of architectural assets found his origins in the study of the building types from which important data such as the historical techniques and architectonical solutions have emerged. This method highlight, in addition to the potentialities of the site, also the features that can be considered as threats, for example, current degradation phenomena and the trigger causes. This facet is very important inside the entire task because it takes in consideration impelling aspect that can be treated as critical or features that potentially could damage the monuments, for example the behavior of buildings under calamitous events such as earthquakes. The analysis highlight also the users' behavior and the way in which the site is used because it is demonstrate that it has effects on the state of conservation of the monuments. For this reason, it is important to show the best practices to use the assets in order to delay the appearance of diseases that can damage it.

The goal is to preserve the area so that the features that characterized the land and that are at the base of his sustainable development will last in the future.

Considering, with a holistic point of view, the weaknesses, the trigger causes and the properties of monuments allows to define a strategy with the goal to enhance and preserve the asset.

Thanks to the use of a GIS software it has been possible, starting from the data coming from the surveys, to develop cartographic bases showing the "system of knowledge" neces-

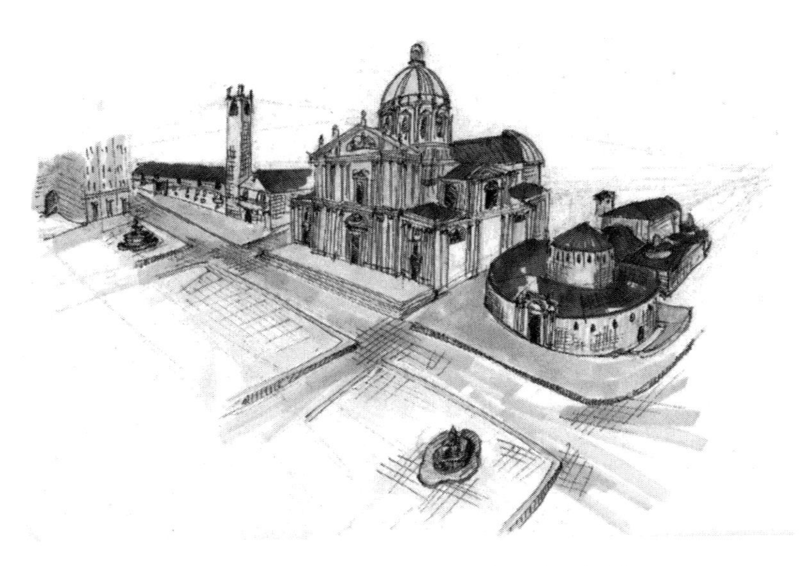

Figure 3. The main architectural elements characterizing piazza Paolo VI.

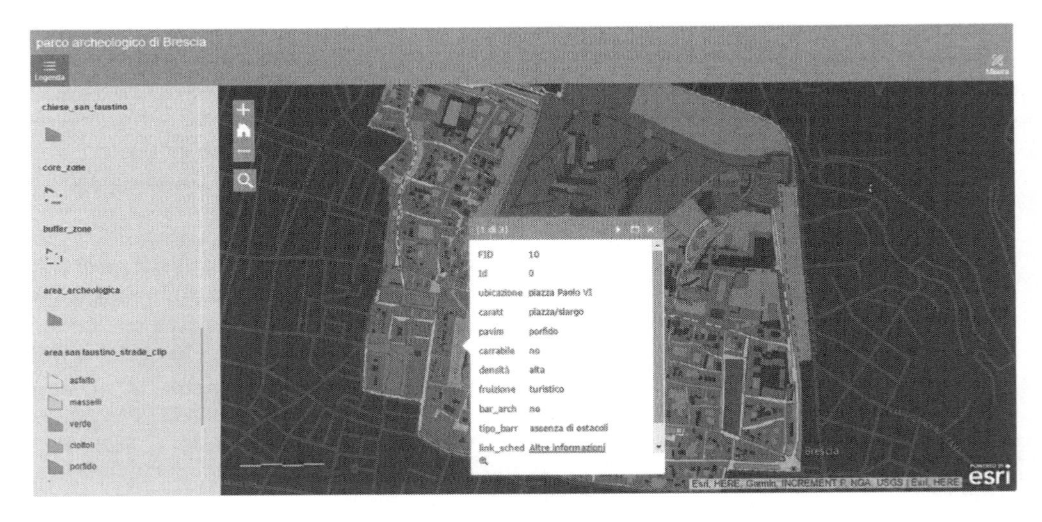

Figure 4. The open spaces database.

sary to develop a project based on the typical features of the examined site. On detail, it has been feasible to show the open spaces features (Fig. 4), such as:

- The open spaces types;
- The cobblestones types;
- The presence of architectural barriers and obstacles and their type;
- Services and infrastructures linked to mobility;
- Pedestrian areas and vehicle accessible streets;

This data has been linked to the ones concerning the users' behavior and how they live in the space. At the end the work shows different maps highlighting:

- How many users there are on every path;
- The main point of interest on every path and their type;

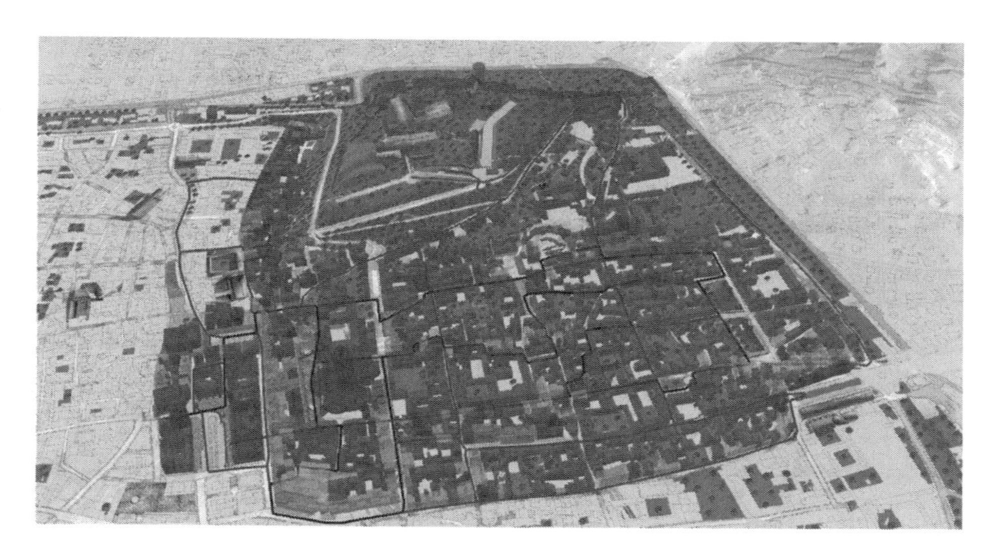

Figure 5. The enhancement project with the network of designed paths that have to highlight the monuments.

Linking this information, the results shows the most used paths and which type of features they have. Going on with this method it was doable to understand which itineraries have the most part of monuments and elements of cultural interest and therefore they are the areas on which is more important to intervene with the aim to promote the use and to enhance the cultural asset of the Archaeological Park.

The open spaces database has the aim to highlight the potentialities and the weaknesses that the area presents at the moment: starting from this analysis is feasible to define criteria that should be at the base of political and planning decisions with the purpose to enhance the historic and artistic asset composing the Park. In this sphere, the GIS turned out to be an important instrument because it has allowed to manage the data flow and his elaboration let to define the topics on which is most important to intervene in the design phase. On another side, the study of preexisting monuments, in particular of materials employed and constructions techniques, allows to guide the designer in the choice of the best and most compatible intervention (Fig. 5).

It is possible to say that the GIS has been used as a "container" because it let to link to the objects composing the maps all the information found in the surveys phase or elaborated during the project. The innovation is linked to the fact that to create sceneries of sustainable development based on the peculiarities of the territory it is necessary that the elements that characterize the land will last in time with the purpose to let the push given to the site by his own elements won't stop in the future. The GIS is suitable for editing web-maps that, thanks to the employment of online platforms, are an important tool useful to diffuse the projects interesting a certain area with the aim to create a shared base in support to them. The web-maps also allow to make known the potentiality offered by a site. These abilities are useful to promote the fruition of the land with the goal to create sceneries of sustainable development.

3 FINAL CONSIDERATIONS

The conveniences that this method offer are a lot and for this reason it can be used for many other future tasks; on detail the developed method let to define in an univocal way the urbanistic background and the existing relations between the site and the monuments inside her. This allows to create the conditions for different designers that are working in the same area but on different monuments to form a homogeneous context and so the result will be a series of requalification actions that will show a univocal image of the site. It is possible to say that the site definition is the starting point to dictate guide lines to supervise punctual interventions on monuments, in particular on the ones exposed to architectural or archaeological limits. With the purpose to create an *unicum* at the urbanistic scale it is necessary that all the stake holders work together. The Public Administration has to be promoter of the guide lines publication, and they have to be shared also with the private sector because every intervention has to be done in a harmonic and coordinated way. The univocal definition of the area allows also to define which are the most compatible design and planning choices. The employment of webGIS platforms let to enlarge the participation of both stakeholders and city users and this allows to collect a higher number of requests and to join up more demands so the given solution will take in consideration the highest number of variables. Starting from the analysis done with the GIS software it has been possible to compare the users' demands and the site features in order to verify the compatibility. In this way, the final design choice will be based on the area qualities and also it will be sheared with the stake holders. To define in an univocal way the context and the assets potentialities is the first step to define the correct urbanistic destinations. This is very important because the first damage cause of monuments is the wrong use. The realized database contains a lot of voices and their compiling allows to take in exam a lot of aspects characterizing the location. From the analysis of the open spaces, guide lines are emerged for the requalification of the Archaeological Park (Fig. 6). The main users took in consideration were the pedestrians in particular families with children so the main topic was to create a safety place that make them feeling comfortable.

In conclusion, if the Public Administration wants to work on the open spaces, she should do:

- Considering the area evolution with the aim to understand the *genius loci* that regulate the origin of the existing streets and the historical relations between the squares and the network of streets;
- Analyzing materials and ancient techniques employed with the aim to develop compatibles projects;
- Studying the interactions between the historic city and the modern innovations in order to avoid conflicts;
- Protecting the historical evidences and to take them in consideration during the services and infrastructures design phase;
- Enhancing the asset giving importance to the open spaces and don't consider them as simple empty places;
- Promoting the creation of clean mobility and sustainable tourism itineraries based on the enhancement and correct fruition of the asset as an instrument to encourage local activities and economy;
- Studying the users' behavior and why they use the asset in a certain way with the aim to design not only the landscape but also its correct fruition;
- Understanding why sometimes the asset is used in a wrong way, developing a holistic view of the problems in order to activate requalification and enhancement policies;
- Promoting the dialog with the private sphere;

Figure 6. The enhancement of piazza del Foro, an example of the application of the guide lines.

Figure 7. Damage caused by humidity and the presence of rivers under the streets, an example of holistic view linking problems and causes.

- Using the street furniture as a tool to guide the users inside the Park and don't place it thinking only about the simple functionality without considering the interactions with the surrounding area;
- Developing alternative solutions trying to understand which one presents the best relationship between functionality and protection needs;
- Promoting policies with the purpose to enhance the asset;
- Using the suitable signage placed in the strategic point and don't place it all over the Park;

A similar method has been used to realize the database with all the features of the architectonical asset characterizing the Park and that allows to define information regarding buildings conservation, the main trigger causes and many other features such as the property and the historical value (Fig. 7). To promote the birth of a common scenario that will be a background for every redevelopment project is necessary that the Public Administration develops good practices on their own buildings with the aim to persuade the private sector to do the same. Obviously, the common background is the Archaeological Park, the excuse for the creation of a big point of interest has to be the justification to start requalification policies at the urbanistic scale. The guide lines must become a tool, useful to facilitate the dialog with the Superintendence, it is important to define a range of planned actions from which the designer can start to define the interventions on buildings.

REFERENCES

Arenghi A., Pezzagno M. (2006). L'accessibilità delle pavimentazioni antiche. In *Pavimentazioni storiche. Uso e conservazione* (pp. 217–222), Edizioni Arcadia Ricerche S.r.l., ISBN 9788895409108.

Bonotti R., Rossetti S., Tiboni M., Tira M. (2015). *Analysing Space-Time Accessibility toward the Implementation of the Light Rail System: The Case Study of Brescia*, Planning Practice and Research, 30(4), pp. 424–442, DOI: 10.1080/02697459.2015.1028254.

Buda A., Coccoli C., Pracchi V., Scala B. (2017) (eds.), *Per una definizione non univoca del concetto di conservazione. Scritti di Gian Paolo Treccani*, Brixia University Press, Brescia.

Carmona M. (2003), *Public Places Urban Spaces. The dimension of urban design*. Architectural Press.

Comune di Brescia (2016). *Piano di Governo del Territorio 2016 approvato*. Available at: http://www.comune.brescia.it/servizi/urbanistica/PGT/Pagine/Variante%202015-PGT-approvato.aspx (accessed May, 15th 2017).
Limongelli M.P., Scala B. (2013) (eds.). *Tra prevenzione e cura: la protezione del patrimonio edilizio dal rischio sismico*, Gangemi editore, Roma.
Robecchi F. (1944), *Le strade di Brescia*, Periodici locali Newton, Roma.
Tiboni, M., Rossetti, S. (2012). L'utente debole quale misura dell'attrattività urbana. *Tema. Journal of Land Use, Mobility and Environment*, 5(3), 91–102.
Tira, M., Rossetti, S., Tiboni, M. (2016). Managing Mobility to Save Energy Through Parking Planning. In *Smart Energy in the Smart City* (pp. 103–115). Springer International Publishing, doi:10.1007/978-3-319-31157-9_6.
Treccani G.P. (1988). *Questioni di patri monumenti: tutela e restauro a Brescia* (1859–1891), Milano.

Town and Infrastructure Planning for Safety and Urban Quality – Pezzagno & Tira (Eds)
© 2018 Taylor & Francis Group, London, ISBN 978-0-8153-8731-2

The queensway of New York city. A proposal for sustainable mobility in queens

C. Ignaccolo
Columbia University, USA

N. Giuffrida & V. Torrisi
Università degli Studi di Catania, Italy

ABSTRACT: In United States of America, road transport and private cars have a considerable role in transport system, areas of disused railways are often remained unused.

The new High Line in NYC is an outstanding example, because it winds between buildings and constitutes a green elevated walk-path with spaces to stay and to relax in a no-green fully urbanized area.

The topic of this paper is a project of conversion of Rockaway Beach Branch Line (RBBL) in Queens into a green cycling – pedestrian path. The thread of this green infrastructure could link the major parks of the area, offering a safer connection with recreation spaces and vibrant commercial strips.

The methodological approach (based on data analysis on GIS) is particularly careful to the interaction with heavy road traffic.

Studies about this project has been drawn up in MsAUD (Architecture and urban design) 2016/17 activities at Columbia University, NYC, USA.

1 INTRODUCTION

The gradual increase in private mobility, dating back to the second half of the last century in western countries, has caused the shutdown of several secondary railway lines which are rarely used and therefore little profitable to any institution, either owner or manager (Guerrieri and Ticali, 2012). In addition to the decrease of rail passenger transport demand due to the improvement of the road system, the other two causes, that have contributed over the last 50 years to this phenomenon, can be sought in the construction of new high performance railway track parallel to the pre-existing one and in the decrease of rail freight transport demand due to the disposal of industrial areas.

Therefore, it is urgent to consider the issues related to inactive railway lines as there are hundreds of thousands of kilometres of inactive railways (Bertolini and Spit, 1998). One estimate is that it costs substantially less to redevelop an abandoned urban rail line into a linear park than to demolish it. Consequently, the disused railways are potential new pathways and the abandoned stations provide available spaces for new activities, supporting sustainable local development and regeneration processes.

According to this, disused railway sites are becoming a focus of redevelopment projects in many European countries and, recently, some former railway lines have been converted into cycling and pedestrian paths. In the USA, where road transport and private cars have a considerable role in transport system, areas of disused railways are often replaced by road layouts. Only recently it is possible to notice a few cases of conversion of railway tracks in non-motorized mobility spaces, especially in urbanized areas. This is due to the fact that issues such as ecology and sustainability have come to the forefront only in recent years, raising awareness and urging cities to promote environmental protection programs, including the convertion of disused railways within the concept of "soft mobility". Some of the norms and initiatives aimed at maintenance or recovery of the disused railway in the USA are the voluntary agreement *Rail Banking* (1983), the no profit organisation Rail to Rail Conservancy (1986), the transport legislation *Intermodal Surface Transportation Efficiency Act* (1991) and the policy statement of Federal Highway Administration *"Design Guidance on Accommodating Bicycle and Pedestrian Travel"* (2000). Currently, a movement is being developed thanks to a "bottom-up" push, which sees the population aggregated in spontaneous organizations that stimulate, provide ideas, collaborate in the creation and management of greenways.

Although the actuality of greenways' concept is nowadays increasing more and more, thinking the greenway as part of a network infrastructure should be one of the main concept to be taken into account in its planning and designing. The planning process should try to provide sustainable landscapes against disintegrating, space decreasing, urban development and uncontrollable change of area use (Ahern, 1995).

In this view the topic of this paper is a project of conversion of Rockaway Beach Branch Line (RBBL) in Queens into a greenway. This study proposes a methodology characterized by a GIS approach to evaluate the need of different kind of interventions for the realization of the greenway and the requalification of its surroundings.

2 GREENWAYS AND RELEVANT BEST PRACTICES

The greenway literature of the past decade consistently names Frederick Law Olmsted as the father of the greenway movement in America (Little, 1990). He developed the idea of *parkway system*, which leads to taking shape of current greenways (Kent and Elliott, 1995).

The influence of the environmental decades on landscape architecture was most prevalent in the academic environments during the 1960s and the 1970s. Lewis' *environmental corridor* concept was used to plan first a major state wide greenway system with a focus on protecting environmentally sensitive areas, or river corridors (Lewis, 1964).

After 1985, greenways were integrated with space and resource management concepts (Mugavin, 2004). They started to have more comprehensive duties: beyond meet people needs and satisfy aesthetical and recreational requirements of city dwellers, they took on a lot of goals such as preserving habitat, reducing flood harms, increasing water quality, protecting historical sites and education.

Nowadays greenways brought together 2 functions: to form open spaces for public and for recreational uses and to ensure the protection and development of natural resources: many countries around the World have tackled these issues in creative and successful ways (Fig. 1).

In recent years an outstanding example of greenway promoted by a bottom-up process is the High Line, a linear park built in Manhattan on an elevated section of a disused New York Central Railroad. In 1999, the nonprofit organization Friends of the High Line was formed by 2 residents of the neighborhood that the line ran through, advocating for the line's preservation and reuse as public open space. The High Line is inherently a green structure: it winds between buildings and constitutes a green elevated walk-path with spaces to stay and to

Country	Project	Length (km)
France (Paris)	Promenade Plantée	4,9
Belgium	RAVeL Réseau Autonome de Voies Lentes	900
Australia	East Gippsland Rail Trail	96
Australia (Sydney)	Goods Line	0.5
United Kingdom	Bristol and Bath Railway Path	24
USA (Missouri)	Katy Trail	390
USA (Chicago)	Bloomingdale trail	4.3
USA (New York)	High Line	2.33

Figure 1. Best practices for the requalification of abandoned railway lines worldwide.

Figure 2. Park's attractions and views of the city from the High Line.

relax in a no-green fully urbanized area. Furthermore, there is a good relationship with some requalified adjacent buildings having a new modified destination of use (Fig. 2a and b).

As great number of studies recommends new approaches to urban and transport planning as solutions to climate change mitigation (Caprì et al., 2016), the High Line landscape functions essentially like a green roof designed to allow the plants to retain as much water as possible. This can be considered a soft approach seeking to raise awareness on how green infrastructures can play a vital role in create climate-resilient development—a role which is currently not sufficient recognised nor integrated into mainstream planning (Inturri, 2011).

3 CONTEXT FRAMEWORK: THE QUEENSWAY

The QueensWay is a project of conversion of a former rail line, LIRR Rockaway Beach Branch (RBB), a 3.5 miles stretch which lies abandoned since 1962. During this time, vegetation have sprouted along the former right of way and illegal dumping has become an increasing problem, with trash and remnants of drug and alcohol use litter the ground (Fig. 3). In 2011, a group of residents living along the former RBB, teamed up to advocate for its conversion into a new linear park, joining in a movement called *The Friends of the QueensWay* (FQW) with the goal of converting the long-abandoned property into a public park. FQW entered into a partnership with *The Trust for Public Land*, the nation's leading non-profit organization working to create parks and protect land for people.

Thanks to the fundings obtained by the State of New York, in 2013 The Trust for Public Land has commissioned the QueensWay Plan to *WXY* and *dlandstudio*, in order to lead an interdisciplinary team to analyze the economic, social, environmental, engineering and transportation dynamics of the site and surrounding area.

The planning approach was based on community involvement, with five large public meetings, 30 workshops and meetings with community groups, and hundreds of stakeholder discussions. The ideas arising from these sessions, as well as the analysis of the site, helped establish the six themes explaining the vision for the QueensWay:

- Connections + Neighborhoods: the QueensWay is seen as a connector to parks, commercial avenues and facilities; it's also a gateway to neighboring communities;
- Ecology + Education: there are 12 schools within a 5 minute walk of the QueensWay; moreover visitors can encounter a variety of environments and learn about plants, geology, stormwater management, and natural habitats for urban wildlife;
- Safety + Comfort: the QueensWay will provide for the needs of all ages and abilities; it will be carefully designed to avoid conflicts between walkers and cyclists. Particular attention will be given to the preservation of privacy for neighbors;
- Play + Health: sport and recreational programs will be developed in partnership with local associations;
- Culture + Economic Development: visitors to the QueensWay will bring new business to commercial activities located in the surrounding neighborhoods; provision of platforms

for performances and public art and the opportunity for adaptive reuse of underutilized buildings will give life to a new cultural offer;

- Care + Stewardship: The community will be engaged through a continued public input process to ensure the park and design meet local needs.

The QueensWay plan divides the park into 6 areas (Fig. 4): 4 integrate activities; 2, called *the passages*, are closer to homes and will be paths for walkers and cyclists.

Figure 3. Vegetation sprouting in the abandoned RBB line.

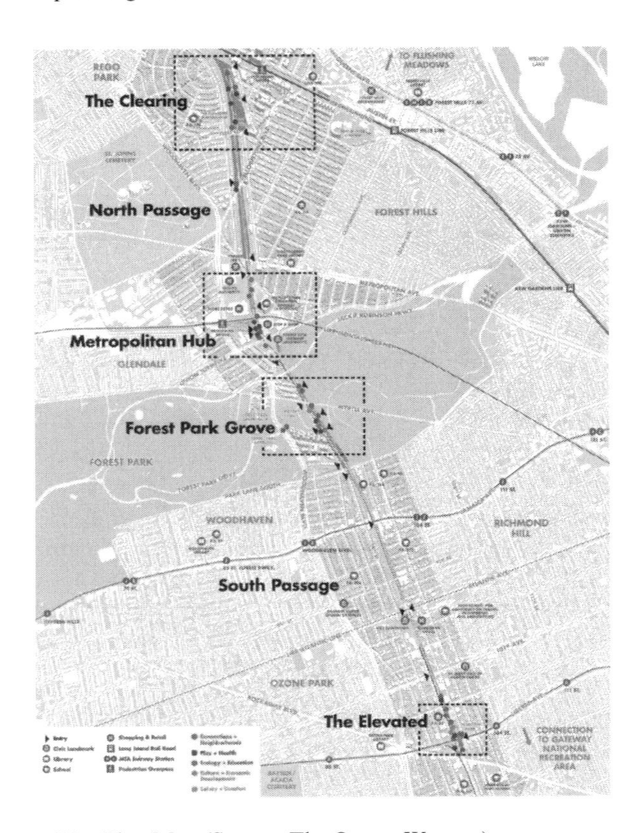

Figure 4. The QueensWay Plan Map (Source: The QueensWay.org).

Figure 5. Closeness of track to private houses.

A particular attention is put by WXY and *dlandstudio* proposal in a design which try to maximize safety and privacy for neighbors (whose houses are close to the track, Fig. 5) while still giving a good permeability and visibility for park users. A proposed solution for pathways that run by private homes is a design with vegetated buffers at the top of the embankment; secure fencing with planting to provide additional screening are put at the property line to physically and visually separate backyards and homes from park users. Moreover, in order to decrease any visual connection between the QueensWay users and adjacent homes, the pathway can be lowered by excavating the embankment. Finally all the main activity spaces will be located close to non-residential amenities, while the two lengths that run by homes (North Passage and South Passage), will be used as a walking and cycle path.

4 METHODOLOGICAL APPROACH

The QueensWay project realized by FQW is a bottom up process involving stakeholder. This paper proposes a GIS approach, based on suitability analysis, useful to evaluate the necessity of the different kind of intervention according to the 6 focus areas defined during the decision process. Suitability analysis describes the search for locations or areas that are characterized by a combination of certain properties.

GIS allows to obtain suitability scores that can be used to determine hot spots regarding each focus area through the intersection of multiple levels of information. In our case, the approach consist of the superposition of 5 different score layers (one for each focus area) constructed through critera which are depending from spatial characteristics evaluated through the use of a GIS software.

$$S_{fa} = c_1 + c_2 + \ldots + c_i + \ldots + c_n$$

where S_{fa} is the score for the specific focus area and i are the n related criteria cocurring to its construction. The selected criteria for each focus are showed in Fig. 6.

A buffer of 1 km from the rail line is taken into account as threshold for the analysis of the surrounding land use. The method assumes that each criterion's values are normalized between 0 and 1 according to the following equation:

$$SN_{fa} = \frac{S_{fa-j} - S_{fa-\min}}{S_{fa-\max} - S_{fa-\min}}$$

where, for each focus area, SN_{fa} is the normalized score, S_{fa-j} is the generic score and S_{fa-min} and S_{fa-max} are respectively its minimum and maximum value. Since the score is a need score a value close to 1 corresponds to an area with more need.

FOCUS AREA	CRITERIA
Connections + Neighborhoods	Number of road crossings Subway stations within 1 km Bus stop within 1 km Number of parks Gateways to adjacent avenues Number of parkings within 1 km Population age
Ecology + Education	Number of schools within 1 km Number of libraries within 1 km
Safety + Comfort	Buffer from neighboring residents' houses Viewshed analysis Number of accessible ramps needed Lighting
Play + Health	Number of sport facilities Emissions from road traffic
Culture + Economic Development	Number of workers Turistic points within 1 km Commercial buildings Number of abandoned facilities within 1 km State of the buildings

Figure 6. Criteria for each focus area.

Figure 7. Viewshed analysis.

Figure 8. Activities analysis.

After the evaluation of each score into a layer, the six layers would be combined/overlaid using raster calculation function to get a composite map showing priority hot spots and areas where intervention is not necessary: the final result of our suitability analysis will be a thematic map showing which locations or areas are more in need for a specific focus area.

5 FIRST RESULTS

This viewshed analysis (Fig. 7) shows which areas are visible from a specific location. Viewshed analysis was performed putting some points on the railroad as observation points. The raster is a DSM (DEM + building eights). The result shows that the future project would guarantee the privacy of people who live nearby the QueensWay infrastructure (especially in

Figure 9. Infrastructure analysis.

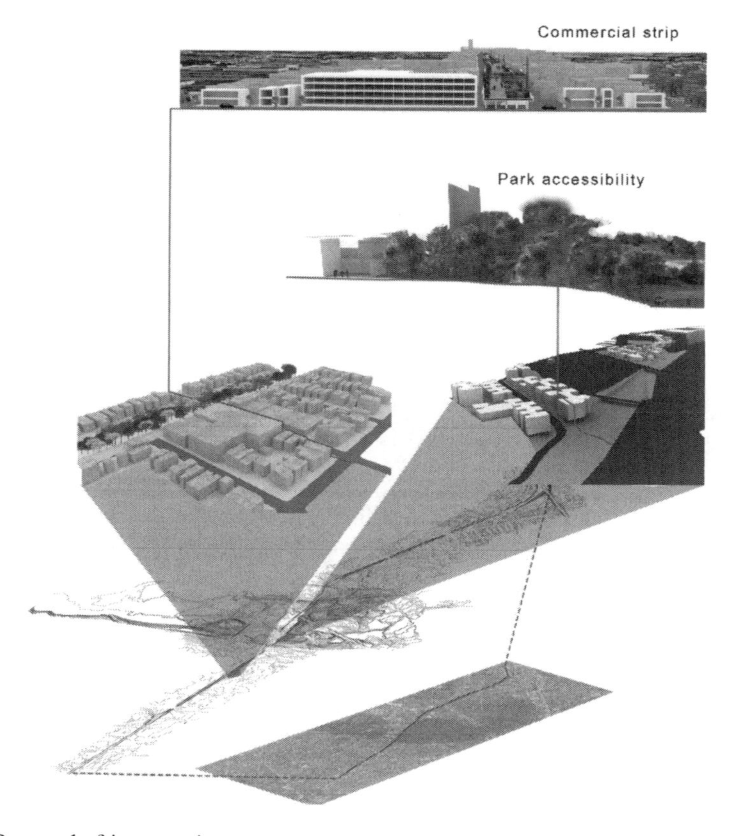

Figure 10. Proposal of intervention.

the Southern part) and, in the same moment, the QueensWay would offer great views of the Forest Park. Fig. 8 shows that QueensWay would be a great link between the built area of Southern Queens and the Forest Park which is not easily accessible nowadays. In addiction a lot of students could use the QueensWay as daily path to reach their schools or other public facilities. In this way there will be also a decrease of traffic congestion, because a lot of commercial buildings are located nearby the former railroad. Last but not least QueensWay could connect two areas of Queens with different Medium Age. The park would be easily reached by Metro thanks to 5 stops (1 in the Northern part, 2 in the middle, 2 in the Southern part) located within a distance of 300 m from the former railroad (Fig. 9). QueensWay would be the only N-S link between the metro stops which are on 3 different lines.

A proposal of intervention is shown in Fig. 10, which highlights the importance of accessibility to Forest Park. Connection to parks and commercial areas is one of the main project aims; the reconversion would expect 1 million annual visitors to the QueensWay, based on similar projects and on the annual number of visits to Forest Park (approx. 900,000), assuming that 250,000 of the visitors will be from outside of Queens bringing new business to local shops and restaurants.

6 CONCLUSIONS AND FURTHER RESEARCH

In the USA, where road transport and private cars have a considerable role in transport system, the disused railway areas are often remained unused, or replaced by road layouts. Only recently more sensibility towards ecology and sustainability have come to the forefront and it is possible to notice a few cases of conversion of railway tracks in non-motorized mobility spaces, especially in urbanized areas.

In this theme, the topic of this paper is a method to improve a project idea of conversion of an abandoned rail line in Queens into a green cycling—pedestrian path. This study proposes a GIS approach, based on suitability analysis, to evaluate the needs of different kind of intervention according to different focus areas. Since community involvement is one of the basis of the proposal project of FQW, in future researches the GIS approach could be integrated with a Multi Criteria Decision Analysis process. This would give the possibility to the community to assign different values to focus areas' criteria and allow the decision maker to obtain priorities of interventions.

REFERENCES

Ahern, J., (1995). Greenways as a Planning Strategy. Landscape and Urban Planning, Vol.33, pp. 131–155.

Bertolini, L., Spit, T., (1998). Cities on Rails—the Redevelopment of Station Areas. E & FN Spon, London.

Caprì, S., Ignaccolo, M., Inturri, G., Le Pira, M., (2016). Green walking networks for climate change adaptation. Transportation Research Part D 45, 84–95.

Guerrieri, M., Ticali, D., (2012). Design standards for converting unused railway lines into greenways. ICSDC 2011, pp. 654–660. http://dx.doi.org/10.1061/41204(426)80.

Inturri, G., Ignaccolo, M., (2011). The role of transport in mitigation and adaptation to Climate change impacts in urban areas, Resilient Cities, Springer Netherlands, pp. 465–478.

Kent, R.L. and C.L. Elliott, (1995). Scenic Routes Linking and Protecting Natural and Cultural Landscape Features: A Greenway Skeleton. Landscape and Urban Planning, Vol. 33, Issues 1–3, pp. 341–355.

Lewis Jr., P., (1964). Quality corridors in Wisconsin. Landscape Architecture Quarterly, Washington, DC, January, pp. 101–108.

Little, C.E., (1990). Greenways for America. Johns Hopkins University Press, Baltimore.

Mugavin, D., (2004). Adelaide's Greenway: River Torrens Linear Park. Landscape and Urban, Planning, Vol. 68, Issues 2–3, pp. 223–240.

Town and Infrastructure Planning for Safety and Urban Quality – Pezzagno & Tira (Eds)
© 2018 Taylor & Francis Group, London, ISBN 978-0-8153-8731-2

Beyond the street: An urban regeneration project for the Porta Milano district in Brescia

M. Tiboni
Università degli Studi di Brescia, Italy

G. Ribolla
Comune di Brescia, Italy

S. Rossetti
Università degli Studi di Brescia, Italy

L. Treccani
Comune di Brescia, Italy

ABSTRACT: The Municipality of Brescia is developing an urban regeneration project, co-financed by the National government and with an overall budget of 47 million euros, called "Oltre la Strada (Beyond the Street)", that interests the area of Porta Milano.

This paper explains how the project was built, starting from the Strategic Urban Plan (PGT) of Brescia and then focusing on the urban design, aimed at redeveloping via Milano as a district and not just as a traffic artery. A focus on the infrastructural and sustainable mobility solutions developed within the Oltre la Strada project, which are mostly linked to the "Living and Walking in Cities" conference theme, is also provided.

1 INTRODUCTION

This paper presents "*Oltre la Strada*": an urban regeneration initiative that is interesting the Italian City of Brescia. After a short introduction on urban regeneration, the paper highlights the efforts made by the Strategic Urban Plan (PGT) of Brescia towards the reduction of soil consumption and the recovery of the existing building heritage (see par. 2). Then, the background, the origins (par. 3) and the contents (par. 4) of the *Oltre la Strada* project are described. Finally, a focus on the infrastructural and sustainable mobility solutions developed within the project, which are the mostly linked to the "*Living and Walking in Cities*" conference theme, is provided (par. 5).

By definition, urban regeneration projects aim at the rehabilitation of impoverished urban neighbourhoods by large scale renovation or reconstruction of housing and public works. Urban regeneration involves comprehensive and integrated actions which seek to resolve urban problems and bring lasting improvements in the economic, physical, social and environmental condition of an area that has been subject to change or offers opportunities for improvements (Roberts, 2000).

But what makes a regeneration project successful? And how should such interventions be pursued and managed?

It is now widely recognised in the literature that successful urban regeneration projects need to ensure affordability, access to facilities and involve local communities and residents (see, i.a., Forrest, 2017; Roberts et al., 2016; Santangelo et al., 2014). Regeneration is a process that takes time, and should be adaptable enough to give residents a genuine voice. At European level, noteworthy urban regeneration experiences include the case study of

Manchester, the King's cross station area in London and the Hammarby Sjöstad district in Stockholm (see, i.a., Bianchini and Parkinson, 1993; Iverot and Brandt, 2011).

Within this background, an urban regeneration project is being developed in Brescia, a city in the North of Italy, with a population of nearly 196,000. Brescia is a municipality with a great amount of emissions per year, and is characterized by the presence of many industries also located within the urban area. The main objective of the regeneration project here described is to make the rundown neighbourhood of *Porta Milano*, in the periphery of the city, attractive and vibrant again, and to find new purposes for underused and neglected spaces in that area.

2 URBAN REGENERATION IN THE STRATEGIC URBAN PLAN (PGT) OF BRESCIA

Like in the most of Italian cities, in the recent years a deep economic crisis affected the real estate market in Brescia. This crisis, together with demographic dynamics (due both to natural balances and migration flows), marked the history of its urban transformations.

Figure 1. Sironi M. – Paesaggio Urbano. This Novecento painting can symbolise the productive function that the Porta Milano neighbourhood had for Brescia.

Figure 2. The location of Porta Milano district.
Source: Municipality of Brescia.

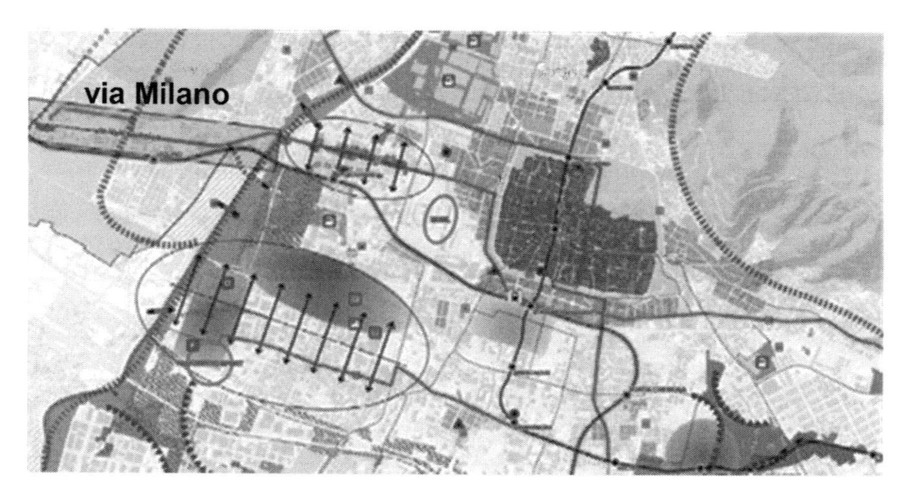

Figure 3. The PGT of Brescia identifies via Milano as an urban regeneration area.
Source: Municipality of Brescia – PGT 2016 tav. V-DP02.

In 2013 the city administration of Brescia launched general modifications to its territorial governance tool, the Strategic Urban Plan *(PGT—Piano di Governo del Territorio)*, focusing its urban policies on the reduction of soil consumption, on urban regeneration and environmental improvements.

The new PGT recognises a central role to the preservation of the environmental assets (water, air and soil), and foresees actions to produce qualitative environmental improvements, and to strength the ecological network. The PGT drastically reduces soil consumption (–51% compared with the previous plan) (source: Comune di Brescia, 2016a), and links the objective of environmental quality with urban regeneration initiatives. Urban Regeneration is applied in the plan through strategies that activate architectural and environmental redevelopment processes of the existing building heritage.

Focusing on Urban Regeneration, the plan, finally approved in February 2016, identifies some areas, which are characterized by a widespread urban and social degradation. According to the plan, those areas are subjected to complex urban regeneration programs: the area of the *Porta Milano* neighbourhood (often simply called *via Milano*, which is the main penetration road to the city for those coming from the West—Fig. 2), is among them.

Particularly, the area of via Milano is part of a larger territorial compendium, that the PGT identifies as "[...] *Parts of the city currently affected by significant divestment phenomena and particular critical situations due to environmental or social degradation, in urban areas characterized by a significant population presence, and to be implemented through complex urban regeneration programs*" [cfr. art. 49 NTA PGT], and for which a number of integrated actions are envisaged (Fig. 3).

3 FROM THE PGT TO "*OLTRE LA STRADA*": AN URBAN REGENERATION PROJECT FOR *VIA MILANO*

Via Milano is identified in the imagination of the city as the whole urban agglomeration overlooking the main traffic artery that connects the western periphery of the city to the historic centre.

There are several factors that contribute to an overall degradation condition of that area.

Via Milano is the first "true" periphery of Brescia, located at the extreme edge of the SIN (*Sito di Interesse Nazionale*—contaminated national site) *Brescia Caffaro*, which is highly affected by contamination and spread by polychlorinated biphenyl (PCB).

Figure 4. Pictures from industrial plants and buildings of Via Milano.

The whole neighbourhood has a history of abandonment of the former industrial sites and, in parallel, is highly characterised by the concentration of non-EU ethnicities and marginality.

Former manufacturing plants (*Ideal Clima, Ideal Standard and Caffaro*) occupy an area of 200,000 m^2, of which 120,000 m^2 are covered by buildings. Most of them are completely dismantled. Only few Caffaro's plants are still active, but very close to disposal. The abandoned industrial area (included in the SIN) extends south of via Milano in direction east/west for almost a kilometre, constituting, together with a predominantly residential building curtain, a single block interrupted transversely only by secondary traffic. To the south the area borders with the Brescia-Iseo-Edolo railway line, that connects Brescia railway station with Iseo Lake and *Valle Camonica*.

The two fronts of via Milano are characterized by a construction curtain that combines different historical thresholds and functional typologies. They are generally characterized by a widespread deterioration, which already led the City Administration to draw up a Recovery Plan in 2010. However, the Recovery Plan was then not realised, due to the real estate crisis and the drastic reduction of public resources.

The historic function of Via Milano, as a traffic penetration axis to the city centre for those coming from the West, has contributed to aggravate the condition of urban degradation, inhibiting the development of high quality district functions for non-motorised users, of urban services, and activities. Furthermore, building deterioration, coupled with a low-cost housing offer, has attracted very low-income users, predominantly of non-EU origin, and fostered social exclusion.

Finally, the abundance of abandoned manufacturing sites makes it difficult to control the accesses, and dismantled plants are therefore used as a shelter without control by inappropriate uses, thus contributing to the dilemma of degradation and delinquency.

Therefore, when in June 2016, the Italian Council of Ministers published the "*Bando Periferie*", a specific call aimed at co-financing projects for urban renewal and safety of the outskirts, the City of Brescia decided to apply with the project "*Oltre la strada*—Beyond

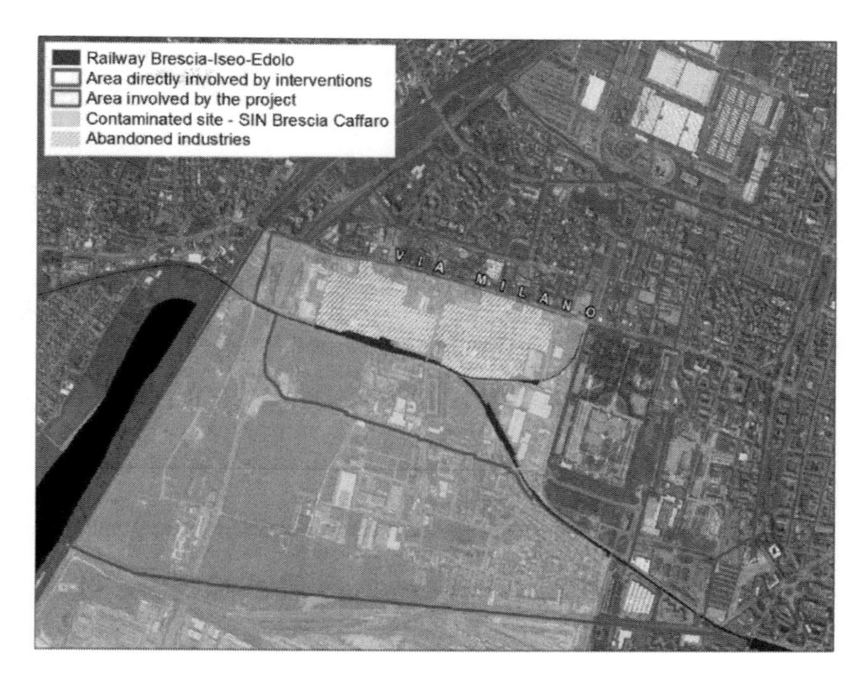

Figure 5. The area of via Milano interested by the urban regeneration project "Oltre la Strada". Source: Municipality of Brescia.

the Street". This project includes interventions on via Milano and on its surroundings, and aims at realising and applying the PGT urban regeneration previsions for that area (Fig. 5).

The borders of the intervention (Fig. 5) are defined by the involvement of those areas that most contribute to the urban degradation of Porta Milano, but also of those contexts that, for different functional reasons, may represent the cornerstones of the process of requalification and development desired by the project. The project *Oltre la Strada* is expected to have a great impact on the city. It involves an overall area where approximately 9,361 families and 19,613 citizens live. And, in the area directly involved by the project interventions (bordered in red in Fig. 5), the 45% of the residents are immigrants (compared with an average of 19,6% in the whole city). The actions of the project heavily focus on the youth, considering that in interested area children are 17.1% (compared with 13.1% of the whole city) and young people between 15 and 34 are 23.0% (compared with 20.1% of the city) (data, updated to 2015, provided by the Municipal Bureau of Statistics).

While drafting the application of the project to the *Bando Periferie*, the city activated strategies to involve public and private stakeholders interested in investing on the project with public interest objectives. *Oltre la strada* is therefore the result of a co-design path between the Municipality of Brescia and a network of partners, public and private entities. There are 14 private partners who have signed agreements with the Municipality of Brescia, identified through a notice that was published on the municipal website in July 2016.

This resulted in a project with an overall budget of 47 million euros, worth of 18 million of co-financing from the government.

4 THE CONTENTS OF "*OLTRE LA STRADA*"

The project "*Oltre la Strada*—Beyond the Street" gathers interventions to support the resilience of via Milano, to enhance quality of life, sustainable mobility, integration, and participatory life in the neighbourhood, and to protect the new and consolidated fragilities.

Figure 6. *Oltre la Strada* project. Infrastructural and urban planning interventions.
Source: Municipality of Brescia.

The intervention also supports many bottom-up initiatives that already exists in this area, but that require the endorsement of the city administration and integrated approaches to grow, consolidate and create greater benefits. Therefore, in less than five years (the project is supposed to end in 2021) the district should be revamped, made vital and attractive.

The project roots on three levels of intervention:

- infrastructural and urban (including mobility) interventions;
- sociocultural approaches, with new dwelling formulas and formulas that combine housing with workspaces for young and creative businesses and for craft workshops;
- a continuous coordination of the project by the city administration, which is first and foremost an active listening of the citizens' needs and of the stakeholders interested in investing in the area.

From the infrastructural and urban points of view (Fig. 6), the foreseen interventions belong to two main clusters:

- **Infrastructures** (in particular via Milano, a road which will have to be identified as the axis of the new urban centre; and the railway line Brescia-Iseo) (see par. 5).
- **Demolished and/or heavily underused buildings** (a former mill, houses of Via Mazzucchelli, Borgo San Giacomo), some of which are characterized by strong marginalization and social degradation (i.e. *via Milano 140* and *Ideal Clima* building).

Infrastructural and sustainable mobility actions of the project are deeply described in the next paragraph (par. 5), while from the Urban Renewal point of view, *Oltre la Strada* is composed by the following interventions:

- over 150 new housing units with moderate and/or contracted housing and the refurbishment of over 200 social housing units for weak users
- three new high-quality designed public spaces (squares) opened along via Milano
- Public-Private Partnerships to match the re-qualifying targets of the new PGT (among others: New Space Theatre IDEAL, Neighbourhood *Mazzuchelli*, Refurbishment of the ancient *Borgo San Giacomo*, a former mill recovery, *via Milano 140*, *Case del Sole*)
- ICT Infrastructures (connection of at least 70% of the project area to the fibre optic network; implementation of an *Internet of Things* radio based network and services, coverage of the 70% of the project area with a new Wi-Fi network; increasing of the area covered by video surveillance).

5 "*OLTRE LA STRADA*": A FOCUS ON THE INFRASTRUCTURAL AND SUSTAINABLE MOBILITY ACTIONS OF THE PROJECT

The overall mobility project for Porta Milano foreseen by the *Oltre la Strada* urban regeneration project seeks a more modern and sustainable accessibility scenario for those places, also making a synthesis of various infrastructural and regulatory initiatives already established.

Furthermore, the Municipality of Brescia is now approving its Sustainable Urban Mobility Plan (SUMP), which is deeply intertwined with the Strategic Urban Plan (PGT—see par. 2) of the city, and therefore with the *Oltre la Strada* project as well.

Today, the whole area of Porta Milano presents many criticalities: Milano is the main access road to the city from the West, characterised by high traffic levels in both directions. *Via Rose*, which is located south of *via Milano* (see Fig. 7) doesn't represent an alternative for vehicles, mainly because there is a railroad crossing.

Oltre la strada aims at solving those criticalities, adopting an holistic approach to mobility issues:

1. Boosting non motorised and soft mobility;
2. Improving Public Transport;
3. Solving criticalities for motorised traffic;

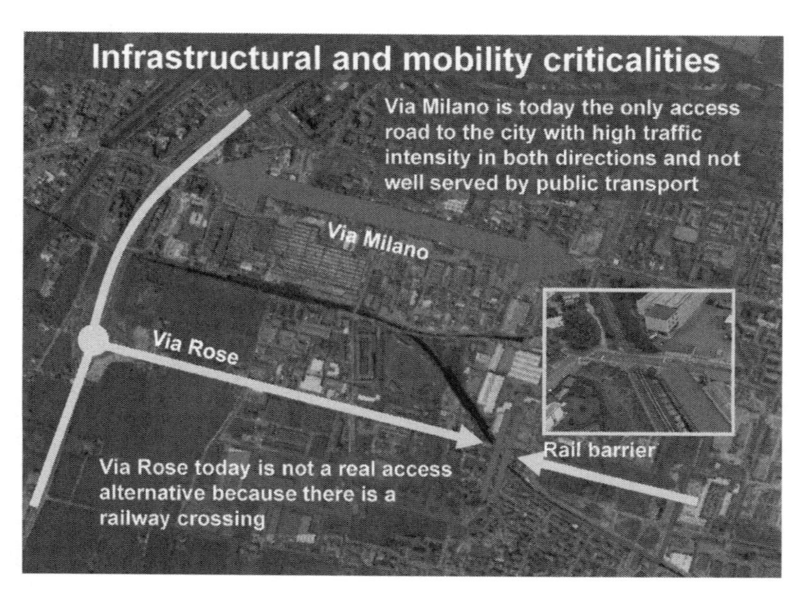

Figure 7. *Oltre la Strada* project. Infrastructural and mobility criticalities in Porta Milano. Source: Municipality of Brescia.

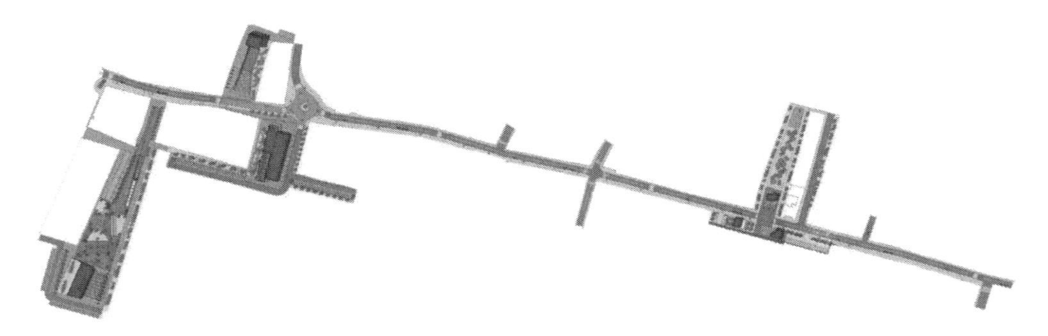

Figure 8. *Oltre la Strada* project. The design of new public spaces along via Milano.

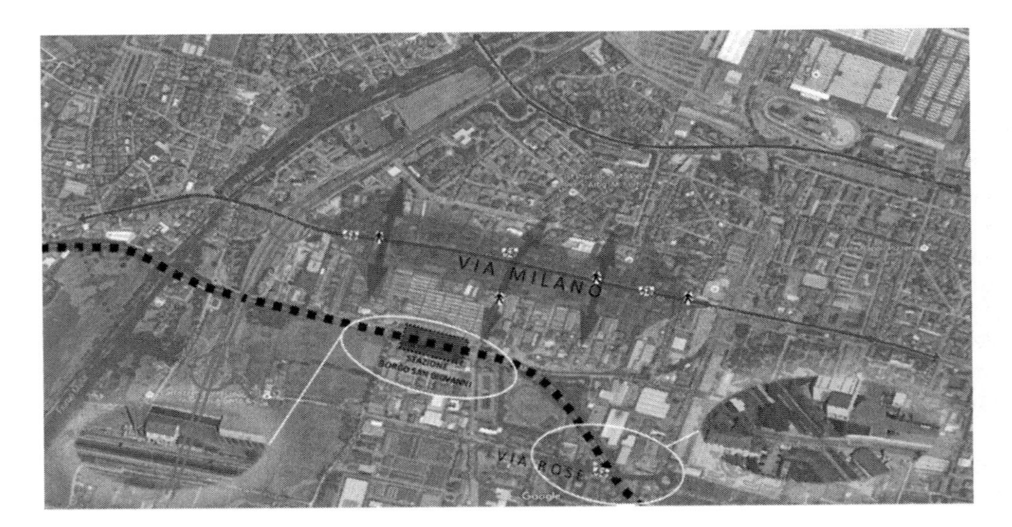

Figure 9. *Oltre la Strada* project. Mobility interventions.

Concerning non-motorised and soft mobility, *Oltre la Strada* set specific targets:

- at least 5 km of new cycling routes will be built throughout Porta Milano;
- the estimated 40% reduction in vehicular transit on Via Milano (compared to the current 20,000 vehicles per day) will facilitate non-motorised road users;
- the reduction of speed limits, especially at squares and in specific road sections to increase road safety for vulnerable users (for an assessment of road safety issues in Italy see, i.a., Rossetti et al., 2014)
- 3 new pedestrian public squares will be opened along via Milano (Fig. 8).

From the Public Transport perspective, the area of Via Milano (as defined by the Oltre la Strada project's perimeter of intervention—see Fig. 5), is very close to the first urban outgrowth of the city centre. But, it is not served by the main public transport line of Brescia, represented by the light metro system active since 2013 (for an assessment of the accessibility levels in Brescia see Bonotti et al., 2015; Tira et al., 2016). Thus, via Milano was excluded from those regeneration processes related to the increased accessibility provided by the light metro.

Therefore, the SUMP envisages both the upgrading of the services of the Brescia-Iseo-Edolo railway line (that goes through the *Porta Milano district*) to promote its use also within the urban area (according to Regional Plans), and the establishment of a strengthful local public transport line along the axis of Via Milano.

Ferrovie Nord (the company that manages the Brescia-Iseo-Edolo railway line), which is partner of *Oltre la Strada*, already started with the Lombardy Region a table for the definition of an agreement aimed at redesigning the service offered by the Brescia-Iseo-Edolo railway line. This represents for Brescia the opportunity to benefit from infrastructural works on the railway line (e.g. the restoration of the unused local station of *Borgo San Giovanni*). The final aim is to upgrade the existing railway line providing a strengthful local public transport option for the western periphery of Brescia. The restoration, putting in security and strengthening of Borgo San Giovanni railway station, that will become "*Brescia—Porta Milano*" represent a pillar of the *Oltre la Strada* project. The existing railway line will be used as a "new" local public transport service for metropolitan use and train frequencies will be increased and cadenced.

Concerning motorised traffic the *Oltre la Strada* project, aim at overcoming the local urban traffic criticalities, eliminating the railroad crossing at Via Rose and design a carriage underpass in order to:

- allow the introduction of suburban services on the historical railway line Brescia-Iseo-Edolo;
- dismantle via Milano from the role of penetration axis, conveying to the new underpass the important traffic connections between the city centre and a new junction on the ring road.

6 FINAL COMMENTS AND RECOMMENDATIONS

To sum up, the paper has presented some (but not all, for a wider description of the single interventions, please refer to Comune di Brescia, 2016b) of the key-features of the *Oltre la Strada* Urban regeneration project for via Milano.

The objectives of the *Oltre la Strada* project are many and involve different typologies of actions: from the enhancement of well-being and quality of life, through the integration and the protection of new or consolidated fragilities in the district, the requalification of public spaces, the infrastructural concerns, the regeneration of former plants and the enhancement of personal services.

The picture of urban, social and environmental criticalities for via Milano necessary requires an holistic and integrated urban regeneration approach. Via Milano issues can not be successfully addressed only by individual actors or through sporadic, specific or thematic interventions: public-private partnerships and a public coordination of the interventions are needed.

Even if the project is now defined and financed, the municipality of Brescia is still putting many efforts in involving as many stakeholders as possible to obtain a permanent and co-designed result in the medium and long term time frames.

As a final recommendation, integral to the success of the project there should be the importance of 'people' in the process, to build trust through face-to-face interactions with citizens: nowadays it is more than ever necessary a focus on the individuals and a re-affirmation of a people-centred planning vision, to create smart strategies for sustainable and inclusive urban environments, and therefore improve the quality of life in our cities (see, i.a., Tiboni & Rossetti, 2012; Pezzagno & Richiedei, 2017).

REFERENCES

Bianchini, F., & Parkinson, M. (1993). Cultural policy and urban regeneration: the West European experience. Manchester University Press.

Bonotti R., Rossetti S., Tiboni M., & Tira M. (2015). Analysing Space-Time Accessibility toward the Implementation of the Light Rail System: The Case Study of Brescia, Planning Practice and Research, 30(4), pp. 424–442, DOI: 10.1080/02697459.2015.1028254.

Comune di Brescia (2016a). Piano di Governo del Territorio 2016 approvato. Available at: http://www.comune.brescia.it/servizi/urbanistica/PGT/Pagine/Variante%202015-PGT-approvato.aspx (accessed May, 15th 2017).

Comune di Brescia (2016b). Oltre la Strada. Relazione di progetto. Available at: http://www.comune.brescia.it/news/2016/settembre/Pagine/Bando-periferie.aspx (accessed May, 15th 2017).

Forrest, A. (2017). We need to talk about urban regeneration. The Guardian, Monday 10 April 2017.
Iverot, S.P., & Brandt, N. (2011). The development of a sustainable urban district in Hammarby Sjöstad, Stockholm, Sweden. Environment, Development and Sustainability, 13(6), 1043–1064.
Paddison, R. (1993). City marketing, image reconstruction and urban regeneration. Urban studies, 30(2), 339–349.
Pezzagno M., Richiedei A. (2017), La conoscenza dei servizi socio-sanitari per una rinnovata attenzione al welfare urbano. Metodiche di censimento, Maggioli.
Roberts, P. (2000). The evolution, definition and purpose of urban regeneration. Urban regeneration, 9–36.
Roberts, P., Sykes, H., & Granger, R. (Eds.). (2016). Urban regeneration. Sage.
Rossetti, S., Tiboni, M., & Tira, M. (2014). Road safety in Italy: An assessment of the current situation and the priorities of intervention. Periodica Polytechnica Transportation Engineering, 42(2), 159–165. doi:10.3311/PPtr.7490.
Santangelo A., Tondelli S., Proli S., Eco-social urban regeneration of residential areas in South East Europe regions—BUILD SEE project results, Urbanistica Informazioni, 2014, 257, pp. 83–87.
Tiboni, M., & Rossetti, S. (2012). L'utente debole quale misura dell'attrattività urbana. Tema. Journal of Land Use, Mobility and Environment, 5(3), 91–102.
Tira, M., Rossetti, S., & Tiboni, M. (2016). Managing Mobility to Save Energy Through Parking Planning. In Smart Energy in the Smart City (pp. 103–115). Springer International Publishing, doi:10.1007/978-3-319-31157-9_6.

Town and Infrastructure Planning for Safety and Urban Quality – Pezzagno & Tira (Eds)
© 2018 Taylor & Francis Group, London, ISBN 978-0-8153-8731-2

GIS-based monitoring and evaluation system as an urban planning tool to enhance the quality of pedestrian mobility in Parma

M. Zazzi, P. Ventura, B. Caselli & M. Carra
Università di Parma, Parma, Italy

ABSTRACT: This paper presents a tool to assess the pedestrian accessibility to local services in residential areas, with special attention to geo-referenced mapping models. This study developed a GIS database, an updated monitoring tool, in order to determine the current offer in terms of pedestrian connectivity in our urban residential areas in general, and more specifically in the case-study of Parma. A series of value judgements on the intrinsic properties of the walkways – in terms of state of maintenance, quality and safety – have been summarized combining different descriptive attributes. This value judgements related to the arcs that make up the shortest path from each dwelling to the nearby public facilities of primary interest. Despite the resistance of our municipalities to consider it a useful tool in decision-making and planning practices, it may play a supporting role in upgrading public services plans or in planning pedestrian network maintenance or future expansion.

1 INTRODUCTION

The paper originates from a wider research activity, started in 2003, concerning the analysis and the evaluation of proximity public services in the urban outskirts of Parma. More authors agree that, in the context of urban and transport planning, accessibility is the essential feature of sustainable mobility (Rossetti, et al., 2015; Handy & Clifton, 2001). This idea is also supported by the Sustainable Urban Mobility Plans, introduced in 2013 by EC Urban Mobility Package; its declared objective is "*improving accessibility of urban areas and providing high-quality and sustainable mobility and transport to, through and within the urban area*" (European Commission, 2013). Many theories and definition of accessibility have been elaborated. Susan Handy and Kelly Clifton suggest an increase in urban planners' interest because accessibility is able to define the potential interaction of residents activities (Handy & Clifton, 2001). Again, the degree of accesibility to urban services determines the number of supplies and opportunities (Geurs, et al., 2012). According to this, the measurment of the quality of pedestrian accessibility becomes very important and, in this sense, the role of GIS tools also becomes crucial.

This paper argues about the real contribution of a GIS tool in supporting mobility planning, thanks to the correlation of alphanumeric data on territorial basis and the identification of significant statistical connections and trend lines. Whilst we find positions that are entirely propositional in terms of monitoring—simulation of possible future scenarios and participatory techniques – (Murgante, 2008), we identify some scepticism in the public administration. In the last few years there have been several contributions from the scientific literature which have employed GIS analysis to evaluate accessibility issues of pedestrian paths. Many of them operate through a raster analysis. A significant investigation, applied to the case study of the San Polo district in Brescia, concerns the application of ET-Geowizard tool for the evaluation of the walking travel times (Rossetti, et al., 2015). Another set of contributions employs vector analysis to examine the issue of disadvantaged users (Xu, 2014), interconnecting data about the presence of barriers (Yairi & Igi,

2006). A Michigan study about measuring the degree of accessiblity to healthcare facilities employs both medologies, raster and vectorial, identifying their specific advantages (Delamater, et al., 2012).

2 OBJECTIVE OF THE PAPER

The general purpose of this paper is to suggest a new GIS-based analysis model able to assess pedestrian mobility in peripheral areas in terms of quality of the paths and time needed to reach the main public services. First at all, this GIS database may have a role as an operative tool to pursue the goal of improving paths' quality, identifying and intervening on the path attributes which cause the impoverishment of safety, easy accessibility and pleasantness of the route. It should also be able to evaluate the integrity of the existing pedestrian infrastructure system, in terms of presence/absence of pavements and crossing paths, and to identify eventual deficiencies.

To improve pedestrian accessibility, both in terms of quality and integrity, is also a significant matter of concern of planning tools related to pedestrian mobility. Therefore, the functional aim of this model is not only to help in an operative way, but also to show its potential in planning for the efficiency of pedestrian mobility.

The GIS-based model can provide remarkable support to the final outcomes of all decisional processes, in either urban policies, mobility plans, or public programmes. Not by chance GIS techniques are now being widespread in the scientific community for what concerns accessibility analysis (Delamater, et al., 2012; Rossetti, et al., 2015; Xu, 2014; Yairi & Igi, 2006). In both cases the purpose is to provide balanced conditions of liveability of the pedestrian paths, to grant the possibility of moving 'from and to' different parts of the city securely, comfortably, and in an attractive way. This matter has a crucial role particularly in peripheral and fringe areas—at a neighbourhood scale—in which the accessibility to the branched and scattered proximity services should be on foot and therefore encouraged.

The specific purpose of the paper is to apply this methodological approach to the pilot case study of *Cittadella*, the second more populated neighbourhood in Parma, where residential use is prevalent.

3 METHODOLOGICAL APPROACH

In the context of this paper, accessibility will be assessed mainly using a GIS database, managed with ArcGIS software, which incorporates various layers of information. The first step to its development has been the design of its structure, combining multiple datasets—pedestrian paths, public services, streets, buildings and housing numbers—and defining correlations and spatial interactions. The relationships which correlate the four indispensable geodatabase's elements have been defined in the logical modelling as shown in Figure 1.

Obviously particular care has been put in the design of the pedestrian paths' dataset which also implies the construction of a pedestrian network as a base map. This network has been developed following a node-and-link structure.

In the graph, all walking trails available on a public space have been mapped: pavements, crossing paths and walkways, especially those which allow users to walk across green areas to reach different parts of the neighbourhood. It takes also into account virtual pavements and virtual crossing paths assuming that pedestrians, in absence of a dedicated path or a marked crossing path, might choose to walk along the street border and to cross the road in proximity of an intersection. This is an indispensable simplification subdued to the analysis model because, generally, people tend to use all the available walking space which is not always public: all accessible private spaces which can make the journey shorter become potential walkways.

Then the constructed graph has been implemented with qualitative and quantitative indicators collected through an accurate urban survey.

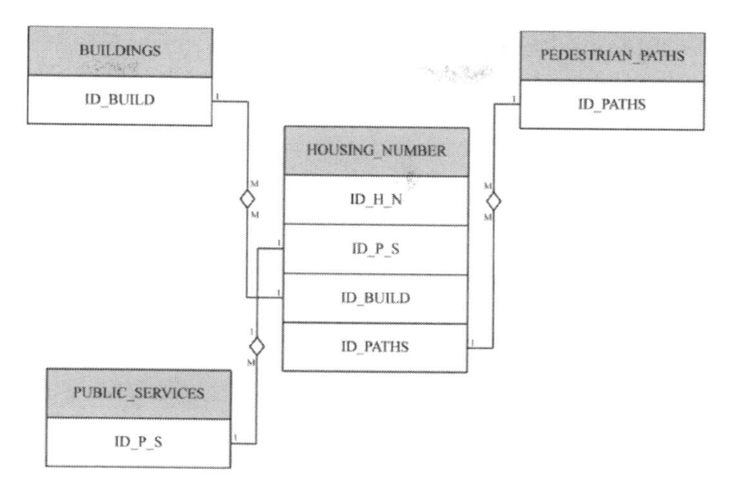

Figure 1. Logical modelling of the geodatabase.

One of the indispensable quantitative attributes is the "cost", i.e. the ratio between the link length and pedestrian speed, estimated at 6 km/h (National Research Council, 2000). The computing of the cost, expressed in minutes, considers also a "slowdown factor" at pedestrian crossings that is slightly different depending on whether a traffic light is present or not.

4 DESCRIPTION

4.1 *Innovative aspects*

The methodological approach is here applied to the pilot case of the urban sector of Cittadella, in the first outskirts of Parma historic core. Through a detailed urban survey, all information has been collected and then implemented in the designed geodatabase. Firstly, the Geographic Information System of Parma municipality, from which the streets, buildings and housing numbers' datasets have been extracted, has been updated with the pedestrian paths' dataset, defining all the possible walkable trails and, in the meanwhile, also the limits to pedestrian mobility. Secondly, public services have been represented in a dedicated dataset, classified in typologies, such as green areas, schools and other public facilities. The case study processing considers three middle schools as potential destinations for residents throughout the neighbourhood.

4.2 *Technical developments*

Each link of the constructed pedestrian network has been associated with a value judgement resulted from the combination of pedestrian paths' qualitative attributes, such as geometry, typology, pleasantness features, state of maintenance and safety standards especially for fragile users. The selection of these attributes has considered the connotations of the specific urban context. Then the "cost" value has been assigned to each link of the network as defined in the methodological approach. In particular the last quantitative attribute is indispensable for the analysis because planning the enhancement of pedestrian accessibility to public services brings in issues of time and distance besides quality.

Thanks to the Arc GIS software's tools dedicated to network analysis, it has been possible to assess either the catchment area of the selected public facilities and the quickest routes to reach them from the point of access of each dwelling unit.

The Service Area analysis can be used to determine the catchment area of one or more facilities within the neighbourhood; in this case study processing, middle schools has been chosen as an example. Each network-based service area includes all the pedestrian links that can be reached within a 3, 5 and 10-minute walk from the chosen facilities, considering also the probable slowdowns caused by street crossing (Fig. 2).

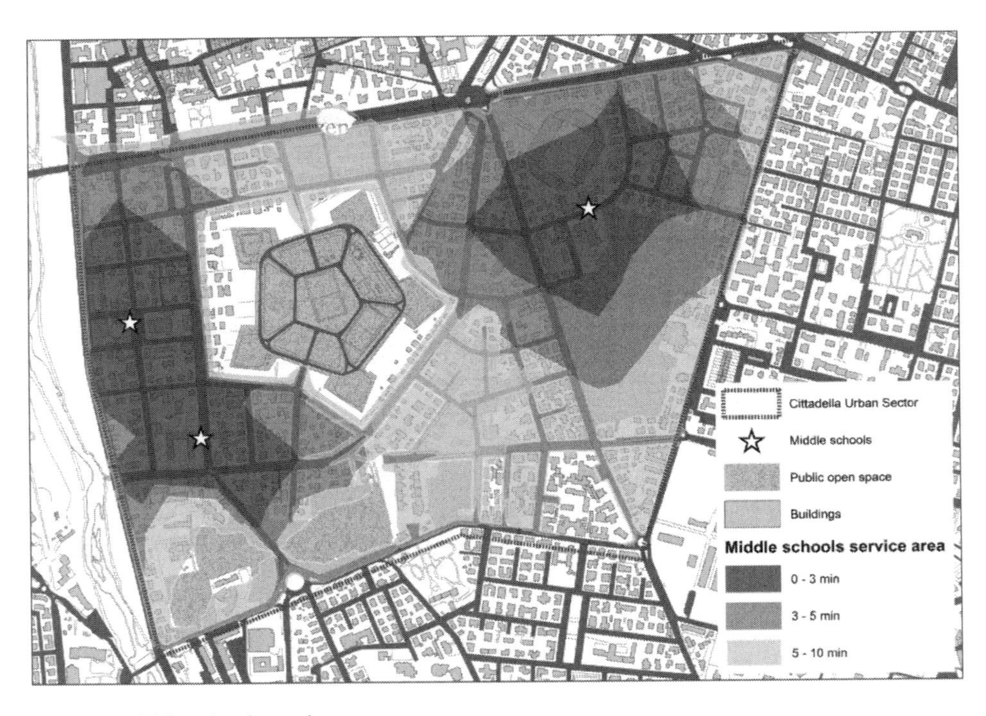

Figure 2. Middle schools service areas.

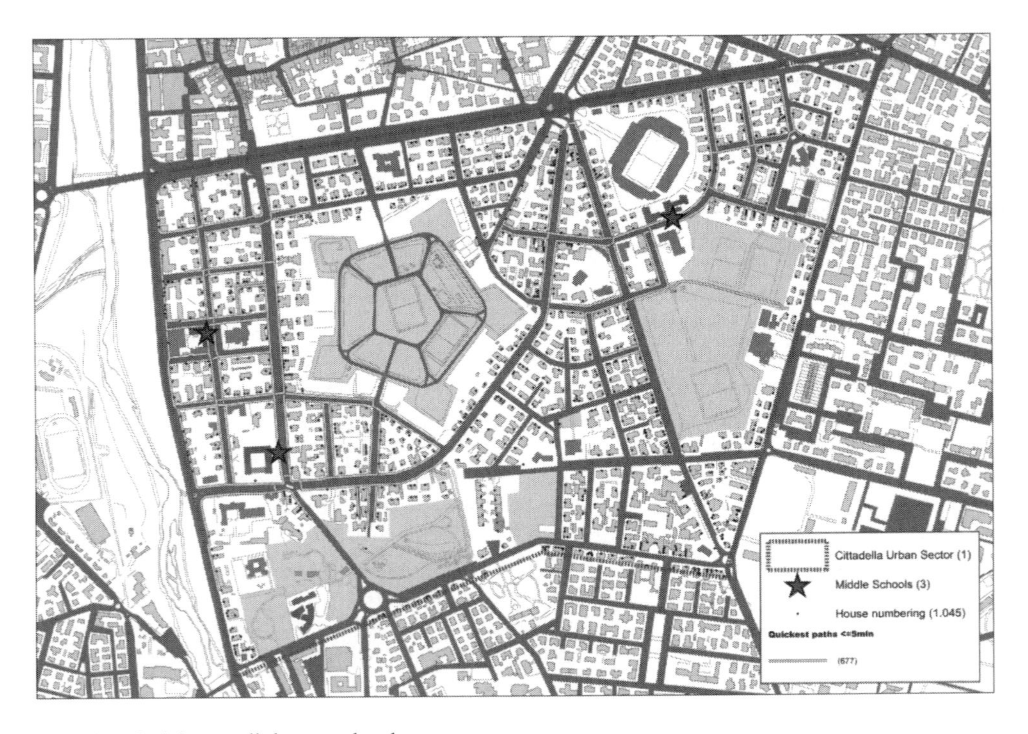

Figure 3. Quickest walk home-school.

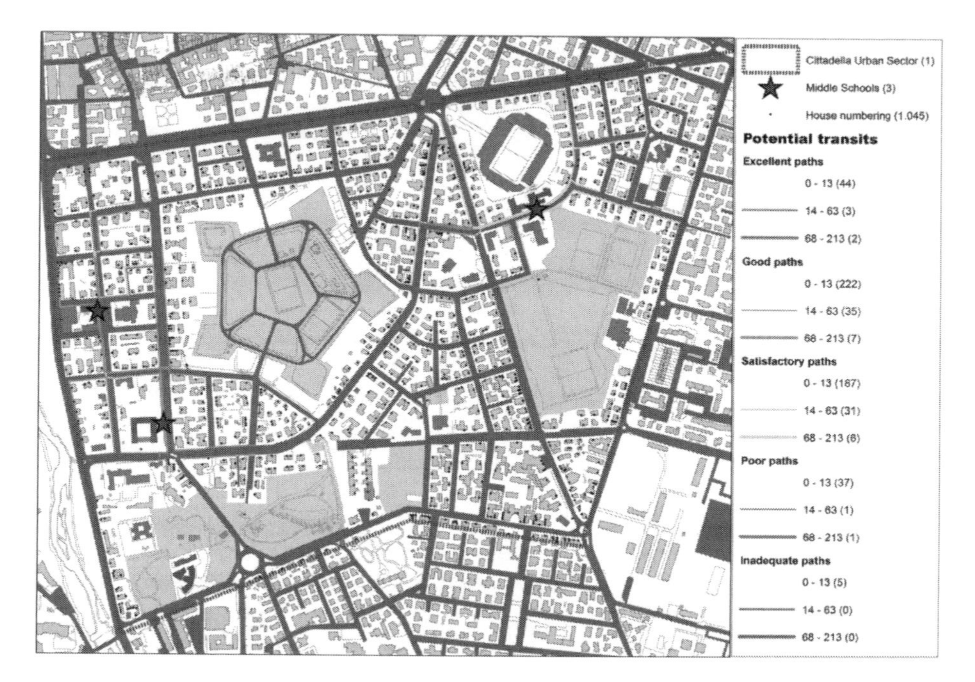

Figure 4. N. of potential transits.

The Closest Facility tool computes the least-cost path, i.e. the quickest walk between dwelling units and public services. In the output, it is possible to identify the paths which are potentially more often travelled, counting automatically the number of links intersections with the quickest walking routes (Figure 3).

5 RESULTS OF THE RESEARCH

Through the GIS model multiple thematic maps have been printed to show the performance of pedestrian mobility related to the timing of pedestrian accessibility. The results of this analysis lead easily to the identification of time zones for the urban pedestrian usability: the service area approach assumes that all locations within the determined zones are accessible in a 3, 5 and 10-minute walk and the assumption is not overestimated because the compute starts from the connectivity of a pedestrian network. This result may also be used as a criterion for housing assessment, measuring the residential value based on proximity thresholds to access public services. Furthermore, this management offers the possibility to give an estimation of all the various urban sectors and therefore to plan public provisions according to the needs.

The GIS model makes it also possible to find trade-offs between the best paths, in terms of safety and pleasantness, and the shortest ones (Figure 4). This can be considered a relevant tool to determine the priority for interventions on pedestrian infrastructures, based on the frequency of use and the potential users involved, e.g. students.

6 CONCLUSIONS

The GIS data model survey singles out a set of indicators apt to define the quality of spaces, the quality of the collective interest equipment and of the walking routes from residential units to public services. Both the quality of the accessibility to public services and of the pedestrian infrastructure can be considered as indicators of a good urban performance of a residential urban sector—according to the objectives of SUMPS. The layering of attributes related to such equipment, the thickening of social functionality gathered in specific places, and the addition of high-quality and qualified routes aggregated in a network, establish a reliable marker of urban centrality and therefore of a presumable quality of the urban settlement.

6.1 *Barriers and drivers*

The present research has the following further possible developments. Firstly, the GIS tool might be improved in order to monitor pedestrian accessibility on a larger scale—urban or metropolitan—establishing a correlation even with public transport networks. This perspective should be able to provide a more complete picture of integrated mobility as a whole.

Secondly, the GIS database could provide support to the PEBA (*Piano di Eliminazione Barriere Architettoniche*). This significant plan operates on the overcoming of architectural barriers in urban areas in order to improve the accessibility for users with disabilities. The description of the barriers proposed by PEBA introduces more interesting qualitative and quantitative indicators (Centro Documentazione Barriere Architettoniche, 2003) that can be used to refine the GIS database:

- Floor paving: presence of narrows, improper transverse slope and slippery, uneven or damaged surfaces, presence of obstructions caused by street furniture and absence of tactile and coloured markers at crossroads;
- Changes in Elevation: presence of stairs or improper longitudinal slope;
- Signage and protections: presence of horizontal signage, acoustic traffic lights, protective barriers and well-lighted pedestrian crossings.

Lastly, the GIS database should be able to integrate the research with a new participative interaction. In fact, through a SMART system set right on the geo-referenced map, the neighbourhood residents, exercising an active role along the lines of the *Citizen Science* project, could be involved directly in monitoring the state of conservation of the trails in real-time. All reports, properly filtered by public officials may constitute, for all practical purposes, a very useful material for updating the database. This GIS database might be used as an instrument for a continuous auditing of the state of the art and its possible future changes. According to this, the model could become a real-time observation, easily accessible by public authorities and citizens. As a result, this 'observatory' could be able to provide a participatory and comparative tool between the municipality and the resident population. Thanks to it, citizens might have an active part in reporting improper uses of pedestrian paths, for example: maintenance conditions, the presence of obstacles, the malfunctioning in signage and lighting, illegal car parking on bike or walking lanes or wrong use of walking trails by bicycles, and any other needs depending on more varied types of users, especially weaker ones.

REFERENCES

Centro Documentazione Barriere Architettoniche ed. (2003), *Linee guida per la redazione del Piano di Eliminazione Barriere Architettoniche.* Regione Veneto, Venezia.

Delamater, P.L., Messina, J.P., Shortridge, A.M. & Grady, S.C. (2012), "Measuring geographic access to health care: raster and network-based methods", *International of health geographics*, Vol.11 No.15, pp. 1–18.

European Commission (2013). *A concept for Sustainable Urban Mobility Plans,* European Commission, Brussels.

Geurs, K.T., Krizek, K.J. & Reggiani, A. (2012), Accessibility analysis and transport planning. Challenges for Europe and north America, Edward Elgar, Northampton.

Handy, S. & Clifton, K.J. (2001), "Evaluating neighborhood accessibility. Possibilities and practicalities", *Journal of transportation and statistic,* Vol.4 No.2, pp. 67–78.

Murgante, B. (2008), "L'informatica, i Sistemi Informativi Geografici e la Pianificazione del Territorio", in Murgante B. (Ed.), *L'informazione geografica a supporto della pianificazione territoriale.* FrancoAngeli, Milano, pp. 7–37.

National Research Council (2000), *Highway Capacity Manual.* TRB, Washington.

Rossetti, S., Tiboni, M., Vetturi, D. & Calderón, E.J. (2015), "Pedestrian mobility and accessibility planning: some remarks towards the implementation of travel time maps", *Journal City Safety Energy,* No.1, pp. 67–78.
Ventura, P. & Zazzi, M. (2014). "L'accessibilità pedonale ai servizi di prossimità nella programmazione urbanistica comunale. Scenari di rete e criticità locali nel caso di Parma", in Giuliani F. (Ed.), *L'utente debole nelle intersezioni stradali.* Egaf, Forlì, pp. 249–268.
Xu, M. (2014), *A GIS-based pedestrian network model for assessment of spatial accessibility equity and improvement prioritization and its application to the spokane public transit benefit area,* Washington State University, Pullman.
Yairi, I.E. & Igi, S. (2006), "Mobility support GIS with universal-designed data of barrier/barrier-free. Terrains and facilities for all pedestrians including the elderly and the disables", in *IEEE International Conference on Systems, Man and Cybernetics, Taipei, 8–11 October 2006,* Curran Associates, Taipei, pp. 2909–2914.
Zazzi, M. (2006), "Localizzazione e accessibilità dei servizi di prossimità nelle periferie della città contemporanea", in *International Conference Living and Walking in Cities. The Underskirts, Brescia-Bergamo, 9/10 June 2005,* Tipografia Camuna, Brescia, pp. 262–278.

Town and Infrastructure Planning for Safety and Urban Quality – Pezzagno & Tira (Eds)
 ISBN 978-0-8153-8731-2

Town planning management: Accessibility to personal care service

M. Pezzagno & A. Richiedei
Università degli Studi di Brescia, Italy

ABSTRACT: The paper presents an expeditious methodology to assess the accessibility of medical, healthcare and social assistance services. It properly focuses on vulnerable city user like elderly and disabled people. The general aim was to evaluate supply and demand of services in order to improve citizen wellbeing in the municipality of Brescia.

All services were georeferenced and the data sheets were collected in a GIS database. The final database is extremely useful because it could be easily updated and it is also transferable. The main outcome of the analyses were thematic maps: they showed clearly e.g. the absence of services in areas whit high percentage of elderly, the kind of services and their distance from public transportation.

In this way municipal government has a scientific feedback for their policies of services creating, services moving and for identifying priority schedule of services and public open spaces maintenance.

1 INTRODUCTION

Research on town facilities planning, starting from the already existing literature on this topic, can lead to interesting debates among the various disciplines such as economics and halthcare management.

The carried out research proposed an interdisciplinary approach to better understand what a family in a middle sized city needs for its welfare. This means putting together knowledge related to offer and demand of facilities and families location within the city, i.e. different neighborhoods charateristics and needs.

This goal opens to connections between specific knowledge new instruments and analysis of the problem set in a transversal way. In Italy the analysis of facility services is a particularly important subject strictly connected to town planning.

In 2001 in Region Lombardy the facilities planning bacame mandatory at the local level (Piano dei Servizi L.R. 1/2001; L.R. 12/2005).

Such care seems to be unique in the Italian situation where planning instruments both strategical and operational do not sufficiently highlight, through a standardized procedure, the dynamics between demand and supply in services to the citizens.

The *Local facilities plan* try to forecast population needs, but most of the existing plans still lack updating of the existing public facilities. Analysis completeness and reference to the private sphere is an essential element to understand the social dynamics linked to facilities.

The overall general knowledge on the exisiting facilies location and on their status is a fundamental starting point for planning and it finds its completeness in the proposed methodology. It integrates public and private healthcare services offer together with their potential demand in residential areas according to specific characteristics of the population (age of people, family needs, etc.). Transferable information and methods could be useful to the expert administrations, to operators, to all operational services and citizens.

2 OBJECTIVE OF THE PAPER

The paper shows the work done by the "town planning team" in evaluating the private and public offer of welfare services in Brescia. The goal of the research is to prompt innovation in the organization and furnishing of services so as to optimize municipal resources and to better wellbeing for individuals and families.

The quality of the offer does not rely exclusively on organization and territorial features, but also, and most of all, on the level of spatial and temporal accessibility. The concept of accessibility refers to the easiness with which a facility can be reached considering the various possibilities of transport (on foot, by bike, using a bus, or a car). Improving accessibility has an impact both on environmental sustainability (Himanen et al., 1994; Rolli and Bruschi, 2005), and on road safety (Godefrooij and Schepel, 2010), besides assuring accessibility to all kinds of users, with particular attention to week users. The facilities taken into consideration in the research deal with children, elderly and disabled people. The research was carried out in three steps.

The methodology of the proposed analysis was set-up and tested to investigate facilities for children and then extended the other two categories. The investigation on desabled facilities was integrated investigating the specific accessibility problems from a visual, motor, auditive point of view. The methodology and the results achieved are presented in the paper.

3 METHODOLOGICAL APPROACH

3.1 *Preparatory phase and technical aspects*

Methodological approach to the re-organization of welfare facilities began with the selection of which facilities were strictly related to welfare goals: i.e. medical, healthcare and social assistance services needed and provided to citizens. To improve welfare services it was necessary to really understand which services could be integrated or grouped together and have a clear idea of their role in the territory, their main and complementary functions and how they had to be managed. Only after this it was possible to classify them in a usefull way. The correct grouping of activities within a category is bound to the considered parameters and the classes themselves. An universal criteria in cataloging needs to avoid subjectivity and personal interpretation which may lead to a wrong or unclear cataloging. As a matter of fact, due to the subjectivity of the criteria adopted by the different municipalities in planning facilities at local level facilities are often catalogued in different sections and, some of them, are not inserted in any category.

So finding a clear objective way of cataloging to be used all over the country seems to be fundamental. It has to be able to describe and differentiate each type of facility and its activity. We decided to use the classification of economic activities proposed by the National Statistic Institute (ISTAT) as it is quite simple to associate and refer a specific activity to an economic category. ATECO is the classification of economic activities proosed by ISTAT in 2007 and it is the l'Italian version of NACE (Nomenclature statistique des Acticites economique dans la Communaute Europeenne) which is the general classification system used to define economic/industrial activities within the European Union. ATECO was devised by ISTAT to account for the specificity of the Italian productive structure and point out activities particularly important throughout the country.

The first result was the introduction of a "universal classification" provided on a single database (catalogue of services and facilities) to be shared and used by all those offices involved in the problem. Univocity was sought not only in the structure but in the contents of classification and in interpretation so as to have a code made up of six figures at most.

It was decided to limit the analyses of facilities for the person vs those that care for the individual physical and psychological health satisfiing the needs and solving the difficulties that people may face in life, that is SAS facilities.

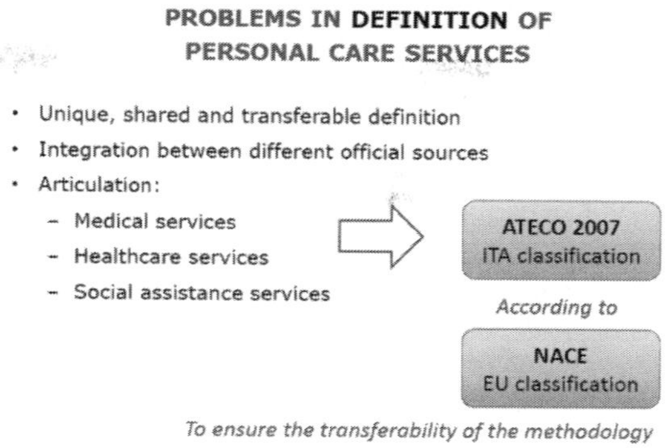

PROBLEMS IN DEFINITION OF
PERSONAL CARE SERVICES

- Unique, shared and transferable definition
- Integration between different official sources
- Articulation:
  - Medical services
  - Healthcare services
  - Social assistance services

ATECO 2007
ITA classification
According to
NACE
EU classification

To ensure the transferability of the methodology

Figure 1. Univocal classification of services and database creation.

From the methodological point of view research was divided in three steps:

- 1st step: "Identification and Classification of services and facilities":
  The Local Urban Plan is the main source to identify public services, but it has to be integrated using data from other databases (for e.g. private services);
  Data harmonization is necessary in order to build up an "unique database"; the focus must be on:
  - updating outdated information;
  - completion of partial information;
  - introducing universal classification criteria according to ATECO 2007 – NACE.
- 2nd step: "Survey and Cataloguing":
  Each service must be verified and/or censored through the following criteria:
  - Location;
  - Management and ownership;
  - Urban contest;
  - Size and urban index;
  - Maintenance of the building;
  - Functionality (opening time, utilities, employees,…);
  - Accessibility (indoor, outdoor).
- 3rd step: "Data analysis and processing"
  GIS database had to be created to:
  - collect survey data;
  - georeference services (using the same rules);
  - obtain new data (through spatial analysis);
  - produce cartography to find out critical issues.

The investigated facilities were divided in three groups: social assistance, social-health assistance and health assistance. Besides this definition it was agreed to consider that some facilities are exclusively dedicated to specific users, for instance R.S.A. (Residenza Sanitaria Assistenziale, i.e. Health Care Residences)[1] that is only for elderly people while other facilities could be used by a wider group of people.

3.2 *Operational phase*

The 1st step was the reclassification of the existing local facilities already scheduled in the plan of Brescia using ATECO identification criteria. The local facilities plan of the Munici-

1. The health care residences, RSA, introduced in Italy in the mid-1990s, are non-hospital facilities which are lodging not self-sufficient people that can not be assisted home and that require specialized medical care by more specialists and extensive health care.

pality scheduled most of the public facilities owned and or managed by the Municipality pinpointing name, location, type of service offered; in few cases specific information on the state of the building and on its accessibility were already set in the database.

The updating of the existing municipal database implied also the introduction of a dedicated column to identify the specific ATECO category.

The 2nd step consisted in the recording the missing welfare services.

The database was integrated step by step through a standard survey card, most of the welfare services were not catalogued in the existing database; during the research a dedicated survey for facilities was set up and the welfare service were recorded using it.

The proposed standard survey card—that applies to any kind of existing facility on the territory—is made up of 11 sections.

1. identifying codes
2. description of the service/facility
3. locating data
4. management
5. analysis of the context
6. description and opinion about the structure
7. functionality of the facility or service
8. access to the facility within the territory
9. accessibility to the facility
10. main architectonic barriers noticed
11. signposting.

In the new database every facility is represented by a single line divided in columns, every group of colunmns is related to one specific section (from 1 to 11).

The investigation on accessibility to *welfare services* (point 8 & 9) collected the information on the area strictly related to the building such as the access to the building (entrance), the parking of the building in which the facility is set and the information related to the public space structure around the facility (Outodoor area: location and equipement of local public transport stops and pedestrian paths). The urban context around the facility was deeply analized within a range of 300 m. For each question of the survey a "quality score" – from 1 (very bad, i.e. not accessible) to 6 (very good) – was introduced in order to have a ranking of the wellness facility.

Such figure has been stemmed from considering accessibility for those users who get to the facility using public transport or those users who live in the surrounding area and need to reach the facility on foot.

In several cases conflicting situations are located just outside the facility. It is in fact possible to notice that these situations of inaccessibility are due to the change of ownership (between public and private). Spaces inside the building in which the facility is located were not investigated.

Furthermore when dealing with facilities for disabled people more specific information was added in relation to the facility itself. The information was collected through checklists focused on the type of disability. These checklists[2] need to be filled in to verify if the facility satisfies the requirements for the disabled category. Each card is characterized by its own identifying code which makes it exclusive within the database, besides the card holds the specific data required to link it unequivocally to the analyzed facility.

The new integrated database can be easily updated, and sensitive data have been reported only in aggregated form.

Whenever it wasn't possible to get the data on the site the missing information was acquired through other official databases (i.e. For the calculation of catastal areas/surfaces the official Topografic Database of Brescia Province were used).

The 3rd step was "Data analysis and processing."

2. It must be underlined that the checklist must be filled in by people who have a knowledge about architectonical barriers and accessibility problems/rooles.

All facilities were georeferenced using a GIS (Geograaphical Information System), so now it is possible to have them on a single map: the areas completely dedicated to a specific facility (monofunctional areas) are identified according to their lot borders (polygon) as they have a single function; while a spot has been considered more useful whenever facilities are located within poli-functional buildings as they haven't got distinct areas in the map. Each shape (polygonal or spot) is linked to the database through the ID code.

4 RESULT OF THE RESEARCH

The methodology to assess the accessibility of medical, healthcare and social assistance services was tested on Brescia facilities.

Localization, quality and accessibility of the existing services were investigated filling the survey for more than 267 welfare facilities:

- 135 specifically for children[3],
- 53 specifically for elderly people,
- 48 specifically for disabled people and
- 31 non-specialized.

Two insights for the facilities dedicated to elderly and disabled users are presented in the paper pointing out the data whenever deemed important for the offer dynamics of the specific kind of facility. A general comment on the collected data follows.

In Brescia the distribution of facilities for elderly people is highly inhomogeneous, also because in most cases they are set in existing buildings. This led to the location of the facility where there was an available building (both public or private) but not where there was highest user request and where there were the best accessibility features.

Table 1. Welfare facilities in Brescia divided in categories and mapped.

	Medical facilities	Healthcare facilities	Social assistance facilities	Other facilities
Children	Hospital (7), Health centre (3), Paedriatician (25)	Swimming pool whit neonatal courses (6)	Residential Care Facility (4), Nursery (45), Foster care facilities or adoption facilities (1), Voluntary social assistence facilities (14)	Kindergarten (60)
Elderly	Residential Care for Elderly and Nursing Home (12)	Old age home (5) Integrated Old people's center (5)	Residence for elderly (9) Old people's center (19)	Other elderly facilities (3)
Disabled people	Clinic (4), Specialized Health centre (3), Physiotherapist (4)	Residential Care for Disabled people (2) Swimming pool (rehabilitate center) (2)	Disable people's center (2) Socio-educational center (1) Self-sufficiency development facilities (1) Work placement facilities (3) Assistence facilities (3) Scholar Assistance (3) Nursery (1) School (3)	Sport activities (3) Volunteers association (2) Associations for the protection of rights (9) Cultural association (2)

3. Children welfare facilities were used to calibrate the investigation methodology.

The parameter SLP[4]/user (i.e. square meters of the facility/number of users per day[5]) shows that the highest area per person is related to R.S.A. facilities.

Facility	Theoric SLP (mq)	N° of daily users	Sq.m. SLP/User
R.S.A.	98.246	1.167	84,19
C.D.I.	12.949	205	63,17
Welfare hotel for elderly	2.534	57	44,46
Small houses for elderly	3.224	66	48,85
Daycare centers	6.340	595	10,66
Other facilities (cultural, etc.)	480	105	4,57
Total	123.773	2.195	56,39

This was expectable as these facilities include health facilities and residences for elderly people. Mini residences are closer to 59 sq.m./user, which is about a standard dimension for a studio flat. C.D.I.s have a bigger dimension compared to Welfare hotel.

For what concerns the dimension of facilities relating it to residing population (31st dec. 2015 elderly aged over 65 and old over 75) and daily users, it is clear that in the Northern part of Brescia there is an increased incidence of facilities in terms of SLP and there is also the higher number of elderly people. In this area there are 5 R.S.A.

In the Southern part of Brescia instead facilities are lacking, while in the historical centre there is a high number of users in relation with the number of the facilities (i.e. a high demand of facilities).

In the Western part of the town there is a very high SLP/user ratio (i.e. more square meters per user).

The day-time centers are those having a more capillary distribution throughout the territory but there is a great disparity between the Northern/Center areas and the Southern/Western ones. In relation to the other facilities for elderly people the San Polo district turns out to be the better served throughout the Municipal territory followed by Centro Storico Nord. The Crocefissa di Rosa, Buffalora, Caionvico, Casazza, San Rocchino, Chiusure, Villaggio Violino, Don Bosco, Folzano, Fornaci, and Porta Cremona districts instead are without any kind of social health and assistance facility for the elderly.

The ratio between the number of facilities and the residing and elderly population per district shows how the high concentration of facilities in the Center (18) coincides with a high rate of elderly residents (28%). The highest number of elderly people is located in the East part of Brescia (32%) though not as absolute value and it has the lowest amount of dedicated facilities (10). The Southern part of the city has the absolute number of elderly residents (more than 10,000) with lower incidence (23% elderly residents) and a very low number of facilities (5).

Bus service is available to access all the facilities and the bus stops are located within a range of 300 meters. No facility is close to the metro line. All the facilities are easily reachable and accessible by car. Pedestrian and cyclist accessibility absolutely need to be improved.

Disabled facilities are also not distributed homogeneously across the territory. During the inspection the 80% of the buildings dedicated to the disabled were considered in good maintenance condition[6].

Regarding the accessibility to the facilities the parameters considered during the survey highlight several problems especially related to the parking areas specifically equipped for disabled people.

Pedestrian accessibility can be considered positive, except for some cases of not adequate width. The access paths to the facility are almost always without necks or obstacles that could reduce the needed width. The pavings are suitable and generally in good conditions.

4. SLP (Superficie Lorda di Pavimento) is the italian acronym for gross floor area.
5. Data related to 2015.
6. The assessment of the buildings conservation state has been focused to the outside only, as entry to the facilities is highly limited by the users' -privacy protection regulations.

Table 2. N. of socio-assistance and health facilities for children, elderly people and the disabled per district in Brescia.

Town area	Name of the district	N. of facilities for children	N. of facilities for elderly	N. of facilities for disabled
Historical center	Brescia antica	5	4	1
	Centro storico nord	3	6	0
	Centro storico sud	8	5	3
	Crocifissa di Rosa	11	0	1
	Porta Milano	2	1	0
	Porta Venezia	9	2	6
Eastern	Buffalora	2	0	0
	Caionvico	1	0	0
	San Polo	9	8	3
	Sant'Eufemia	3	2	0
Northern	Borgo Trento	7	2	1
	Casazza	2	0	1
	Mompiano	5	5	8
	San Bartolomeo	4	4	0
	San Rocchino	7	0	2
	Sant'Eustacchio	2	1	1
	Villaggio Prealpino	2	1	1
Western	Chiusure	3	0	3
	Fiumicello	3	2	4
	Primo Maggio	1	1	0
	Urago Mella	4	1	0
	Villaggio Badia	3	3	0
	Villaggio Violino	3	0	0
Southern	Chiesanuova	4	1	0
	Don Bosco	4	0	1
	Folzano	3	0	0
	Fornaci	2	0	0
	Lamarmora	7	2	5
	Porta Cremona	13	0	5
	Villaggio Sereno	3	2	2
	Total	**135**	**53**	**48**

Figure 2. Services for elderly vs. density of age 65+ citizens.

Figure 3. Services for elderly vs. density of vulnerable families with at least one age 65+.

	Evaluation parameter	Percentage
Accessibility close to the facility	Served by metro	40%
	Served by bus	98%
	Served by urban bus	28%
Parking for users	Facilities with internal parking for users	73%
	Facilities with external parking for users	89%
	Facilities without parking	0%
	Facilities with parking reserve for disabled	75%
Access routes to the facility	Facilities in 30 km/h Zone	0%
	Facilities reachable with pedestrian routes correctly dimensioned	92%

A critical point is represented by the directional signage. It presented problems both in the identification of the facility and in orientation along the belonging areas. Signposting is often lacking or poor mainly in the case of privately owned buildings. Besides there never are well marked suitable devices to ask for assistance in accessing the building specially if lifting equipment is needed or just to lead the user.

As per the reserved parking spaces there is a rather good situation for what concerns the connection between the parking area and the facility. Only rarely it has been noted that the connections were not adequate or connected to a pedestrian route leading to a service entrance. A very small number of facilities have reserved parking spaces with dimensional characteristics not complying with Italian regulation.

The accessibility to the service using public transport is within a range of 300 meters, but covered bus stop are often missing. The available space for transport waiting, ascent and

Figure 4. Services for disabled people classified by category of disability.

descent is generally critical, while suitable information devices are instead always present at all stops.

5 INNOVATIVE ASPECTS AND CONCLUSIONS

The innovative aspects of this approach mainly concern the completeness of the method: all kinds of facilities have been considered using ATECO classification and most importantly it doesn't exclude privately owned or association managed facilities (as instead it happens in the town plan). The used census database is easy to update and the univocal identification of the facilities (ID code) allows to modify and update all the data. Besides the information acquired through the survey can be transferred to those agencies which may need them (municipality or local health service).

FINAL REMARKS

ATECO-NACE identifies all services, not just public ones

GIS Database easy to update (+100 services)

Collection of information based on the type of disability

Georeferenced information useful to

- Identify priorities and evaluate relocation of services
- Map supply and demand of services

Municipal government has
DATA for DECISION
a scientific feedback for policies of services

The structuring of the census through two kinds of filings, a static one for all facilities and a variable one according to the specific function (or type of service) allows to collect the highest number of information possible and to compare it extensively according to the investigation needed. Information georeferencing has various positive consequences: it allows to directly connect facilities supply and demand (the resident population also according to age or other features); it allows to find out facility-overloaded or lacking districts and decide possible relocations; it allows to prioritize maintenance tasks in publicly owned facilities and it gives an operational contribution also in pointing out architectonical barriers area around the facilities.

REFERENCES

Godefrooij, T. and Schepel, S. (2010). Co-benefits of Cycling-inclusive Planning and Promotion. Interface for Cycling Expertise. World Bank, GRSF, contract 7152334.

Himanen, V., P. Nijkamp and J. Padjen (1992). Transport Mobility, Spatial Accessibility and Environmental Sustainability, Research-Memorandum 1992–53 December 1992, p. 9, 20–28.

Rolli, G.L. and Bruschi, M. (2005). Accessibility and environmental quality in vulnerable historical sentre: l'Aquila. In: Busi R. and Pezzagno M. (eds.) Living and walking in cities. Town and infrastructure planning for saferty and urban quality for pedestrian, p. 159–172.

Vulnerable road users

Town and Infrastructure Planning for Safety and Urban Quality – Pezzagno & Tira (Eds)
 ISBN 978-0-8153-8731-2

The city of images. Urban mobility policies and extra-small tactical projects for promoting quality of urban life of people with ASD (Autism Spectrum Disorder)

T. Congiu, V. Talu & G. Tola
Università degli Studi di Sassari, Sassari, Italy

ABSTRACT: The paper focuses on the rarely discussed topic of the relation between the city and people with ASD (Autism Spectrum Disorder).

It has become incresingly important to think that it is necessary to 'broaden' the research perspective by investigating contribution of urban mobility policies and urban design to the enhancement of the quality of life of people with ASD.

The paper describes a recent and ongoing research aimed at defining integrated urban mobility policies and extra-small tactical projects for promoting and providing the real opportunity for people with ASD of "using" their everyday city.

In the first we provide a framework for illustrating the commonly recurring problems that people with ASD face in their daily life when they interact with the urban environment.

In second the paper is supported by an in depth analysis of existing contributions (researches and projects) and an exchange with different experts.

1 INTRODUCTION

The research focuses on the rarely discussed topic of the relationship between the city and people with ASD [Autism Spectrum Disorder], with the specific aim of promoting individual «urban capabilities» (Talu 2013, 2014; Blečić, Cecchini, Talu, in press) by increasing the autonomy and safety of walking across the city at a neighborhood scale, thus making the access to relevant urban spaces and services possible, also to this vulnerable group of citizens.

The knowledge gained so far on autism catch up continuously, since both the definition of autism as well as its causes and its diagnostic framing have been and are being subjected to continuous integrations and remodeling[1].

The most recent disorder's definition[2] includes autism within a *spectrum* or *continuum of severity*[3] characterized, with great variability, from persistent deficits in mutual social communication and social interaction, from the presence of restricted interests and patterns of repetitive behaviour, hypo and hyper reactivity to sensory stimuli, and comorbidity[4].

Furthermore, the incidence of Autism Spectrum Disorder is steadily increasing. According to the European Commission, today autism is the most widespread developmental disability (Giofrè, 2010).

The increasing incidence of the disorder is mainly due to the change of criteria and diagnostic categories (Plauchè J. e Meyers, 2007), namely to an improvement of the diagnosis that currently considers as autistic individuals that would not have been considered as such before.

1. Cf. Diagnostic and Statistical Manual of Mental Disorder (1968, 1987, 1994, 2013).
2. Cf. Diagnostic and Statistical Manual of Mental Disorder (DSM 5) (2013), APA.
3. Vivanti G., (2010), p. 24.
4. Comorbidity is the combined presence of more neurological disorders beyond autism (Bartolomeo and Cerquiglini, 2010).

A monitoring made in 2010 in eleven States of USA on children of eight years of age revealed that 1 child on 68 (1,5%) was autistic, with an increase of 119.4% compared with 2000 (CDC, 2016)[5].

Nowadays there is no cure for autism (Vivanti, 2010) and despite the progress achieved in the early diagnosis that allows to begin therapeutic treatments in a timely manner, reaching adulthood in many cases means a deterioration aggravated by the loss of a set of support services guaranteed only until the majority.

Thinking of an accessible urban context means to reconsider the spatial and functional urban priorities according to all inhabitants to provide even people with ASD the necessary facilities to access and "use" urban spaces.

2 OBJECTIVE OF THE PAPER

Starting from the development of a cognitive framework on the relationship between the city and people with ASD, the research aims at defining a set of policies (of mobility) and urban extra-small tactical projects (flexible, modulable, and low cost) for promoting "urban capabilities" of people with ASD, with particular reference to the ability and opportunity of walking independently and safely across the city to access to relevant spaces and services in the urban environment.

Specifically, the aim of the research is to provide implementing tools to improve the recognizability of urban sites and services through structured paths in order to facilitate their reaching.

Since autism is a *spectrum* (it manifests in a wide variety of ways in different people), it has been essential to identify the *behavioral constants* that characterize the relationship between the autistic person and the environment, in order to define the spatial and sensory requirements needed to improve the urban accessibility level and quality.

3 METHODOLOGICAL APPROACH

The methodological process followed during the research may be divided into four distinct phases extremely related to each other.

First and foremost, it has been necessary to know and understand the Disorder, focusing on all those daily needs that can be translated in spacial actions.

Particular attention was given to the interaction individual-environment, since people with ASD tend to establish a conflictual relationship with the surrounding environment; in fact, their behavioural problems are not just mere consequences of a different functioning of their central nervous system (CERPA, 2016), but also a response caused by an altered sensory perception.

The *sensory experiences of autism* (Bogdashina, 2011) depend not only on the six senses receptors, but also on the placement's disorder in the *spectrum* range. In fact, these are distributed according to a scale of intensity that mainly distinguishes the sensory experience in two perceptual categories: *hyposensitivity* and *hypersensitivity*.

In the case of *hyposensitivity* the individual perceives a significantly lower quantity and intensity of stimuli and informations, while, in the case of *hypersensitivity*, she/he over-perceives, with several resulting problems for the brain's elaboration process (Delacato, 1974).

This is one of the most relevant aspects of the research because most of the times the sensory perception affects negatively the daily lives of people with autism spectrum disorder (Pellicano, Dinsmore, Carman, 2013).

5. See Christensen D. et al., Prevalence and Characteristics of Autism Spectrum Disorder Among Children Aged 8 Years – Autism and Developmental Disabilities Monitoring Network, 11Sities, United States 2010. MMWR Surveillance Summaries. 2014; 63: 1–22.

This first phase of the research lead us to identify the *"atypical urban functionings"* of the people with autism spectrum disorder.

The second phase, through the literature review and the examination in depth of several realized projects, allowed us to identify a set of spatial requirements for the design of autism friendly urban environment able to meet the spatial needs of people with autism.

Current researches and applications aimed at exploring the role of spatial configuration as a means for improving the autonomy of people with ASD, exclusively focus on the definition of criteria for the design of *closed, separated, private* spaces, *devoted* only to people—mainly children—with ASD, (i.e. assisted living residences, day care centers and schools, healing gardens) (Beaver 2003, 2006; Brand 2010; Gaudion e McGinley 2012; Herbert 2003; Linehan 2008; Mostafa 2008; Sachs e Vincenta 2011; Wilson 2006).

Currently, only one research has tried to explore the urban scale with the aim of defining urban paths well connected to relevant and specific urban spaces and services to support and increase autonomy of adult people with ASD (Decker, 2014).

Beside this activity, we have organized several meetings with different experts: parents of the association ANGSA Sassari Onlus, special needs teachers of the primary school (located in the neighborhood interest by the application), and neuropsychiatrists. The exchange with the parents (interviews and questionnaire) lead us to "fine-tuning" the definition of three *"enabling urban requirements"* for urban public spaces accessible and livable also for people with ASD.

The last phase of the research focused on an application in the city of Sassari, in the neighborhood where is located the primary school "V° Circolo Didattico", necessary to translate the research's results on a real spatial dimension.

This phase has been really useful because allowed us to face with a complex and problematic reality that, due to its structure and localization within the urban context, hinders the actual possibility of walking (not only for people with ASD). One of the major is related to the vehicular traffic that especially in Via Gorizia (the street where the main entrance of the school is located) is really intense and unsafe.

The design of the actual road section determines a priority vehicular space. Pedestrian areas are limited to uncomfortable and tight sidewalks in which there are several types of obstacles (bins, bumps, light posts, trees, etc.) that reduce the already little available space.

Sidewalks are also delimited by vehicular parking areas that create space-use conflicts: most of the time cars are parked on the pedestrian crossings and on the sidewalks, on the bus stop areas, on the disabled parking without any permission reducing visibility, accessibility, safety and making pedestrian accessibility very difficult or almost impossible.

4 DESCRIPTION

Starting from the *"atypical urban functionings"* three reading keys has been developed aimed of identifying urban policies and extra-small tactical projects.

I. The first one considers the problem of sensory overload and its management in the urban context. Studies, research and the several interviews have reported traffic noise and veicholar flow as the main obstacles for walking or, in this specific case for getting to school, therefore to increase accessibility, autonomy and safety the first necessary condition is related to the adoption of sensory mitigation measures to reduce vehicular noises. Referring to the Barcelona Urban Mobility Plan (2013–2018)[6], the first macro-action

6. The Barcelona Mobility Plan focus on the reorganisation of urban mobility defining macro-blocks (*superillas*) whose vehicle access is restricted and regulated by a set of policies that reduce the speed at 10 km/h and hinder the passing traffic because of the introduction of one-ways. Bus and private cars traffic is diverted to main roads outside the macro-block. See: http://mobilitat.ajuntament.barcelona.cat/es/plan-de-movilidad-urbana/presentacion.

concerns the introduction of traffic regulation policies at neighborhood scale for reducing acoustic impact and vehicular speed and for improving pedestrian accessibility and safety. This define pedestrian priority macro-blocks where car parking is limited to the access areas or to a dedicated sites.

Through physical and perceptive resistance solutions (raised pedestrian crossings, extension of sidewalks at intersections and crossing or the extension of the whole sidewalk reducing parking areas, greenery) passing traffic is discouraged providing at the same time more security.

II. The second, however, refers to the importance of the use of images and in general of visual supports (pictures and maps) because people with ASD think in pictures (Grandin, 2001) and understand the surrounding environment using pictures that represent objects and actions.

A vertical signage with images, supported by a horizontal coloured strip, is a key element to facilitate orientation and recognition of urban places and services. The horizontal signage is defined by two colours indicating a direction and a change, infact colour is an important orientation device also used in the visual supports of the therapeutic treatment for people with ASD.

For example the colour red appears in several pictures of the Augmentative and Alternative Communication[7] to symbolize the "stop" action.

III. Lastly, the need to respect and follow a specific routine is necessary for people with ASD because they are extremely creatures of habit and is very hard for them to accept changes in the normal ritual of daily actions. Faced with change they can react with anger crises and aggressive attitudes (Bartolomeo and Cerquiglini, 2010). In their everyday life time planning of the activities is indispensable. This need occurs through the structuring of "social stories" consisting of objects images, photographs or words (it depends on the severity of the disorder) that illustrate to the person the ordered succession of the activities to be done during the day and the sequence of the single micro-actions to carry out in order to accomplish a more complex one.

IV. This type of signage should be located at significance places and in general before embarking on a path to getting to a defined site and before carrying out an action such as crossing the street. Providing the sequence is essential because guarantees them to be prepared for what's going to happen.

5 RESULT OF THE RESEARCH

The application in the city of Sassari have been carried out in a part of the neighborhood in which, beside the primary school of "V° Circolo Didattico", are located a secondary school, a high school and a nursing home (Fig. 1).

The first action, essential prerequisite in order to develop the subsequent ones, is the definition of a low speeds block to reduce auditory sensory stimuli and to discourage passing traffic (Fig. 2).

This measure is combined with the integration of traffic calming solutions such as raised pedestrian crossing (Fig. 3).

Parking area is limited to the access point of the block.

Physical and sensory accessibility and autonomy are facilitate by the design of "*enabling path*", properly supported whose starting are located in the access points (Fig. 4).

---

7. Augmentative and Alternative Communication involved the communication skills by learning a symbolic system alternative to the verbal code composed of pictures, written words and objects (Vivanti, 2010). During childhood many people with ASD don't develop verbal language and use images as a support for communication and the development of associative thinking (Grandin, 2001).

Figure 1. The application area.

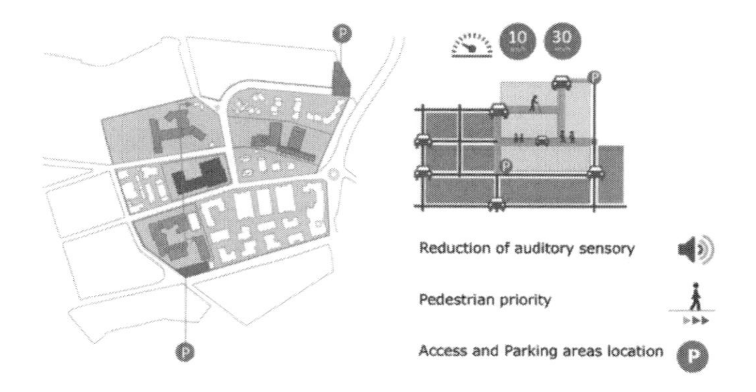

Figure 2. The low speeds block.

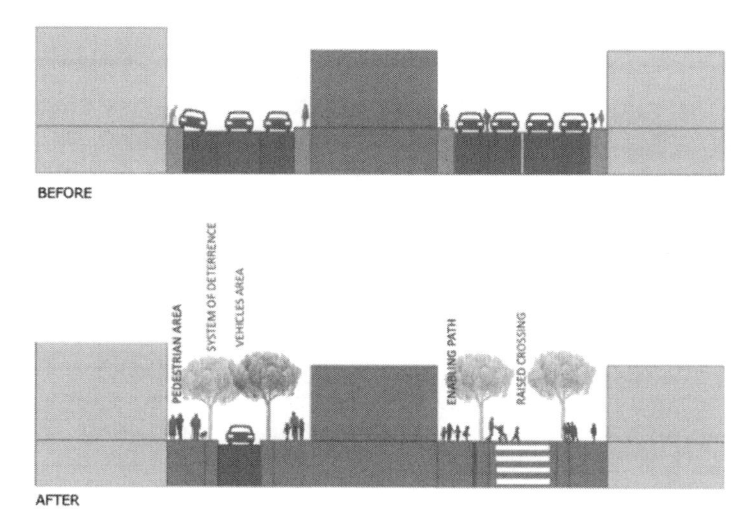

Figure 3. Traffic calming solutions.

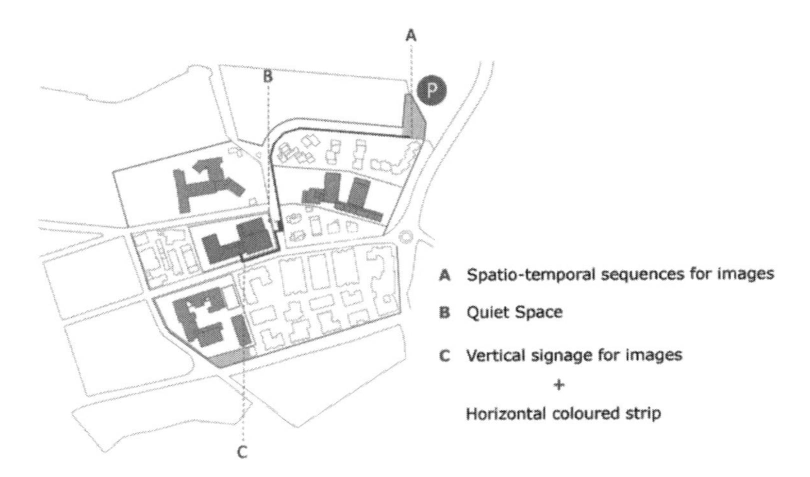

Figure 4. The enabling path.

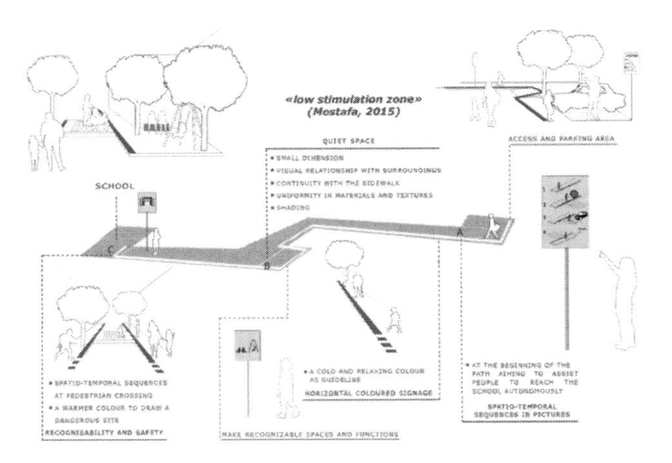

Figure 5. Structure and design of the enabling path.

The requiring of reduce the sensory overload is also solved providing the *enablig path* with "Quiet Spaces" (Fig. 5), protected spaces able to maintain a good visual relationship with the context and with a particular focus on sensory quality (cadence depends on path's length and on the sensory stimulation level) aimed at defining a "*low stimulation zone*" (Mostafa, 2015). These spaces have to comply specific requirements: continuity with the sidewalk, small dimensions to avoid crowding and confusion, protection, visual relationship with the surroundings, shading, uniformity and continuity in the use of materials and textures.

Quiet Spaces are located on the sidewalks in the parking areas subtracted to vehicles.

The need to communicate for images and thus the use of visual supports is solved with vertical signage for images and with a horizontal coloured strip (Fig. 5). To avoid confusion only two colours are used; a colder and relaxing one (blue) that act as a guideline along the path and a warmer one (red) to draw attention when occur a change or a dangerous situation.

The *routine* requirement is met by the integration of a signage that introduce *spatio-temporal sequences* (Fig. 5) to support the carrying out of complex actions. This sequences infact are localized at the beginning of the paths to assist the reaching of the school and at relevant and dangerous sites as the pedestrian crossing.

6 CONCLUSIONS

The proposals and the first results of the research show that thinking of a City for Images means consider not only people with ASD but consider and intercept also negated and unrecognized spatial needs and desires of many other vulnerable inhabitants such older people, children, woman and in general a group of people who function in an atypical way (Terzi, 2011).

Promoting the autonomy of people with autism spectrum disorders through the adoption of mobility policies and micro tactical urban projects become an opportunity for promoting and improving the quality of life of the entire city.

Furthermore, the significant presence of this group of inhabitants that, according to the experts and prediction, will continue to grow and the almost total lack of attention to this demand at the urban scale mark the need and the duty to focus attention on a real ignored demand.

Clearly, it is not possible thinking to deprive city of their complexity to satisfy each of this need, but people with ASD deserve to find, at least in defined parts of the city, public spaces that accommodate and support other forms of functionings.

As Decker writes:

"Individuals with autism add an additional level of health to cities[8]"

REFERENCES

American Psychiatric Association (APA), (1968). Diagnostic and Statistical Manual of Mental Disorders, Second Edition, American Psychiatric Association.

American Psychiatric Association (APA), (1987). Diagnostic and Statistical Manual of Mental Disorders, Third Edition Revised, American Psychiatric Association.

American Psychiatric Association (APA), (1994). Diagnostic and Statistical Manual of Mental Disorders, Fourth Edition, American Psychiatric Association.

American Psychiatric Association (APA), (2013). Diagnostic and Statistical Manual of Mental Disorders, Fifth Edition, American Psychiatric Association.

Bartolomeo S., Cerquiglini A. "Il quotidiano dei soggetti autistici: definizione dei programmi terapeutici e di socializzazione e fabbisogno di servizi", in Giofrè F., (2010). Autismo. Protezione sociale e architettura, ALINEA editrice.

Beaver C., (2003). Breaking the mould. Communication. 37(3):40.

Beaver C. Designin Environments for Children and Adults with ASD, AUTISM SAFARI. 2nd WORLD AUTISM CONGRESS & EXHIBITION, Città del Capo, 30 Ottobre – 2 Novembre 2006.

Blečić I., Cecchini A., Talu V., (in press). Approccio delle capacità e pianificazione urbana. Camminabilità come strumento di promozione della qualità della vita urbana degli abitanti svantaggiati in Archivio di studi urbani e regionali.

Bogdashina O., (2011). Le percezioni sensoriali nell'autismo e nella Sindrome di Asperger, uovonero edizioni.

Brand A., (2010). Living in the Community. Housing Design for Adults with Autism. Helen Hamlyn Centre – Royal College of Art.

Centers For Disease Control And Prevention (CDC), (2016). Community Report on Autism. From the Autism and Developmental Disabilities Monitoring (ADDM) Network. CDC.

CERPA. I bisogni di persone con disabilità intellettive e disturbi dell'apprendimento: quali opportunità per progettare l'inclusione e sostenere l'autonomia, Bologna, 28 Aprile 2016.

Christensen D. et al., (2014). Prevalence and Characteristics of Autism Spectrum Disorder Among Children Aged 8 Years – Autism and Developmental Disabilities Monitoring Network, 11Sities, United States 2010. MMWR Surveillance Summaries. 63: 1–22.

Decker E.F., (2014). A city for Marc. An inclusive urban design approach to planning for adults with autism. Kansas State University.

8. Decker E.F., (2014), p. 34.

Delacato C., (1996). The Ultimate Stranger: The Autistic Child, Academic Therapy Publications, Novato, 1974. Trad. it. Alla scoperta del bambino autistico, The Ultimate Stranger, Armando Editore.
Gaudion K., Mcginley C., (2012). Green Spaces. Outdoor Environment for Adults with Autism. Helen Hamlyn Centre of Design – Royal College of Art.
Giofrè F., (2010). Autismo. Protezione sociale e architettura. ALINEA Editrice.
Grandin T., (2001). Pensare in immagini e altre testimonianze della mia vita autistica, Erickson.
Herbert B.B., (2003). Design guidelines of a therapeutic garden for autistic childre N. Loyola University.
Linehan J., (2008). Landscape for Autism. Guidelines and Design of Outdoor Spaces for Children with Autism Spectrum Disorder. University of California.
Mostafa M., (2008). An architecture for autism: concepts of design intervention for the autistic user. International Journal of Architectural Research. 2: 189–211.
Mostafa M., (2015). Architecture for Autism: Built Environment Performance in Accordance to the Autism ASPECTSS Design Index. Design Principles and Practices. 8: 56–71.
Sachs N., Vicenta T., (2011). Outdoor Environments for Children with Autism and Special Needs. Implications. 9: 1–8.
Terzi L., (2011). "What metric of justice for disabled people? Capability and disability", in Brighouse H. e Robeyns I., eds., Measuring justice. Primary goods and capabilities. Cambridge University Press.
Vivanti G., (2010). La Mente Autistica. Le risposte della ricerca scientifica al mistero dell'autismo. Omega Edizioni.
WHO. Autism spectrum disorders & other developmental disorders. From raising awareness to building capacity, Geneva, 16–18 Settembre 2013.
Wilson B.J., (2006). Sensory garden for children with autism. University of Arizona.

Town and Infrastructure Planning for Safety and Urban Quality – Pezzagno & Tira (Eds)
 ISBN 978-0-8153-8731-2

Safety evaluation of urban streets with mixed land use: Examining the role of early stage of planning

V. Gitelman
Israel Institute of Technology, Haifa, Israel

E. Doveh
Technion Statistical Laboratory, Haifa, Israel

D. Balasha
Israel Institute of Technology, Haifa, Israel
Traffic and Transport Engineering Consultancy, Haifa, Israel

ABSTRACT: This study investigated the relationships between road infrastructure and land use characteristics and the level of road safety on mixed land-use streets in Israel seeking for ways to prevent unsafe solutions at early stages of planning, such as detailed town plans (DTP).

The majority of streets were with two-way traffic, straight and flat, with 70%–100% of commercial frontage, multi-level buildings and with DTP classified as "planned for mixed land-uses".

The model for total injury crashes showed that higher number of crashes are associated with a higher percentage of commercial frontage, more midblock crosswalks and higher traffic volumes. Similarly, stronger positive relations with pedestrian crashes were found for higher percentages of commercial facades.

The study findings supported the assumption that the initial street planning as mixed land-use affects its safety level, where higher extent of the mixed land-uses on the street was associated with an increased crash risk.

1 INTRODUCTION

In Israel, similar to other countries, more than 70% of injury crashes and above 60% of fatalities and serious injuries occur in urban areas (RSA, 2016). Many studies showed the associations of the number of crashes with variables describing urban environment, such as street types and land uses, and road design characteristics (e.g. Wedagama et al., 2006; Ewing and Dumbaugh, 2009; Ukkusuri et al., 2012). It can be agreed that design stages define road characteristics, which, in turn, affect safety performance of streets. Hence, one would expect that road design practices would account for safety considerations, starting from an early stage of urban planning, that is not always the case.

In Israel, the statutory planning process is dictated by the *Design and Construction Law*, which determines a hierarchy of plans, including: national outline plan, county outline plan, local outline plan, and a detailed town plan (DTP). A lower level plan is derived from the higher levels; when a plan is approved it becomes a secondary legislation. The DTP determines the obligatory conditions for urban infrastructure planning, including road network. The transportation conditions are dictated by the areas dedicated to roads ("road rights"); the road designers should share these areas between traffic lanes, pedestrians and parking zones. In general, safety considerations are not included in the statutory planning process. Such issues can be considered only at the stage of a detailed road design prior to building but then the designer usually applies a local interpretation without a vision of the whole picture. Moreover, safety faults at early planning stages may result in building roads/areas with embedded safety problems. Thus, including safety considerations in the DTP may prevent unsafe solutions at a later stage.

The process of involving safety considerations at the design stage is known in the literature as *safety audits*, with proven safety contributions (e.g. Macaulay & McInerney, 2002). However, safety solutions suggested at the design stage should be based on evaluation studies (Elvik et al., 2009), which examined the relationships between urban road design and built environment characteristics, traffic exposure and road safety.

2 OBJECTIVE OF THE STUDY

Urgent problems in the design practices in urban areas are associated with planning streets with mixed land-use. Such streets serve both residential and other land-use purposes, e.g. commerce, business and public; they are common in the centers of Israeli towns. The road functions of such streets are mixed as they should satisfy a combination of both mobility and access demands. Pedestrians present a significant share of road users on such streets, that, together with high traffic volumes, creates multiple pedestrian-vehicle interactions and safety problems. The risk of pedestrian injury on such streets in Israel is higher compared to other urban roads (Gitelman et al., 2012). This study investigated the relationships between road infrastructure and land use characteristics, and the level of road safety on mixed land-use streets in Israel seeking for ways for preventing unsafe solutions at early stages of urban planning, such as DTP.

3 RESEARCH APPROACH

As urban streets with mixed land-uses were selected road sections where commerce, business and public buildings composed the majority of properties along the street. For each street, a wide range of land use, road infrastructure, vehicle and pedestrian traffic characteristics was collected, through detailed field surveys. Crash data on the total injury, severe and pedestrian crashes that occurred in five years were extracted, for each section, from the Central Bureau of Statistics' files. The information on the DTPs of the streets was found in the municipality archives and geographic systems of the local authorities. Using the plans, the road rights' width and the lengths of non-residential land uses (shops, business, public buildings, parks, etc.) were measured, on both street sides. The ratio of mixed land-uses out of the total street length was estimated; when it was over 50%, the street was defined as *planned* for mixed land-uses.

Using the study's database, the relationships between the street characteristics and crashes were analyzed by means of adjusting multivariate regression models for the number of crashes on the street sections. The models were fitted to total injury crashes and pedestrian crashes. As crashes are subject to over-dispersion (Elvik et al., 2009), negative binomial regression models were applied, in the form of:

$$E\{acc\} = \exp\sum_i [\beta_i x_i]$$

where $E\{acc\}$ is the expectancy of crash numbers on the road section; x_i – street characteristics, β_i – model coefficients; the street length was used as an *offset*. The model parameters were estimated using the *glm.nb* function of the MASS R library (Venables and Ripley, 2002). Both *forward* and *stepwise* strategies were attempted for model development, using a minimum of *Akaike information criterion* to select the best model. To identify the street characteristics with stronger impacts on crashes a complementary analysis by means of binary trees was applied, using the *rpart* function of R.

4 RESULTS

4.1 *Characteristics of street sections with mixed land uses*

For the study's purposes, detailed data were collected for 88 street sections with mixed land-uses, in 25 towns. Table 1 summarizes their main characteristics. The data showed that the streets examined were of three road types: dual-carriageway two-way, single-carriageway two-way and single-carriageway one-way; some streets of the latter type were with a dedicated bus-lane. Fig. 1 provides examples of typical views of such streets in Israeli towns.

The majority of streets were with two-way traffic, straight and flat, with 70%–100% of commercial frontage, multi-level buildings and a DTP classified as *planned* for mixed land-uses. Most of the streets had marked pedestrian crossings and bus stops, and considerable volumes of pedestrians and vehicle traffic: with hourly averages of 500–1300 passing vehicles, 800–920 walking pedestrians and 140–240 crossing pedestrians. At the same time, most sites had neither sidewalk extensions along the street nor pedestrian fences for preventing unauthorized crossings (74% and 85% of the total), and were characterized by a mix of various types of junctions (signalized, un-signalized, roundabouts). In total, in five years, 624 injury crashes were reported on the study sections, of which 57% occurred at junctions and 11% were severe (fatal + serious) crashes. In general, existing traffic arrangements did not convey

Table 1. Characteristics of street sections with mixed land uses (N = 88).

a – Road and land use characteristics.

Characteristic	Categories (% of the total set)
Road type	Dual-carriageway (44%), single-carriageway two-way (31%), single-carriageway one-way (19%), single-carriageway with bus lane (6%)
% of commercial facades	Below 50% (12.5%), 50%–70% (17%), 70%–100% (70.5%)
Type of housing	Multi-level buildings up to 8 stores (92%), others – one/two family homes and multi-level buildings with over 8 stores (8%)
Planned land use purposes – DTP	Mixed uses on both sides (61.4%), mixed uses on one side (19.3%), others (19.3%)
Junction types	Un-signalized only (15%), un-signalized and roundabouts (19%), all types (66%)
No of midblock crosswalks	0 (17%), 1 (28%), 2 (15%), 3 (21%), more (19%)
No of bus stops	0 (20.5%), 1 (26%), 2 (27%), 3 (12.5%), more (14%)
Horizontal alignment	Straight section (71%), slight curve (22%), sharp curve (8%)
Vertical alignment	Plain (53%), slight grade[2] (39%), strong grade (8%)

b – Length, traffic and crashes.

Road type	Mean length m	Vehicle traffic*	Walking pedestrians*	Crossing pedestrians*	No of injury crashes#	% of severe crashes	% of pedestrian crashes	% of crashes at junctions
Dual-carriageway	489	1289	801	222	306	12%	33%	63%
Single-carriageway two-ways	321	763	823	140	145	12%	36%	42%
Single-carriageway one-way	314	492	922	238	58	12%	26%	47%
Single-carriageway + bus lane	897	1148	830	192	115	6%	20%	67%

*Hourly average. #Total, in five years.

a – a dual-carriageway two-way street

b – a single-carriageway two-way street

c – a single-carriageway one-way street

Figure 1. Examples of typical streets with mixed land uses in Israel.

a clear message as to pedestrian versus vehicle priorities in the streets, whereas traffic calming elements for vulnerable road users were generally missing. Not surprisingly, 31% of crashes on these streets were with pedestrian injury.

4.2 *Models for predicting crashes*

Table 2, a–b presents the explanatory models fitted to total and pedestrian crashes on the study streets. Among the explanatory variables for total injury crashes were found the characteristics of the road layout (roadway width and geometry), speed variance, frequency of midblock pedestrian crosswalks, type of housing and the extents of planned and actual non-residential land uses along the street. The model showed that an increase in the number of total crashes is associated with a higher percentage of commercial frontage, more pedestrian crosswalks on the street, higher traffic volumes, increased variance in the travel speeds, "other" types of housing (i.e. different from the common multi-level buildings), and when

Table 2. Explanatory models for crashes on streets with mixed land uses.

a – Total injury crashes (deviance explained = 40.1%).

Variables	Estimate	Std. Error	z value	Pr(>|z|)
(Intercept)	–6.53	0.94	–6.97	0.000
Commercial facades: 50%–70%	1.17	0.32	3.67	0.000
Commercial facades: 70%–100%	0.93	0.31	2.98	0.003
Standard deviation of travel speeds#	0.08	0.05	1.49	0.136
Vertical alignment: slight grade	–0.49	0.17	–2.89	0.004
Vertical alignment: strong grade	–0.51	0.32	–1.60	0.110
Total width of roadway#	–0.07	0.02	–4.25	0.000
No of crosswalks#	0.18	0.06	2.96	0.003
Log (Hourly vehicle traffic)#	0.30	0.16	1.79	0.073
Share of street length planned with mixed residential and commercial uses#	0.95	0.28	3.38	0.001
Indicator of total planned mixed uses*	–0.72	0.23	–3.19	0.001
Type of buildings: others	0.50	0.28	1.78	0.076
Share of street length planned with open public space#	1.11	0.69	1.61	0.107
Horizontal alignment: slight curve	–0.19	0.18	–1.03	0.302
Horizontal alignment: sharp curve	–0.54	0.28	–1.92	0.055

b – Pedestrian crashes (deviance explained = 47.1%).

Variables	Estimate	Std. Error	z value	Pr(>|z|)
(Intercept)	–6.58	0.60	–11.02	0.000
Share of street length planned with mixed residential and commercial uses#	1.81	0.29	6.20	0.000
Hourly vehicle traffic#	0.001	0.00	5.46	0.000
Commercial facades: 50%–70%	1.54	0.42	3.69	0.000
Commercial facades: 70%–100%	1.12	0.43	2.61	0.009
Share of public transport in total vehicle traffic#	2.94	1.11	2.65	0.008
Junction types: un-signalized and roundabouts	–0.83	0.41	–2.05	0.041
Junction types: all	–1.80	0.43	–4.18	0.000
Share of street length planned with open public space#	3.03	0.84	3.61	0.000
Total width of roadway#	–0.04	0.02	–2.29	0.022
Width of sidewalks#	0.07	0.03	2.29	0.022
Presence of pedestrian fencing: on one sidewalk	–0.20	0.47	–0.43	0.671
Presence of pedestrian fencing: on both sidewalks	–37.43		0.00	1.000
Presence of pedestrian fencing: on the median	0.33	0.34	0.98	0.329
Standard deviation of travel speeds#	–0.10	0.06	–1.77	0.077

*Equal to "1" when, in total, more than 50% of mixed land uses were planned, at least on one side of the street. #Numeric variable.

longer shares of the street land uses were originally planned as residential with commerce or as open public space[1].

Conversely, a decrease in the number of crashes was associated with presence of grades and curves along the street, wider roadway and when in total the street was planned for mixed land-uses on the majority of its length[2]. As a whole, the model indicated a positive link between wider planned and actual mixed land uses on the street and total crashes, whereas the land use indicators served as stronger crash predictors than road design features.

1. Most impacts are significant with $p < 0.01$, the impact of vehicle traffic – with $p < 0.1$, speed variance and share of open public space – with $p < 0.15$.
2. 3%–7%

In the model for pedestrian crashes, the explanatory variables included the road layout (widths of roadway and of sidewalks), intersection types, vehicle traffic, percentage of public transport, presence of pedestrian fences, variance of travel speeds and the indicators of actual and planned non-residential land uses along the street[3,4]. However, the impacts of roadway and sidewalk widths, vehicle traffic and speed variance were minor, the impacts of pedestrian fences – inconsistent. Stronger positive relations with pedestrian crashes were found for higher percentages of commercial facades, longer shares of the street land uses that were planned as residential with commerce or as open public space and higher presence of public transport in the vehicle traffic. On the other hand, fewer pedestrian crashes are expected on streets with various intersection types compared to streets with un-signalized junctions only.

Similarly, a complementary binary tree model for predicting pedestrian crashes indicated a positive link between the share of the street land uses planned for residential with commerce uses and crash frequencies.

In general, the models showed that the early-stage planning of the street for mixed land-uses increased the crash risk for pedestrians. The study streets were with high actual levels of vehicle traffic and pedestrians, while the presence of road design elements intended to improve pedestrian safety, such as wider sidewalks or pedestrian fences, were not found to decrease pedestrian risk considerably.

5 CONCLUSIONS AND FUTURE DEVELOPMENTS

The study findings supported the assumption that the early street planning for mixed land-uses affects its safety level, where, in general, we found that a higher extent of mixed land-uses was associated with worse safety levels of the street. However, the streets with mixed land uses are essential for urban texture as they stimulate city developments. Hence, appropriate traffic arrangements should be developed for such streets, for attaining better safety. Such solutions are not readily available in the current guidelines for urban streets' design (Guidelines, 2008).

In a brain-storming discussion with experts in town planning and traffic engineering, the main directions of further thinking were suggested as follows:

- To minimize the potential conflicts between vehicle traffic and pedestrians and safety problems, streets with mixed land-uses should be prevented from turning into arterials of passing traffic.
- Safety issues should be examined at the DTP stage, by means of safety audits. Particular attention is required to preventing continuity of traffic flow and providing space for safer solutions at junctions.
- For junctions, the possibilities of building roundabouts should be considered, with respective space planned. Alternative solutions for left-turns can be examined, based on traffic forecasts. Visibility issues while approaching junctions should be checked in the DTP, due to implications on the reserved "road rights".
- The priority of pedestrians on the streets with mixed land-use should be stated and supported by road design settings, such as: one traffic lane per direction; on two-way streets, a physical separation between the directions, partly with a raised median; and extensive use of traffic calming measures.

REFERENCES

Elvik R., Hoya A., Vaa T. and Sorensen M. (2009). The handbook of road safety measures. 2nd edition. Emerald.

---

3. Most impacts are significant with $p < 0.01$, the impact of horizontal alignment – with $p < 0.15$.
4. Most impacts are significant with $p < 0.01$, the impacts of layout characteristics and pedestrian fences are significant with $p < 0.05$, speed variance – with $p < 0.1$.

Ewing R. and Dumbaugh E. (2009). The built environment and traffic aafety: a review of empirical evidence. Journal of Planning Literature 23(4), 347–367.
Gitelman V., Balasha D., Carmel R., Hendel L. and Pesahov F. (2012). Characterization of pedestrian accidents and an examination of infrastructure measures to improve pedestrian safety in Israel. Accident Analysis & Prevention 44(1), 63–73.
Guidelines for urban streets' design (2008). Ministry of Transport and Road Safety, Jerusalem.
Macaulay J. & McInerney R. (2002). Evaluation of the proposed actions emanating from Road Safety Audits. Austroads Inc., Sydney.
RSA (2016). Road safety trends in Israel. Road Safety Authority, Jerusalem.
Ukkusuri S., Miranda-Moreno L.F., Ramadurai G. and Isa-Tavarez J. (2012). The role of built environment on pedestrian crash frequency. Safety Science 50, 1141–1151.
Venables W.N. and Ripley B.D. (2002). Modern applied statistics with S. Fourth edition. Springer.
Wedagama D.M.P, Bird R.N., Metcalfe A.V. (2006). The influence of urban land-use on non-motorised transport casualties. Accident Analysis and Prevention, 38, 1049–1057.

Town and Infrastructure Planning for Safety and Urban Quality – Pezzagno & Tira (Eds)
 ISBN 978-0-8153-8731-2

An identification of infrastructure measures to improve elderly pedestrian safety in towns in Israel

V. Gitelman, R. Carmel & F. Pesahov
Israel Institute of Technology, Haifa, Israel

ABSTRACT: The purpose of this study was to identify the patterns of road accidents that most commonly cause injury to elderly pedestrians in Israeli towns, and to recommend infrastructure solutions that may contribute to reducing elderly pedestrian injury.

A sample of 15 towns with a high prevalence of elderly pedestrian accidents was selected to characterize sites where elderly pedestrian accidents typically occur. Accident sites in these towns were mapped and information on 476 sites where 580 elderly pedestrian accidents occurred was collected.

The field surveys focused on the areas with higher concentrations of elderly pedestrian accidents in the towns that were selected using the Average Nearest Neighbor examinations and space-distance analysis.

Based on a combined consideration of literature findings and accident site characteristics the study suggested infrastructure solutions for improving elderly pedestrian safety at various types of urban sites.

1 INTRODUCTION

Elderly pedestrians are a vulnerable group of road users, in many countries (OECD, 2001; DaCoTa, 2012). On the one hand, walking is a more essential mode of transport for older people than for other age groups, due to giving up driving. On the other hand, older people suffer more road trauma due to deterioration in their physical and mental ability that brings them to higher involvement in road accidents and to more severe consequences of such. Fig. 1 presents pedestrian injury rates by age groups, in Israel, illustrating a higher risk for elderly pedestrians related to other age groups. The data from the Central Bureau of Statistics (CBS) accident files in Israel show that elderly pedestrian accidents account for over a fifth of all pedestrian accidents, whereas the severity of elderly pedestrian injury is higher compared to the overall severity of pedestrian injury: e.g., 31% of fatal and serious cases in elderly pedestrian accidents versus

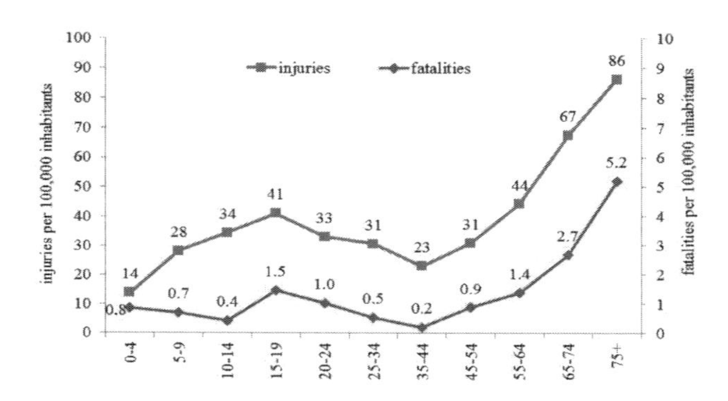

Figure 1. Pedestrian injury rates by age groups, in 2012, in Israel.

22% in all pedestrian accidents, in 2006–2010. Annually, over 680 elderly[1] pedestrians are injured in road accidents in Israel, of which 40 are killed and over 160 are seriously injured (RSA, 2016).

In addition, population ageing is one of the growing issues in the developed countries with the expected effect of an increased involvement of older road users in accidents, thus, requiring attention from a decision-making, technical and research viewpoints (Hakkert and Gitelman, 2014). Infrastructure conditions have to be adapted to the limitations of older pedestrians (Oxley et al, 2006). However, such solutions are not readily available and should be fitted to local needs based on research studies.

2 OBJECTIVE OF THE STUDY

The majority of accidents involving elderly pedestrians occur in urban areas: e.g., 97% of injury and 74% of fatal pedestrian accidents, in 2012. Thus, the purpose of this study was to identify the patterns of road accidents that most commonly cause injury to elderly pedestrians in Israeli towns, and to recommend infrastructure solutions that may contribute to reducing elderly pedestrian injury. The study was commissioned by the National Road Safety Authority.

3 RESEARCH APPROACH

The methodological approach of the study included three components. First, a review of international literature was conducted focusing on the distinct characteristics of elderly pedestrian accidents and infrastructure solutions suggested for improving elderly pedestrians' safety. Second, urban sites where elderly pedestrian accidents typically occur were identified and explored statistically in terms of engineering settings, road and urban environment characteristics. At this step, the whole list of Israeli towns with over 10,000 residents was examined and ranked according to the values of several indicators: the number of elderly pedestrian injuries (in five years), the share of elderly out of the total pedestrian injuries and the share of elderly in the city population. A set of higher ranked cities was selected, for which all elderly pedestrian accidents (in five years) were mapped[2]. The accident locations were characterized in terms of traffic engineering settings, urban road types and land uses, and then summarized to provide the leading patterns.

Third, field surveys were conducted at sites of elderly pedestrian accidents and at a set of non-accident sites, in the same cities, to identify the infrastructure characteristics and deficiencies that may be associated with accident occurrences and, consequently, to suggest infrastructure measures that may prevent such accidents. The areas with higher concentrations of elderly pedestrian accidents, for the survey, were selected using the Average Nearest Neighbor (ANN) examinations and space-distance analysis (Mitchell, 2005). The ANN indicator was estimated for each city, based on the distances between the accident points, in order to verify that the distribution of accident sites was not random and, hence, areas with higher accident numbers can be identified. Such areas were defined when shared zones between the "circle environments[3]" of various points were observed. At each site, in the selected areas, detailed infrastructure characteristics were collected, hourly traffic and pedestrian volumes were estimated and travel speeds were measured, in morning hours (8–13)[4]. Similar examinations were conducted at the control sites—with similar traffic settings but no elderly pedestrian accidents. The findings were summarized for typical urban settings, indicating common infrastructure deficiencies at the accident sites and features related to the absence of accidents at the control sites. Finally, based on the literature findings and the field surveys,

1. Ages 65+ were used for the definition of "elderly", in the current study.
2. Based on the CBS files.
3. With a radius of 50 m for most locations, except for accidents at mid-block crosswalks where the radius of 100 m was applied.
4. Speeds were measured by a speed gun, for a sample of 30 free-flow vehicles; excluding signalized junctions and roundabouts.

infrastructure solutions for improving elderly pedestrian safety were suggested, for each type of sites.

4 RESULTS

4.1 *Findings from the international experience*

The literature review included summaries of international experience (e.g. OECD, 2001; Whelan et al, 2006; DaCoTa, 2012) and research papers (e.g. Oxley et al, 2006; Leden et al, 2006). The literature showed that elderly pedestrian accidents mostly occur during regular walking; close to home, shopping or leisure centers; on straight road sections or at junctions; in day hours. The accident risk increases at areas with higher demands on the abilities of crossing pedestrians such as roads without median, wide roads with many traffic lanes, heavier traffic and mixed road user activities. Various measures for improving elderly pedestrian safety were indicated, mostly by designing infrastructure to suit the functional limitations that accompany the ageing process, such as slower walking, longer reaction time, difficulties in visual search, estimating gaps to vehicles, etc.

Among the infrastructure measures that may be particularly beneficial for elderly pedestrians, the most common ones were: speed reduction measures; separating or restricting motor traffic in areas with increased pedestrian activity; reducing the complexity of the road-crossing process; and shorting the distance from one road side to the other. Table 1 summarizes elderly pedestrian related infrastructure measures that were found in the literature. For the majority of them, safety efficiency was shown not in terms of accident reductions but in positive behavior changes, e.g. reduced vehicle-pedestrian conflicts, lower travel speeds, better pedestrian conspicuity, etc. They all improve crossing conditions for pedestrians. However, only a

Table 1. Infrastructure measures for improving elderly pedestrian safety in urban areas, from the literature.

On street sections	At intersections
1. For elderly:	1. For elderly at junctions/signalised crosswalks:
– use of large and bright signs	– green light extension for pedestrians
– different sidewalk texture near crosswalks	– adding voice signals
	– adding tactile devices
2. At mid-block crosswalks:	2. At crosswalks:
– proper signing	– refuge islands in the middle
– pedestrian-activated pavement marking	– sidewalk extensions
– fencing on sidewalks and medians for guiding pedestrians to crosswalks	
– refuge islands in the middle	
– sidewalk extensions near crosswalks	
– traffic lights' installation	
3. Traffic calming:	3. Traffic calming:
– limiting travel speeds	– blocking selected directions of travel
– narrowing lanes	– (one-lane) roundabouts
– reducing lanes	– raised junctions
– visual narrowing of the street	
– speed humps	
– raised crosswalk	
– horizontal shifts of the route	
– change in pavement texture	
4. Improved lighting	4. Improved lighting
	5. Grade-separation for pedestrians
	6. Traffic lights: cancellation of shared vehicle-pedestrian green; advanced green for pedestrians

few measures are specifically fitted to elderly pedestrian needs, while the majority are related to pedestrian safety, in general.

4.2 *Characterization of elderly pedestrian accident sites*

A sample of 15 towns with a high prevalence of elderly pedestrian accidents was selected, representing a mix of various types of sizes, geographic areas and socio-economic levels of Israeli cities. Accident sites in these towns were mapped providing information on 476 accident sites where 580 elderly pedestrian accidents occurred.

The sites were classified according to six types of urban infrastructure settings: signalized junctions, non-signalized junctions, non-signalized midblock pedestrian crosswalks, road sections without crosswalks, roundabouts and others. Summarizing the characteristics of accident sites we found that the majority of elderly accidents occurred on road sections without crosswalks (26%), non-signalized junctions (22%) and signalized junctions (22%). The most common type of streets with the accident sites was collector roads, e.g. dual-carriageway with two lanes per travel direction (47%) and single-carriageway (30%); the most common areas—with mixed land uses, i.e. both commercial and residential purposes (67%), situated in town centers (24%) and their proximity (25%).

Furthermore, a cluster analysis indicated groups of towns with various patterns of accident distributions across various types of settings, e.g. with prevalence of elderly accidents on road sections, with the majority of accidents on roundabouts and on sections without crosswalks, etc. Such results can be useful for creating pedestrian safety programs in various cities.

4.3 *Field survey's results*

The analysis of ANN indicators showed that in 14 towns (out of 15) the accident sites' distribution was not random but clustered ($p < 0.01$). Hence, the areas with higher accident numbers were identified in each town and field surveys were conducted at 92 accident sites and 21 control sites, in total. The sites were divided into five types as defined above, excluding "others".

The characteristics of sites with higher concentrations of elderly pedestrian accidents indicated that they are mostly associated with:

- non-signalized and signalized junctions (38% and 27% of the total set, respectively);
- sites with medium traffic volumes[5] (49%);
- sites with medium and high volumes[6] of crossing pedestrians (54% and 28%);
- collector roads (dual-carriageway roads with two lanes per direction – 59%, and single-carriageway roads – 26%);
- areas with mixed uses (55%) and residential areas with attractions for elderly pedestrians (22%).

In addition, high travel speeds, with 85th percentile speed over 50 km/h, were observed at a substantial share of accident sites such as non-signalized junctions and sections with mid-block crosswalks (at 23% and 44% of such sites, respectively), whereas the speed levels at the control sites were lower.

Following a detailed engineering survey, typical infrastructure deficiencies at the accident sites and the features possibly related to the absence of accidents at the control sites, were summarized for each type of sites. Fig. 2 provides examples of common infrastructure deficiencies observed at sites with concentrations of elderly pedestrian accidents.

For signalized junctions, the leading deficiencies included: lacking a "green wave" for pedestrians that cause pedestrian groups on the median and crossings on red (see Fig. 2, a–b), and lack of fencing on medians resulting in unauthorized pedestrian crossings near junctions (found at 7 and 4, out of 25 accident sites). In addition, long waiting times for pedestrians, lack of marked crosswalks on all sides of the junction, narrow crossing areas,

5. An average of 100–200 entering vehicles, in 5 minutes, at a signalized junction, 50–100 – at other sites.
6. Correspond to an average of 5–20 and over 20 crossing pedestrians, respectively, in 5 minutes.

Figure 2. Examples of infrastructure deficiencies observed at accident locations.

etc. were observed. Among infrastructure solutions for improving elderly pedestrian safety at signalized junctions, were suggested:

- an earlier appearance of the pedestrian signal in shared pedestrians' and right-turning vehicles' "green" (at 14 out of 25 junctions);
- guiding pedestrians to crosswalks by means of fences on medians or sidewalks (at 7);
- reducing waiting time for pedestrians (at 7);
- an exclusive "all green" for pedestrians at junctions with high pedestrian volumes (at 2);
- extending crossing areas (at 2), etc. Pedestrian fencing and a "green wave" for pedestrians were found at some control sites.

At non-signalized junctions, among leading deficiencies were: lack of marked crosswalks on all sides of the junction; insufficient junction conspicuity (Fig. 2, c); lack of refuge islands/ sidewalk extensions in the crosswalk (Fig. 2, d) (found at 15, 10 and 8, out of 35 accident sites). Thus, among possible remedial measures can be:

- improved signing—adding overhead flashing lights for crosswalks (at 15 out of 35 junctions);
- arranging crosswalks on all sides of the junction (at 11);
- arranging refuge islands and/or sidewalk extensions for the crosswalks (at 10);
- traffic calming measures (at 9), etc.

At mid-block non-signalized pedestrian crosswalks among common deficiencies were: lack of refuge islands/sidewalk extensions in the crosswalk; visibility problems due to illegal parking; insufficient crosswalk's conspicuity; lacking signing/marking, etc. Among safety improving solutions were suggested:

- installation of a raised pedestrian crosswalk;
- adding overhead flashing lights;
- arranging refuge islands/sidewalk extensions for crosswalks, etc.

5 CONCLUSIONS

In this study, a systematic approach was applied for a detailed examination of accident sites where elderly pedestrians are typically injured in Israeli towns. Typical characteristics of such sites and common infrastructure deficiencies were identified based on data analyses. The results showed that the main elderly pedestrian problems belong to streets and junctions with mixed land uses, situated in town centers and their proximity.

Based on a combined consideration of literature findings and accident site characteristics, infrastructure solutions were suggested for improving elderly pedestrian safety at various types of urban sites. In addition, we believe that treatment of selected sites only is not sufficient and that a system-wide approach should be promoted for reducing elderly pedestrian injury in towns. This approach should diminish the conflict areas between pedestrians, especially elderly ones, and vehicular traffic, in general, by means of removing through vehicle traffic from the pedestrian activity areas, separation between pedestrian movements and vehicle traffic in time and in place, and introducing traffic calming measures in the pedestrian activity zones.

REFERENCES

DaCoTA (2012). Pedestrians and Cyclists, Deliverable 4.81 of the EC FP7 project DaCoTA.

Hakkert, A.S., Gitelman, V. (2014). Thinking about the history of road safety research: Past achievements and future challenges. Transportation Research Part F, 25, 137–149.

Leden, L., Gårder, P. & Johansson, C. (2006). Safe pedestrian crossings for children and elderly. Accident Analysis and Prevention, 38 (2), 289–294.

Mitchell, A. (2005). The ESRI Guide to GIS Analysis, Volume 2. ESRI Press.

OECD (2001). Ageing and Transport, Mobility needs and safety issues. Organisation for Economic Co-operation and Development, Paris.

Oxley, J., Charlton, J., Corben, B., and Fildes, B. (2006). Crash and injury risk of older pedestrians and identification of measures to meet their mobility and safety needs. 7th International Conference on Walking and Liveable Communities, Melbourne, Australia.

RSA (2016). Road safety trends in Israel. Road Safety Authority, Jerusalem.

Whelan, M., Langford, J., Oxley, J., Koppel, S., and Charlton, J. (2006). The elderly and mobility: A review of the literature. Monash University Accident Research Centre, Melbourne.

Town and Infrastructure Planning for Safety and Urban Quality – Pezzagno & Tira (Eds)

Using multiple correspondence analysis to improve safety in interaction between road transit and public spaces

N. Giuffrida, M. Ignaccolo & G. Inturri
Università degli Studi di Catania, Catania, Italy

ABSTRACT: Multiple Correspondence Analysis (MCA) is a statistical technique used to present the relative closeness of the categorical variables from any dataset by analyzing data in the form of numerical frequencies and represent them graphically. Since pedestrian accidents data can be represented as transactions of multiple categorical variables, MCA can be a considered a good methodology to analyze the relationship among such variables.

This paper is the study of application of MCA to analyze safety in the interaction among urban road transit and pedestrians. MCA is applied to a Bus Rapid Transit (BRT) line in the city of Catania, operating since 2013; during those years there were no accidents, but a series of traffic conflicts have happened; that's the reason why authors proceed with the application of the methodology taking into account the occurrence of the so called "near misses".

1 INTRODUCTION

Data from Italia statistic institute reveal that in 2015, 173.892 road accidents occurred in Italy, resulting in 3.419 deaths and 246.050 injured. For the first time, the number of deaths increased after 15 years (since 2001). The increase in the number of deaths involved, in particular, motorcyclists (+7.2%) and pedestrians (+ 4%), for which this is the second consecutive increase. Every 100 pedestrian accidents recorded 3.07 deaths and 104.51 injured. In particular, in 2015, the city of Catania was among the 14 major municipalities in Italy selected for mortality/road accident.

Most of the time in urban environments, pedestrians are provided with exclusive or protected routes, but sometimes they can interact with various modes of transport. In particular, it problems related to the interaction of pedestrians with public collective transport are not negligible. In order to prevent pedestrian accidents, it's necessary to answer questions about their circumstances and the characteristics of people involved; it is also important to consider the relationships between these different criteria and to identify distinct categories of accidents (Fontaine and Gourlet, 1995).

MCA appears to be an interesting method for analyzing data captured through a survey; specifically, it's used to locate ordinary associations between categorical variables observed for a sample of individuals.

The objective of the paper is to explore the methodology of the MCA as a tool to identifiy the most important variables and their relationship in the occurrence of accidents. In order to achieve this, a case study regarding the city of Catania will be analyzed with, reference to interaction between an express bus line and pedestrians.

2 STATE OF THE ART

The use of MCA for pedestrian accidents analysis began with a study conducted by Fontaine (1995), in which a classification of pedestrians involved in road accidents in France was conducted, identifying different type of users; in this research the MCA is presented as a basis for in-depth analysis to improve the understanding of accidents and propose suitable actions.

Factor et al. (2010), by applying MCA to a wide database of official data about road accidents that took place in Israel and socio-economic data, showed that different social groups tend to be involved in different type of accidents; in particular drivers belonging to low-income social group tend to be involved in more serious accidents.

Dan and Sun (2014) used MCA to analyze accident involving pedestrians and road vehicles in Louisiana, with the aim to identify important contributing factors and their degree of association. Their study highlighted several type of groups of drivers and pedestrians that are closely related to typical kind of accidents, but also particular geometric and environment characteristic of the road. The findings of the research are considered helpful to traffic safety professionals in determining the hidden risk association group of variables in fatal single-vehicle crashes.

In our study we will focus on the application of MCA in the analysis of safety in the interaction between public transport and pedestrians, in lack of significant data on accidents, by using data from a survey on traffic conflicts previously conducted.

3 RESEARCH APPROACH

MCA (Benzécri, 1979), is an exploratory data analysis method used for detecting and representing underlying structures in a categorical data set. It's used to analyze a set of observations/individuals described by nominal variables (Abdi and Valentin, 2007). Its purpose is to make graphically clear the relationships between categories, between individuals and between individuals and categories, projecting their profiles into subspaces of reduced dimensionality. It's a particularly suitable method for multidimensional analysis of qualitative data obtained through a survey of a sample of individuals. However, it requires the starting data to be organized into a matrix Z of indicators such as individuals/categories, with a number of i rows equal to the number of individuals, and j columns equal to the number of active categories. Each element of Z is 0 if the category is rejected by the individual, or 1, if the category is chosen by the individual. This type of encoding applied to Z is defined as "disjunctive" because the categories provided by each variable are exclusive (only one can be chosen by each individual), and "complete" because a category must be chosen necessarily.

Map is the main output of the MCA, where the associated points (categories) are relatively close. In addition, the nearest points to the average value are near the origin of the map, while the farthest are located at the margins of the map.

In order to obtain map of variables, MCA focuses on some fundamental geometric concepts evaluated through the following steps:

1. Evaluation of masses, which correspond to the weights assigned to the point. Considering Z matrix, the first step to evaluate masses is to compute marginal frequencies. For each individual, row marginal frequencies are calculated by adding the chosen categories (which coincides with the Q variables):

$$z_{i,\cdot} = \sum_{j=1}^{J} z_{ij} = Q \quad \text{con}\, \mathrm{i} = 1,2,\ldots \mathrm{I}$$

   Similarly, column marginal frequencies are calculated by adding the individuals that chase the category:

$$z_{\cdot,j} = \sum_{i=1}^{I} z_{ij} \quad \text{con}\, \mathrm{j} = 1,2,\ldots \mathrm{J}$$

   The overall frequency is given by:

$$z = \sum_{i=1}^{I} z_{i,\cdot} = \sum_{j=1}^{J} z_{\cdot,j} = IQ$$

Columns and rows' masses $\bar{r}_j$ and $\bar{c}_j$ are obtained dividing marginal frequencies by overall frequency; vectors of rows and columns' masses are then evaluated:

$$\bar{r} = (\bar{r}_1\, \bar{r}_2 \ldots \bar{r}_j \ldots \bar{r}_J)^T$$

$$\bar{c} = (\bar{c}_1 \bar{c}_2 \ldots \bar{c}_i \ldots \bar{c}_I)^T$$

2. Profile gives the coordinates on the graph of the point. They should be obtained by dividing each of Z elements by the sum of columns (or rows); since it's not a numerical matrix an assignment process is adopted. In the case of column's profile, the value 1/z is assigned to each column j if the category is chosen by the individuals, 0 if the category is refused. In this way the matrix C of columns' profile is obtained and the column's profile cj can be represented:

$$c_j = (c_{1j} c_{2j} \ldots c_{ij} \ldots c_{Ij})^T$$

The mass $\bar{r}_j$ is attributed to cj, and it results to be directly proportional to zj: this means that the most chosen categories have a bigger mass.

At the same way by assigning the value 1/Q to the Q categories that the individual i chooses and 0 to the refused ones we obtain the matrix R of rows' profile and the row's profile ri, to which the mass $\bar{c}$ is attributed. Rows profiles are weighted equally, since individuals have the same mass, so they're considered in the same manner.

3. Centorid is the weighted average position. From a geometrical point of view, cj profiles represents a cloud of J points with variable mass in the I-dimensional space RI, with centroid in their medium point $\bar{c}$, while ri profiles represents a cloud of I points with constant mass in the J-dimensional space RJ, with centroid in their medium point $\bar{r}$. The 2 spaces are equipped with a orthogonal reference system in which similar profiles are represented as close points. Since the orthogonal reference system doesn't guarantee the comparison between the profiles, a new orthonormal reference system for each cloud is sought, with origin in the centroid of the cloud, and configured in a way to maximize the dispersion of profile projections on the axes.
4. Distance is the proximity between two points. In RI the distributional distance between ci and cj is defined as:

$$d^2(c_i, c_j) = \sum_{i=1}^{I} \frac{(c_{ij} - c_{ijr})^2}{\bar{c}_I} = I \sum_{i=1}^{I} (c_{ij} - c_{ijr})^2$$

So, the 2 categories profiles coincide when both categories are chosen by the same individuals and their distance increases with the number of individuals choosing one and not the other category. This means that a category chosen by a few individuals will be far from the others. Distance of cj from 0I of the global reference system RI is:

$$d^2(c_J, 0_I) = \sum_{i=1}^{I} \frac{(c_{ij} - 0)^2}{\bar{c}_I} = I \sum_{i=1}^{I} c_{ij}^2 = I z_j \left(\frac{1}{z_j}\right)^2 = \frac{1}{z_j}$$

Distance of cj from the centroid $\bar{c}$ of the cloud f all J profiles is:

$$d^2(c_J, \bar{c}) = \sum_{i=1}^{I} \frac{(c_{ij} - \bar{c}_i)^2}{\bar{c}_I} = I \sum_{i=1}^{I} \left(c_{ij} - \frac{1}{I}\right)^2 = \frac{1}{z_j} - 1$$

It follows that the profile of a category more resembles the average profile as its frequency is bigger, which means that the profiles of the most frequently selected categories are located close to the centroid, while those of the rarely chosen categories are far away.

5. Inertia is the weighted sum of squared distances to centroid. The inertia of cj with reference to 0I of the global reference system RI is:

$$In_{0I}\left(c_j\right)=\overline{r}_J d^2\left(c_J,0\right)=\frac{z_j}{IQ}\frac{I}{z_j}=\frac{1}{Q}$$

The overall inertia of J profiles of the cloud is:

$$In_{0I}\sum_{j=1}^{J} In_{0I}\left(c_J\right)=\sum_{j=1}^{J}\frac{1}{Q}=\frac{J}{Q}$$

The inertia of cj with reference to the centroid is:

$$In_{\overline{c}}\left(c_j\right)=\overline{r}_J d^2\left(c_J,\overline{c}\right)=\frac{z_j}{IQ}\left(\frac{I}{z_j}-1\right)=\frac{1}{Q}-\overline{r}_J$$

so it increases if the number individuals who have chosen it decreases, up to a maximum of 1/Q. This relationship reveals that only mass of a category can influence the orientation of the axis. This means that the first factor axes are conditioned almost exclusively by rare categories. Consequently, it's appropriate to eliminate categories that are too rare, which have little significant phenomena.

Variables appear implicitly in the MCA through their categories. The cloud of the profiles of the categories consists of the sovraposition of Q sub-clouds, as much as variables, all with the same centroid. This property is preserved in category projections on the axes. It follows that the abscissa of the projections does not all have the same sign. As a result, the factor axes oppose all J categories and, in particular, the Jq categories of each variable q.

The inertia of the categories of each variable is proportional to Jq, thus, the contribution of each variable to the total cloud inertia increases linearly as the categories increase.

The total inertia of the profiles' cloud depends only on the number of categories and variables:

$$In_{\overline{c}}=\sum_{q=1}^{Q} In_{\overline{c}}(q)=\frac{1}{Q}\sum_{q=1}^{Q}(J_q-Q)=\frac{J}{Q}-1$$

The total inertia of the two clouds of profiles (of categories and individuals) are the same.

6. From a mathematical point of view the problem of the research of the factorial axes which consent to maximize the dispersion of the profiles' projection can be formalized for columns's profile cloud RI and for rows' profile cloud RJ like this:

$$CR^T u_a=\lambda_a u_a \quad \text{with} \quad u_a{}^T D_c^{-1} u_a=1$$

$$R^T C v_a=\mu_a v_a \quad \text{with} \quad v_a{}^T D_r^{-1} u_a=1$$

where λ_a and μ_a are eigenvalues and ua and va are the correspondent eigenvectors, with rank a = 1,2, …, A with origin in the centroid of the two profile cloud. The size of the subspace that contains the cloud of J points is at most:

J1+(J2–1)+…+(Jq–1)+…+(Jq–1) = J–Q+1

The eigenvectors and the eigenvalues will also be the same number.

Considering now the cloud of the cj profiles in the RI space with the source in $\overline{c}$, the eigenvalue corresponding to $\overline{c}$ is equal to 0, so at least we will find non-null A = J–Q eigenvalues.

From the mathematical point of view, the solution of the problem is to extract the eigenvalues and eigenvectors of the matrix RTC of order JxJ and rank A, which is also the number of non-null eigenvalues to which the eigenvectors match. Such eigenvalues have values of less than 1 and are all different from each other.

7. The coordinates of the profiles on the axes detected by the eigenvectors are obtained by projecting the profiles on the axes and they are the elements of matrices G and F:

$$G = C^T D_{\bar{\bar{c}}}^{-1} U$$

$$F = R^T D_{\bar{r}}^{-1} V$$

By reducing these matrices to the first A*<A columns, a "smoothing" of the geometric configuration of the cloud is performed, so it can be said that the method allows obtaining a small number A* of ga factors, which are nothing more than synthetic variables directly linked to the original variables Q. The relative contribution of a category to the inertia of the axis is defined as the ratio between the inertia related to the origin of the projection g of its axis profile and the overall inertia λa on the axis; it measures the 'effort' made by the category to attract the axis itself and allows to identify the most effective categories. The relative contribution of a variable q is also useful for interpreting a factorial axis, and is calculated as sum of the relative contributions of its Jq categories; the comparison of contributions allows to identify the variables that have contributed more to the definition of the axis and thus to give it meaning.

Maps are final objective of MCA, since they considerably ease the data perception; 2 type of maps can be obtained:

- Symmetric maps are those where the observations and categories are scaled in principal coordinates.
- Asymmetric maps are those where the categories are scaled in principal coordinates and observations in standard coordinates (categories' asymmetric maps) and vice versa (observation's asymmetric maps). Additionally, if 2 observations are displayed in the same direction as a category vector, the observation that is the furthest in the category vector direction is more likely to have selected that category of response. The axes simultaneously oppose the categories of the same variable and the categories of all the variables. Therefore, if 2 categories of the same variable are close, it means that the two segments of individuals have made quite similar choices, relative to the categories of the other active variables. Instead, if 2 categories of different variables are close, it means that the 2 categories were almost always chosen by the same individuals. In practice, the proximity of different categories of variables is interpreted as association between categories, because the same individuals have frequently associated them in their choices, and the proximity between categories of the same variable is interpreted as the similarity of behavior of different individuals.

4 CASE STUDY

4.1 *Territorial framework and data acquisition*

BRT1 is a bus express line which connects a park-and-ride facility and the city center; it has a length of almost 13 km, most of it running on protected right of way (by curb or marking) and a total of 20 stops and 1 station. In a survey of 2014, data analysis of the situations was carried out for conflicts between BRT1 and pedestrians. Along the route there are 84 pedestrian crossings identified by zebra markings: 35 located at vehicle crossings; 20 located at a BRT1 stop; 29 located on along the line. On the basis of the survey, 3 sites, representing each of the 3 types of crossings were chosen. For each of the crossings selected, further highlighting was carried out to identify the major hazardous behaviors adopted by pedestrians. A database of 115 individuals and 5 variables (Fig. 2) was created, with the variable related categories (Fig. 1).

Variable	Categories				
Sex	Male (M)			Female (F)	
Age	Adult (A)			Children (B)	
Pedestrian crossing type	Close to stops (FS)	Along the line (LI)		Vehicle crossing (IN)	
Time	morning	afternoon		evening	
Event cause	pedestrians cross outside authorized space without curb	pedestrians cross outside authorized space with curb	pedestrians cross in a rush to take the bus	pedestrians cross with lack of attention	other

Figure 1. Variable related categories for database.

— ID	sex	age	time	type of crossing	cause
i1	M	B	morning	FS	1
i2	M	B	morning	FS	1
i3	M	B	morning	FS	1
i4	F	B	morning	FS	1
i5	F	B	morning	FS	1
...	...	...	...	...	...
...	...	...	...	...	...
...	...	...	...	...	...
i114	F	A	evening	LI	1
i115	M	A	evening	LI	1

Figure 2. Database example.

ID	Sex		Age		Time			Type			Cause				
	F	M	A	B	M	A	E	FS	IN	LI	1	2	3	4	5
i1	0	1	0	1	1	0	0	1	0	0	1	0	0	0	0
i2	0	1	0	1	1	0	0	1	0	0	1	0	0	0	0
i3	0	1	0	1	1	0	0	1	0	0	1	0	0	0	0
i4	1	0	0	1	1	0	0	1	0	0	1	0	0	0	0
i5	1	0	0	1	1	0	0	1	0	0	1	0	0	0	0
...	...	...	...	...	...	...	...	...	...	...	...	...	...	...	...
...	...	...	...	...	...	...	...	...	...	...	...	...	...	...	...
...	...	...	...	...	...	...	...	...	...	...	...	...	...	...	...
...	...	...	...	...	...	...	...	...	...	...	...	...	...	...	...
i113	0	1	1	0	0	0	1	0	0	1	1	0	0	0	0
i114	1	0	1	0	0	0	1	0	0	1	1	0	0	0	0
i115	0	1	1	0	0	0	1	0	0	1	1	0	0	0	0

Figure 3. Disjoint table.

4.2 *MCA application*

For the application of MCA we used the XLSTAT program that allows to run a series of statistical analysis in Microsoft Excel environment.

	F1	F2	F3	F4	F5	F6	F7	F8	F9	F10
Eigenvalue	0.41	0.33	0.24	0.22	0.21	0.20	0.14	0.11	0.09	0.06
Inertia (%)	20.41	16.44	12.22	11.12	10.38	10.01	6.79	5.42	4.46	2.75
% cumulative	20.41	36.85	49.08	60.20	70.58	80.59	87.38	92.79	97.25	100.00

Figure 4. Eigenvalues and Inertia percentage.

	F1	F2	F3	F4	F5	F6
Corrected inertia	0.068	0.026	0.003	0.001	0.000	0.000
Corrected inertia (%)	50.014	19.170	2.278	0.580	0.067	0.000
% cumulative	50.014	69.184	71.462	72.043	72.109	72.109

Figure 5. Corrected eigenvalues and Inertia percentage.

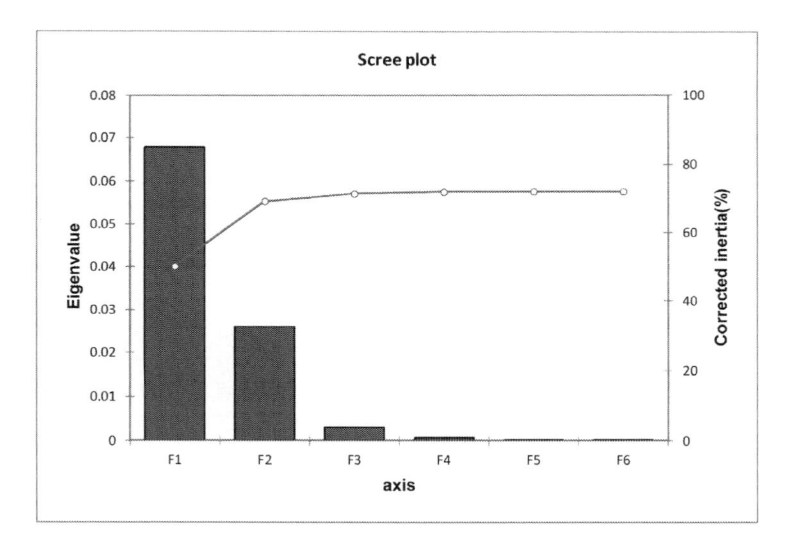

Figure 6. Scree plot.

From the analysis of the observation of individuals the software derives the disjoint table (Fig. 3).

The distribution of inertia on the axes is the next step. In our case, the number A of non-null eigenvalues depends only on the number of categories and the number of variables: $A = J–Q = 15–5 = 10$. Total inertia, eigenvalues and percentages of inertia are then evaluated (Fig. 4).

Total inertia it's equal to:

$$In_{\bar{c}} = In_{\bar{r}} = \frac{J}{Q} - 1 = \frac{15}{5} - 1 = 2$$

The average inertia on an axis is:

$$\lambda_a = \frac{1}{Q} = \frac{1}{5} = 0,20$$

	F1	F2	F3	F4	F5
sex-F	0.434	0.054	0.081	0.920	0.285
sex-M	-0.259	-0.032	-0.048	-0.550	-0.170
age-A	0.263	-0.205	-0.161	0.061	-0.091
age-B	-1.515	1.184	0.930	-0.353	0.525
morning	-0.958	0.636	0.423	0.235	-0.118
afternoon	1.061	0.307	0.200	0.232	0.084
evening	-0.130	-0.951	-0.629	-0.473	0.032
type FS	-0.111	0.427	-0.342	0.033	0.035
type IN	1.473	0.252	1.378	-1.307	-0.138
type LI	-0.370	-1.531	0.437	0.545	-0.045
C1	-0.695	-0.434	0.167	0.157	0.209
C2	0.613	0.481	-0.499	-0.384	0.181
C3	-0.373	1.186	-0.684	1.302	-4.747
C4	1.599	-1.305	3.400	-0.698	-1.417
C5	1.629	0.769	0.643	1.574	1.108

Figure 7. Principal coordinates (variables).

	F1	F2	F3	F4	F5
sex-F	0.679	0.094	0.163	1.951	0.626
sex-M	-0.405	-0.056	-0.097	-1.165	-0.374
age-A	0.411	-0.358	-0.326	0.130	-0.200
age-B	-2.372	2.065	1.881	-0.748	1.152
morning	-1.500	1.109	0.856	0.498	-0.260
afternoon	1.660	0.536	0.404	0.492	0.185
evening	-0.204	-1.659	-1.272	-1.003	0.070
type FS	-0.174	0.744	-0.692	0.069	0.076
type IN	2.306	0.440	2.787	-2.770	-0.303
type LI	-0.579	-2.669	0.885	1.157	-0.099
c1 no	-1.088	-0.756	0.338	0.334	0.458
c1 yes	0.960	0.838	-1.010	-0.815	0.398
c2-no	-0.583	2.068	-1.383	2.762	-10.418
c2-no	2.503	-2.275	6.877	-1.479	-3.111
c2-yes	2.549	1.341	1.300	3.338	2.433

Figure 8. Standard coordinates (variables).

	Weight	Weight (relative)	F1	F2	F3	F4	F5
sex-F	43	0.075	0.034	0.001	0.002	0.285	0.029
sex-M	72	0.125	0.021	0.000	0.001	0.170	0.018
age-A	98	0.170	0.029	0.022	0.018	0.003	0.007
age-B	17	0.030	0.166	0.126	0.105	0.017	0.039
morning	38	0.066	0.149	0.081	0.048	0.016	0.004
afternoon	39	0.068	0.187	0.019	0.011	0.016	0.002
evening	38	0.066	0.003	0.182	0.107	0.066	0.000
type FS	79	0.137	0.004	0.076	0.066	0.001	0.001
type IN	12	0.021	0.111	0.004	0.162	0.160	0.002
type LI	24	0.042	0.014	0.297	0.033	0.056	0.000
C1	57	0.099	0.117	0.057	0.011	0.011	0.021
C2	46	0.080	0.074	0.056	0.082	0.053	0.013
C3	4	0.007	0.002	0.030	0.013	0.053	0.755
C4	4	0.007	0.044	0.036	0.329	0.015	0.067
C5	4	0.007	0.045	0.013	0.012	0.077	0.041

Figure 9. Contributions (variables).

This is the value chosen as empirical threshold to determine the number of axes to take. In our case we should take the first 6, but the first 2 explains 36.85% of inertia and none of the other explains more than 16.45%, so we don't need to consider them.

Greenacre (Greenacre et al. 2006) proposed a correction of inertia to give a better idea of the quality of the maps. Fig. 5 shows that the method based on the correct inertia gives the 69.18% with the first 2 axes. Corrected inertia percentage are displayed in the scree plot of Fig. 6 to locate the "elbow" of their descending sequence.

Fig. 7 and Fig. 8 show the main coordinates used to represent projections in maps, and the standard coordinates used to represent the vertices-observation projections in asymmetric maps.

The categories that most influence the axis calculation are the ones that have the largest contributions (Fig. 9).

5 RESULTS

Since in symmetric maps there's no a specific interpretation of the distance category-observations, we just consider asymmetric map (Fig. 11) in which it's possible to study the way observations are positioned relative in relation to the category vectors.

Results show that there are no significant gender differences for pedestrians in adopting risky behaviors. Clouds of categories are indicated in the Fig. 11 and are furthly explained in Fig. 10.

5.1 *Possible intervention measure*

On the basis of the results it's possibile to improve strategies for the selection of countermeasure. Based on the elements of the three combinations, some examples of best practice from European and International countries have been taken into account to be implemented to improve safety in the different crossing configurations.

- *Channeling*: a barrier system for pedestrian traffic channeling could be introduced at the sidewalk in order to orientate pedestrians to a safe place crossing route (Fig. 12).
- *Additional traffic light*: the introduction of additional lights located at a lower position can call the attention of vehicles and users approaching at the intersection (Fig. 13).
- *Identification of pedestrian crossing without zebra*: if the pedestrian strips are traced on the private lane of the BRT, pedestrian could wrongly think that they have priority on BRT, even when they don't (Fig. 14).

COMBINATION	CATEGORIES
1: accidents on a pedestrian crossing located close to a stop occur in the morning and involve kids crossing the street in a rush as they see the bus at the stop or in approach.	Age B
	Morning
	C3
	Type FS
2: accidents on a pedestrian crossing located close to a vehicular crossing occur in the afternoon and involve people crossing out of authorized spaces, with the presence of curb and stopping on the reserved lane during the cross.	C2
	Afternoon
	Type IN
	C5
3: accidents on a pedestrian crossing located along the line occur in the evening and they involve adults crossing out of authorized spaces, without the presence of curb.	C1
	Evening
	Type LI

Figure 10. Categories' combinations.

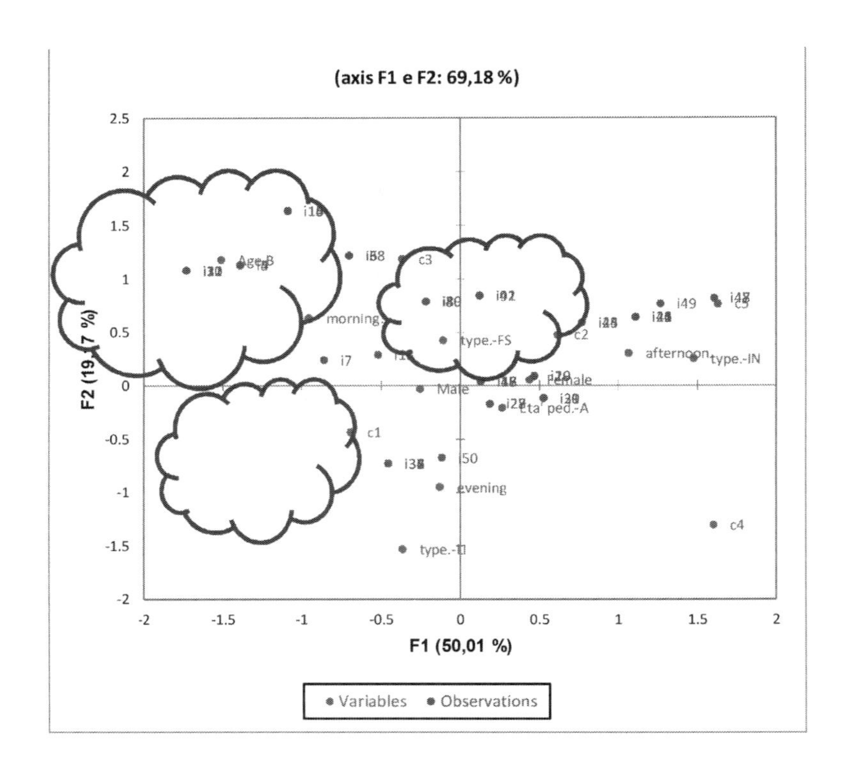

Figure 11. Variables asymmetric graph.

Figure 12. Channeling in Turin.

- *Pedestrian traffic light*: Due to the presence of a vehicle traffic light at the crossing intersection prior to the crossing, the presence of a pedestrian traffic light could draw more attention on pedestrians crossing and regulate some of the risky behaviors;
- *Markings and sign for forbidding crossing on unauthorized location* (Fig. 15);
- Road pavement treatment: Flooring and other surface treatments could be used for the crossing section affected by vehicular traffic (Fig. 16);
- Look both ways panel: this sign is designed to caution pedestrians crossing about the possibile presence of bus in both direction (Fig. 16).

In Fig. 17 it's possible to see which intervention is recommendable to use any of the interventions proposed.

Figure 13. Traffic light with additional lights in Santa Cruz de Tenerife.

Figure 14. San Francisco.

Figure 15. San Diego (image from https://goo.gl/snDfGr).

Figure 16. Manchester (https://goo.gl/2QZl6x).

Measure	Comb. 1	Comb. 2	Comb. 3
Barriers	x		
Additional lights on traffic light	x		
Crossing without zebra	x	x	x
Road pavement treatment		x	
Crossing ban on unauthorized route		x	
Pedestrian traffic light		x	x
Barriere			
Channeling		x	x
Sign warning of presence of transit in both direction			x

Figure 17. Coupling between interventions and combination.

6 CONCLUSIONS

The results from this study show that MCA is a good option to extract useful information from accident data. In particular results provide a basis for knowledge of pedestrian accident patterns that can characterize the BRT1 line in Catania through the associations of specific factors. From these results, appropriate pedestrian safety interventions can be implemented.

One of the limitations is that since the initial selection of variables was based on an earlier survey, some interesting variables were probably not explored. More in-depth analysis of the variables could be the subject of future research. If the database would be more complete, the MCA will probably generate more significant correlations between the variables.

REFERENCES

Abdi H, Valentin D (2007). Multiple correspondence analysis. In Salkind NJ (ed) Encyclopedia of measurement and statistics. Sage, Thousand Oaks.

Benzécri JP (1979). Sur le calcul des taux d'inertie dans l'analyse d'un questionnaire. Cahiers de l'Analyse des Données, 4, 377–378.

Das S, Sun X, Yair G, Mahalel, D. (2014) Exploring Clusters of Contributing Factors for Single-Vehicle Fatal Crashes Through Multiple Correspondence Analysis. TRB 93rd Annual Meeting Compendium of Papers, Washington DC.

Factor R, Yair G, and Mahalel, D. (2010). Who by Accident? The Social Morphology of Car Accidents. Risk Analysis 30(9).

Fontaine H, Gourlet Y, & Ziani A. (1995). Les accidents de piétons, analyse typologique. Rapport INRETS No. 201.

Fontaine H. (1995). A Typological Analysis of Pedestrian Accidents. 7th, workshop of ICTCT, Paris.

Greenacre M and Blasius J (2006). Multiple Correspondence Analysis and Related Methods. Chapman & Hall/CRC, FL.

Town and Infrastructure Planning for Safety and Urban Quality – Pezzagno & Tira (Eds)
© 2018 Taylor & Francis Group, London, ISBN 978-0-8153-8731-2

New mobilities and territorial complexity: Is the promotion of sustainable mobility risk-free for cities? The case of Strasbourg, France

E. Propeck-Zimmermann, S. Liziard & A. Conesa
University of Strasbourg, Strasbourg, France

J. Villette
University of Normandy Caen, Caen, France

R. Kahn
University of Strasbourg, Strasbourg, France

T. Saint-Gerand
University of Normandy Caen, Caen, France

ABSTRACT: The intensification of urban public policies in favor of sustainable mobility appears legitimate regarding the energy transition, but could it not generate any unwanted side effect? An exploratory spatial and statistical analysis approach linking a set of socio-urban environments, sustainable mobility facilities and accidentology indicators has been implemented at the Eurometropolis of Strasbourg scale. Its aim is to study the risk of increasing socio-spatial inequalities linked to the spatial selectivity of developments in favor of sustainable mobility. This work is part of a multidisciplinary RED research project aimed at producing knowledge on the emerging risks of sustainable mobility policies and fostering their integration into public policy.

1 INTRODUCTION

The sustainable mobility belongs to the global development model of territories. It aims to promote the forms of the mobility that are more environment-friendly and to meet the entire population needs under the economic viability constraint. In this context, the travel patterns have evolved significantly, and particularly at the urban areas scale. They have been diversified and oriented towards soft, public or shared transport modes, which are encouraged by the urban planning and incentive policies in order to enhance the eco-mobility.

However, we are questioning whether the intensification of the urban public policies in favor of a sustainable mobility, which is legitimate regarding the energy transition, is virtuous. In parallel, we are interrogating whether the development of a new organization of mobility, which can be accompanied by infrastructures and new practices, is really devoid of negative effects, in case we consider the cities as socio-spatial systems. If not, we are seeking to know which domains are affected by these negative effects and whom they affect. The RED multidisciplinary research project, in which this work is carried out, aims to provide knowledge on the emerging risks of sustainable mobility policies and to foster their integration into the public action.

From a geographical point of view, several questions arise. The first one is how the sustainable mobility facilities are distributed across the agglomerations. The second one is to which spaces and populations they benefit. And the third question is how the type and spatial distribution of crashes evolve, particularly those that involve vulnerable users.

To answer these questions, spatial and statistical analysis methods are applied to the Eurometropolis of Strasbourg (EMS) using a spatial and detailed database. This Eurometropolis is pursuing a very voluntarist policy in promoting the sustainable mobility, especially by according a significant place to the soft mobility.

Within our work, the sustainable mobility is, first, briefly contextualized in its exercise framework (the territorial complexity and current development model). And then, a global approach, which consists in linking a set of socio-urban environments, sustainable mobility managements, and accidentology indicators, is presented. In fact, our objective is to identify new forms of risks induced by the sustainable mobility in Eurometropolis of Strasbourg.

2 SUSTAINABLE MOBILITY AND TERRITORIAL COMPLEXITY

At the metropolitan scale, many factors which are derived from the configuration of the territory, influence the traffic conditions and their associated risks: morphologies and structures of developed areas, organization of buildings (Engel, 1986; Millot, 2003; Fleury, 2012), continuity, density, and network mesh, etc. All of these elements make up a system (Millot and Brenac, 2001) and influence the number, frequency, location, to some extent the severity of the crashes, and the concerned population. The organization of use of the public spaces (parking, meeting areas, etc.) and the distribution of users should, also, be taken into account.

The contemporary policies of the sustainable mobility and associated developments (pedestrian areas, zones 30, cycling networks, and public transports, etc.) modify the socio-technical system. Moreover, the spatial configurations, traffic conditions, behaviors, and accidents are dynamically reconfigured across the whole territory which reveals new forms of risks and vulnerabilities.

In France, the Urban Solidarity and Renewal act of 2000, which is supplemented by the Environment Grenelle laws (2009–2010), aims to enhance the legal effect of the urban mobility schemes. Its purpose is to integrate the urban quality and environmental protection into the economic development and urbanization choices in order to give a new priority to other transport modes than the car. This regulation has been occurred in most large urban agglomerations through reducing the car designed places in certain urban areas, promoting soft modes, and developing public transport, especially in a greenfield sites, while improving the accessibility by cars at the "metropolis of flows" scale (Dupuy, 2006; Offner, 2006; Orfeuil, 2008; Reigner, Hernandez, Brenac, 2009; Brenac, Reigner, Hernandez, 2013).

We wonder whether the dual objective of the economic and ecological performance (accessibility/attractiveness) lead to a selective processing of spaces, accompanied with a risk of an increase in the socio-spatial inequalities between the strategic urban territories, "showcases" of the sustainable mobility (the competitive urban centers and economic areas), and other more ordinary territories.

3 MANAGEMENT POLICY OF THE SUSTAINABLE MOBILITY IN EUROMETROPOLIS OF STRASBOURG

In Strasbourg agglomeration, the management policy of the sustainable mobility has been, mainly, implemented since 1990s. Thus, the urban transport plan (PDU) of 1992 prohibits the access and parking of cars in the city-center. After the pioneering examples of Grenoble and Nantes, this pedestrianization has been accompanied by the inauguration of the tram in 1994, resulting in a highly mediated political confrontation. It symbolizes the rebirth of the modern tramway which has affected most of the major French cities between 1990 and 2000. Nowadays, it is remarkable for its extent (second in France behind Lyon), its mesh (six lines and 42 km of rails), and its cross-border character since the 28th of April, 2017.

The specificity of Strasbourg is the development of the cycling policy, partly inspired from its Rhine neighbors (Basel, Karlsruhe, and Freiburg). It has been initiated since 1978 with the two-wheeled master plan (the latest version dates back to 2011), and beefed up between 1990 and 2010. Nowadays, this strategy allows Strasbourg to have more than 580 km of cycling network and 19000 bows (approximately, 500 are built every year). Since 2010, a self-service

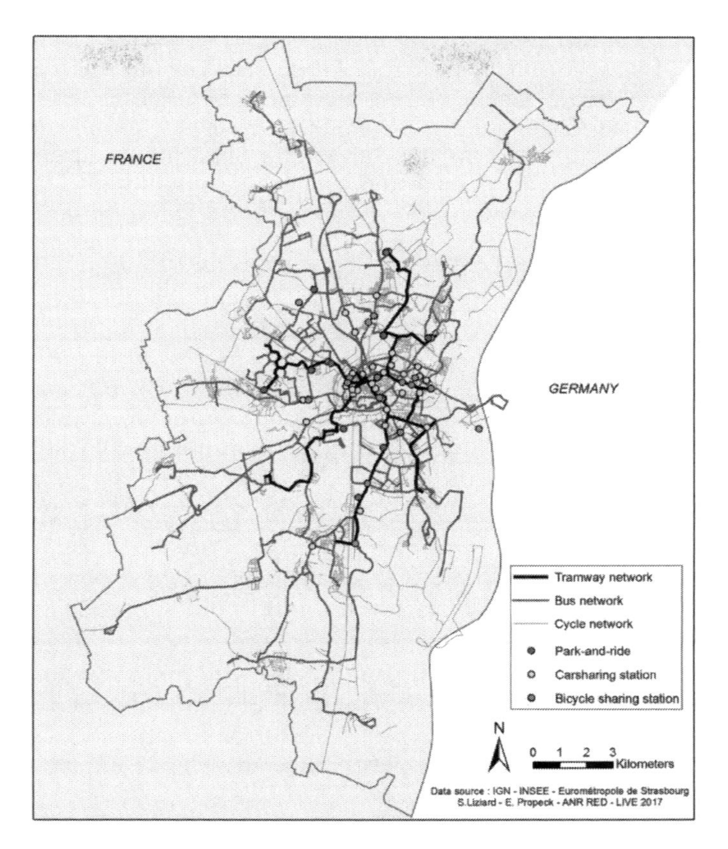

Figure 1. Sustainable mobility in Eurometropolis of Strasbourg.

bicycle system pursues this policy. Nevertheless, there are still a number of difficulties in the implementation of such policy (the saturation and conflicts with pedestrians, etc.). In addition to that, since 2002, the EMS provides a car-sharing service. The overall situation of the sustainable mobility developments is presented in Figure 1.

The map highlights a gradient of the sustainable mobility management from the center to the periphery. It raises the problems related to the periurbanisation, as in the last PDU of the EMS (2012): the periphery to periphery trips, the active modes in the second-ring suburb, and the limits of the cycling network, etc. These challenges are reflected in the following objectives: "Acting, in priority, on the peripheries in order to re-balance the mobility issues", "Expanding the range of the mobility services, and using each mode of transport according to its area of relevance" (PDU 2012).

Multiple scale analysis seems to be important, particularly the strategic scale of the agglomeration, in order to highlight the induced effects that are sometimes unsuspected, of the adopted policies. An original approach, based on the elaboration of a socio-urban typology, has been developed. It aims to study the socio-spatial inequalities of the sustainable mobility facilities and behaviors, as well as the related crashes. With respect to that, we focus our objective in providing a knowledge that will inform stakeholders and support them in their strategic choices.

4 SOCIO-SPATIAL SELECTIVITY OF SUSTAINABLE MOBILITY MANAGEMENTS: METHODOLOGICAL FRAMEWORK

The identification of the risks due to the unequal treatment of the urban spaces, which result from the spatial selectivity of the urban developments designed to the sustainable mobility, requires working at the agglomeration level. Such level represents the strategic scale of the public action. It involves linking the physical and socio-economic characteristics (buildings, infrastructure, population, motilities, etc.), technical objects of the sustainable mobility (pedestrian zones, zones 30 (zone where the speed is restricted to 30 km/h), bicycle stations, public transport in the greenfield site, etc.), and traffic accidents. A statistical and spatial analysis framework under GIS is set up to understand these relationships.

4.1 *Implementation of the spatial database*

The spatial database is organized into six themes according to a conceptual model which is elaborated using the HBDS (Hypergraph Based Data Structure) method (Fig. 2).

The data cover the entire territory of the Eurometropolis of Strasbourg which accounts for 28 communes that sums up to 314 km^2, till the end of 2016. Most of the data is provided by the Eurometropolis (PLU, social housing, economic activities and public services, transport infrastructures and associated facilities and crashes). The data on the population characteristics and its mobility practices is provided by the INSEE (General Population Census 2012 of the National Institute of Statistics and Economic Studies). They are complemented by the data from the IGN concerning the geographic reference system (GEOFLA, BD TOPO 2013, and Orthophoto 2011).

4.2 *Development of a socio-urban typology*

The socio-urban structure is the basis of any planning policy. The developed approach is based on a prior implementation of a fine socio-urban structure typology of the Strasbourg agglomeration. This typology will serve as a reference for the subsequent analyses. According to that, the objective is to see how the mobility facilities, mobility behaviors, and crashes are organized according to this structure.

The study is based on 45 indicators related to the population (age, household size, socio-professional composition, etc.) and urban environment (land cover, buildings, housing, and road network). The data is processed based on a mesh of 200 m, with respect to the finest division of the INSEE. At first, a data construct is necessary in order to measure all the 45 indicators at the mesh scale (such as the area of each land cover type). For certain socio-economic data (such as the number of people per socio-professional category) which are only available in a broader division (IRIS of INSEE), the values are estimated with reference to the proportion of inhabitants or households in the mesh.

From the constituted indicators, the typology is established by using a multiple correspondence analysis which is followed by an agglomerative hierarchical clustering. As a result, the meshes sharing many modalities are gathered in a cluster, without a spatial contiguity constraint.

Figure 3 shows a center-periphery logic and variations within the different suburbs' rings. These variations can be related to the urban structure (dense suburbs and housing blocks, both close to the center) and/or socio-demographic characteristics (distinction between the 3 clusters of the second-ring suburb).

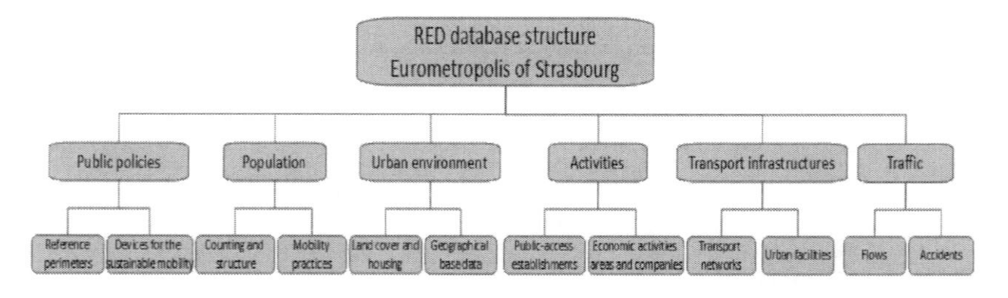

Figure 2. Database structure.

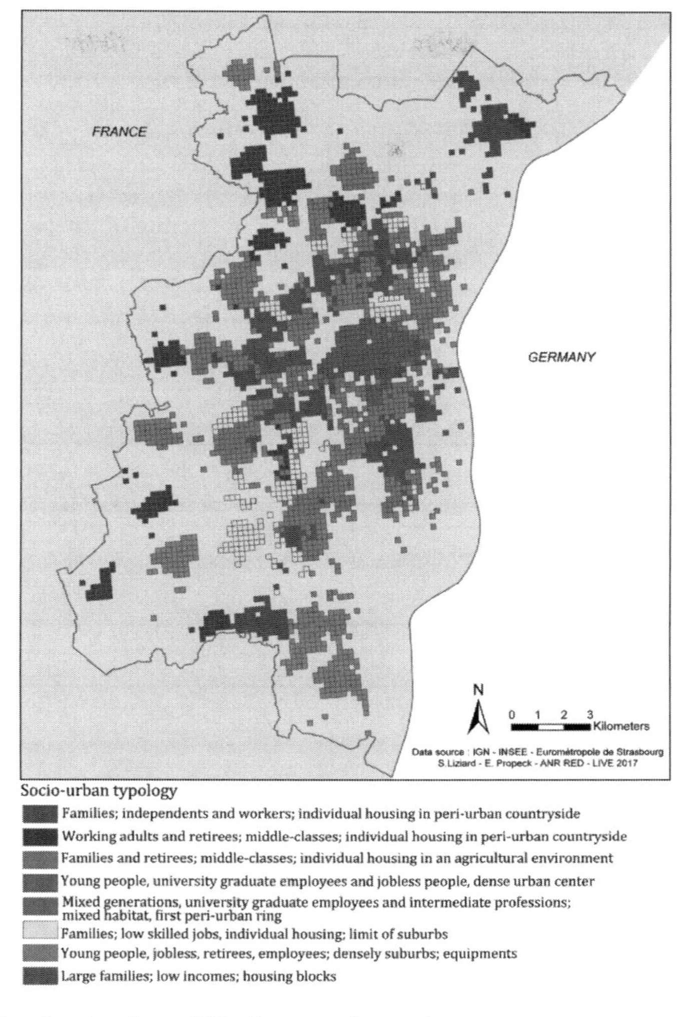

Figure 3. Socio-urban typology of Strasbourg agglomeration.

5 LINKING TERRITORY, SUSTAINABLE MOBILITY FACILITIES AND INSECURITY

The socio-spatial disparities in mobility infrastructures and behaviors and road traffic accidents are analyzed according to the clusters of the socio-urban typology previously elaborated.

5.1 *Socio-spatial inequalities in the mobility facilities and behaviors*

The methodology consists in examining how the mobility facilities and behaviors variables are differentiated based on the socio-spatial typology clusters. The selected variables are presented in the Figures 4a, 4b, and 4c. They allow the socio-urban typology clusters to be gathered into three different profiles. Within a given cluster, the values from the most cleaving variables are highly different from the average value over the entire study area. This variation is determined using the test value which allows to compare the mean values and to represent a possible over or under-representation of a variable in one or many clusters.

The Figure 4a shows that the second-ring suburb (blue, dark purple and pink clusters) is highly specialized in the automobile. It is characterized by a large share of households that possess at least two cars. In addition to that, it contains a large share of people going to work by car. Whereas, the public and soft transport modes are under-represented. The profiles are very similar between the three clusters. Furthermore, the public transport, soft modes, and self-service are under-represented.

Conversely to that, the Figure 4b indicates that the share of people going to work by car is relatively low in the dense urban center (red cluster), suburb (orange), and blocks housing (brown) clusters. Additionally, the households tend to possess only one car. Beside, these three clusters, especially the block housing cluster, present a high share of people that use public transports to go to work. Nevertheless, only the dense center and suburbs clusters are distinguished by the use of the soft transport modes (biking and walking), whereas only the center offers the self-service mobility modes (Velhop stations for bicycles and Citiz stations for cars). Thus, the results show that the profile of the center, suburbs, and housing blocks clusters is oriented towards the public and soft mobility modes, with variable intensities. Furthermore, it is important to remind that the share of people going to work using public transports is much higher in the housing blocks cluster than in the dense urban center and suburbs clusters.

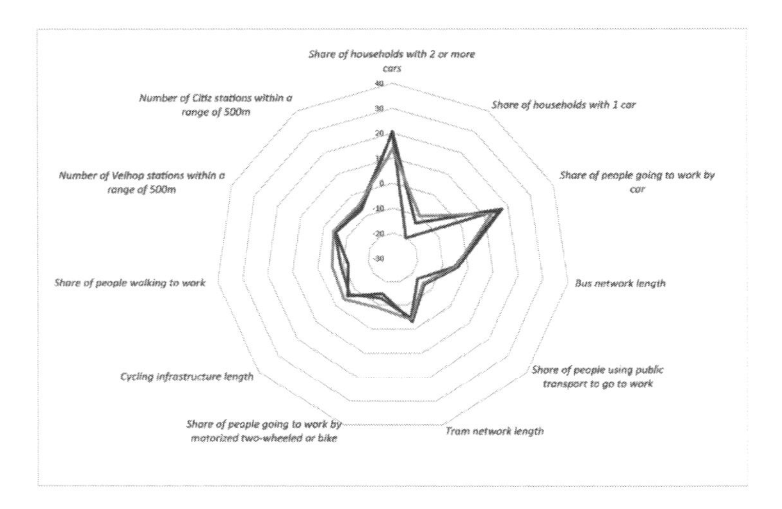

Figure 4a. Mobility profile within the clusters of the suburban second-ring.

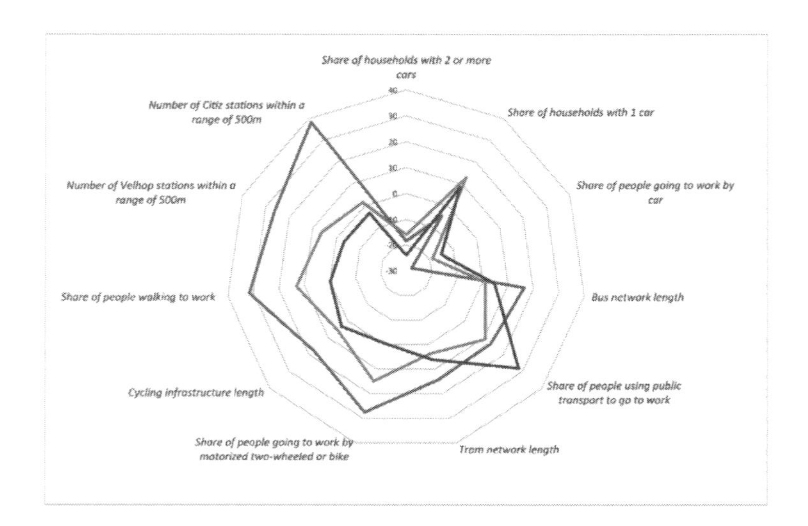

Figure 4b. Mobility profile within the dense urban area (center, suburbs, and housing blocks).

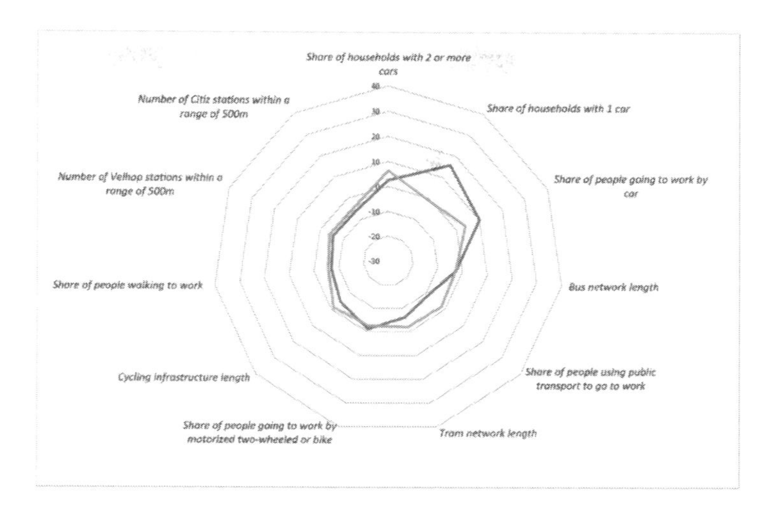

Figure 4c. Mobility profile within the clusters of the suburban first-ring.

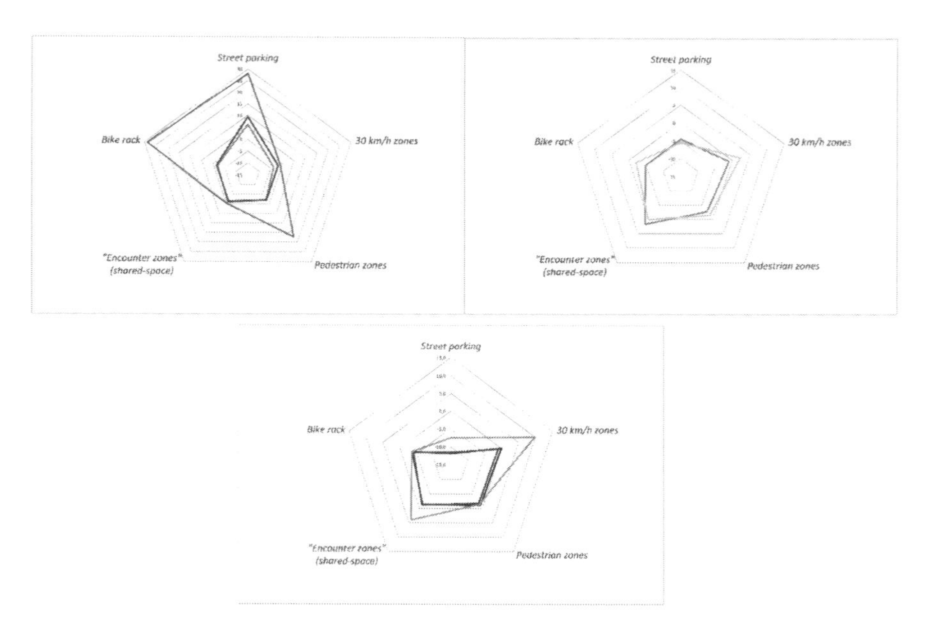

Figure 5. Clusters differentiation within different mobility profiles.

However, the public transport service supply in the housing blocks (the length of the tramway and bus networks) is lower or similar compared to the dense urban center and suburbs.

The Figure 4c shows that the first-ring suburb (green and yellow clusters) is more homogeneous regarding the different mobility modes. The yellow cluster is particularly homogeneous. It presents no cleaving variable. The green cluster, which is characterized by an over-representation of the automobile and an under-representation of the public transport, differs slightly from the yellow one. As illustrated in the first profile (Fig. 4a), the soft modes and self-service are under-represented in both clusters.

Other variables relevant to the mobility arrangements more accurately determine the clusters differentiation within the same profile. The analysis of these variables shows that the dense urban center (red) is different from suburbs (orange) and housing blocks (brown) because of a strong presence of bicycle networks, parking spaces, and pedestrian areas.

Besides, the pink cluster of the second-ring suburbs, corresponding to families and retirees living in individual housings within an agricultural environment, differ from the other two clusters of this suburbs ring because of the presence of zones 30 and meeting zones (Fig. 5).

The adopted approach is, gradually, refining the analysis of the urban specializations and risks of widening social inequalities risks that linked to unequal treatment of the urban spaces. It, also, examines the socio-spatial disparities of crashes among different mobility modes.

5.2 *Socio-spatial disparities in road traffic crashes*

The number of road traffic crashes and their evolution among different mobility modes differ according to the socio-urban typology clusters. The data on road crashes victims correspond to the crashes occurring on public roads and involving at least one vehicle that causes at least one victim who needs a medical care. These data, which is collected by the law enforcement officers (police or gendarmerie) intervening at the crash site, is stored on the "BAAC File". This file is administered by the National Inter-ministerial Road Safety Observatory.

In general, the number of crashes for all modes is characterized by an explicit decrease. It is more regular at the national level than the Eurometropolis of Strasbourg level. Indeed, at the national level, the highest decrease rate is observed between 2002 and 2003 (Fig. 6). During this period, the road safety has been raised to a major national cause level by the French president Jacques Chirac. On one hand, this period has led to an increased control, training, and sensitization of drivers. On the other hand, it has led to the deployment of speed control radars.

At the local level, the road safety policy has been launched in February 2003 with different communication actions undertaken by the city and Eurometropolis of Strasbourg. Accordingly, a real decline of the number of crashes is observed between 2003 and 2004. In 2008, the number of crashes increased then steadily declined until 2015. This decrease has become a national trend. It has become even more obvious at the local level in recent years.

The analysis of crashes, with respect to each transport mode, shows that this particular evolution between 2003 and 2008 corresponds particularly to the motorized modes –cars and motorized two-wheeled vehicles– whereas, it is, almost, unseen for pedestrians and buses (Fig. 7a to 7e).

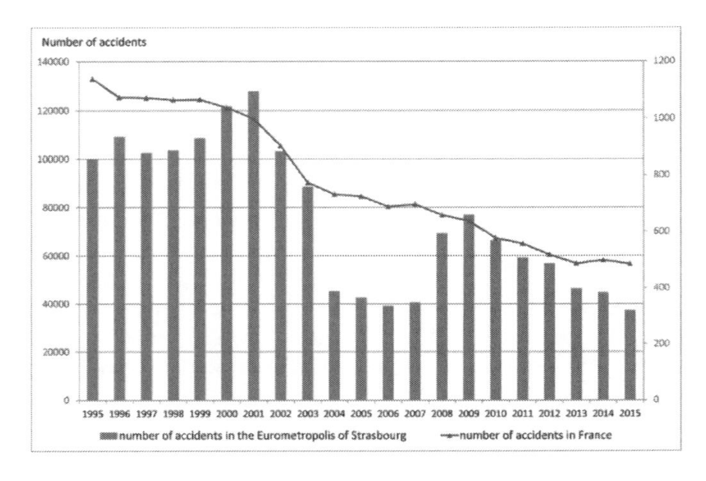

Figure 6. Evolution of the number of road accidents in France and Eurometropolis of Strasbourg between 1995 and 2015. (Source: National Interministerial Observatory for Road Safety, SIRAC—Eurometropolis of Strasbourg).

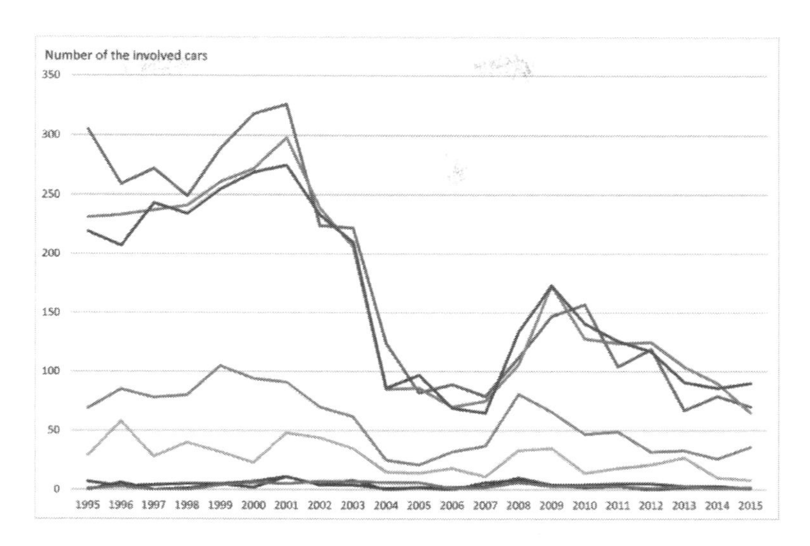

Figure 7a. Evolution of the number of the cars involved in road accidents per cluster.

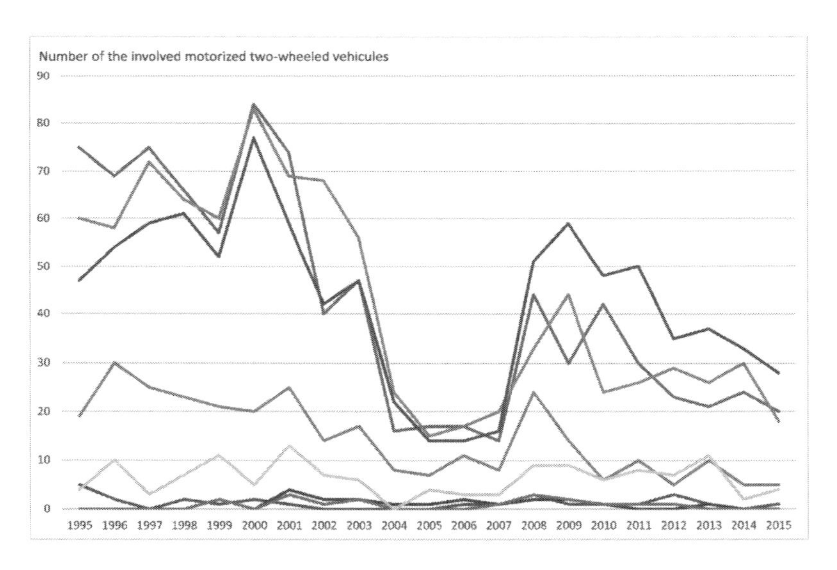

Figure 7b. Evolution of the number of the motorized two-wheeled vehicules that are involved in road accidents per cluster.

The distribution of the number of crashes per cluster reflects the organization center-periphery. In addition to that, for all modes, the number of vehicles and people involved in the central clusters, where traffic is higher, is more important than in the periphery. However, this difference declines thanks to an important decrease in the number of crashes in the central clusters. Still, the absence or low number of individuals in the clusters of the second peri-urban crown pushed us not to consider them in the Figures 7a to 7e.

Moreover, the hierarchy between the clusters has evolved since 2007 for several modes. It is the case of cars whose quantity has become more important since 2008 in the suburbs (orange) and housing blocks (Brown) than in the urban center (red) (Fig. 7a).

The number of crashes involving motorized two-wheeled vehicles in the large housing blocks exceeds the observed values in the urban center and suburban areas. These evolutions in the clusters hierarchy concern the first periurban crown. The number of motorized two-wheeled vehicles decreased in the green cluster which is characterized by a mix of different

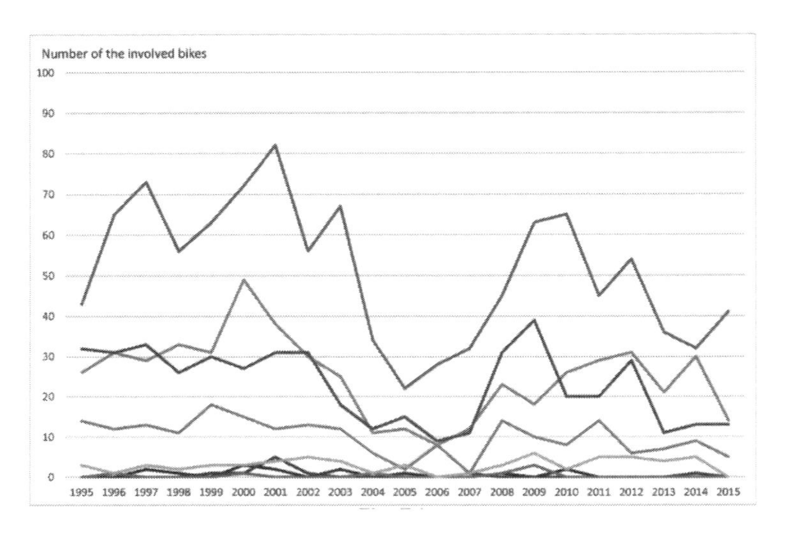

Figure 7c. Evolution of the bikes involved in road accidents per cluster.

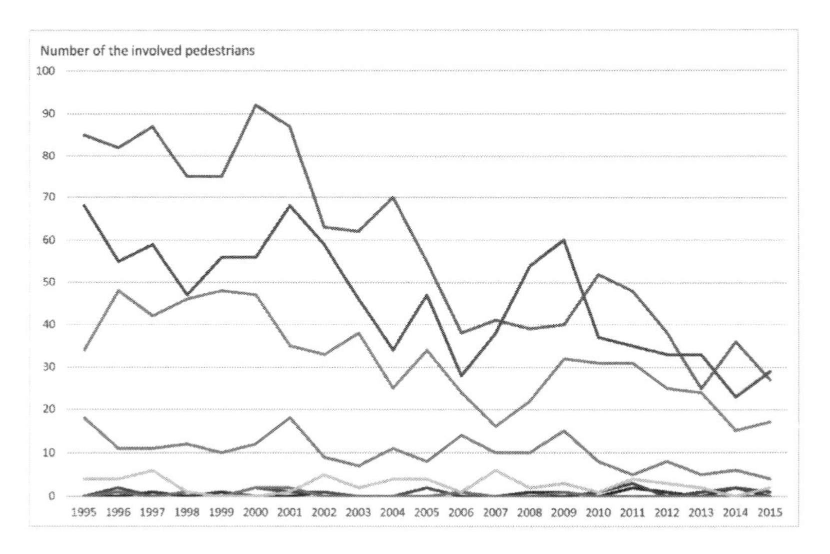

Figure 7d. Evolution of the number of the pedestrians involved in road accidents per cluster.

generations and housings. Consequently, it joins the values of the yellow cluster that consists in low-rise housing and whose values have remained stable (Fig. 7b).

Figure 7c shows that the overall downward trend which is observed for bicycles is much less pronounced than cars and motorized two-wheeled vehicles. This trend is linked to the increase of cycling habits in the Eurometropolis of Strasbourg, whose share from home-to-work trips has increased from 5.9% in 1997 to 7.6% in 2009. It reached 15% in 2015 in the more restricted area of the city of Strasbourg.

Moreover, the decrease in the number of involved bicycles is less obvious in the suburbs and large housing blocks compared to the urban center. The differences, previously identified, between these clusters, in terms of pedestrian areas and cycle tracks, can explain these evolutions.

Figure 7d illustrates that the number of pedestrians involved in crashes is higher in the large housing blocks than the suburbs, throughout the study period. This suggests a particular

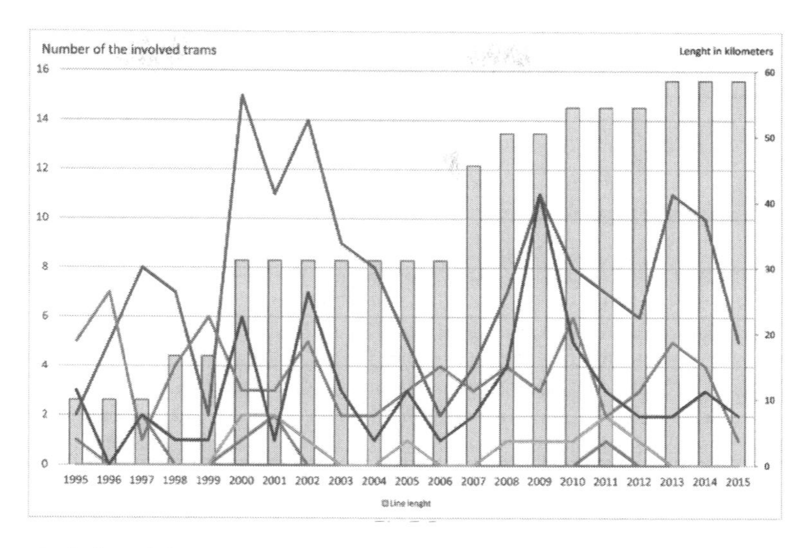

Figure 7e. Evolution of the tramway network and the number of trams involved in road accidents per cluster.

vulnerability of this transport mode in large housing blocks, as in the case of the motorized two-wheeled vehicles, over the second half of the study period.

Concerning the public transport, the small numbers of the involved tramways and buses do not allow an accurate interpretation of their variations. Nevertheless, following the opening of new tram lines, we observe phases of increase in the number of the involved trams in accidents that caused road traffic victims, within 2000, 2007, 2013 (Figure 7e). For buses, the different clusters do not exhibit a trend of evolution, except a decrease in the urban center, probably thanks to the tram service.

6 CONCLUSION AND PERSPECTIVES

The GIS and undertaken exploratory approaches have shown their interest in establishing diagnoses at the urban agglomeration strategic scale. They have allowed the analysis of the increasing social and spatial inequalities risk induced by the sustainable mobility policy. In addition to that, the integration of the social and economic dimension allows to contextualize the public action. In parallel, a similar application is carried out on Caen and Lille agglomerations in order to compare different urban contexts (type, size, and situation).

New perspectives are expected to evaluate and spatially locate the evolutions that are concomitant of the sustainable planning in different urbanized sectors. The evolution of the crashes will be particularly characterized by "a before and an after" diagnoses at the local (itinerary of a tramway and zones with calm traffic, etc.) and metropolitan global scales. Such analysis faces the complexity of the circulation system which is made up of interactions, regulations and readjustments loops that operate continuously by the users and the urban developers at various scales. It, also, faces the lack of adequate data in comparing the risk situations "before and after" an urban planning operation. In fact, if accidents data are often available over a long time period, the urban developments are rarely recorded. Furthermore, if the flow data exist, they are often fragmentary (CERTU, 2009). Despite these difficulties, empirically observing the effect of any action on the level of risk seems necessary to understand the functioning of the system and to assist in the definition of the strategies for an overall risk reduction in the territory. Such an analysis is undertaken in Strasbourg in order to improve knowledge of the impact of the tramway on the accidentology, especially on vulnerable users.

For further analyses, new perspectives consist in developing a more functional approach in analyzing the risks of unequal treatment of the urban spaces which are induced by the sustainable mobility. The implementation in GIS of an operational concept such as the "spatial ergonomics", defined by T. Saint-Gérand (2002), permits an evaluation of the conditions of access to resources needed by the population everyday and wherever and according to different mobility modes. As matter of fact, developing a new spatial analysis tool, which integrates the functional complexity, is the subject of the RED research project.

Spatial ergonomics is defined as the ability degree of a territory to meet its population needs at the minimum cost and effort (distance, time, money, security, comfort). A level of spatial ergonomics encompasses two criteria. The first criterion is the global access effort to the closest resources. It depends on the spatial distribution of resources (shops, health care, education, leisure, and public services, etc.) and the access means to that. The second criterion consists of resources and access alternatives within an acceptable distance (resources varieties, transport modes choice, and itineraries, etc.). This ability degree determination involves the modeling of a spatial, organizational, and costs constraints system. These constraints control the access to resources at different scales for different mobility modes.

A synthetic indicator, which is calculated according to different modes at each location, allows an evaluation of the impact of the policies of sustainable mobility on the potentialities of access to resources. The overall level of ergonomics, at every point in the space, is related to the socio-urban environments typologies, which are previously analyzed. It characterizes the socio-spatial dissimilarities. The goal of such analysis is to identify the forms of organization and functioning that ensure an ergonomics access to the optimal resources for the population.

The whole of the approach undertaken—linking multiple indicators of the urban environments, land planning, and crashes by exploratory spatial and statistical analysis, modeling and implementation in a GIS of the spatial ergonomics concept, put into perspective with the socio-urban environments—is targeted. The purpose is to reveal, at the strategic scale of the metropolis, the socio-spatial risks induced by the sustainable mobility policy, which may remain ignored, while they pose real questions. This research can provide a help and new insights in the diagnostics and improve the definition of urban development projects.

ACKNOWLEDGMENTS

This research is supported by the National Research Agency: ANR RED project (2014–2018).

REFERENCES

Brenac T., Reigner H., Hernandez F. (2013), « Centres-villes aménagés pour les piétons: développement durable ou marketing urbain et tri social ? », *Revue Recherche Transports Sécurité*, 29, 271–282.

Dupuy G. (2006), *La dépendance à l'égard de l'automobile*, Prédit, Paris, La documentation française.

Engel U. (1986) "Risk figures for road users in different urban street categories". 5ème congrès international de l'ATEC, "*L'insécurité routière*", 9–13 juin 1986, Paris, proceedings vol. 3, 17 p.

Fleury D., (2012), *Sicureza e urbanistica. L'integratione della sicurezza stradale nel governo urbano*. Gangemi Editore 254 p.

Millot M. (2003) *Développement urbain et insécurité routière: l'influence complexe des formes urbaines*. Thèse de doctorat en Transport de l'Ecole Nationale des Ponts et Chaussées, Paris, 414 p.

Millot M., Brenac T. (2001) « Analyse qualitative et modélisation de l'influence des caractéristiques des voiries urbaines et de leur environnement sur les phénomènes d'accidents. Une première approche ». BLPC, n° 232 Mai-juin 2001, pp. 5–18.

Offner J.M. (2006), *Les Plans de Déplacements Urbains*, Paris, La Documentation française.

Orfeuil, J.P. (2008), *Mobilités urbaines. L'âge des possibles,* Les Carnets de l'info, 254 p.

Reigner H., Hernandez F., Brenac T. (2009), « Circuler dans la ville sûre et durable: des politiques publiques contemporaines ambiguës, consensuelles et insoutenables », *Métropoles*, n°5.

Saint-Gérand T., (2002) « *S.I.G.: Structures conceptuelles pour l'analyse spatiale* » thèse d'H.D.R., Université de Rouen, 2002.

Boosting and planning soft mobility

Town and Infrastructure Planning for Safety and Urban Quality – Pezzagno & Tira (Eds)
© 2018 Taylor & Francis Group, London, ISBN 978-0-8153-8731-2

Innovations in cycling mobility for sustainable cities

E. Conticelli, A. Santangelo & S. Tondelli
Università di Bologna, Bologna, Italy

ABSTRACT: Reducing pollution and greenhouse gas emissions generated by transport is a great challenge for any sustainable city aiming to preserve the health of its citizens. Bicycles could represent an alternative to private car, especially over the short distances. Thus, the main challenge is to make cycling mobility a more appealing means of transport.

The aim of the paper is to describe how innovations can enhance cycling mobility in a framework of drastic reduction of air pollution and GHG emissions.

The paper identifies possible mobility contexts which represent fields of innovation, characterized by different mobility needs and target groups where the introduction of innovative solutions can boost cycling mobility. The contexts are: dedicated paths and technological devices, bike-sharing solutions, trainings & educational programs, cargo bikes.

Describing less explored fields of research, the paper aims at giving a contribution to make cycling a greener and funnier alternative to motorized means of transport.

1 INTRODUCTION

Although in the last decade GHG emissions due to car traffic have decreased, since 2014 transport emissions have started to grow again, and are 20.1% above 1990 levels (TERM, 2016), thus maintaining transport as the second main emitting sector after energy supply (EEA, 2014). This trend will not allow to achieve the targets fixed by the EU for 2030. Moreover, the introduction of more innovative and cleaner motorized vehicles, such as electric vehicles, does not seem to be sufficient to entirely address all the car traffic issues: i.e. traffic congestion which is strictly related to private road transport and it is destined to grow (EC, 2016).

The promotion of walking and cycling and the development of efficient public transports may be the only feasible, convenient alternatives to reduce levels of car use (EC/UN-Habitat, 2016; Banister, 2008), to encourage active travels and to make cities healthier and more liveable. Public transport is the backbone of a competitive low carbon transport system, but cannot satisfy all mobility needs because of its rigidity, discouraging people to use it. Moreover, making public transport systems more efficient frequently means to allocate heavy investments for renewing infrastructures and vehicles. Conversely, walking and cycling are sustainable, flexible, low-cost and more active travel modes that respond to both environmental and social needs and contribute to increase physical activity (Leonard, 2014).

Since more than 30% of the passengers trips made by cars in Europe cover distances of less than 3 km and 50% are less than 5 km, cycling could represent a serious alternative to private car (Dekoster and Schollaert, 1999). Moreover, cycling may be very versatile, thus covering all kind of trips made alone or carrying goods or other passengers, such as kids, serving people of all ages and social backgrounds, especially within smaller urban areas where public transport are scarce and local trips are relatively short (Pucher and Buehler, 2012). Bicycle can play a role also over longer distances, especially when bicycle paths serve commuting trips, when fast, comfortable and safer cycling services are offered. Finally, cycling mobility can be an effective solution also for freight delivery, especially in central areas, covering the last miles of supply chains (Russo and Comi, 2012).

To this aim, the introduction of different kinds of innovation in EU cities could determine a positive impulse for a wide diffusion of cycling mobility (Vettori, 2016). Social and technological innovations may effectively support the investigation of mobility needs and attitudes, change behaviour and the appeal of the bike, make cycling safer and better integrated within a comprehensive urban transport policy, actively involving stakeholders and target groups in the definitions of policies and actions to be adopted.

2 OBJECTIVE OF THE PAPER

While innovations in public and personal motorized transport are frequently analysed and mentioned, relatively little attention has been devoted to innovation in cycling mobility so far. In this framework and assuming that cycling is one of the most sustainable modes of transport, (Kenworthy, 2006), the paper aims to describe how innovations can effectively boost cycling mobility in the framework of drastic reduction of air pollution and GHG emissions.

3 METHODOLOGICAL APPROACH

The paper firstly identifies four main mobility contexts representing four fields of innovation in cycling mobility, characterized by different mobility needs and target groups, where the introduction of new solutions and approaches can boost bike mobility. Within these contexts, innovative measures and solutions—both currently in action or at the study level—able to stimulate a significant shift towards cycling mobility are analysed and discussed. Finally, the paper concludes by discussing the outcomes of introducing innovations for the promotion of cycling mobility and opens to possible further innovation scenarios to be explored in the future. The ambition is to suggest less unexplored fields of research in order to make cycling a real greener, healthier and funnier alternative to motorized means of transport.

The identified innovation fields, addressed as the most promising contexts of innovation, involving a wide range of target groups and being able to mainstream cycling mobility for all and for every purpose, are:

- Cyclists priority and safety;
- Bike sharing schemes;
- Behavioural change;
- Urban logistics.

These four innovation fields as well as the solutions provided in the next paragraphs might not be completely exhaustive, but they are representative of the main issues to be addressed for encouraging an actual and long lasting shift towards cycling mobility.

4 CYCLISTS' PRIORITY AND SAFETY

Numerous and diverse are the factors influencing bicycle use, but safety, longer travel time and higher efforts needed for travelling seem to be very crucial factors for choosing bicycle, especially for commuting journeys (Heinen et al, 2010). In this framework, new solutions mainly dedicated to commuters for travelling in a safe and pleasant environment should be sought in order to increase the use of bicycles. To this aim, pedelecs and e-bikes, but also automated traffic light priority to bicycles and smart wheels represent new solutions for reducing travel time and efforts, while the improvement of bicycle infrastructures and networks through the development of fast dedicated cycling routes can increase safety and comfort of commuters but also of other users.

4.1 *Electric bicycles*

In the field of e-bikes, important progresses have been undertaken during the last decade, thus increasing their diffusion all over Europe (CONEBI, 2016). The motorization of the bike enables cyclists to cover longer distances and/or shorter travel duration, being an efficient solution especially for commuters and for older people (*Dufour D., Ligtermoet & Partners*, 2010), and having the potential to actually reduce car mileage (Kämper et al, 2016).

Innovative and alternative solutions are continuously at the study both for making pedelecs more effective and user-friendly and for converting traditional bicycles in hybrid ones. Interesting examples are prototypes of electric wheels, such as the Copenhagen wheel[1] or the Freeduck wheel by Ducati[2], which can substitute the traditional wheels for assisting cyclists during their daily trips, without changing the bicycle.

4.2 *Traffic light priority systems*

Reducing travel time caused by delays at traffic lights is another issue influencing the use of bicycle, especially for commuters. New interesting solutions have been experimented in the last few years in the city of Aarhus in Denmark, going on along the positive experience of the green waves of Copenhagen[3]. As partner of a European project[4] devoted to encourage smart solutions for cities, the municipality has engineered a popular intersection to be responsive to bikers, turning the light green in their favor thanks to a RFID tag that triggers a sensor. The first results of this pilot show good performances of the system, which has never caused traffic problems or pushbacks from cyclists or drivers.

4.3 *Cycle paths and networks*

Innovations that give priority to and improve the safety of cyclists are not limited to the bicycle itself, but concern also infrastructures. The concept of fast cycling routes is relatively new (ECF, 2014) and comprises "high standard bicycle paths reserved for cyclists for fast and direct commuting over long distances" (Sørensen, 2012, p. I). Fast cycling routes are mainly spread in Northern European countries, such as the Netherlands, Denmark and Germany, where people are traditionally keen to travel by bike. They are generally located in densely populated urban areas serving people who commute 5 kilometres or more to their destination, but also leisure cyclists, and respond to 5 main criteria: directness; attraction; comfort; safety; connection (Bicycle Research, 2010).

The introduction of this kind of infrastructures in terms of reduction of car use and air pollution has already shown some benefits. In Germany, for instance, the first stretch of the biggest bicycle highway in the world—100-km long, connecting ten cities and four universities, serving the largest urban agglomeration in Germany—was opened in December 2015 in the Ruhr region. Together with the booming popularity of the electric bike, the highway is expecting to lead to a new era of cycle commuting in Germany (RR, 2014): when completed, the network will remove 50,000 cars from the road—with an associated annual reduction of 16,000 tonnes of CO_2 emissions, attracting daily between 1,000 and 4,000 of cyclists on each section.

5 BIKE SHARING SCHEMES

Public bike sharing systems (BSSs) have becoming increasingly popular over the last decade. More than 400 schemes exist in Europe, mostly in Italy, Spain and France (ECF, 2015). They represent an important innovation in the cycling sector as well as for public transport, since these systems are open to all citizens for daily mobility (Parkes et al, 2013; ECF, 2015) and are conceived in close relation with other public transport services.

There have been three generations of bike-sharing systems over the past 45 years (DeMaio, 2009) characterized by increasing innovations in their management. The first generation was implemented in Amsterdam in 1965 and was characterised by no payment or security func-

1. https://superpedestrian.com/the-copenhagen-wheel.
2. http://www.freeduck.it/.
3. http://www.copenhagenize.com/2014/08/the-green-waves-of-copenhagen.html.
4. http://www.radical-project.eu/.

tions, which determined its collapse in a short time. The second was firstly launched in Copenhagen (DeMaio, 2009), where bikes could be picked up and returned at specific locations with a coin deposit, but the anonymity of the users caused the failure of this type of programmes as well. When the BSSs were smartened with a variety of technological improvements (smart card access; automatic docks and stations; real time information; "first 30 minutes for free", etc.) the third generation began (DeMaio, 2009; ECF, 2015). This last generation have created a larger cycling population, increasing the number of cities operating a BSS from 13 in 2004 to 855 in 2014 (Fishman, 2016) raising bike mode share between 1.0–1.5% in cities with pre-existing low cycling use (DeMaio, 2009). Nowadays increasing innovations in bike sharing systems and alternative solutions, even for smaller cities, are emerging, making bike sharing a more and more promising way of moving around the city.

5.1 *The 4th generation of bike sharing schemes*

Innovation in BSSs is on the go, paving the way to the fourth generation of bike-sharing schemes. It will be characterized by important improvements, based on extensive electrification of shared bikes and use of smartphone applications for real-time updates, (ECF, 2015), solar-powered and movable docking stations, improved distribution of bikes, new business models (Parkes et al, 2013; DeMaio, 2009; ECF, 2015), in order to make BSSs more effective and reliable.

An interesting example of more advanced BSSs is the Call-A-Bike service, runned by Deutsche Bahn in several German big cities. The scheme is designed for one way trips and does not rely on fixed racks but rather on GPS technology and georeferencing which enables a "floating" bicycle access (Shaheen et al., 2012; Parkes et al., 2013). On the same line is Cycle. land, the UK's first peer-to-peer bike-sharing company launched in Oxford in 2006 which enables registered travellers to borrow bikes from local cyclists.

5.2 *Bike sharing schemes in small towns*

Despite bike sharing schemes have gained wide success in medium and large cities (more than 200,000 inhab.), in small cities they could be even more important, representing a useful addition to existing public transport systems. In fact, in rural areas and in small towns with up to 100,000 inhabitants, Public Transport (PT) is often not as well developed as in larger cities, therefore BSSs can complete or even substitute PT (OBIS, 2011). The cases of the rural south Bohemia (CZ) (OBIS, 2011) and Lower Austria (MOG, 2014) show that regional BSSs linked to railway stations have shown good performances in terms of attractiveness of bike sharing for tourism and even for commuting, increasing the synergies between small cities (OBIS, 2011).

As these systems continue to expand in number and size, interoperability among them is becoming more and more necessary (Midgley, 2009). This is what is happening in the cities around Valencia, where 12 municipalities have been developing an integrated bike sharing system with the aim to extend the service to the inner metropolitan ring of Valencia, comprising 30 municipalities, and to connect the system with the one already present in Valencia. The ambition is indeed to integrate and encourage sustainable trips inside the whole inner metropolitan area. A first pilot station was installed reaching more than 4,500 uses in 4 months.

6 BEHAVIOURAL CHANGE

Shifting towards sustainable mobility requires that people radically change the way they make decisions on their daily travels, but people are generally reluctant to alter their behaviour (Banister, 2008). At the same time, public acceptability drives political acceptability, which is a necessary prerequisite for change and for being actions take place (Banister, 2008). Therefore, pushing a long-lasting behavioural change means first of all dealing with people's mobility habits, giving the opportunity, ability and motivation to change them (Leonard, 2014; Poiesz, 1999).

New methods to investigate mobility needs and attitudes and to change mobility behaviour of road users are required to encourage people to actually choose the bike as a means of transport, thus enhancing its appeal. Special efforts should be dedicated to children and older adults, who are the most vulnerable people especially for what concerns safety but also the ones who may benefit more from this means of transport.

6.1 *Bicycle trainings and educational programmes*

Involving pupils in order to make their school environments more cycling friendly and make cycling safer, may encourage the use of bicycles influencing young people's perception of bicycle as a valuable and funny means of transport and may convince parents to leave kids bicycling to school as well. The city of Reggio Emilia in Italy has a decennial tradition in promoting bicycle culture with 200 km of cycle paths, 20% of travels made by bicycle and decennial investments. Since 2003, the city has been experimenting the "Bicibus", consisting in a group of pupils going and coming back to school by bike together with adult volunteers (e.g. parents, grandparents, teachers) along predefined and signalled paths. Currently the service involves around 7% of children of 18 primary schools. Bicibus has produced positive effects in terms of road safety, environmental conditions, change of mobility habits, children independence and people relationships. The availability of parents and relatives both for proposing the service to their kids' schools and for escorting the children has ensured the success of the measure in the long run and represents a valuable example of how social innovation processes can boost innovation.

6.2 *Gamification and reward-based competitions*

Other innovative approaches for encouraging a sound shift towards the use of bicycle, especially among youngsters and new generations of cyclists, might be ascribed to gamification and reward-based competitions. Frequently emotional factors such as fun, personal engagement or team spirit are more influent than factors like health benefits or environmental reasons (Millonig et al., 2016). In addition, gamification is now incorporated into many running and cycling mobile phone applications where users collect points for the miles they monitor or for reaching a specific target (Coombes and Jones, 2016).

Taking part of these competitions often means also taking part of data gaining processes which might be relevant for that bodies who plan urban transports and infrastructures. Crowdsourcing for data collection made by enthusiastic and motivated users, as in the case of bicyclists, has a huge potential in augmenting the standard data collection procedures by including the requirements of otherwise marginalized groups of users (Misra et al., 2014) and giving real time and capillary information.

6.3 *Urban logistics and the use of cargo bikes for personal and commercial purposes*

Using the bicycle for carrying goods has many advantages over the motorized vehicles in urban areas, where traffic restrictions and congestion are frequent and the distances are relatively short. The project Cyclelogistic (Reiter et al., 2014) has estimated that 49% of all trips in the European urban areas are partly related to transport of goods and are made by motorized vehicles, and 51% of them could be potentially shifted to bicycle. Within this potential the project esteemed that 1/3 is due to commercial logistic trips (businesses & service and delivery) and 2/3 to private logistic trips (commuting, shopping, leisure). In this framework, some interesting fields of innovation may be identified.

6.4 *Cargo-bikes for commercial logistic trips*

Using cargo-bikes for delivering goods helps covering the last mile trips within dense urban areas. This new delivery system is commonly based on the creation of micro-logistic centers placed in the outskirts of the inner city where goods are carried by trucks coming from the main logistics centers located in suburban areas and redistributed by cargo-bikes within the city centre.

While these models are mainly focused on businesses, other innovative and more customer-oriented solutions have been implemented. In Valencia, ENCICLE has developed a cargo bike delivery system which acts in the e-commerce framework. Customers buy on-line and can easily plan the time and place of delivery, which is made by bike.

A sort of integration of these two systems may produce an advancement, towards a more articulated logistic service: through ICT-based management systems the micro-logistic center could be also used as delivery point for businesses where to store the wares already sold out and to deliver them through cargo-bikes, allowing customers to shopping by foot, by bike or by public transport, without caring about their volume or weight.

6.5 *Cargo-bikes for private logistic trips*

Cargo-bike for private uses is currently experiencing a revival in cities all over Europe and many new and different models have been appearing, but it has not yet penetrated the bike market on a significant level because of high costs, lack of parking space and rare use (Eltis.org, 2014; Willems and Rodtheut, 2013). In this perspective, a cargo bike sharing system could represent a valuable solution, aiming at improving access to and use of cargo bikes for a wide range of diverse user groups. In Ghent, different cargo-bike sharing schemes have been experimented: on the one side, neighbourhood cargo-bikes shared among neighbours and, on the other side, the introduction of cargo-bikes in the existing bike sharing scheme.

7 RESULT OF THE RESEARCH AND CONCLUSIONS

The paper has mapped some current innovations and trends, where a clear commitment of diverse stakeholders and user groups in terms of boosting cycling mobility is leading to a cultural change.

The described examples have shown that social and technological innovations are hardly separable but they claim each other, and their mix seems to strengthen the action of the broad public for planning and promoting cycling as a more appealing means of transport. Innovations with a predominant non-technological nature and mainly focused on the behavioural change aspect, such as gamification, reward-based competitions or peer-to-peer bike sharing schemes, needs technological infrastructures and services for working. At the same time, technological innovations (such as smart bikes, new generations of bike and cargo-bike sharing schemes) may be proposed to users open to experiment new forms of mobility and lifestyles. In some cases, technological innovations have stimulated social innovations, such as in peer-to-peer bike sharing schemes, while in other experiences social innovation gets first, like in bike training programmes for children, which may include gamification activities supported by ICT tools and apps.

This mutual integration of technological and social innovations is an important ingredient of a systemic process for effectively boosting cycling mobility adopted by policy-makers as well as user groups and providers; another key element is to design innovations through inclusive co-design and co-creation process by involving all the relevant stakeholders and target groups (children, youngsters, commuters, older adults, businesses), starting from the real mobility needs.

Innovations cannot radically change the way a city can move but may be a trigger for change, creating a fertile environment. A more structured change comprising new organisational and governance concepts, changes in planning processes, and promoting integrated transport-land use policies may be achieved through a strong commitment of public policies (transport policies, land-use policies, urban development policies, housing policies, environmental policies, taxation policies and parking policies) which should be open as well for defining a long-term strategy based on the promotion of soft mobility, bringing cycling into new smart and inclusive city design.

REFERENCES

Banister D., (2008). The sustainable mobility paradigm, Transport Policy, 15, 73–80. DOI:10.1016/j.tranpol.2007.10.005.

Bicycle Research, 2010. Radschnellwege (Bicycle Express Routes). Berlin.
Confederation of the European Bicycle Industry, (CONEBI, 2016). European bicycle market. 2016 edition. Industry and market profile.
Coombes, E. and Jones, A., (2016). Gamification of active travel to school: A pilot evaluation of the Beat the Street physical activity intervention. Health & place, 39, pp. 62–69.
Dekoster J., Schollaert U., (1999). Cycling: the way ahead for towns and cities, European Communities.
DeMaio P., (2009). Bike-sharing: History, Impacts, Models of Provision, and Future. Journal of Public Transportation, 12 (4): 41–56. DOI: http://dx.doi.org/10.5038/2375-0901.12.4.3. Available at: http://scholarcommons.usf.edu/jpt/vol12/iss4/3.
Dufour D., Ligtermoet & Partners, 2010, PRESTO. Cycling policy guide. Cycling infrastructures (https://ec.europa.eu/energy/intelligent/projects/en/projects/presto).
Eltis.org: Public cargo-bike-sharing in Ghent (Belgium) (2014), http://www.eltis.org/discover/case-studies/public-cargo-bike-sharing-ghent-belgium.
European Commission (EC, 2016). EU Reference Scenario 2016: Energy, transport and GHG emissions —Trends to 2050, Publications Office of the European Union, Luxembourg.
European Commission, United Nations Human Settlements Programme (EC/UN-HABITAT, 2016). The state of European Cities 2016. Cities leading the way to a better future. Publications Office of the European Union, Luxemburg.
European Cyclists' Federation, (ECF, 2014). Factsheet. Fast Cycling Routes: towards barrier-free commuting.
European Cyclists' Federation, (ECF, 2015). A European Roadmap for cycling – ECF proposal.
European Environment Agency (EEA, 2013). A closer look at urban transport. TERM 2013: transport indicators tracking progress towards environmental targets in Europe. EEA Report No 11/2013, European Environment Agency.
European Environment Agency (EEA, 2014). Total greenhouse gas emissions trends and projections. http://www.eea.europa.eu/data-and-maps/indicators/greenhouse-gas-emission-trends-6/assessment/#total-greenhouse-gas-emissions-trends-and-projections.
Fishman E., (2016). Bikeshare: A Review of Recent Literature, Transport Reviews, 36:1, 92–113. DOI: 10.1080/01441647.2015.1033036.
Heinen E., van Wee B., Maat K., (2010). Commuting by Bicycle: An Overview of the Literature, Transport Reviews, 30:1, 59–96, DOI: 10.1080/01441640903187001.
Kämper C., Helms H. and Jöhrens J., (2016). Modal Shifting Effects and Climate Impacts through Electric Bicycle Use in Germany. Journal of Earth Sciences and Geotechnical Engineering, vol. 6, no. 4, 2016, 331–345.
Kenworthy, (2006) The Eco-City: Ten Key Transport and Planning Dimensions for Sustainable City Development. Environment & Urbanization, 18(1).
Leonard C. (ed.), (2014). Civitas, Innovative Urban Transport Solutions. Civitas makes the difference. How 25 cities learned to make urban transport cleaner and better, ICLEI – Local Governments for Sustainability, Freiburg.
Midgley, P. (2009). The Role of Smart Bike-sharing Systems. In: Urban Mobility. Journeys. May. 23–31.
Millonig A., Wunsch M., Stibe A., Seer S., Dai C., Schechtner K., and Chin R. C.C., (2016). Gamification and social dynamics behind corporate cycling campaigns. Transportation Research Procedia, 19, 33–39.
Misra A., Gooze A., Watkins K., Asad M., and Le Dantec C. A., (2014). Crowdsourcing and Its Application to Transportation Data Collection and Management. Transportation Research Record: Journal of the Transportation Research Board No. 2414, 1–8.
Move on Green (MOG, 2014). Good practices collection on sustainable mobility in rural EU. http://www.euromontana.org/wp-content/uploads/2014/07/mog_good_practices_collection.pdf, (accessed 05.04.2017).
OBIS, (2011). Optimising Bike Sharing In European Cities. A handbook.
Parkes S. D., Marsden G, Shaheen S. A. and Cohen A. P., (2013). Understanding the diffusion of public bikesharing systems: evidence from Europe and North America, Journal of Transport Geography, 31, 94–103. DOI: 10.1016/j.jtrangeo.2013.06.003.
Poiesz T. B. C., (1999). Gedragsmanagement, Waarom mensen zich (niet) gedragen. [Bahavorial Management, Why people (do not) behave], Immerc bv.
Pucher J. and Buehler R., (eds) (2012). City Cycling, MIT Press, Boston.
Regionalverband Ruhr, (RR, 2014). Machbarkeitsstudie Radschnellweg Ruhr RS1. http://www.rs1.ruhr/fileadmin/user_upload/ RS1/pdf/RS1_Machbarkeitsstudie_web.pdf (accessed 04.04.2017).
Reiter K., Wrighton S., and Rzewnicki R., (2014). Cyclelogistics. Moving Europe forward. D7.1 A set of updated IEE Common performance indicators including their baseline and assumptions for extrapolation. www.cyclelogistics.eu.

Russo F., Comi A., (2012). City characteristics and urban goods movements: A way to environmental transportation system in a sustainable city. Procedia—Social and behavioral sciences, 39, pp. 61–73. http://dx.doi.org/10.1016/j.sbspro.2012.03.091.
Shaheen, S.A., Martin, E., Cohen, A., Finson, R., (2012). Public Bikesharing in North America: Early Operator and User Understanding. Mineta Transportation Institute. MTI Report 11–26, San Jose State University. <http://transweb.sjsu.edu/PDFs/research/1029-public-bikesharing-understanding-early-operators-users.pdf> (accessed 03.04.2017).
Sørensen M. W. J., (2012). Sykkelekspressveger i Norge og andre land. Status, erfaringer og anbefalinger, [Bicycle express routes in Norway and other countries—Status, experiences and recommendations]. TØI rapport 1196/2012, Transportøkonomisk Institutt, Oslo.
Vettori M. P., (2016). Cycling city project: strategie e tecnologie delle infrastrutture per la mobilità sostenibile. Il caso di Copenhagen, TECHNE, 11, 66–73.
Willems, F. & Rodtheut, S. (2013): Can cargo bike sharing exist in Berlin? http://www.unserlastenfahrrad.de/UnserLastenfahrrad_Final_report.pdf.

Town and Infrastructure Planning for Safety and Urban Quality – Pezzagno & Tira (Eds)
 ISBN 978-0-8153-8731-2

Sustainable mobility in the functional mix of the urban project

M. Francini, S. Gaudio, G. Mercurio, A. Palermo & M. Viapiana
Università della Calabria, Rende, Italy

ABSTRACT: The current transport system, based mostly on road traffic and the use of private vehicles, impacts the environment, the health and the economy. To overcome these, it is necessary to move towards sustainable models of mobility. These can be achieved by adopting an appropriate and strategic land-use planning approach at the urban scale.

The aim of the paper is to verify the transferability of the criteria identified in the European project PAES ("Housing Policy for Sustainable Construction", financed by "Erasmus+" Programme) to small and medium-sized Calabrian cities. The PAES project studied the good practices in promoting efficient land use models. These aimed at linking sustainable mobility goals with territorial modelling and public space enjoyment goals.

In order to verify the application of these criteria, the single city is assimilated to "neighbourhood" dimension and a new governance model based on sharing the investments and the services is assumed.

1 STATE OF THE ART

An efficient mobility system leads to a better quality of life and better availability and functionality of cities. The efficiency is real only if it goes hand in hand with sustainability.

It is therefore necessary to design a sustainable mobility that will reconcile the right to movement with the need to reduce negative externalities (namely the variety of external effects that are also sustained by those who do not cause them).

In the urban area, the need of a sustainable transport system is leading to the identification new models of demand and movement management. Furthermore, the emphasis is moving from the construction of new infrastructures to the optimisation of existing ones. In this respect, urban planning plays an important role because its task is the creation of better settlement patterns and the adoption of a series of measures and policies. The actions of mobility rationalization and improvement with the objective of sustainability have become an opportunity for urban regeneration. This occurs through integration policies between urban transformation governance and mobility governance in the existing city, with the aim of driving and controlling the effect of positive externalities on the urban areas.

In recent years, the research has focused its attention on sustainable neighbourhoods. The neighbourhood represents a slice of urban reality with a size large enough to appeal to sustainability criteria, which are transferable to the city. Furthermore, this level is also linked to quality of life; a sustainable neighbourhood's design causes communities to develop with environmental considerations and social and economic goals, in a balanced perspective.

The approach is conceptually holistic: it necessitates an integrated design that is not only linked to transport infrastructures, but also to the built heritage, to public spaces and to other urban aspects.

Urban planning has to work particularly to promote efficient models of land use, with the aim of linking sustainable mobility goals with territorial modelling and public space enjoyment goals.

In some European States, the relevance of the problem has led to attempts seeking to systematize the possible range of measures, to exchange and spread "best practices" lists, to test innovative solutions of urban mobility and city structure organization.

The work described in this contribution starts from the results of a European project PAES ("Housing Policy for Sustainable Construction", financed by "Erasmus+" Programme—in "KA2—Cooperation for Innovation and the Exchange of Good Practices Strategic Partnerships for higher education" Action). The partners of this project were: the Department of Civil Engineering of the University of Calabria (Italy), the Technical University of Iasi (Romania), the Technological Education Institute of Serres (Greece), the Technical University Kosice (Slovakia) and the Transylvania University of Brasov (Romania).

Particularly, the project has demonstrated the supremacy of northern EU Member States in sustainable design processes at the neighbourhood level. Some powerful examples are several new German and English neighbourhoods created for families[1]. Their organization model of urban mobility is based on a clear distinction between large urban express roads and neighbourhood streets, characterized by a local low-speed mobility and high standards of accessibility, distribution, security and practicability.

The project highlighted that these neighbourhoods, from the mobility point of view, are the result of a quality urban planning in which the enhancement of the potential of built-up areas is crucial.

The neighbourhoods analysed were all planned with a setting based on the strong presence of public spaces, which can be used, by a large number of citizens. They are characterized by low motor density caused by means alternative forms of transport to the car, from the relaunch of public transport and from the introduction of several services such as *car sharing*, *car pooling* and *bike sharing*.

In these experiences, urban planning has placed the importance of the city shape and the role of public spaces at centre stage. In fact, they show that a compact urban shape is a determining factor to promote sustainable mobility. It allows the concentration of services and functions, reducing movements that the citizens have to execute to meet their needs.

Furthermore, these experiences highlight the need to increase urban development near public transport services. This is achieved by encouraging mixed use zoning and integrating residential, commercial, services and recreational functions, which are accessible with public transport or on foot and by bicycle (for medium-short distances) in the same neighbourhood.

In Italy, these policies are asserting more slowly. Among previous experiences, it is possible to remember that of Bolzano. Here the integration between the Casanova neighbourhood and the city centre has been guaranteed through the realization of cycle-pedestrian paths and public transport lines; near to them, an internal road system is planned for a limited traffic flow, thanks to deliberately winding roads.

The enhancement of the public transport system through the realization of a new railway station has taken place simultaneously to these interventions; it is an important opportunity to activate a "metropolitan railway" project and a strong element of identification and a fundamental junction for mobility throughout the neighbourhood. Also in the Le Albere neighbourhood in Trento, the focus was the mending of the area with the existing urban fabric; for this purpose, the neighbourhood has been linked to the historic centre thanks to three subways. Furthermore, a functional mix (commercial activities, residences, service industry and recreational areas) was created to allow revitalization of an area that had been marginalized for years and ensure it a balanced and coherent development with the rest of the urban fabric. The quality of life in the neighbourhood has also increased thanks to the presence of bicycle and pedestrian paths, lanes, squares and green spaces.

Among the causes of poor implementation of sustainable mobility policies, there is generally a cultural delay on the issues of urban environment quality, pollution and security; problems linked to programming and financing of transport in urban areas are added to it.

1. For example Riesefeld neighbourhood in Freiburg, Germany, or Bedzed neighbourhood in London, England.

2 OBJECTIVE OF THE PAPER

Starting from these considerations, the research details the issue of small and medium-sized cities. The goal is to check how the best practices already experienced in neighbourhoods of wider urban realities can also be exported to contexts in which the neighbourhood level overlaps with the same urban dimension, to be included in a wider territorial system.

For a long time, these realities seemed to be exempt from the transport problems of big cities, being characterized by less pollution and congestion and by a better quality of public spaces and services to citizens. However, in recent decades, they have also been affected by the systematic use of private vehicles generally associated with the dispersion of urban activities and the vast territorial dispersion that have made the public transport very uncompetitive. In this sense, the small and medium-sized cities can be an effective scope for the quality of urban space policies based on enhancing the role of public transport and sustainable mobility networks[2]. In the small and medium centres, even more so than in consolidated urban areas, public transport has to become the armour on which to reorganize the widespread city[3].

A possible resolution of the problem is suggested by Community policies[4]. These highlight the importance of strengthening the territorial dimension, as well as adopting an integrated, coherent and global approach that involves all sectors, administrative levels and territories.

Therefore, in the case of small and medium territorial dimensions the research addresses the issue not only at the level of a single city, but also in a wider territorial context, in which the single city is considered in a "neighbourhood" dimension. The research is therefore geared towards new governance models aimed at a sustainable use of the territory that considers both the inner urban layout, which is no longer understood as a closed and self-referential system, and the wider territorial reality. Urban sustainability characteristics must be reflected, therefore, on a territorial perimeter that goes far beyond the administrative boundaries, creating territorial and performance ties with adjacent urban fabrics.

3 DESCRIPTION

The development of research is related to small and medium Calabrian urban realities (representing more than 80% of the regional total) characterized by difficult economic, social and cultural situations and extremely small dimensions (less than 5.000 inhabitants). In particular, the scope of the study coincides with a territorial district located in the upper Cosenza Tyrrhenian coast and consists of 13 municipalities[5]. The choice of the area was determined by the transformation dynamics of this territory that in recent decades have been strongly tied to the spread of residential settlements. This has caused serious impacts on the overall functionality of the residential areas, as well as an exponential increase in environmental negative externalities, mobility infrastructures and overall quality of the urbanised area.

The complex nature of this process requires the adoption of a perspective that takes account of all components and the definition of integrated policies coherent with the aim of sustainable mobility. In this regard, the research highlights the need for a thorough review of urban planning tools. In particular, in order to give greater effectiveness and sharing to strategic policies, it was necessary to promote a new communal urban planning season more linked to the design of Municipal Structural Plans drawn up in associate form.

These tools, already provided in the Calabrian Regional Urban Law[6], today have been used little and often badly. The experiences gained so far have distorted the meaning the legislator sought to give to this tool. It has been interpreted not as an opportunity to systematize

---

2. Vittadini, 2006.
3. Di Vito et al., 2008.
4. Leipzig Charter, Toledo Declaration and Territorial Agenda 2020.
5. Aieta, Laino Borgo, Laino Castello, Mormanno, Orsomarso, Papasidero, Praia a Mare, San Nicola Arcella, Santa Domenica Talao, Santa Maria del Cedro, Scalea, Tortora e Verbicaro.
6. Law n.19 of 16/04/2002 "Rules for land protection, government and land use".

the resources and energies of neighbouring municipalities, but as a means of attracting the financial incentives provided by the same law. The result is that these Plans form as a summation of municipal urban planning tools between which there is no form of integration and sharing.

It is therefore necessary to ensure that Municipalities are more motivated to plan jointly a unitary structure of territorial organization and, at the same time, to share sustainable scenarios of urban and territorial development. The goal must be to make truly collective and strategic choices—in relation to the location of mobility infrastructures, productive environments and supra-municipal public services and activities—to be made explicit within individual territories.

Only in this way is it possible to plan public and private investments more efficiently and to respond to functional and performance needs of networks and public services more easily.

By assigning a more cogent role to the associated Plan, the research strategy is based on a territorial organization inspired by a polycentric settlement model that allows integrated actions able to build relations and connections between the coastline, the hilly intermediate zone and the mountain interior area of the upper Cosenza Tyrrhenian coast. This polycentric model is based on a rationalization of activities, functions and a transport system on an inter-municipal scale.

In the model, the planning occurs at a supra-municipal level through an Associated Structural Plan. In particular, it is aimed at dividing the public services and the commercial activities between all the concerned municipalities, in addition to residential and productive assets. At the same time, it has the aim of the containment of building expansion—through the reuse and regeneration of the existing built heritage in dense urban areas, and the realization of new urban enlargements only around the public transport nodes—and at promotion of the integration and functional mix to limit the creation of mono-functional zones[7]. It is therefore necessary to reorganize the territory by assigning different functions to different parts, in relation to the specific infrastructural and environmental conditions. Each territorial element thus contributes to the formation of a single organism with a higher urban and environmental quality and with a consequential better quality of life[8].

At the base of the strategy, there is an organic distribution of activities and functions closely linked to the redesign of the public transport system. In this reorganization of mobility, the interactions between the local infrastructure systems and between the single urban centre and the territorial context are fundamental[9].

At the first level of this mobility scheme there is the public transport system on a supra-municipal scale; it represents the bearing structure of the sustainable mobility system at a territorial scale and it fulfils the task of ensuring higher order connections. Consistent with these needs, the priority goal is the rationalization and the systemization of the different types of public transport on the territory (national and local railways, extra urban bus lines), as well as the improvement of intermodality and the quality of service (frequencies, speeds, served centres, coincidences).

At the second level, there are the integrated systems of sustainable mobility within the single residential zone. They are aimed to ensure the protection of urban inland areas (in particular the many valuable historical dwellings) and the areas of ecological and tourism interest (mainly coastal and river bands). Low-impact forms of mobility are expected, which have been tested in the studied experiences, motivated by the strong presence of modal interchange infrastructures and the prediction of dissuasive measures. These include ecological vehicles for public transport and the distribution of goods, pedestrian walkways through a capillary network of routes linked to services and public transport networks, trails and bicycle paths in a separate roadway and connected directly to the attraction poles of the urban scene, "traffic calming" interventions that can aid the coexistence of pedestrian, bicycles and cars.

---

7. Marletto et al., 2015.
8. Delpiano, 2006.
9. Fratini, 2013.

4 CONCLUSIONS

There are many difficulties moving in this direction. The need to translate environmental sustainability goals into operational measures aimed at enhancing the quality of the territory must be reconciled with the goal of building and ensuring fairness and effectiveness in the vast territorial policies.

The strategic approach requires addressing organizational and institutional issues to ensure consistency between local and sectoral initiatives with an impact on the territory, as well as organicity in localization choices. Horizontal and vertical coordination is needed: Municipalities must cooperate with other administrative levels and consolidate collaboration with other cities to share the required investment and services on a wider territorial scale.

In order to meet these goals, "fixed" coordination mechanisms must be integrated with other "flexible" ones to ensure dialogue and cooperation between the territorial and administrative levels, as well as among sectors concerned by urban development.

In this process, it is crucial to overcome the tensions between different interests and to reach a compromise between conflicting goals and divergent development models. It is important to have a shared view, which, however, must take into account the differences between urban realities: the diversity of development processes, territorial dimensions, demographic and social framework, cultural and economic activities.

Therefore, a supportive behaviour and a tangible territorial cohesion are necessary in which the quality of spatial, programmatic and functional synergies plays a fundamental role, determined by both the adjacent urban realities and the territorial context.

The research reconciles these needs by resorting to the institute of "territorial equalization". In particular, the stipulation of perequotic agreements is expected between the Municipalities involved in the distribution of charges and benefits derived from the specific urbanistic localization of the interventions and activities.

The breadth of the territorial and administrative dimension increases the political, and technical, complexity of the operation, and the difficulty to reach agreements[10].

However, this step is necessary above all to minimize competition compared to the need to make similar urbanistic choices to attract settlements on their territory by each Municipality. Ultimately, Municipalities are not competing with each other, but they compete in the achievement of the same goals.

REFERENCES

Cecchini D., Castelli G. (a cura di) (2016), Scenari, risorse, metodi e realizzazioni per città sostenibili, Gangemi Editore.

Delpiano A. (2006), "Una città di comuni", in Environnement. Ambiente e Territorio in Valle d'Aosta, n.33.

Di Vito G., Follesa F., Murmura L. (2008), "Un programma di area vasta per i piani locali", Atti convegno SIU.

D'Onofrio R., Talia M. (2015), La rigenerazione urbana alla prova, Franco Angeli Editore.

Fratini F. (2013), "I quartieri sostenibili di Friburgo", in Urbanistica Informazioni n. 248.

Gerundo R., Fasolino I. (2006), "Procedure perequative in ambito intercomunale per l'attuazione degli strumenti urbanistici previgenti", in Francini M. (a cura di), Modelli di sviluppo delle aree costiere e rurali a elevata strutturazione storica, Centro Editoriale e Librario Unical.

INU Emilia-Romagna (2008), La perequazione territoriale nella disciplina urbanistica, Provincia di Bologna.

Marletto G., Trepiedi L. (2015), "Vocazioni territoriali e politiche per la mobilità sostenibile. Il caso di Terni", in EyesReg, Vol. 5, n. 3.

Vittadini M.R. (2006), "Problemi di trasporto nelle città medio-piccole", in Environnement. Ambiente e Territorio in Valle d'Aosta, n. 33.

10. Gerundo et al., 2005.

Town and Infrastructure Planning for Safety and Urban Quality – Pezzagno & Tira (Eds)
© 2018 Taylor & Francis Group, London, ISBN 978-0-8153-8731-2

Cities alive: Towards a walking world

C. Fraticelli
Arup, Italy

ABSTRACT: The paper describes the research developed by Arup, “Cities Alive. Towards a walking world”.

This explores the walkability concept which goes beyond the good design of sidewalks and street-crossings which guarantee the “ability to walk” for citizens. It promotes more liveable, healthy streets for all and emphasises the idea of a multisensory, interactive and social urban experience. The negative effects of heavy automative use are significant, therefore intrinsic to the success of cities and the quality of life they offer is how people move around within them.

The reasearch intends to demonstrate the role that walking could have in reversing cars’ negative effects and developing more liveable, sustainable, healthy, safe and attractive cities. Thus, 80 international case studies are invastigated.

Learning from international best practices, the research provides a useful guide in order to inspire politicians, designers, planners, engineers, consultants and technical specialists to plan long-term strategies and promote innovation.

1 INTRODUCTION

With nearly 70% of the world’s population set to live in urban areas by 2030, the quality of life experienced by the urban population will determine our global future.

It is increasingly cities, in fact, more than national governments, that have the power to tackle climate change, promote the global economy and deliver prosperity[1]. Moreover, with the advent of social media and the growth of the global middle class, urban citizens have nowadays greater power than ever to stop or start projects, choose where they live, work and invest and demand a better quality of life.

And it’s not by chance that at a global scale, citizens are now reclaiming their street as public space. Intrinsic to the success of cities and the quality of life they offer is how people move around within them and experience their places.

Undoubtedly, the 20th century was the century of cars. They colonised the space of every-day human life and the legacies of the traffic-dominated planning era are still clearly visible in cities worldwide. The main issue, according to what Lewis Mumford wrote half a century ago, is that “the right to have access to every building in the city by private motorcar, in an age where everyone possess such vehicles, is actually the right to destroy the city”.[2]

Current challenges, such as climate change and urban inequalities, are finally casting a light on the necessity to plan a sustainable urban development and put walking at the centre of the urban discourse.

The concept was explored firstly in the 60s by a critical movement led by Lewis Mumford, William H. Whyte and Jan Gehl, which began questioning car dominance, driven by the concern for the decline in the human-focused approach to urban projects. However, the rising awareness of the key role of walking in the urban debate and the decline of car culture, at least in the Western world, are relatively recent phenomena.

1. 2014 Revision of the World Urbanization Prospects, by United Nations. 2014. Available from: http://esa.un.org/unpd/wup/Publications/Files/WUP2014-Report.pdf.
2. Lewis Mumford in J. Speck, “Walkable City”, 2012, pag.78.

Today, studies indicate that in North America, Japan, Australia and European countries we may have reached 'peak car' – the apex at which car ownership, licence ownership and the distance driven per vehicle level off, and then turn down.[3] Cities around the world are beginning to realise that by getting more people on foot, in tandem with reducing the number of cars, they will have healthier, happier citizens and thriving streets and public spaces.

The report makes the case for policies that encourage walking to be placed at the heart of all decisions about the built environment and to design physical activity back into our everyday lives by incentivising and facilitating walking as a regular daily mode of transport. Walking, after all, is the lowest carbon, least polluting, cheapest and most reliable form of transport, and is also a great social leveller. Having people walking through urban spaces makes the spaces safer for others and, best of all, it makes people happy. As argued by Enrique Peñalosa, Mayor of Bogotá, *"As a fish needs to swim, a bird to fly, a deer to run, we need to walk, not in order to survive, but to be happy"*[4]

The aim of this research was to analyse and communicate trends, benchmarks, data and benefits of shaping walkable cities, in order to inspire politicians and professionals.

The research process was focused on providing a shared common ground, based on the definition of a comprehensive Framework (Fig. 1) consisting of a broad range of 50 Drivers, 50 Benefits, 40 Actions and 80 Case studies, able to recognise (using qualitative and quantitative references) the creation of walkable environments as a fundamental tool for developing more liveable, sustainable, healthy, safe and attractive cities.

In order to reach the goal, the research was based on a multi-disciplinary approach, involving also relevant external contributors and leaders in the field: Gehl Architects (Europe), Project for Public Space (Americas), Hong Kong University (East Asia), and the City of Auckland (Australasia). Learning from international best practices, the research aims at providing a useful guide in order to plan long-term strategies and promote innovation.

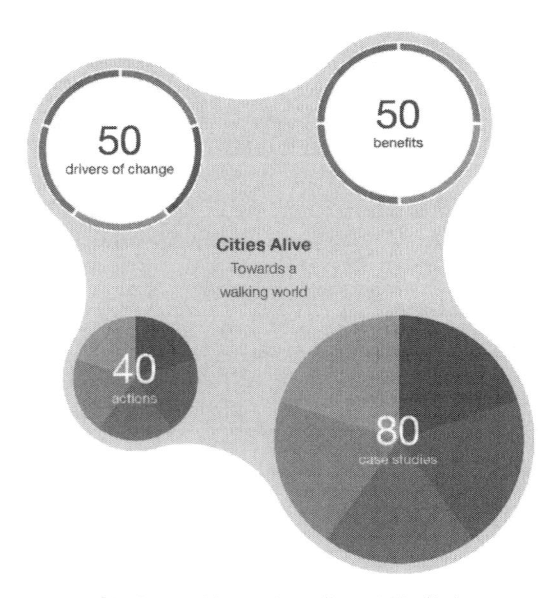

Figure 1. Towards a walking world framework.

3. Are we reaching "peak travel"? Trends in passenger transport in industrialized countries, by Adam Millard-Ball and Lee Schipper. 2010. Available from: http://www.tandfonline.com/doi/abs/10.1080/01441647.2010.518291.

4. Enrique Peñalosa in J. Speck, "Walkable City", 2012.

2 DRIVERS

Moving towards a walking world requires, first of all, a cultural shift. Several cities have started to take steps on this direction guided by a set of social, technological, economic, environmental and political trends which are rapidly and intensely changing our society: from the development of new demographic patterns, characterised by a progressively ageing population and families shrinking, which are fundamentally transforming the way in which people want to live and interact in cities to the advent of new forms of autonomous mobility, which are on the rise and in the near future may radically change the way we commute and bring to a decrease in road infrastructure demand. All those drivers are signals of future cities' changes:

- **Social** trends: Car ownership, Demographic patterns, Digital lifestyles, Emotional cityness, Fortress cities, Public health, Sustainable behaviours, Urban inequality, Urban spontaneity, Urban stress;
- **Technological** trends: Autonomous vehicles, Availability of sensors, Big data, Communication and sharing, Constant connectivity, Digital Information, Free public Wi-Fi, Gamified incentives, Interactive street furniture, Quantified-self;
- **Economic** trends: Congestion costs, Digital Economy, Genius hub, Glocalism, Health costs, Recession, Sharing economy, Tourism, Unemployment, Urban regeneration;
- **Environmental** trends: Active transportation, Air pollution, Climate change, Decarbonisation, Energy consumption, Green infrastructure, Heat Island, Land use patterns, loss of biodiversity, transport safety; and
- **Political** trends: City competitiveness, Collective consciousness, Green politics, Leadership, Micro-solutions, Policy integration, Privitisation, Public space, Stakeholder engagement, Urban resilience.

3 BENEFITS

Considering the potential benefits, investing in walkable public spaces should be a no-brainer. With a broad range of environmental, social and economic benefits—such as reducing air pollution and traffic congestion, fostering social cohesion and addressing physical and mental health problems, attracting private investments, bringing vibrancy to local shops and providing opportunities for communities—the adaptation of public space to encourage walking has become a growing urban development trend. A rising consciousness around the fundamental role of public space, in fact, is leading cities to update out-dated regulations based around cars and parking in favour of a more holistic view of mobility and access.

Therefore, through a literature review and a series of multidisciplinary workshops held in different geographic areas by an Arup global specialised team, the research highlighted a list of 50 benefits of walking that should be achievable in most contexts and demonstrable by quantitive and qualitative measurements. The findings rely on the definition of areas of benefits, with a series of secondary indicators, identified as follows:

- **Social** benefits such as health and wellbeing, safety, placemaking, social cohesion and equality;
- **Economic** benefits including city attractiveness, the local economy, urban regeneration, and cost savings;
- **Environmental** benefits to do with virtuous cycles, ecosystem services, liveability and transport efficiency; and
- **Political** benefits associated with leadership, urban governance, sustainable development and planning opportunities.

To achieve the shift from car-centric to human-scale cities, the development of an evidence-based methodology is a fundamental step to establish a shared global recognition. Evidence, in fact, can be one of the best way to influence decision-makers.

For instance, it is globally recognized that social interaction between people is a fundamental feature of a thriving urban experience. Promoting walkable environments intensifies the use of public space, raising the frequency and quality of informal social interactions between people. That happens because walkability can dramatically boost opportunities for people meeting, sharing and mixing, planting the basis for the creation of stronger communities.[5] As argued by William H. Whyte: "What attracts people most, it would appear, is other people"[6]. That is demonstrated, for example, by Donald Appleyard, Professor of Urban Design at the University of California, which conducted, during the late 1960s, a comparison of three streets identical in every dimension except for levels of traffic ('light', 'moderate' and 'heavy').[7] He wanted to show that the mere presence of cars can crush the quality of social life in neighbourhoods. His research demonstrated that residents of streets with light car traffic volumes had three times as many friends and twice as many acquaintances as those living on a street with heavy car traffic.[8] Moreover, looking at an Irish study, it was discovered that residents living in walkable neighbourhoods, exhibit at least 80% greater levels of social capital than those living in car-dependent ones and are more likely to know—and therefore to trust—their neighbours and get involved in local issues and politics[9].

"Strenghetening community identity" and "Fostering social interaction" are just examples of the varied set of benefits identified in the report. The fifth mayor re-election of Miguel Anxo Fernandez Lopez in the Spanish city of Pontevedra, after the decision to pedestrianize the whole city in 1999, represents, for instance, the demonstration of how walking can also build and raise public consensus[10]. Or even, the 300% increase in employment, after the redevelopment of Temple Bar District in Dublin[11] or the 335% increase in Barcelona's annual visitors in the last 20 years[12], are clear evidences of how boosting soft mobility and promoting a walkable public space policy, can "Boost prosperity", "Enhance city branding and identity" and "Promote tourism". Studying the New York Highline case study, then, can highlight further economic incentives. The creation of the iconic linear pedestrian park, funded with only $115 m of public investments, has generated, in fact, over $2bn in private investment surrounding the park, attracting 5 million visitors a year, creating 12,000 new jobs and doubling the property value in the neighbourhood[12]. And there are enormous environmental benefits. Even a single day without traffic can bring huge benefits. In September 2015, for instance, the Paris' Journée Sans Voiture cut levels of nitrogen dioxide by 40% in parts of the city.[13]

4 ACTIONS

How to start? Moving towards a walking world requires actions. Giving consistency to the large spectrum of benefits identified needs the definition of a holistic walkable strategy (Fig. 2).

5. Walking towards walking stronger communities, by Chiara Fraticelli. 2016. Available from: http://thoughts.arup.com/post/details/577/walking-towards-stronger-communities.
6. Whyte W. H. (1980). The Social Life of Small Urban Spaces. Washington, D.C. The Conservation Foundation.
7. Revisiting Donal Appleyard's Livable Streets, by Streetfilms. 2011. Available from: https://vimeo.com/16399180.
8. Appleyard D., Gerson M. S., Lintell M. (1981). Livable Streets. University of California Press. 364 pages.
9. Social Capital and the Built Environment: The Importance of Walkable Neighborhoods, by Kevin M. Leyden. 2003. AJPH, vol. 93, no. 9, pp. 1546–1551. Available from http://www.jtc.sala.ubc.ca/reports/leyden.pdf.
10. Pontevedra, come si vive in una città senza auto (e senza smog)?, by Michele Cocchiarella. 2016. WIRED. Available from: http://www.wired.it/attualita/ambiente/2016/01/18/pontevedra-smog-linquinamento-banditi-15-anni-rischio-misure-immediate/?utm_source=facebook.com&utm_medium=marketing&utm_campaign=wired.
11. The pedestrian pound. The business case for better streets and places, by Living Streets. 2014. Available from: http://www.livingstreets.org.uk/media/1391/pedestrianpound_fullreport_web.pdf.
12. Walkonomics: the High Line effect, by Demetrio Scopelliti. 2015. Available from: http://thoughts.arup.com/post/details/429/walkonomics-the-high-line-effect.
13. Paris car ban cut harmful exhaust emissions by up to 40 per cent, by Caroline Mortimer. 2015. The Indipendent. Available from: http://www.independent.co.uk/environment/paris-cuts-harmful-no2-exhaust-emissions-byup-to-40-per-cent-after-banning-cars-for-a-day-a6679686.html.

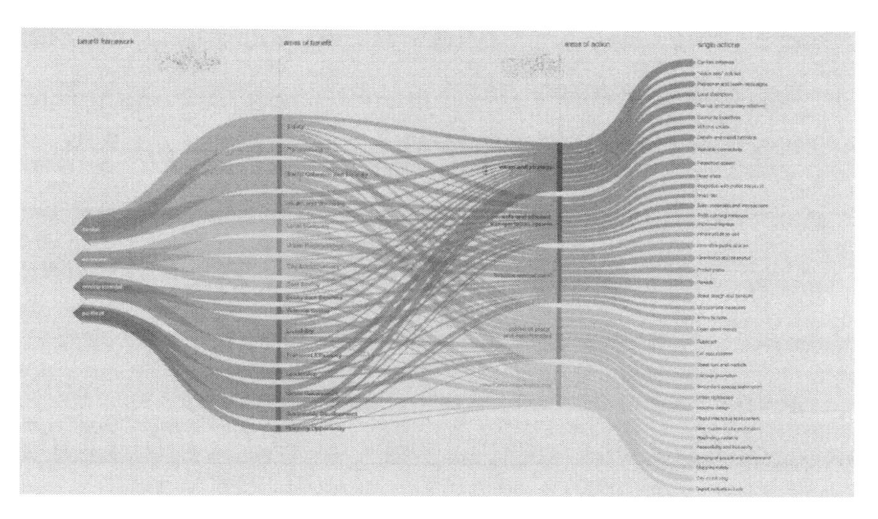

Figure 2. From benefits to actions: the diagrams summarises the interrelation between walkability's benefits and suggests implementable actions.

Figure 3. The walkable city shown in the illustration gives a lively overview of the actions proposed and the way they could considerably change our cities.

The Walkable City (Fig. 3), in fact, is a city that puts people first and shapes itself in accordance to its citizens' needs and desires. Therefore, to address the complexity of the urban issues, a kaleidoscopic set of actions and policies is required, diversified both by nature and by dimension. These 40 actions concern:

- *visions and strategies*: series of urban policies, involving city plan policies and innovative interventions, promoting a diffused walkable approach to the city;
- *safe and efficient transportation systems*: interventions that operate on a city's infrastructure providing an improved street network;
- *creating liveable environments*: set of actions that affects the urban quality, re-designing public space on the basis of pedestrians' priorities;
- *sense of place and community*: strategies and proposals to encourage the active and emotional participation of citizen in everyday urban life; and
- *smart and responsive cities*: technological tools and innovative approaches contributing to a city's monitoring and evaluation.

This varied set of actions provides a useful guide for decision-makers in order to plan long-term strategies and envision new frontiers of development.

Table 1. 40 actions.

Visions & strategies	Safe and efficient transportation systems	Creating liveable environments	Sense of place and community	Smart and responsive cities
car free initiatives	walkable connectivity	infrastructure re-use	open street events	playful interactive environment
vision zero policies	pedestrian streets	innovative public spaces	public art	new modes of city exploration
pedestrian & health campaigns	road share	street design & furniture	DIY opportunities	wayfinding systems
local champions	integration with public transport	pocket parks	street fairs & markets	accessibility & inclusivity
pop-up & temporary initiatives	road diet	parklets	heritage promotion	sensing of people & environment
economic incentives	safe crosswalks & intersections	greenways & blueways	redundant spaces reallocation	Mapping safety
virtuous cylces	traffic calming measures	microclimate measure	improved urban nightscape	city monitoring
density & mixed functions	improving signage	active facades	inclusive deisgn	digital evalution tools

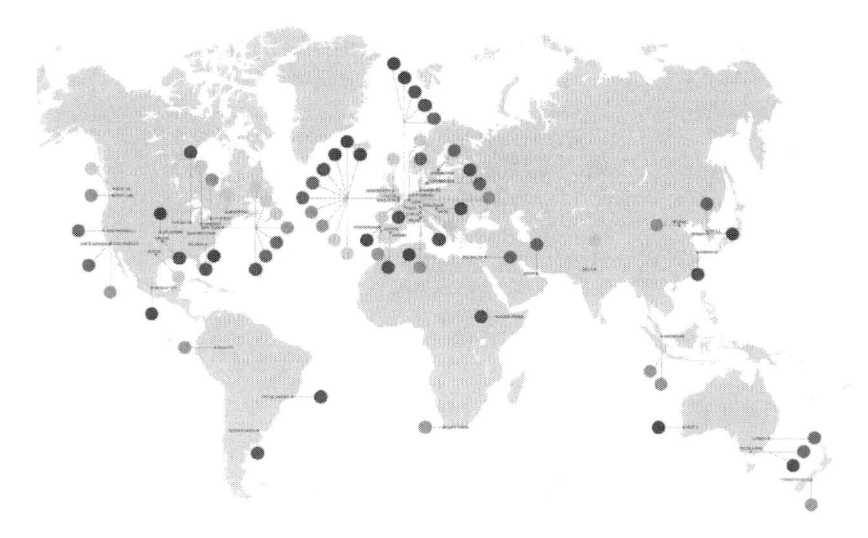

Figure 4. Case studies map.

The 5 areas of action have been articulated in order to provide a comprehensive guide of action.

5 CASE STUDIES

Finally, a wide collection of emerging ideas and case studies represents a fundamental tool to capture the state of the art of walkable cities and aim to inspire planners, designers, decision makers to provide outstanding solution towards a walking world.

In order to demonstrate the practicality of the actions highlighted, a wide database of examples has been gathered by a global team of Arup experts and summarised in a series of 80 case studies from across the world (Fig. 4).

In particular, referring again to the capacity of walking in strenghtening communities, as demonstrated by the proliferation of 'do-it-yourself' crowd funded initiatives, across the world community-led projects are on the rise to maintain or improve aspects of daily life. That's what happened, for instance, in Ghent, when a group of residents suggested the idea of building a network of temporary car-free zones. In May 2015, 22 of Ghent's busiest streets were closed to traffic and converted into Living Streets for ten weeks, featuring pop-up parks and bars, helping local people to play, relax and socialise[14]. The project achieved a huge popular success.

This case study is part of a varied catalogue of best practices that depicts a wide global scenario and emphasizes international champions. Among those: a masked superhero defending the pedestrian rights through the streets' of Mexico city[15]; the redevelopment of Mariahilfer Strasse[16], in Wien, into a pedestrian-friendly shared space with a central 'car free' zone and specifically-designed paved surfaces, street side furniture and trees. The creation of parks and public spaces built over a recessed section of freeway as in the case of "Klyde Warren Park" in Chicago[17]. The infrastructure re-use of a disused rail corridor in Sydney, with the creation of a new green urban corridor, "The Goods Line", a landscaped pedestrian and cycle route that connects existing cultural, educational entertainment areas via 500 m of new public space[18]. The use of hundreds of fibres for an interactive light and sound installation in order to encourage public interaction and the exploration of urban space for DUNE project, developed by Studio Roosegaarde[19]. The creation of a web and app-based neighbourhood assessment tool that allows user to view any location's "walkable score", based on average distance to a user-defined set of amenities[20].

Those actions are just the beginning of an urban walking revolution.

This report wants to shine a light on walking and demonstrate through evidence and case studies that a walkable city is a better city for everyone and putting walking first will keep our Cities Alive.

Some cities have started to create a vision and strategy for walking, recognising it as a transport mode in its own right, as well as an important part of almost all trips, whether by car, bus, train or bicycle. However, more walking champions are needed to help make that change. As argued by Jan Gehl, *"We are realising that if you have people walk and bicycle more, you have a more lively, more liveable, more attractive, more safe, more sustainable and more healthy city."* So, *"[...] what are you waiting for?"*[21]

Link to the research:

http://publications.arup.com/publications/c/cities_alive_towards_a_walking_world

REFERENCES

Speck J. (2013) Walkable City: How Downtown Can Save America, One Step at a Time. North Point Press. p. 320.

Mumford L. (1981). The Highway and the City. Praeger. p. 246.

Gehl, J. (1987/2011). Life between buildings; Using public space. London:Island Press. p. 200.

Whyte W. H. (1980). The Social Life of Small Urban Spaces. Washington, D.C. The Conservation Foundation. p. 125.

Jacobs J. (1961). The Death and Life of Great American Cities. New York: Random House. p. 472.

14. Belgian Streets Got Rid Of Cars And Turned Into Beautiful Parks This Summer, by A. Peters. 2015. Co. Exist. Available from: http://www.fastcoexist.com/3050299/belgian-streets-got-rid-of-cars-and-turned-into-beautiful-parks-this-summer.
15. Mexico's masked hero making streets safe – for pedestrians, by AFP. 2015 Available from: http://tribune.com.pk/story/926286/mexicos-masked-hero-making-streets-safe-for-pedestrians/.
16. Available from: http://www.mariahilferstrasse.at/.
17. Available from: https://www.klydewarrenpark.org/.
18. Available from: http://aspect.net.au/?p=384
19. Available from: https://www.studioroosegaarde.net/project/dune/info/.
20. Available from: https://www.walkscore.com/.
21. Jan Gehl in London: no city for cyclists, by Peter Walker. 2013. Available from: https://www.theguardian.com/environment/bike-blog/2013/dec/03/london-cycling-provisions-laughable-bike-blog.

Appleyard D., Gerson M. S., Lintell M. (1981). Livable Streets. University of California Press: Berkeley, USA. p. 243.
Leyden K. M. (2003). Social Capital and the Built Environment: The Importance of Walkable Neighborhoods. AJPH, vol. 93, no. 9, pp. 1546–1551.

SITOGRAPHY

2014 Revision of the World Urbanization Prospects, by United Nations. 2014. Available from: https://esa.un.org/unpd/wup/publications/files/wup2014-highlights.Pdf.
Are we reaching "peak travel"? Trends in passenger transport in industrialized countries, by Adam Millard-Ball and Lee Schipper. 2010. Available from: http://www.tandfonline.com/doi/abs/10.1080/01441647.2010.518291.
The walking city, by Demetrio Scopelliti. 2015. Available from: http://doggerel.arup.com/research-roundup-the-walking-city/.
Active transportation and Real Estate. The next frontier, by the Urban Land Institute. 2016. Available from: http://uli.org/wp-content/uploads/ULI-Documents/Active-Transportation-and-Real-Estate-The-Next-Frontier.pdf.
As traffic deaths soar, cities pursue lower speed limits to eliminate fatalities, by Luz Lazo. 2017. Available from: https://www.washingtonpost.com/local/trafficandcommuting/as-traffic-deaths-soar-cities-pursue-lower-speed-limits-to-eliminate-fatalities/2017/02/25/6f86e614-f216–11e6-a9b0-ecee7ce475fc_story.html?utm_term=.5968efb37a26.
Walking towards walking stronger communities, by Chiara Fraticelli. 2016. Available from: http://thoughts.arup.com/post/details/577/walking-towards-stronger-communities.
Revisiting Donal Appleyard's Livable Streets, by Streetfilms. 2011. Available from: https://vimeo.com/16399180.
Pontevedra, come si vive in una città senza auto (e senza smog)?, by Michele Cocchiarella. 2016. WIRED. Available from: http://www.wired.it/attualita/ambiente/2016/01/18/pontevedra-smog-linquinamento-banditi-15-anni-rischio-misure-immediate/?utm_source=facebook.com&utm_medium=marketing&utm_campaign=wired.
The pedestrian pound. The business case for better streets and places, by Living Streets. 2014. Available from: http://www.livingstreets.org.uk/media/1391/pedestrianpound_fullreport_web.pdf.
Belgian Streets Got Rid Of Cars And Turned Into Beautiful Parks This Summer, by A. Peters. 2015. Co.Exist. Available from: http://www.fastcoexist.com/3050299/belgian-streets-got-rid-of-cars-and-turned-into-beautiful-parks-this-summer.
Mexico's masked hero making streets safe—for pedestrians, by AFP. 2015. Available from: http://tribune.com.pk/story/926286/mexicos-masked-hero-making-streets-safe-for-pedestrians/.
Jan Gehl in London: no city for cyclists, by Peter Walker. 2013. Available from: https://www.theguardian.com/environment/bike-blog/2013/dec/03/london-cy.

Town and Infrastructure Planning for Safety and Urban Quality – Pezzagno & Tira (Eds)
 ISBN 978-0-8153-8731-2

Re-shaping a post-seismic re-construction district through cycling infrastructures. The case of Monterusciello

R. Gerundo
Università degli Studi di Salerno, Fisciano, Italy

C. Gerundo
Università degli Studi di Napoli "Federico II", Naples, Italy

ABSTRACT: Among non-motorised transport mode, which are often considered as vital elements of sustainable transport systems, the bicycle is substantially faster than walking and is more flexible than public transport. That's the reason why the number of policy initiatives to promote the creation of bike paths within urban areas has grown considerably over the past two decades, as part of the search for more sustainable transport solutions, as well as for recreational purposes. The paper deals with the efforts put in place to enhance urban quality and lifestyle in Monterusciello, a suburban district of Pozzuoli, a medium-size town near Naples (Italy), highlighting the role of urban projects aimed at increasing cycling mobility.

1 INTRODUCTION

Non-motorised transport modes are often considered as vital elements of sustainable transport systems. Their emissions of pollutants and noise, and the accident risks they pose for other road users are very low. Thus, a high share of non-motorised transport modes would certainly contribute to a more attractive urban environment (Rietveld et al., 2004).

Among non-motorised transport mode, bicycle is substantially faster than walking and, in addition, is more flexible than public transport due to its 'continuous' character eliminating waiting and scheduling costs.

Nevertheless, cycling attractiveness is strongly affected by the quality of infrastructure and traffic safety, as well as physical conditions, such as weather and surface slope.

For these reasons, the number of policy initiatives to promote the creation of bike paths within urban areas has grown considerably over the past two decades as part of the search for more sustainable transport solutions, as well as for recreational purposes.

Policies to promote cycling vary from country to country due to the wide variation of geographic, climatic, cultural, and economic contexts among countries, likewise the level of national involvement in cycling policy is highly variable.

Although the commitment of the national level can have a significant impact on cycling implementation, measures to promote bicycle use are most-efficiently designed, overseen and implemented by local authorities (ECMT, 2004). Municipalities can adopt push and pull policies in order to promote cycling, by making competing modes more expensive and improving its attractiveness. In the first case, active price-setting policies of car parking cost can be pursued, or bike-sharing plan can be implemented. In the second one, measures should be put in practise to reduce travel time and accident risks. These targets should be easily achievable by reducing the number of stops or hindrances and improving safety through bike lanes or cycle tracks creation.

Municipal provision of bike lanes and bike-hire schemes in cities like Paris, Stockholm, London, Milan or Brussels are proven to encourage urban residents to adopt sustainable modes of transport (URBACT II, 2015). In Italy, north-south gap is evident in cycling performance too: in southern cities, the provision of bike lane is two times lower than the Italian average (Legambiente, 2015).

Cycling infrastructure is essential to convert people to the merits of bike as an everyday means of transport. But a high level of cycling, as occurs, for instance, in the Netherlands, can be obtained not only through a comprehensive cycling infrastructure network (including bike lanes and separated paths), but also with some measures focused on bicycle parking facilities, egress trips for train services and access and egress trips for slower modes of public transport (Martens, 2007; Pucher et al. 2008).

But cycling infrastructures are not the only prerequisite to ensure a high percentage of trips performed by bicycle, since urban landscapes can generally affect cycling too. For instance, many scholars has pointed out that multi-polar and mixed-use urban structure with high residential and employment density helps to reduce the amount of home-work traffic in and out of the city centre and makes much of the major destinations reachable by cycling (Meng et al., 2014). Well-connected streets, small city blocks, and close proximity to retail activities are other aspects that were shown to induce non-motorized transport, as well as various exogenous factors, such as topography, darkness, and rainfall, which have far stronger influences (Cervero et al., 2003).

It is quite obvious that cycling improvement in historical european urban areas should chiefly be pursued through adequate policies to push vehicles away from roads, while bike-friendly project and planning criteria can be more easily be applied to new settlements.

With respect to existing modern urban tissues, having a consolidated structure, relevant grid modification are technically and financially hard to put in practice. Likewise, a rearrangement of urban functions is barely achievable since empty spaces within this areas should be preserved from new residential and/or commercial building construction, in order to prevent soil sealing and provide facilities and leisure spaces as parks, green areas, gardens. In this case, the most efficient measures regard the creation of bike lanes and cycle tracks, along the edge of the existing carriageways. Although this works are generally low-cost, nowadays municipalities hardly allocate resources to them, since they prefer to invest the few public funds they can manage into works to enhance motorised transport modes, such as road pavement repair or new paths creation.

In this context, is essential for public administration to take the funding opportunities available at European Union level, in order to finance radical intervention on urban mobility and built a bicycle friendly community.

2 TARGET OF THE PAPER

The paper deals with the efforts put in place to enhance urban quality and lifestyle in Monterusciello, a highly populated suburban district of Pozzuoli, a medium-size town near Naples (Italy), highlighting the role of urban projects, financed by european programmes and public funds, aimed at re-shaping its urban features and people's attitude to non-motorised transport modes, through new cycling infrastructures.

A description of Monterusciello will be provided and the sequence of events that led to its urban layout will be explained in detail, bringing out reasons and effects of the bicycle & pedestrian-unfriendly project criteria used when the district was planned and built. Moreover, a detailed review of all the projects and ideas devised over the time concerning cycling mobility enhancement in Monterusciello will be produced.

3 THE NEW TOWN OF MONTERUSCIELLO AMONG URBAN AND SOCIAL DECAY AND REGENERATION POLICIES

Monterusciello is the youngest public new town in Europe. It was built after the intense bradiseismic phenomenon occurred in Pozzuoli in 1983 that led to the rising of the soil level up to 180 centimetres.

The peak of the bradiseismic crisis in Pozzuoli happened in October 1983 when almost the whole populated area, including the historical city centre, its surroundings and the waterfront, was evacuated and about 30.000 people remained homeless. As a consequence, the Ministry of Civil Protection decreed urgent provisions to let the Municipality of Pozzuoli approve a Public Housing Extraordinary Plan (PHEP), in order to build about 4.000 apartments for the evacuees in the rural and non-urbanized area of Monterusciello. The Plan was developed by the University of Naples at the end of 1983 and all the dwellings were built between 1984 and 1986. The new district should have been built with a regular grid layout, wide roads, large green spaces and various urban facilities (schools, theatre, market, churches, sports centres etc.) (Fig. 1).

Within the general framework of public housing plans, Monterusciello can be considered the most relevant housing development built far from the historical city core in a non-urbanized area, as a new and functionally independent settlement.

All the buildings, urbanization works and amenities planned by the PHEP of Monterusciello should have been entirely built by public authorities but, although almost all the rural lands had already been expropriated, most of the urban facilities were not created, leaving about 50 hectares of rural empty spaces interspersed within the district layout.

Moreover, the district was served by two roadway junctions and a train station. It contributed to enhance the accessibility of Monterusciello, together with the wide road and regular grid design. Regretfully, relevant distances between functions within the district and gentle slopes in some part of the area have always made walking difficult as a mode of transport. This criticality concerning travel within the district was not even overcome through cycling mobility because high speed of vehicles, given by the great streets width, and the lacking of bike lanes, have always exposed cyclists to serious risks. In addition, the great dependency of dwellers on private vehicles in order to reach the ancient town, where social activities were and still are chiefly carried out, have fostered a sense of estrangement that hindered the integration of the community with the new town.

Nowadays Monterusciello district hosts about 20,000 residents with low-income, a high level of unemployment, and is characterised by large unused spaces. Problematic social conditions are combined with a difficult urban environment, in particular for the character of emptiness, anonymity, and decay of the common spaces.

For these reasons, the local government have been struggling in the recent years to concentrate public funds and project ideas on the issue of Monterusciello urban regeneration.

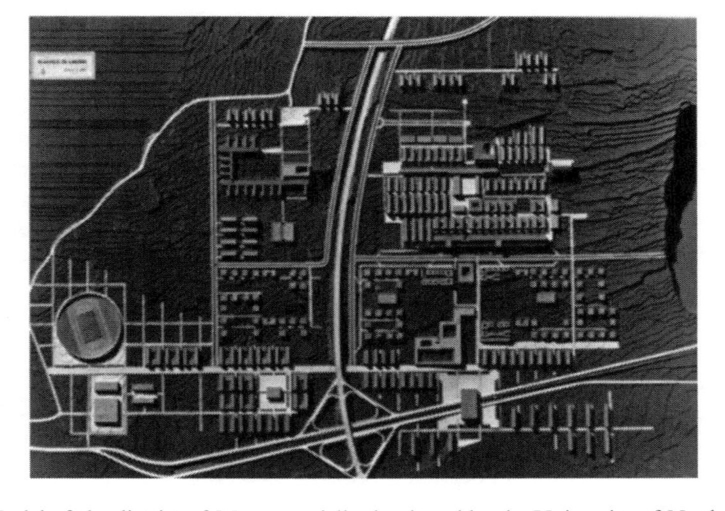

Figure 1. Model of the district of Monterusciello developed by the University of Naples.

4 BIKE-DRIVEN URBAN REGENERATION PROCESSES IN MONTERUSCIELLO

4.1 *Twenty years attempts of cycling mobility promotion*

Starting from few years later the creation of Monterusciello, the local government of Pozzuoli have been trying to adopt measure to increase cycling mobility within the district. At the beginning of the nineties, an initial proposal of bike lane was designed, even if only in the recent times it has been improved, formally adopted and send to the Metropolitan City of Naples in order to achieve regional funds. This project concerns the creation of 4 kilometres of cycle track at the side of the most important urban roads with the aim to connect the main urban facilities, such as schools, public offices, sport centre, and the train station (Fig. 2).

Moreover, the Municipality of Pozzuoli has recently approved a memorandum of understanding with the Metropolitan City of Naples and four other Municipalities, in order to gain government loans to buy 168 bikes, half of which power-assisted, and to build 28 bike stations, some of which in Monterusciello. This measure tends to lay the foundations for the promotion of cycling as an alternative to public and private motorized mobility for morning commutes.

4.2 *The Monterusciello Agro City project*

In April 2016 the Municipality of Pozzuoli submit a project to the first call for Urban Innovative Actions (UIA), an initiative managed by the European Commission and financed by the European Regional Development Fund (ERDF) program. The project, called Monterusciello Agro City (MAC), was selected, together with other 18 proposals, among almost 400 submissions.

MAC project seeks to reduce urban poverty in the district—where poverty is understood in the social and economic sense and within the physical environment—triggering a process of economic, entrepreneurial, and social development, together with the improving of the urban environment.

Thirty hectares of public abandoned non-urbanized open areas, corresponding with the ones where facilities should have been built according to the PHEP, will be transformed into farmland, developed with the innovative techniques of permaculture in order to spearhead an economic process and urban growth as a means to combat poverty (Fig. 3).

Permaculture is a method, developed in the seventies by Bill Mollison and David Holmgren, to plan and manage humanized landscapes (Mollison, 1991). The word permaculture originally referred to "permanent agriculture" but was expanded to stand also for "permanent culture", as it was understood that not only agricultural aspects were integral to a truly sustainable system.

Figure 2. Map of the cycle track project and visual simulation of the intervention.

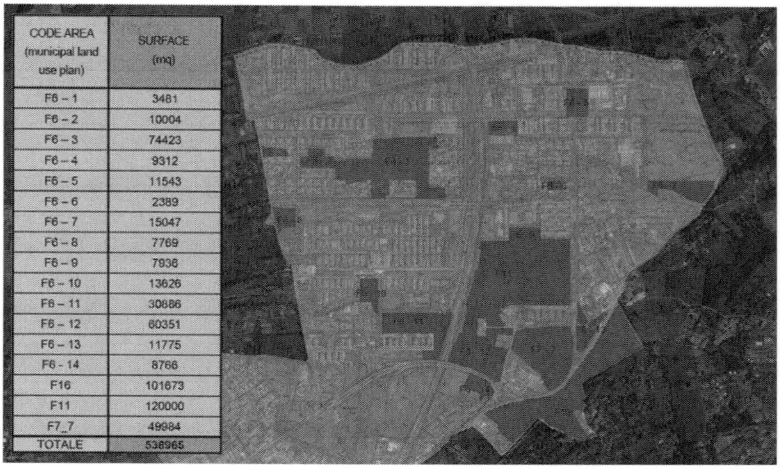

CODE AREA (municipal land use plan)	SURFACE (mq)
F6 – 1	3481
F6 – 2	10004
F6 – 3	74423
F6 – 4	9312
F6 – 5	11543
F6 – 6	2389
F6 – 7	15047
F6 – 8	7769
F6 – 9	7936
F6 – 10	13626
F6 – 11	30886
F6 – 12	60351
F6 – 13	11775
F6 - 14	8766
F16	101673
F11	120000
F7_7	49984
TOTALE	538965

Figure 3. Areas within the district of Monterusciello involved in MAC project.

Permaculture can be defined as an integrated system of landscape planning principles, based on the application of features and patterns observed in natural ecosystems, able to let the landscape satisfy community basic needs, such as food, fibres and energy, and, at the same time, be resilient, rich and permanent like natural ecosystems are.

The final target of MAC project, which will be financed to the tune of 5 million euros[1], is to create through permaculture principles a landscape made of urban areas and agriculture land connected by a productive thread. A circular system will be developed through the implementation of organic agriculture within the borough, creating working opportunities and training, enhancing the networking between people and the local business, in harmony with the territory tradition and driving forward new contemporary innovative systems. These actions will have a positive impact on the hydrogeological stabilization, preservation and increase of soil life, preservation of the biodiversity. It will consequently increase the qualitative and quantitative yield of agricultural crops (Fig. 4).

Furthermore, the MAC project will increase the job opportunities in the agricultural industry developing and managing an educational programme, made of 3 cycles of courses concerning *biointensive agriculture*, *ethical production and rural marketing*, *business innovation and agricultural-business*. With regards to rural production, MAC project intends to lay the foundations for the development of new enterprises, supporting the creation of three start-up and a 'farm-to-table' network to promote bio-intensive products cultivated in Monterusciello and sell them in the existing local market.

MAC project will also act on the quality of the urban spaces: architecture interventions will activate spaces within the existing and un-used public buildings for two laboratories and an Incubator Centre.

One of the key action will be the creation of a bike path, as well as walkways and seating areas, all to be set along the agriculture areas overlooking the greenery.

The bike path will be connected to the other interventions programmed by Local Municipality in order to create an efficient network of bike lanes, parallel to the roadways, for inter-district travels. The main purpose of this infrastructure is to respond to four different demands:

1. connect all the district facilities, such as schools, public offices, parks, regional railway stations, bus stops, and future bike stations;

1. The 80% of the total amount of the project cost (4 million euros) will be financed by ERDF program, while the remaining 20% (1 million euros) will be co-financed by the Municipality of Pozzuoli and other 6 partners: University of Salerno, Coldiretti and Confagricoltura (farmers trade unions), L'Iniziativa (association for social development and support), FORMIT (private Foundation experienced in scientific research, technical support, analysis) and Agrocultura (private enterprise experienced in permaculture).

Figure 4. Visual simulation of landscape changes MAC project intends to achieve for rural abandoned areas in Monterusciello.

2. ensure an easier accessibility to the district for non-motorised people;
3. create new point of observation and enjoyment of the regenerated agricultural urban landscape;
4. increase the road safety by a speed reduction through the shortening of the lane width.

5 CONCLUSIONS

The paper yearns to be a review of best practice that local governments should put in practice to pursue urban regeneration, acting on the promotion of cycling mobility. The case study shown has turned out to be particularly meaningful since Monterusciello is a publicly owned new town, which was planned and build thirty years ago with no attention to cycling mobility. Its layout design caused a high dependency of people on private vehicles to travel and commute within the district. Furthermore, the incomplete implementation of the PHEP led to lacking of urban facilities, presence of huge abandoned uncultivated spaces, urban and social decay.

The actions activated up to the present have assumed cycling as an invariant feature of urban mobility policies: bike lanes and station projects in Monterusciello will not only ensure most efficient connections between main urban facilities and transport hub (train and bus station, parking areas) but also be part and drivers for urban regeneration process, lying in new agro-urban landscape creation and promotion.

Future demanding challenges consist of coordinating and optimizing efforts, ideas, designs and funds coming from different projects in order to maximize the benefits for the community.

REFERENCES

Cervero R., Duncan, M. (2003). Walking, cycling, and urban landscapes: Evidence from the San Francisco Bay Area. American Journal of Public Health, 93, 1478–1483.

ECMT—European Conference of Ministers of Transport, (2004). National Policies to Promote Cycling. OECD.

Martens K., (2007). Promoting bike-and-ride: The Dutch experience. Transportation Research Part A: Policy and Practice, 41. 326–338.

Meng. M, Koh P.P., Wong Y.D., Zhong Y.H. (2014). Influences of urban characteristics on cycling: Experiences of four cities. Sustainable Cities and Society, 13. 78–88.

Mollison B. (1991). Introduction to permaculture. Tagari Publication, Australia.

Pucher J., Buehler, R. (2008). Making cycling irresistible: Lessons from the Netherlands, Denmark and Germany. Transport Reviews, 28, 495–528.

Rietveld P., Daniel V., (2004). Determinants of bicycle use: do municipal policies matter?. Transportation Research Part A. 531–550.

URBACT II capitalisation, (2015). Sustainable regeneration in urban areas. URBACT II programme.

Integrated land use and transport planning. Methodological approaches and case studies

Town and Infrastructure Planning for Safety and Urban Quality – Pezzagno & Tira (Eds)
 ISBN 978-0-8153-8731-2

The Sustainable Urban Mobility Plan (SUMP) of Naples: An example of a rational and participated transportation planning process

A. Cartenì
Università degli Studi di Napoli Federico II, Naples, Italy

M. De Guglielmo
Università degli Studi di Salerno, Fisciano, Italy

N. Pascale
Comune di Napoli, Naples, Italy

M. Calabrese
Università degli Studi di Napoli Federico II, Naples, Italy
Università degli Studi di Salerno, Fisciano, Italy

ABSTRACT: A Sustainable Urban Mobility Plan (SUMP) is a strategic (long period) transportation plan aimed to improve welfare of peoples and workers living in the city and in its surrounding area. The SUMP moves the focus from traffic (vehicles) to people toward a sustainable mobility jointly with the quality of life. According to the recent European Guidelines, this plan has to be built on existing planning practices and after considerating: the integration, the participation, and evaluation.

In absence of a detailed national guideline implementing SUMP, the aim of this paper was to apply an innovative theoretical decision-making approach for the development of the Naples (Italy) SUMP. This case study is interesting because the city is the third largest Italian city and the metropolitan area with the highest population density.

In 2016, the city has ratified a first formalization of acts: 'the analysis of mobility system' and 'the statement of strategic objectives'.

1 INTRODUCTION

The complexity of decision-making processes in transportation planning has long been recognized in the relevant literature, together with the need to "open up" such processes and broaden the consensus around alternative courses of action (see for instance Manheim et al., 1972 and Suhrbier et al., 1987, as well as the far-sighted chapter dedicated to choices in transportation in Manheim, 1979). It is assumed that the decision-making process has, or should have, some form of "rationality" and that quantitative tools, i.e. statistical analyses and mathematical models, play a central role in it. Against this background, transportation system analysis and planning are seen mostly as activities based on the design and simulation of alternative projects and the assessment of priorities. However, such assumptions are often not satisfied in real-life complex cases. Transport-related decisions can be a-rational or "sub-optimal" with respect to stated, formal objectives, and still "rational" in the context of a wider, less defined set of contrasting objectives including maximizing consensus and/or minimizing opposition to proposed solutions. From another angle, plans and projects often impact on multiple and contrasting interests in a complex institutional setting, and result from decision-making processes involving several actors, both public and private. There is a vast literature on "planning failures" in transportation (see, e.g. Hall, 1980; Winston, 2000; Button, 2005; Flyvbjerg et al., 2005; Bartholomew, 2007; Knoflacher, 2007; Lemp and Kockelman, 2009).

The basic assumption of this paper is that the quality of the decision-making process is a key factor for a "successful" planning, and that the quality of the decisions critically depends on how the process is structured. Planning and designing transportation systems should expressly be recognized as managing complex, multi-agent decision-making processes in which political, technical and communication abilities should all be involved in order to design solutions which are technically consistent and, at the same time, maximize stakeholder consensus.

A "desirable" transportation planning process should combine the potential benefits of rational decision-making and stakeholder engagement. By "desirable" we mean a process leading to decisions which are "as efficient as possible" given the circumstances and, once adopted, reduce the risks from unavoidable uncertainties both in the context and in the technical analyses. To this end, the decision-making process should be *transparent*, i.e. decisions should be justifiable and retraceable, and *participated*, i.e. decisions should be shared by the largest possible number of decision-makers and stakeholders.

According to the recent European Guidelines (Wefering et al. 2014), a SUMP is a strategic (long period) transportation plan aimed to improve welfare of both people and workers living in the city as well as in its surrounding area. Compared to "traditional" urban transportation plans, SUMP moves the focus from vehicles to people with significant advancements in the direction of a sustainable mobility jointly with the quality of life, overcoming a too sectorial vision of transport without interactions between interventions.

Recent literature reports several case study of cities that have developed a SUMP (or part of them): see e.g. the case of Dresden in Lindenau and Boehler-Baedeker (2014) or Krakow, in Minchenej and Zwolinski (2016). An advisory group of European Commission have developed also a set of recommendations, reported in May et al. (2017), to give support to European governments and their local authorities and encourage them to learn from each other to develop these plans.

However, less investigated is the developing/applying of a unique theoretical approach as a standard (guideline) of a such complex planning process. A new comprehensive approach for making a rational transportation planning, was proposed in Cascetta et al. (2015). In this paper, the authors proposed an innovative planning process based on "three legs", better explained in the following, which transforms the transport planning process into a complex interaction between three main categories of actors: decision-makers, stakeholders and professionals.

Starting from these considerations, and in absence of a detailed national regulation, the aim of this paper was the application, for the first time in literature, the innovative rational decision-making approach proposed by Cascetta et al. (2015) at urban scale. Precisely this theoretical decision-making approach was applied to the Naples (Italy) Sustainable Urban Mobility Plan (SUMP).

The paper is organized as follows. In the next section, the main elements of a decision-making process are summarized; afterwards the application to the indicated case study is illustrated and discussed. Finally, the main conclusions and some further perspectives are reported.

2 THE DECISION-MAKING PROCESS FOR A RATIONAL TRANSPORTATION PLANNING

In 2015, Cascetta et al. proposed an innovative, also called "a new look", decision-making process for rational transportation planning (see Fig. 1, for a schematic representation) based on three parallel and intertwined processes or "three legs": i) a rational decision-making process; ii) stakeholder engagement process; iii) the use of quantitative methods. The bounded rational model assumes that actors are still goal-oriented, but they implicitly take into account their cognitive limitations in attempting to achieve those objectives. Furthermore, since different decision-makers are involved in the process, it is very likely that their objectives are diverse and possibly contrasting, not to mention those of the variety of stakeholders involved in the process.

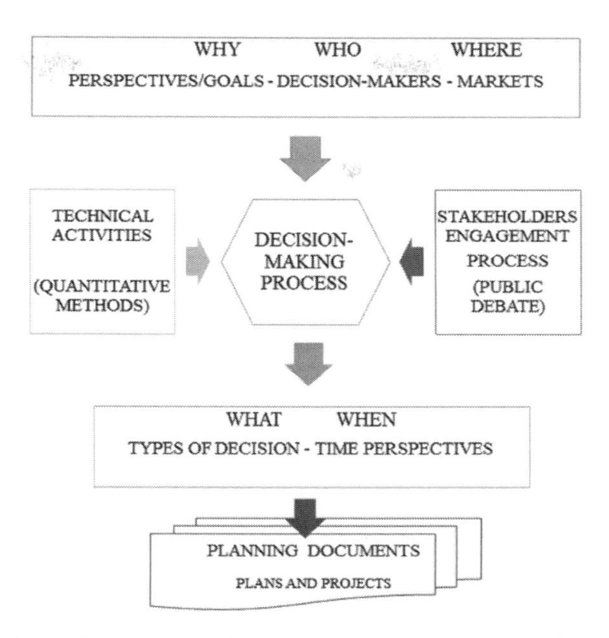

Figure 1. The decision-making processes for a rational transportation planning.

The model aims to find a satisfactory compromise rather than the optimal value of an abstract, multi-objective, function. The bounded rational decision-making model is intrinsically dynamic, with several feedback loops adapting the "solutions" to their ability to satisfy objectives and constraints until reaching a "satisfactory" level of compliance. Stakeholder Engagement (SE) or Public Engagement (PE) can be considered the process of involving stakeholder concerns, needs and values in the transport decision-making process (Kelly et al., 2004). It is a two-way communication process that provides a mechanism for exchanging information and promoting stakeholder interaction with formal decision-makers and the project team. The overall goal of engagement is to achieve a transparent decision-making process with greater input from stakeholders and their support for the decisions that are taken. Stakeholder engagement (SE) may have different potential roles in decision-making models. Different levels for SE can be integrated within the cognitive rational approach. Decisions are still rational, i.e. they are based on the comparison of alternative courses of action (plans/projects), with regard to their expected effects. However, decisions are generated by exploring a (limited) number of alternatives until a solution, which satisfies decision-makers and as many "major stakeholders" as possible, is reached. The comparison of alternative solutions is arguably the most critical phase. A (wide) public consensus is required to define the "satisfactory" alternative, i.e. that with a higher probability of being implemented. This decision stage can be supported by the stakeholder consultation and participation, where different engagement tools can be adopted depending on the specific nature of the decision to be made.

In the proposed framework, quantitative methods, i.e. statistical analysis and mathematical models able to reproduce the transportation system performance and simulate the effects of possible alternative configurations of the system, together with domain-specific capabilities in designing alternative solutions, have a dual role. The first role is to analyse current system performance and formulate and assess alternative projects/options, in later stages. This role is well-established in the literature and in the professional practice of transportation system analysts and planners. The second, and more innovative role, is the contribution to SE by providing information for interaction among decision-makers, process coordinators and stakeholders. This function can be useful from the very beginning of the process, as it helps to develop a common understanding of the current performance and deficiencies of the system, based on data and analysis and not only on impressions and points of view. Analysis of current conditions may help identify groups subject to particular benefits/losses as possible stakeholders.

3 THE APPLICATION CASE STUDY: THE SUMP OF NAPLES

The case study, proposed in this paper, concerns the SUMP of Naples and it appears to be interesting because of the specificity of Naples, described in the following. The SUMP process, characterized by participation and consultation, has been included by ISPRA (the Italian Institute for Protection and Environmental Research) among the best Italian practices (see Faticanti et al., 2016).

Working[1] on its SUMP for the last two years, the City of Naples has made a very complex plan that, recently, has had a first formalization by means of the approval of two acts: the *Analysis of mobility system* and the *Statement of strategic objectives*.

As reported in the following, the process has an intrinsic dynamicity and the objectives, defined in early stages, can be reformulated until the achieving of a mobility system as it should be thought today.

SUMP's vision has as core a shared mobility system, definition that includes both traditional local public transport systems and bike, car and taxi sharing services. To this core, it will be necessary to connect and integrate both private and cycle-pedestrian mobility. Such mobility system requires an intensive use of telematics technologies, in particular for the dematerialization of mobility payments and for data sharing.

These first parts of the plan were carried out, in the absence of a detailed national regulation on this topic, on the practical basis of European Guidelines and on the above described theoretical "three legs" approach as illustrated in Fig. 2 and following described.

3.1 *Technical activities*

In the last 20 years, the city of Naples has developed several transport planning tools (Piano Comunale dei Trasporti, 1997; Programma Urbano dei Parcheggi; 1999; Piano della Rete Stradale Primaria, 2000; Piano delle 100 Stazioni, 2003; Piano del Traffico Urbano, 2002–2004; Piano d'Azione per l'Energia Sostenibile, 2014; Variante generale al PRG, 2004; Variante per la zona occidentale, 1998). Thanks to these tools, design and construction of important works for the city became possible: for example, the Metro Line 1 and its art stations are known throughout the world. However, as all sectorial planning documents, also the planning tools of Naples had a consistency limited to related policy areas.

For deep analysing of the reference strategic context, the local plans were not sufficient: the attention had to be directed to other levels of planning (regional, national[2] and European plans) with aim to identify, for the city of Naples, strategic objectives and targets that were aligned with the previous indicated upper levels.

In the first phase, statistical analyses are considered as able to reproduce the transportation system performance; their results, in the following of the process, will be the basis for using simulation tools. From these analyses, Naples results to be the third largest Italian city and metropolitan area with the highest population density (8,250 inhabitant/km^2, year 2011). This specificity, that has a negative impact for private transport system, could be an encouragement to develop effective public transport policies.

The city of Naples presents: i) a number of cars per inhabitant less than that of Milan and Rome, but similar to in Turin; ii) the highest density of vehicles in Italy: 4,500 veh./km^2 (compared to 3,700 veh./km^2 in Milan and 1,500 veh./km^2 in Rome). In addition, the city has the world' s largest historical centre (UNESCO World Heritage Site, thanks to a Greek/Roman central part) that is totally inadequate to so large transit of cars.

1. According also to tools and information available on Eltis website: http://www.eltis.org.
2. See, e.g., documents available at: http://www.mit.gov.it/sites/default/files/media/notizia/2017-04/Allegato_MIT_AL_DEF_2017.pdf; http://www.mit.gov.it/sites/default/files/media/notizia/2016-07/Strategie%20per%20le%20infrastrutture_2016.pdf.

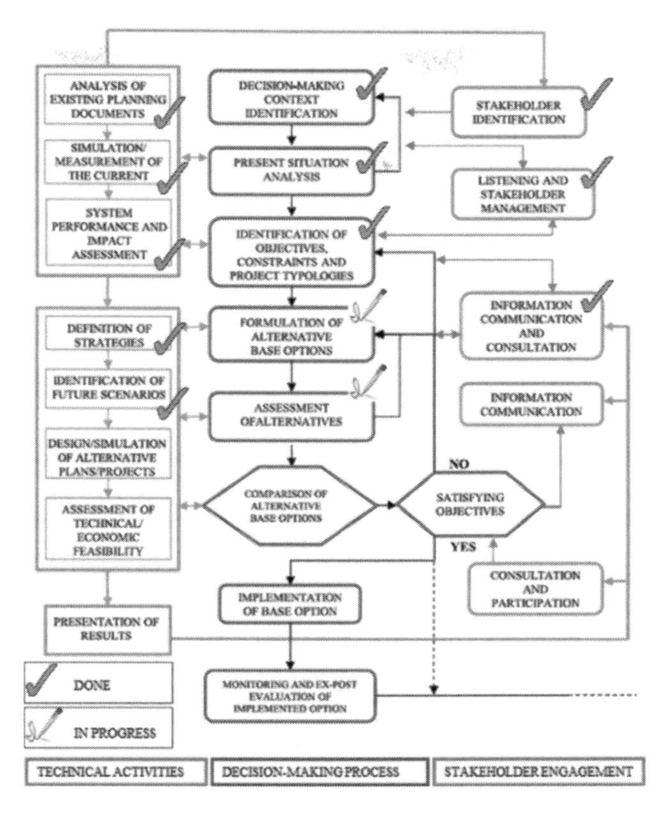

Figure 2. The decision-making process for SUMP of Naples with indicated phases already performed and phases still in progress (based on schematic representation of an overall transportation decision-making model reported in Cascetta et al. 2015).

Currently, since its urban population is decreasing and, on the contrary, the number of inhabitants in metropolitan area are increasing, the mobility towards the city centre is increasing in last years. In fact, the 60% of trips that occur every day in Naples occurs within the city centre, while the remaining 40% are exchange trips from the metropolitan area to the centre. The 42% of these trips are carried out by private transport while the remaining 58% is divided between rail (18%), bus (16%) and other types of mobility (24%).

The current mobility system, focused on vehicles, has costs in terms of:

- *congestion*: as well known, it is associated with time lost in traffic and fuel consumption. For the metropolitan area of Naples, this cost was estimated in 1–2 billion €/year while just for the city of Naples between 300 and 600 million €/year (Ambrosetti, 2012);
- *road accidents*: this cost, that has a global relevance (see e.g. the targets to contain this huge social cost in Horizon 2020), for city of Naples reaches190–250 million €/year (MIT, 2012);
- *pollution*: 20–25% of pollutant emissions in the atmosphere derive from the transport sector and represent about a quarter of those gases and particles that, in urban areas, put at risk the public health;
- *improper use of public space*: monetary estimates are not available, but we have a simple idea of its relevance just thinking that cars, with the same number of passengers, are about 5 times the volume of public transport by buses, about 10 times by a tram, and about 20 times by subway.
- Finally, it can be summarized that the current mobility system has, for the city of Naples, an annual social cost of 490–850 million €/year.

3.2 *Decision-making process*

The decision-making context identification is linked with the phase of stakeholders' identifications.

The analysis of current state has been divided into three levels: technical analysis of the mobility system main indicators, comparison with stakeholders and comparison with other similar territories.

In a first step of plan definition, the objectives (within 10 years) have been identified as follows: i) encourage the use of collective transport; ii) improve safety mobility; iii) reduce pollutant emission; iv) rationalize the road system; v) encourage bicycle and pedestrian mobility; vi) make intelligent the mobility system; vii) reorganize the parking system.

Because of the intrinsic dynamicity of the process, as participatory process, the objectives, defined in early stages, appear to have need of a reformulation in light of continuous advancements. It was so introduced a new objective: viii) optimize urban logistics. In addition, it was understood that the objectives are more and it is not appropriate to list them since they are interrelated as shown in Fig. 3. In fact, the shared areas have great relevance. Three examples can be summarized: 1) the purchase of new buses involves both the *reduction of pollutant emissions* and the *incentive to use public transport*; 2) the installation of intelligent traffic lights, to the two previous objectives, also adds to *make mobility system intelligent*; 3) control and management of tunnels involves *intelligent systems* and increased *mobility safety*.

The projects developed by SUMP can be proposed with alternatives. In particular, this approach has been used for the new lines of collective transport serving the east (San Giovanni) and the west (Bagnoli) areas of the city. Today, these are areas present relevant dismantled industrial activities that will be interesting by a complete redefinition in the next few years. For the east area, SUMP has proposed different routes and different technologies (trams and BRTs) to efficiently serve new settlements. For the west area, different intervention scenarios have been identified involving the extension of existing metropolitan lines (M6 or M2) or the construction of a dedicated line.

3.3 *Stakeholder engagement*

The phase of identification has involved several stakeholders. For saving space, according to the groups indicated in the European Guidelines, they can be summarised as follows (more details on the city of Naples website)[3]:

- *Government/ Authorities*: national and local authorities by means of representatives of the Ministry of Transport, of provincial and regional authorities; European Investment Bank; Municipal Police;

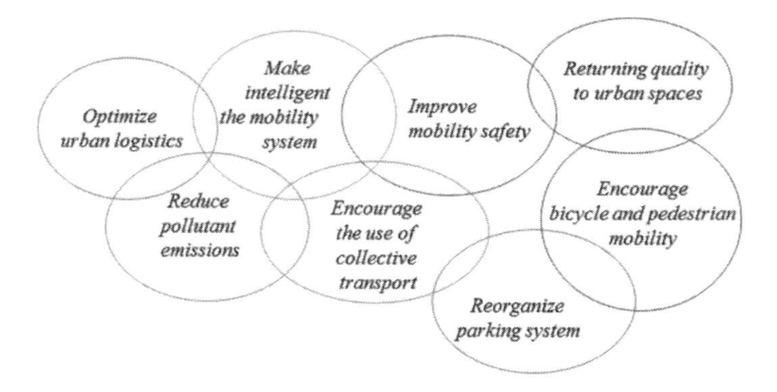

Figure 3. The objectives of SUMP for city of Naples and their interrelations.

3. http://www.comune.napoli.it/flex/cm/pages/ServeBLOB.php/L/IT/IDPagina/28525.

- *Business/Operators*: transport operators as ANM, CTP, Consorzio Unico Campania, Tangenziale di Napoli S.p.A., etc; sectorial economic operators;
- *Communities/Local Neighbourhoods*: transport users' associations and environmental associations;
- *Others:* research institution (University of Naples Federico II), professional associations (Engineers order of Naples).

In event of meetings, the stakeholders are invited to explain their specific points of view, their needs and their objectives.

In addition to public meetings and debates, it was realized a dedicated page for all citizens, within the website of city of Naples[4], with all documents freely downloadable and continuously updated.

4 CONCLUSIONS

The aim of this paper was the application, for the first time in literature, of the innovative rational decision-making approach proposed by Cascetta et al. (2015) at urban scale. The development of the Naples SUMP is not ended. The main activities were implemented (e.g. definition of aims, constraints, strategies, public engagement), but much remains to be done (e.g. implementation and monitoring). In particular, also the topic of "hedonic quality" (e.g. aesthetic, beauty, design) in public transport (e.g. including the effect of the "art" Metro Line 1) will be explicitly take into account.

REFERENCES

Bartholomew K., (2007). Land use-transportation scenario planning: promise and reality. Transportation, 34, 397–412.

Button K., (2005). Market and Government Failures in Transportation, Elsevier.

Cascetta E., Cartenì A., Pagliara F., Montanino M., (2015). A new look at planning and designing transportation systems as decision-making processes; Transport Policy 38. pp. 27–39.

Faticanti M. et al. (2016) Qualità dell'ambiente urbano—XII Rapporto ISPRA Stato dell'Ambiente XX/16 pp. 697–698 Available at http://www.isprambiente.gov.it/public_files/XII_RAPPORTO_AREE_URBANE/XII_RAU_Volume_completo.pdf.

Flyvbjerg B., Skamris Holm M.K., Buhl S.L., (2005). How (In)accurate Are Demand Forecasts in Public Works Projects?: The Case of Transportation. Journal of the American Planning Association, 71(2), 131–146.

Gardner J., Rachlin R., Sweeny A., (1986). Handbook of strategic planning, Wiley, New York.

Hall P., (1980). Great Planning Disasters. University of California Press.

http://www.mit.gov.it/mit/mop_all.php?p_id=12919.

Knoflacher H., (2007). Success and failures in urban transport planning in Europe—understanding the transport system, Sadhana, 4, 293–307.

Lemp J.D., Kockelman, K.M., (2009). Understanding and Accommodating Risk and Uncertainty in Toll Road Projects: A Review of the Literature. Transportation Research Record, 2132, 106–112.

Lindenau M., Boehler-Baedeker S., (2014). Citizen and stakeholder involvement: a precondition for sustaible urban mobility. Tranportation Research Procedia 4, pp. 347–360.

Manheim M., J. Suhrbier E., Deakin L.A., Neumann F.C., Colcord Jr. A. T. Reno Jr., (1972). Transportation Decision-Making: A Guide to Social and Environmental Considerations, NCHRP Report 156.

Manheim M.L., (1979). Fundamentals of transportation systems analysis. MIT Press, Cambridge, Massachusetts, USA.

May A., Bochler-Baedeker S., Delgado L., Durlin T., Enache M., van der Pas J., (2017). Appropriate national policy frameworks for sustainable urban mobility plans. European Transport Research Review. Vol.9, Issue No. 1, Springer Publishing.

Minchenej M., Zwolinski T., (2016). Development of Sustainable Urban Mobility Planning for City of Krakow. 5th IEEE conference on Advanced Logistics & Transport (ICALT).

4. http://www.comune.napoli.it/flex/cm/pages/ServeBLOB.php/L/IT/IDPagina/28525.

MIT Italian Ministry of Transport, (2012). Studio di valutazione dei Costi Sociali dell'incidentalità stradale.
Municipality of Naples (2017). www.comune.napoli.it. Last access in May 2017.
Suhrbier J. H., DeakinE., Loos J., (1987). The Role of the public in Public/Private Partnerships. Institute of Transport Engineers Annual Meeting, New York, USA.
The European House-Ambrosetti, Finmeccanica (2012). Smart Mobility, Muoversi meglio per vivere meglio.
Wefering F., Rupprecht S., Bührmann S., Böhler-Baedeker S., Rupprecht Consult—Forschung und Beratung GmbH, (2014). Guidelines. Developing and Implementing a Sustainable Urban Mobility Plan. Document prepared for the European Commission, 2013.
Winston C., (2000). Government Failure in Urban Transportation. Fiscal Studies, 4, 403–425.
www.ambrosetti.eu/wp-content/uploads/Ricerca-Finmeccanica.pdf.

Town and Infrastructure Planning for Safety and Urban Quality – Pezzagno & Tira (Eds)
 ISBN 978-0-8153-8731-2

Relationship between mobility and urban form in contemporary new town planning. Notes for a comparative perspective

P. Ventura, A. Montepara, M. Zazzi, M. Cillis, B. Caselli & M. Carra
Università degli Studi di Parma, Parma, Italy

ABSTRACT: This paper presents an in-depth analysis to identify good planning practices for mobility and public spaces which are connected together and are considered quality factors in the urban design. Specifically the analysis of displacements at the urban and regional scale must necessarily investigate not only the plurality of attractors, but also a great number of mobility structures that determines a higher level of accessibility and polarization around which the main flows of goods and people gravitates.
Through three interpretative key-words – the relationship between mobility and urban form, intermodality and the urban micro-landscape – a comparative analysis on some case studies has been carried out, with particular attention to the New Towns built in the last twenty years. This overview is useful to prefigure which are the possible scenarios of innovation for the New Towns, both in future ones, keeping in mind that they will mostly be centred in developing countries and in future developments of existing urban areas.

1 INTRODUCTION

This paper is framed in a research program on New Town planning that has been involving the University of Parma research team in the Laboratory of Urban Planning since 2006 (Ventura, et al. 2006–2017).

The *New Town* investigation has been a great field of study yesterday (Unwin, 1909; Howard, 1898; Merlin, 1971; Frampton, 1992; Stern, et al., 2013) as well as today.

The concept of *New Town*, e.i. a comprehensively planned settlement, involves an intentional action for controlling and designing both its birth and its form, hence most of the time New Towns have a regular morphology from which the goals of each planner and their composition rules emerge. These characters distinguish them from "spontaneous" settlements, developed in an unplanned and irregular manner as a "sum" of individual wills or as an outward diachronic expansion of built-up areas (Lavedan, 1952).

The ultimate intent of a *New Town* plan is certainly to programmatically tend toward an optimal placement of human settlements in relation to their functions (Stübben, 1907), and to increase and facilitate the movement of people and goods via public transit. In this logic, the spatial arrangement of transportation infrastructure also determines the spatial aspect of a urban settlement, that is the urban form.

In fact, *New Town* plans have been giving their contribution to the innovation of urban mobility through time: starting from the British new suburbs of the mid-nineteenth century (Hall, 2002), set on the theories of Ebenezer Howard's *Garden City*, where mobility is based on the connection with the 'parent city' via public transportation, to the Radburn pattern (Stein, 1957), in application of the principle of separating vehicles from pedestrians, or even in the introduction of the *precinct* and traffic protection measures in residential areas (Buchanan, 1958).

The *New Town* is closely linked to economic, demographic, and infrastructural development even after World War II, when, answering the great demographic growth and subsequent demand for new housing, Western Countries became a particularly rich laboratory for new settlements planning.

Only a few decades later the construction of new cities seems to answer the pressing problems of congestion of the increasingly populous metropolitan areas, e.g. the case of Paris *Villes nouvelles* or the new Dutch cities destined to host part of Amsterdam's inhabitants. Here, new ways to improve public transportation and to increase the vitality of public space were experienced. In French *Villes nouvelles* in Seventies, for example, the central public space, where a regional commercial centre together with public services and institutions are located, is kept only to pedestrians and bicycles, while roads and railways are underground.

In recent years, the optimization of urban mobility, according to sustainability goals, is part of a global perspective aimed at reducing pollution and improving light mobility in new cities by increasing the efficiency of intermodal transport.

However, this debate, though still open, seems to have shifted its focus to the territories which are rapidly growing such as Middle East, East Asia and the South African regions. Over the last twenty years, at least 240 *New Towns* have been built there. The Asian continent holds the primacy with about 60% and the African continent with about 33%. By comparison, the phenomenon in Europe (4%) and in the United States (2%) has become almost completely inconsistent (INTI, 2017).

2 OBJECTIVE

The general aim of this paper is to investigate the relationship between mobility and urban form in contemporary planned urban communities.

In this sense, it has a double goal: to compare recent new town planning cases and to single out specific planning methods regarding mobility. The research also seeks to state whether it is possible or not to prefigure a new scenario for overcoming previous *New Town* typologies—green cities, factory towns, leisure towns—in view of the changes occurred in the ideas and customs of a population and in public transport innovation.

The specific objective of this paper is to explain how the analysed examples can be classified in specific interpretative models that integrate a view of the urban shape and spatial arrangement with the characteristics of the infrastructure system design. The research undoubtedly leads to identify trends that can be particularly useful in scheduling future redevelopments or transformation processes of the existing cities public realm.

3 METHODOLOGICAL APPROACH

A multi-tiered analysis has been conducted on a large sample of cases in a 50-year timeframe, since 1960, to operate a comparative study, both quantitative and qualitative, among different *New Towns*. This comparison is based on the following set of descriptive parameters: a) dimensional elements (population, surface and density); b) characteristics of the road system and sustainability, especially related to the compactness degree of the settlement; c) characteristics of the infrastructural and public transport system.

3.1 *Infrastructure system and transport*

For each selected city, urban and suburban connections among geographically or economically close urban contexts have been analysed. The interaction among multiple factors (different business environments, the functioning of their private and public transportation systems and its degree of integration with slow mobility) has been taken into account.

3.2 *Size and compactness*

Cities have been also classified according to their size, measuring their population, territorial extension and density. Density is an important parameter because it is also an indicator of compactness, an indispensable character that helps cities to reduce their automobile dependency, to encourage slow mobility and to optimize the use of collective transport systems. The target density of the sample New Towns, calculated using the original plans target population, has then been compared with the current population density based on the actual number of inhabitants.

3.3 *Design of public space and road system*

Three major categories of road systems can be singled out in *New Town* contemporary planning. The first one consists of a predominantly or exclusively homogeneous isotropic fabric, usually on a regular orthogonal basis, as in the antiquity (some contemporary examples are Le Corbusier's Ville Radieuse in 1930 and Ville Contamporaine in 1922, Magnitogorsk in 1930, Islamabad in 1950). Within this category we can also identify less frequent cases of linear development fabric (Ciudad Lineal,1905; Rush City Reformed, 1930) and non-orthogonal fabric with hexagonal grids (Toulose Le Mirail, 1960) or with a triangular equilateral base (GR Le Ricolais). Also slightly isotropic fabrics wrapped to fit the landscape, as Milton Keynes (1965), Brasilia (1960) which can be also classified as linear one, Chandigarh (1950), suit this category.

The second category corresponds to the monocentric (Afula 1925, Littoria, 1932) or mixed orthogonal and multi-center radial system (New Beograd, 1948), typical of large cities, which then evolves into complex and hierarchical networks.

Finally, the third corresponds to a more or less complex curvilinear system, intentionally irregular and organic, able to exploit the orography and to obtain precious landscape effects (Garden Cities).

4 DESCRIPTION

4.1 *Innovative aspects*

A synoptic prospect of data has been set up. It allows, through an interpretative reading: a) to highlight the peculiarities or critical issues of urban design choices linked to the relationship between the infrastructure system and the primary function of a city, its geographic position and its urban form; b) to observe the transformation processes that have taken place since the founding of the city; c) to make it easier to better understand the possible implications of strategic polices for road and transport planning.

4.2 *Research*

The availability of services and facilities within an urban area—city, town or suburb—highly depends on the speed and ease with which citizens can reach them moving inside and outside the city limits.

The importance of speed has introduced a change in *New Towns* planning—and also in the traditional city—pointing to the prevalence of rapid transport modes (Glaeser, et al., 2001).

Different attitudes have been recognized starting with the rise of car-centric *New Towns*, where public transport becomes less important (e.g. Seaside, Laguna West and Celebration), to new urban developments investing in great public infrastructures apt to support institutional/financial uses (Putrajala, Astana), mass flows around specialised consumer related facilities (Milton Keynes, Meixi Lake, Makuhari) or commuter settlements that can offer on one hand a fast suburban connection with employment centres, and on the other an inner accessibility provided by walking and cycling paths (Ørestad, Lelystad). The road structure design in this last case consists, as in the past, of a search for a geometric regularization around a town centre which is both a departure point towards near major settlements, and the pedestrians arrival point inside the city.

Obviously, urban densities are not uniform: some *New Towns* are set on more compact, mixed-use patterns apt to develop sustainable mobility solutions (Duany, et al., 2000) with a relatively high density, from 50 inh/ha to at least 150 inh/ha. On the other hand, where we find a very low density, often below 50 inh/ha, the use of private cars, for travelling inside and outside the city limits, is encouraged.

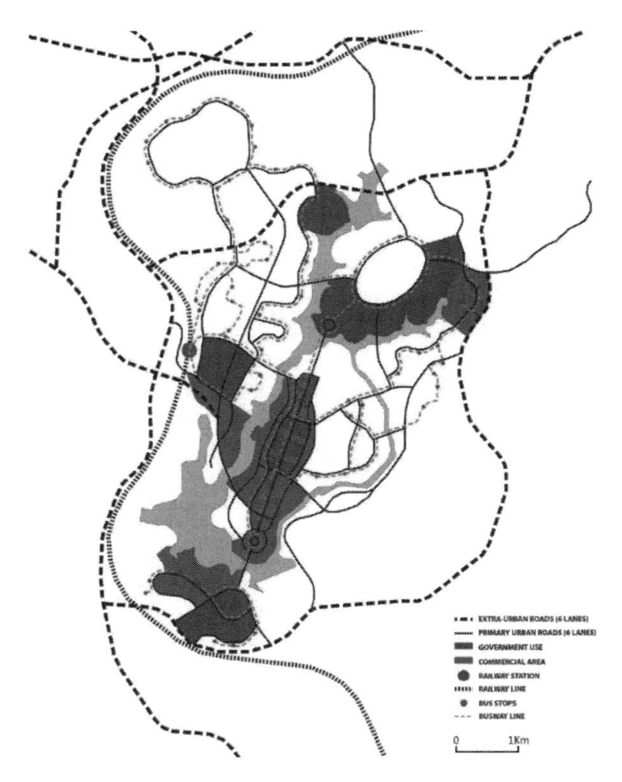

Figure 1. Putrajaya, Malaysia (1995).

5 RESULTS OF THE RESEARCH

By comparing a sample of about 50 case studies, the combination of the collected data has permitted to summarize five interpretative models that take into account the main trends in recent *New Towns* plans, identified from the point of view of their degree of autonomy or dependence, of their specific function, of their gravitational area, and their inner and outer flow motions.

5.1 *Hub city*

The *Hub city* plays a primary role in the infrastructure network and it maintains either a "leadership" function, as in the case of capital cities e.g. Astana, Putrajaya and Nairobi, or a polarizing function thanks to the intersection of multiple public transport networks (e.g. King Abdullah Economic City, Euralille).

Particular reference is made to the main infrastructure network nodes—ports, airports, railway stations—which require a special location in the urban system due to their role of urban condenser of numerous flows of people and goods. As for the flows of people, a rapid urban and suburban interconnection is very important. In Putrajaya (1993) and Astana (1998), for example, this need is granted by a multi-modal public transport system (light rail, transit bus and tramway). In the two mentioned cases, public transport lines converge in specialized urban areas, such as CBDs and government headquarters, or in intermodal hubs.

In King Abdullah Economic City (2005) and Aerotropolis, two realities with an economic system related to logistics activities, the primary gateways for the movement of goods are the International Ports and Airports.

The Hub city layout is often very complex, mixed and irregular, also due to substantial diachronic additions, but it always shows a larger axis passing through the main infrastructure

Figure 2. Milton Keynes, England (1967).

nodes. All of the sample cases, attributable to this model, have a high density between 75 and 332 inh/ha; the densest city is Putrajaya.

5.2 *Consumer city*

The *Consumer City* is a reality where there is a great deal of services, jobs and consumer goods, and where tertiary activities (industry, cultural and commercial activities) are predominant. These activities centre around a large and attracting core for consumers and workers, such as CBDs or generic city centre dominated by the presence of a shopping mall. Even the traditional industrial city (now overcome in an European context) might be considered as a *Consumer City* because of its ability to attract flows of goods and workforce.

This model is characterized by an intermodal infrastructure system that facilitates rapid accessibility (Glaeser, et al., 2001): there is at least one primary infrastructure (in most cases a railway) and a large number of options for public transport on roads or rail.

In general, the road structure here has a regular geometry (orthogonal isotropic or linear layout), sometimes with variation on the theme such as in Milton Keynes (1967), but always *"in accordance with the hypothetical interests of a consumer society"* (Frampton, 1992). The urban density and compactness degree are set to medium or high values to encourage pedestrian and bicycle accessibility to public services and public transports. In the sample analysed, this category is certainly predominant in terms of number of cases that can be attributed to it.

5.3 *Commuter town/suburb*

The *Commuter Town/Suburb*, as a natural evolution of the Garden City concept, is a residential satellite settlement crossed by a suburban public transport line (railway, tramway or

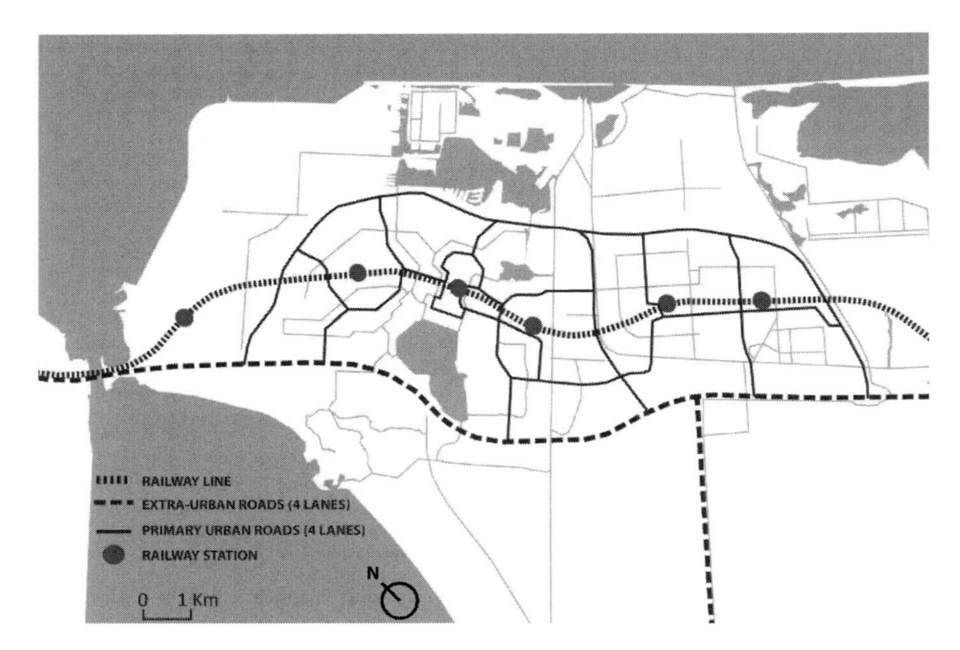

Figure 3. Almere, Netherlands (1977).

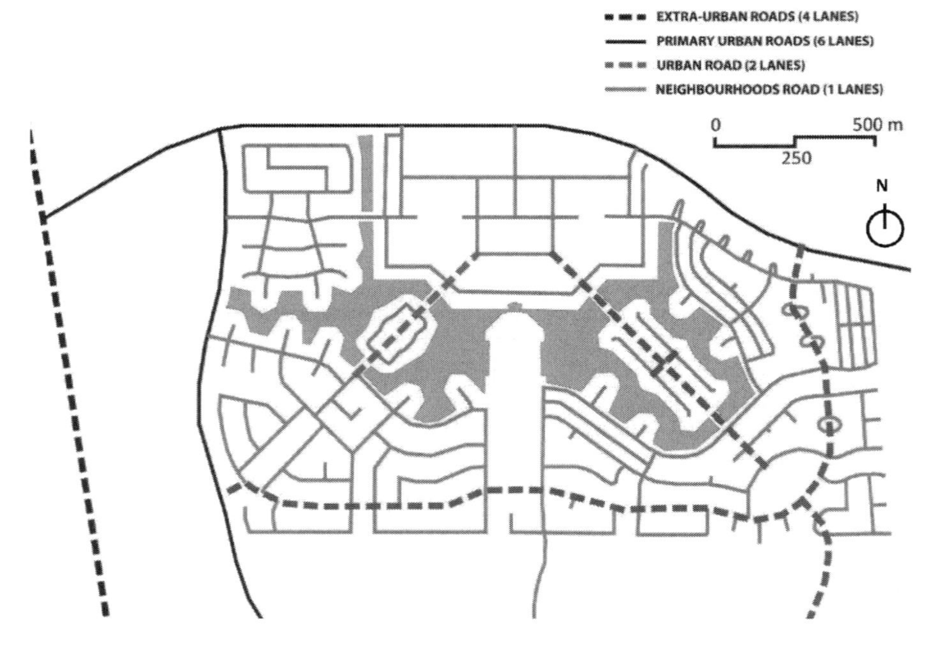

Figure 4. Laguna West, USA (1989).

subway) which grant connection with the 'heart' of the nearest metropolitan core, full of services and job opportunities.

In this sense we can point to the North American model of the *Transit Oriented Development*, which prefigured a compact and accessible city that maximizes the amount of public facilities within walking distance and well connected to the outside thanks to a collective transit system. (Transport Oriented Development Institute, 2016). Sengkang New Town (1970), one of Singapore's satellite cities, is an example of TOD and it adopted both an underground line (North East Mass Transit Line) and an above-ground tram line (Light Rail Transit) as means of transport towards Singapore.

Unlike the Garden City curvilinear layout, *Commuter Towns* follow an isotropic orthogonal system, such as Lelystad (1967), and in some cases linear layouts, such as Ørestad (1997). The residential function, including home-related public services and large green spaces, involves medium and low densities: from 40 inh/ha (Lelystad) to 65 inh/ha (Ørestad). An exception is the case of Pujiang City (2001), Shanghai's satellite city, with a density of 385 inh/ha.

5.4 *Car suburb*

Despite the common goals of increasing sustainability and optimizing public urban mobility, some *New Towns* focus the design of their mobility system on road infrastructures. At best, a sort of collective transport service is to be found, although it is totally inadequate due to long travel times (such as buses in New Cairo, 2009).

The predominance of automobile traffic and the lack of a public transport system is mainly found in North American New Towns, e.g. Seaside (1979), Laguna West (1989) and Celebration (1996).

Though, when *Car Cities* are designed with an irregular and curvilinear layout, often an interesting spin arises concerning the separation of pedestrian mobility from vehicular mobility: cars are forced to run more slowly inside the residential neighbourhoods, thanks to the curved mesh and the narrow road section lined with tree rows as in Laguna West. Anyway *Car Cities* generally have a very low population density, in the order of 28 inh/ha (Laguna West) or 20 inh/ha (Celebration); this definitely favours private transportation and vehicular movement.

6 CONCLUSIONS

This paper confirms that planning from the scratch success is extremely dependent on mobility planning. One can easily single out, along with the previous categories, three different planning attitudes, at least according to the geographical context. In America—despite the elaboration of TOD's theories—the *New Towns'* development "*fails the sustainability test because it is not served by public transportation and is as completely car dependent as any other suburb*" (Hall, 1994). In Europe, where settlements keep a higher density, the attitude towards an efficient public transport system and compact cities prevails. Lastly, in Africa and Asia, the trend is very different. In Far East countries, where urban densities are still high, smart transportation planning principles are affected.

The hierarchical model based on the neighbourhood unit is still applied and affects the hierarchical model of both the road and transport systems. Particularly it affects the pedestrian and cycle paths at the neighbourhood scale and public transport systems that connect the neighbourhood to workplaces or to the railway station. The neighbourhood can have different size depending on urban density ranging from about 50 to 100 ha and population densities from 100 to 400 inh/ha or even more. The grid of main urban, always oversized, thoroughfare collector roads in residential areas can range most of the time from 250 ha (Lelystad) to about 80 ha (Vallingby and Milton Keynes).

6.1 *Barriers and drivers*

The present research can be improved through the analysis of particular phenomena that have been highlighted in some uncommon cases. Firstly, the case of "dormitory towns and suburbs", that are completely lacking attractors centre and work opportunities and are poorly served by public transport (Senri,1962 and Tama, 1965).

Secondly, the case of those *Consumer Cities* which have failed to fulfil their function. This is the case for many recent Chinese New Towns, such as Yujiapu, a financial city classified today as "Ghost town". Another example is Halle-Neustadt (1960) which has now turned into a shrinking city.

An important driver for new researches pertains the quantitative evaluations regarding the characteristics of urban road systems (size compared to traffic) and the search of higher densities, as low densities are the danger of current urban developments, which in the outskirts most of the time are declining and sometime lower than 30 inhabitants per hectare.

An important contribution to this research can be derived from the processes of self-renewal realized by many aged European New Town that nowadays have become "Upgraded cities", such as Aarhus. There are even more innovative projects, where the infrastructure is no longer based on transports but on information technology instead, such as PlanIT Valley (2008), which is globally connected through the web.

REFERENCES

Buchanan, C. D., 1958. Mixed Blessing, The Motor in Britain. Bristol: Leonard Hill.

Duany, A., Plater-Zyberk, E. & Speck, J., 2000. Suburban Nation. The Rise of Sprawl and the Decline of the American Dream. New York: North Point Press.

Foletta, N., 2014. ITDP. [Online] Available at: https://www.itdp.org/wp-content/uploads/2014/07/22.-092211_ITDP_NED_Desktop_Houten.pdf [Accessed 23.05.2017].

Frampton, K., 1992. Modern architecture a critical history. III ed. London: Thames and Hudson.

Giorgieri, P. & Ventura, P. eds., 2007. Strada, strade. Teorie e tecniche di progettazione urbanistica. Firenze: Edifir.

Glaeser, E. L., Kolko, J. & Saiz, A., 2001. Consumer city. Journal of Economic Geography, Vol. 1, pp. 27–50.

Hall, P., 1994. Squaring the circle: can we resolve the Clarkian paradox?. Environment and Planning B, N. 21, pp. 79–94.

Hall, P., 2002. Cities of Tomorrow, an Intellectual History of Urban Planning and Design in the Twentieth Century. III ed. Oxford: Blackwell.

Howard, E., 1898. To-morrow. A Peaceful Path To Social Reform. London: Swann Sonnenschein Publishers.

INTI, 2017. International New Town Institute. [Online] Available at: http://www.newtowninstitute.org/ [Accessed 04.01.2017].

ITS Institute, 2017. Bus Rapid Transit Technologies: Assisting drivers operating buses on road shoulders. [Online] Available at: http://www.its.umn.edu/Research/ProjectDetail.html?id=2001046 [Accessed 23.05.2017].

Kaufmann, R., 1926. Planning of jewish Settlements in Palestina. The Town Planning Review, 11.12(2).

Krier, R., 1979. Urban space (Stadtraum). Foreword by Colin Rowe. London: Academy editions.

Lavedan, P., 1952. Histoire de l'urbanisme. Paris: Henri Laurens Editeur.

Merlin, P., 1971. Le città nuove. Bari: Laterza.

National BRT Institute, 2017. Bus Rapid Transit. [Online] Available at: https://nbrti.org/ [Accessed 23.05.2017].

Russel, A., 2016. Car-centric data encourages car-centric transportation planning—Mobility Lab. [Online][Accessed 23.05.2017].

Stein, C. S., 1957. Towards new towns for America. II ed. Cambridge(MA): MIT Press.

Stern, R. A. M., Fishman, D. & Tilove, J., 2013. Paradise planned. The garden suburb and the Modern City. I ed. New York: The Monacelli Press.

Stübben, J., 1907. Der Städtebau. Stuttgart: Alfred Kröner.

Transport Oriented Development Institute, 2016. [Online] Available at: http://www.tod.org/home/about.html [Accessed 04.01.2017].

Unwin, R., 1971. La pratica della progettazione urbana. II ed. Milano: Il Saggiatore.

Ventura, P., Zazzi, M., Caselli, B., Carra, M., Damianakos D., & Cillis, M., 2006–2017. Urbanistica Paesaggio e Territorio. [Online] Available at: http://www.urbanistica.unipr.it/ [Accessed 23.05.2017].

Town and Infrastructure Planning for Safety and Urban Quality – Pezzagno & Tira (Eds)
© 2018 Taylor & Francis Group, London, ISBN 978-0-8153-8731-2

Integration of SUMPs (Sustainable Urban Mobility Plans) and SEAPs (Sustainable Energy Action Plans): An assessment of the current situation in Italian small and medium-sized cities

F. Morea, L. Mercatelli, S. Alessandrini & I. Gandin
AREA Science Park, Trieste, Italy

ABSTRACT: Sustainable Urban Mobility Plans (SUMPs), Sustainable Energy Action Plans (SEAPs) and Sustainable Energy and Climate Action Plans (SECAPs) are the main tools promoted by the European Union to foster local sustainable mobility, energy, climate change adaptation and mitigation policies.

The paper reports the outputs of a study on the current situation of SUMPs and SEAPs/SECAPs in Italian small and medium-sized cities (50.000 to 350.000 inhabitants). Detailed data on the plans developed to date (or lack thereof), the status of each plan as well as the need to review existing plans are provided. Furthermore, the study focuses on the concept of "harmonization of SUMPs and SEAPs" and detailed data are presented on the barriers hindering harmonization.

Finally, the paper identifies a set of Italian cities that have at least a SUMP, SEAP or SECAP, and have a potential to harmonize planning, implementation and monitoring of their sustainable energy and mobility plans.

1 INTRODUCTION

Sustainable Energy Action Plans (SEAPs), Sustainable Energy and Climate Action Plans (SECAPs) and Sustainable Urban Mobility Plans (SUMPs) are the main tools promoted by the European Union to foster sustainable mobility, sustainable energy production and consumption, climate change adaptation and mitigation policies at the local level.

Based upon the European Union's 2020 Climate and Energy Package, focussing on emissions cuts, renewables and energy efficiency with the main purpose of mitigating the impacts of climate change, the Covenant of Mayors initiative, launched by the European Commission in 2008, has brought global challenges to the local level. Local authorities have been developing SEAPs through a standardized methodology and selecting a set of actions to reduce carbon emissions and increase energy efficiency and production from renewable sources. This way, local authorities have committed to reducing CO_2 emissions by a minimum of 20% by the year 2020. According to the Covenant of Mayors' portal, more than 39% of Italian municipalities have signed the Covenant, representing almost 50% of the signatories. However, although 94,5% of Italian signatories have developed a SEAP, only 23,2% of the municipalities with a SEAP have also monitored their plans' results. This suggests that SEAPs are scarcely implemented in Italy. The twelve Italian larger municipalities (>250.000 inhabitants) are the best performing ones: they have all signed the Covenant and developed a SEAP, and half of them have monitored their plans. Focussing on municipalities with between 50.000 and 350.000 inhabitants, 111 municipalities signed the Covenant (83,4%), 107 developed a SEAP (96,4% of the signatories) and 31 have monitored their plans' results, which suggests that only 27,9% of SEAPs are actually implemented.

With indicators showing good progress and 2020 approaching, in 2014 the European Commission decided to raise the stakes and set new ambitious targets for 2030 with the 2030 Energy Strategy: a 40% cut in greenhouse gas emissions compared to 1990 levels; at least a 27% share of renewable energy consumption; at least 27% energy savings compared with the business-as-usual scenario. In this framework, the new Covenant of Mayors for Climate and Energy proposes an evolution of SEAPs into SECAPs, tackling both climate change mitigation (as in SEAPs) and adaptation (assessment and reduction of major climate risks and vulnerabilities, water management, urban planning). In Italy, 88 municipalities have signed the new Covenant and 6 of them have developed a SECAP. Moreover, 46 more municipalities have added to their SEAPs climate change adaptation objectives.

Introduced with the 2011 White Paper on Transport 'Roadmap to a Single European Transport Area—Towards a competitive and resource efficient transport system' and the 2013 Urban Mobility Package, SUMPs are strategic plans based upon a long-term vision, with the main goal to provide integrated solutions to transport and mobility needs of people and goods, guaranteeing technical, economic, environmental and social sustainability. In Italy SUMPs are considerably less spread than SEAPs for two reasons: firstly, SUMPs are newer, secondly SUMPs are usually presented as useful tools primarily for medium to large-sized functioning areas (>100.000 inhabitants), whereas SEAPs can be successfully developed also for small to medium-sized municipalities. According to the dedicated national online observatory (*'Osservatorio PUMS'*) as of 21 March, 2017, only 70 Italian municipalities have, or are in the process of developing, a SUMP. Those with a population of between 50.000 and 350.000 inhabitants are 48, of which 5 have had their SUMP approved by the City Council, 7 have a SUMP with a formal political adoption, pending the approval by the City Council, and 36 are drafting their SUMPs. In the year 2000, long before the introduction of SUMPs, Law 340 of 24 November 2000 introduced the concept of PUM (italian acronym for "Piano Urbano della Mobilità" i.e. Urban Mobility Plan) in the Italian system, addressed to satisfy mobility needs while reducing environmental pollution, noise, energy consumption, and enhancing road safety through a mix of reduced-impact regulatory, infrastructural and management solutions. Individual or aggregations of municipalities, provinces and regions were called to develop PUMs for areas with at least 100.000 inhabitants. According to the national online observatory, 47 PUMs were passed by municipalities from 2001 to 2015, of which 43 have a population of between 50.000 and 350.000 inhabitants.

PUMs/SUMPs and SEAPs/SECAPs are typically developed by different departments in local authorities, often missing the opportunity to exploit synergies and economies of scale deriving from their harmonization and alignment with urban planning.

2 OBJECTIVE OF THE PAPER

The paper reports the outputs of a study on the current situation of PUMs/SUMPs and SEAPs/SECAPs in Italian small and medium-sized local authorities (with less than 350.000 inhabitants), providing detailed data on the plans developed to date (or lack thereof), status of each plan as well as on the need to review existing plans. Furthermore, the study focuses on the concept of harmonization of SUMPs and SEAPs and the barriers hindering the plans' harmonized development and implementation.

3 METHODOLOGICAL APPROACH AND/OR RESEARCH APPROACH

The main sources of data are public databases on SEAPs, SECAPs, PUMs and SUMPs. Specifically, data on SEAPs and SECAPs are collected by the Covenant of Mayors Office and made available through the official website. Data on Italian municipalities and population have been provided by the National Institute for Statistics (Istat). The source of data on PUMs and SUMPs is the dedicated national online observatory (*'Osservatorio PUMS'*).

Following the SIMPLA project's approach [8], the analysis devotes particular attention to municipalities with a population between 50.000 and 350.000 inhabitants.

Further specific information on the state of development and implementation of SEAPs, SECAPs, PUMs and SUMPs and their integration (or lack thereof) has been collected through questionnaires sent to all Italian cities in the chosen range.

Table 1. PUM/SUMP status.

No mobility plan	The municipality does not have a (sustainable) urban mobility plan
PUM only	The municipality has adopted a PUM (not a SUMP)
SUMP in progress	The municipality is drafting a SUMP
SUMP implemented	The municipality's SUMP is being implemented

Table 2. SEAP/SECAP status.

No SEAP/SECAP	The municipality does not have a SEAP or a SECAP
SEAP	The municipality has a SEAP (target: –20% CO_2 emissions in 2020; may also contain additional adaptation measures)
Joint SEAP	The municipality has a joint SEAP (carried out collectively by a group of neighbouring local authorities)
SECAP	The municipality has a SECAP (target: –40% CO_2 emissions in 2030; + adaptation measures)
Joint SECAP	The municipality has a joint SECAP (carried out collectively by a group of neighbouring local authorities)

Table 3. Presence of PUM/SUMP and/or SEAP/SECAP.

No SUMP or SEAP	The municipality does not have a SUMP, nor a SEAP
SEAP/SECAP, no PUM/SUMP	The municipality has a SEAP/SECAP but does not have a PUM/SUMP
PUM/SUMP, no SEAP/SECAP	The municipality has a PUM/SUMP but does not have a SEAP/SECAP
SEAP/SECAP and PUM/SUMP	The municipality has both a SEAP/SECAP and a SUMP

The stage of development/implementation for SUMPs has been classified on a scale from 0 to 3 according to the following Table 1:

The stage of development/implementation for SEAPs and SECAPs has been classified on a scale from 0 to 4 according to the following Table 2:

The stage of implementation of both SEAPs/SECAPs and PUMs/SUMPs has been classified on a scale from 0 to 3 according to Table 3:

4 RESULTS OF THE RESEARCH

Information on Italian municipalities' population, integrated with data reported in the Covenant of Mayors' monitoring catalogue and the national observatory on SUMPs has been used to further investigate the spreading of SEAPs/SECAPs and PUMs/SUMPs and their stage of development.

Given the national-level of collected data, the first analysed aspect is the spreading of sustainable energy and mobility plans in Italy. Information has been used to create a high-resolution map (see supplementary materials available online at www.simpla-project.eu) where it is possible to check simultaneously the development of SEAPs/SECAPs and PUMs/SUMPs for each municipality. Given the presence of some densely covered areas, an an analysis of the spreading of plans in Italian regions has been carried out. For each region, the proportion of municipalities with a SEAP/SECAP and a PUM/SUMP have been obtained and results reported in Fig. 1.

As regards the presence of PUMs/SUMPs, some regions have significant coverage, as for instance Umbria (with 5 municipalities, or 5% of the total), Puglia (with 12 municipali-

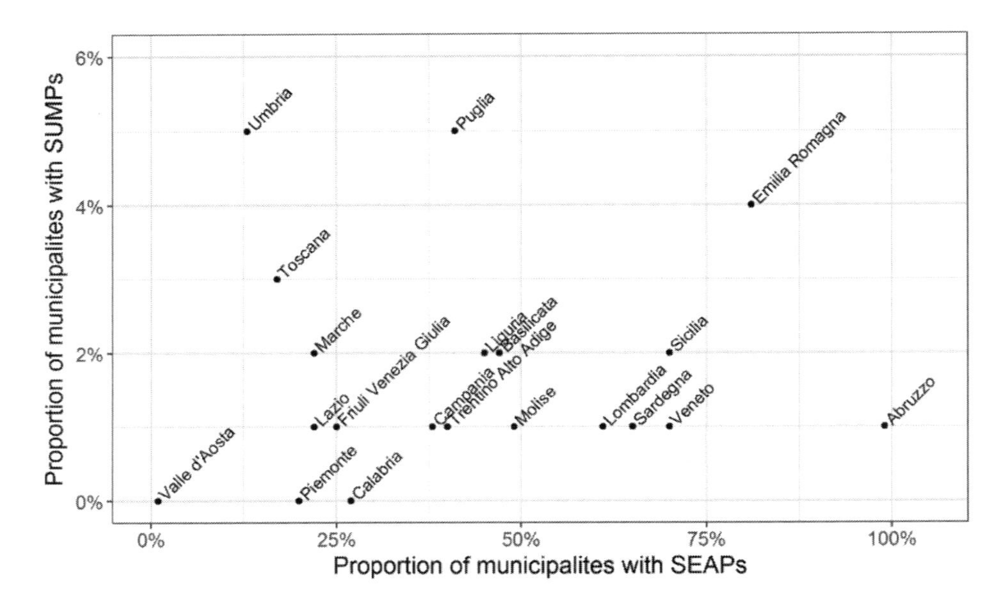

Figure 1. SEAPs/SECAPs and PUMs/SUMPs in Italian regions. Each point of the graph represents one of the 20 Italian regions. The X-axis reports the percentage of municipalities with a SEAP/SECAP in the region, while the Y-axis reports the percentage of PUMs/SUMPs. For visualization purposes, the two axes' scales are different.

ties, or 5%) and Emilia Romagna (with 12 municipalities, or 4%). As for SEAPs/SECAPs, the regions with the highest coverage are Abruzzo (with 302 municipalities, or 99%), again Emilia Romagna (with 274 municipalities, or 81%) Veneto (with 408 municipalities, or 70%) and Sicily (with 272 municipalities, or 70%).

A deduction can be made that those regions have implemented strategies to promote the development and adoption of the plans.

An interesting overview of SEAPs/SECAPs and PUMs/SUMPs adoption can be obtained by crossing the size of municipality involved—and thus the population benefitting from them and the stage of development of SEAPs/SECAPs and PUMs/SUMPs. The Italian population has been devided into 6 classes based upon the size of the population living in the urban area and the distribution of sustainable energy and mobility plans analysed. As reported in

Fig. 2, the coverage of SEAPs/SECAPs is reasonably comparable across the 6 classes (from 63% for 50–100 K to 100% for >350 K). However, given the distribution of the Italian population, mainly living in small and medium centres, the majority of individuals who do not benefit from the effects of a SEAP/SECAP live in small and medium centres. Performing a similar analysis on PUMs/SUMPs, we observe more extreme differences in the spreading of plans. Almost the entirety of urban centres with less than 100 K inhabitants (7,941 out of 8,000) did not reach the earliest step of development of a mobility plan. On the other hand, larger municipalities (>100 K) seem to have very often adopted a mobility plan. Because of that, only 17% of the population for municipalities in the 100–350 K class and 13% for municipalities in the >350 K class do not benefit from the effects of a mobility plan. Overall, the proportion of Italian population benefitting from a SEAP/SECAP is 66%, while for PUMs/SUMPs the percentage decreases to 25%.

As regards the progress of sustainable energy and mobility plans, from the available information we observe a general low level of implementation. Based on the data reported in, only 23% of the municipalities that have developed a SEAP have also provided a monitoring plan. Moreover, only 2% of the centres is developing a SECAP, the new version of the plan, establishing more ambitious goals using a broader system analysis. The trend seems to be confirmed by data on SUMPs. Out of the 96 municipalities with a mobility plan, 68 have adopted the SUMP, but only 16 municipalities have a plan passed by the local council.

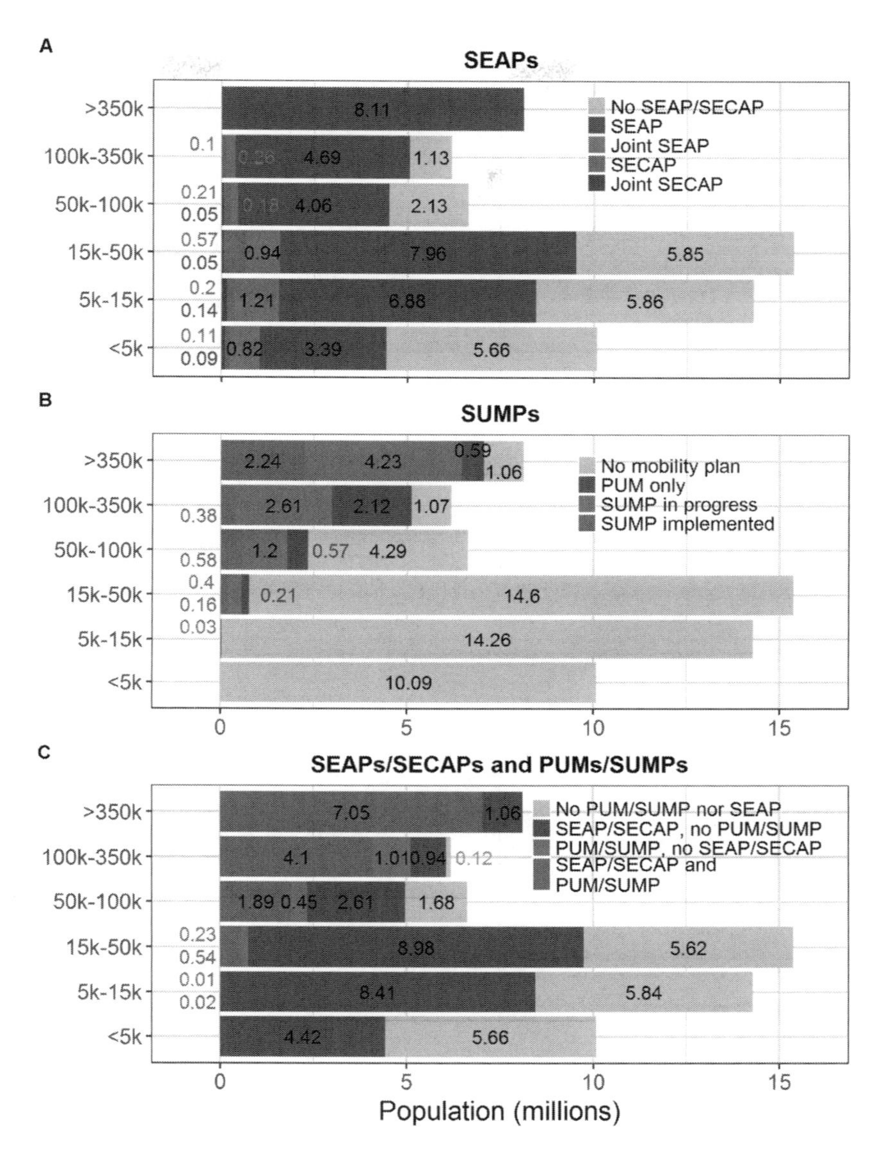

Figure 2. Italian population living in municipalities having SEAPs/SECAPs and PUMs/SUMPs. For each graph, the population has been divided into 6 classes depending on the size of municipalities (x-axis). The Y-axis reports the total number of inhabitants for each class. The number of inhabitants is reported for each bar segment (measured in million). Plot A and B represent the stage of development respectively for SEAPs/SECAPs and PUMs/SUMPs, while plot C reports the overlapping of the plans.

Given the moderate spreading of SUMPs, the number of municipalities that have both a PUM/SUMP and a SEAP/SECAP is also limited: 74 out of 8000, that is 0,9% of the municipalities. However, in terms of people involved it corresponds to 22% of the Italian population for which the analysis of integration of the plans is highly relevant. Moreover, this number is likely to increase. As reported in Fig. 3, a peak of new SEAPs was recorded in the year 2013 with 895 municipalities producing new plans, yet data show that plans have been developed also over the last 3 years (146 in 2015, 268 in 2016 and 14 in the first months of 2017 respectively).

Data shown in Fig. 4, resulting from a survey based upon a single question submitted to the 133 Italian municipalities with populations between 50k and 350k inhabitants, show an overview on SEAPs, SECAPs and SUMPs status and their integration in the cities that par-

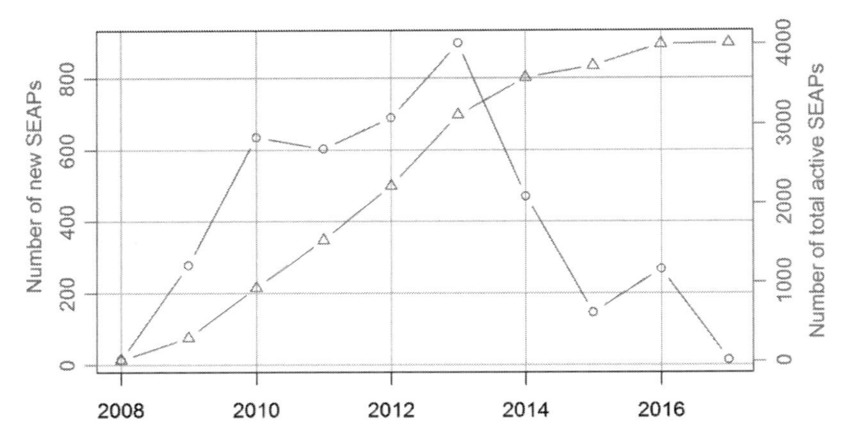

Figure 3. Development trend for SEAPs up to March 2017. The blue line represents the number of new plans developed for each year, while the orange line shows the overall number of plans, including previous years, that is the total number of active plans.

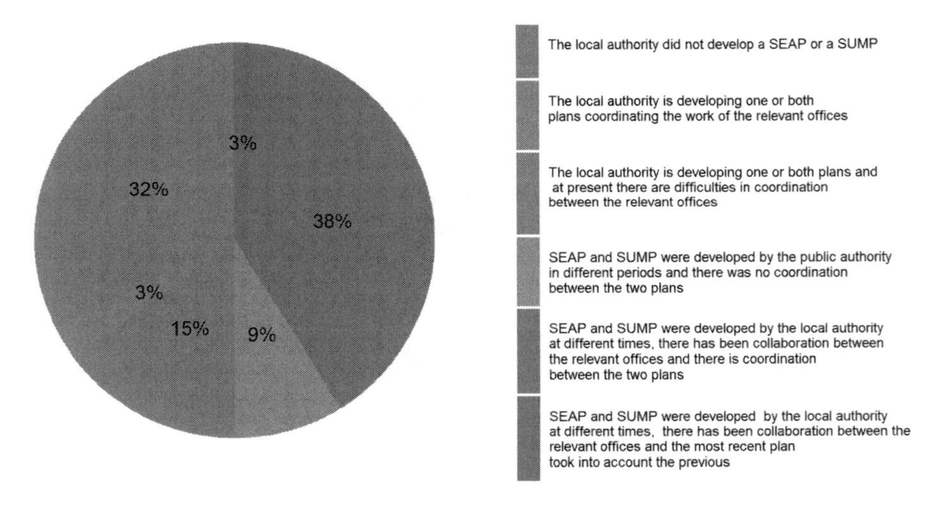

Figure 4. SEAPs and SUMPs status survey results.

ticipated in the survey. As a result, it is possible to observe two predominant situations: the lack of both plans or the presence of both plans where each has been developed taking into account the other with collaboration between departments in the local authority during the design and implementation phases.

5 CONCLUSIONS

The paper describes the Italian SEAP/SECAP and PUM/SUMP scene, highligting development and implementation status in relation to the number of inhabitants of each municipality. Data from ISTAT (the Italian National Institute of Statistics), the Covenant of Mayors Office and the Osservatorio PUMS (Italian national observatory on mobility plans) were collected, analyzed and presented in an aggregated form.

From the analysis of results and looking at data at a regional level, it is possible to state that SEAP development, implementation and monitoring methodology has been strongly supported and well defined both by the Covenant of Mayors' Office (CoMO) and by national and regional coordinators, resulting in a very significant number of plans passed by local councils and approved by CoMO. The SUMP methodology and support strategy has so far not yet led to the same results. Integration between SEAPs/SECAPs and SUMPs can be considered as a step forward to bring sustainable mobility planning and sustainable energy planning closer. The SIMPLA project, with the drawing up of specific guidelines for harmonization between SEAPs and SUMPs, and with training and coaching activities dedicated to local authorities, does exactly that, aiming to support the development and update of sustainable mobility and energy plans.

5.1 *Barriers and drivers*

Based on the above results, significant barriers can be observed, the most relevant being lack or fragmented availability of multidisciplinary skills in public authorities at all levels, for the production of multi-sector planning tools allowing the achievement of an integrated approach to develop sustainable energy and mobility plans.

5.2 *Further research*

Small municipalities have the capacity to develop complex plans (e.g. SEAPs and SECAPs) addressing sustainable energy and climate mitigation and adaptation, but do not seem to be as effective in tackling sustainable mobility. This highlights the need to further investigate appropriate methodologies and support schemes for sustainable mobility in small municipalities.

ACKNOWLEDGMENTS

The research leading to these results has been funded by the European Union's Horizon 2020 research and innovation programme under grant agreement n° 695955.

SIMPLA Sustainable Integrated Multisector Planning

http://www.simpla-project.eu/en/

Special thanks to the Covenant of Mayors Office and to Endurance Italia—Osservatorio PUMS for providing the data required for the purposes of this article.

REFERENCES

European Commission, (2008). 2020 Climate and Energy Package. (web page visited on April 2017). https://ec.europa.eu/clima/policies/strategies/2020_en.

Covenant of Mayors for Climate and Energy. (Web page visited on April 2017). http://www.covenant-ofmayors.eu/about/covenant-of-mayors_en.html.

European Commission, (2014). 2030 Climate and Energy Framework. (web page visited on April 2017). https://ec.europa.eu/clima/policies/strategies/2030_en.

European Commission, (Brussels, 28.3.2011 COM(2011) 144 final). White Paper—Roadmap to a Single European Transport Area—Towards a competitive and resource efficient transport system. http://eur-lex.europa.eu/legal-content/EN/ALL/?uri=CELEX:52011DC0144.

European Commission, (Brussels, 17.12.2013 COM(2013) 913 final). Communication from the Commission to the European Parliament, the Council, the European Economic and Social Committee and the Committee of the Regions—Together towards competitive and resource-efficient urban mobility.

ENDURANCE ITALIA—Osservatorio PUMS, (2016). (web page visited on April 2017). http://www.osservatoriopums.it/.

Fresner J., Krenn C., Morea F., Mercatelli L., Alessandrini S., (2017). Guidelines for the harmonization of energy and mobility planning. (STENUM GmbH, Austria, AREA Science Park, Italy). http://www.simpla-project.eu/en/news/simpla-guidelines-1st-release/.

Italian National Institute of Statistics, (web page visited on April 2017). http://demo.istat.it/.

Town and Infrastructure Planning for Safety and Urban Quality – Pezzagno & Tira (Eds)
© 2018 Taylor & Francis Group, London, ISBN 978-0-8153-8731-2

Investigating the transit-orientation of existing urban environments along four railway corridors in Italy

R. Papa, G. Angiello & G. Carpentieri
Università degli Studi di Napoli Federico II, Naples, Italy

ABSTRACT: A urban environment with high densities, mixed and diverse land uses, located within an easy-walkable distance to railway stations increases the use of public transport.

This work presents a GIS-based analysis of the urban environment along four recently-developed/extended railway corridors: the Milan Line 3, the Turin Line 1, the Rome Line B and the Naples Line 1. The analysis aims to quantify the degree of transit orientation of these four corridors. Twelve built environment indicators are presented and evaluated for the selected case studies.

Results show a higher degree of transit orientation for Line 3. The Naples line 1 is characterized by a high level of population density and street connectivity. While the metro B presents the highest performances, the Turin Line 1 presents lower population and job density.

The analysis provides interesting insights for policy formulations aimed at increasing transit use trough transit corridors and station areas re-development.

1 INTRODUCTION

Investments in urban rail systems has increased rapidity over the past few decades in Italy. These systems have been generally planned as instruments to solve transport and environmental problems associated with the extensive use of private cars such as congestion, air pollution and environmental damage.

A considerable body of professional and academic research has analysed the factors behind success and failure of urban rail systems. In broad overview, the evidence leads to one compelling conclusion: while transit fares, safety and quality have a significant impact on transit ridership (Redman et al., 2013), it needs to be acknowledge that, in the long run, the success of most urban rail policies is highly dependent on the existence of a supportive built environment (Ryan and Frank, 2009; Ewing and Cervero, 2010; Lindsey et al., 2010; Sung and Oh, 2011). Built environment, i.e. the human-made part of our physical surroundings, can indeed provide the necessary (albeit not sufficient) conditions for shaping transit use, without which, urban rail policies would have limited to no effect (Suzuki et al., 2013).

Built environment characteristics can be worked out (within a certain extent) directly by urban planning instruments such as zoning, building standards, revitalization programs or site-specific interventions. Thus, an analysis of built environment characteristics along urban rail corridors could provide interesting insights for policy formulations aimed at increasing transit use trough urban planning.

2 OBJECTIVE OF THE PAPER

In this paper, we present some preliminary results of a research project aimed at assessing the built environment along transit corridors in the largest Italian cities. In particular, our attention here is on four recently-developed or extended transit corridors: the Milan Line

Table 1. Main features of the selected railway corridors.

City	Line	Number of stations	Lenght (km)
Milan	3	21	18,5
Turin	1	21	13,2
Rome	B	26	23,5
Naples	1	18	18

Table 2. Data, formats, and sources.

Data	Format	Source
Census tracks	Spatial	ISTAT
Population	Non spatial	ISTAT
Jobs by sector	Non spatial	ISTAT
Street network	Spatial	OpenStreetMap
Stations' location	Spatial	Cities' open data portal
Building	Spatial	Italian Ministry of the Environment

3, the Turin Line 1, the Rome Line B and the Naples Line 1. Our main aim is to conduct a spatial analysis along these four corridors in order to understand, quantify and visualize their degree of transit orientation. By "*degree of transit orientation*" we mean the capacity of an urban environment to support transit use, encourage walking and reduce car dependency.

3 DATA AND CONTEXT

The main feature of the railway corridors investigated in this paper are reported in Table 1. We select these corridors according to the following criteria: i) they were opened or extended in the same period; ii) they connect the city center with the outlying parts of the city and iii) they have a similar number of stations.

In order to compute the indicators detailed in the next section, we combined spatial data in GIS format and non-spatial data (statistical data) from different sources, as reported in Table 2.

4 METHODOLOGICAL APPROACH

Based on an extensive literature review, we identified four spatial factors that support transit use, encourage walking and reduce car dependency:

- *Density* refers to the number of urban activities located within a walkable distance from a railway station. High densities are important to support high-frequency transit service and to foster lively, walkable communities. Thus, high density development is an indication of high travel demand and, possibly, high transit patronage (Sung and Oh, 2011);
- *Diversity* is a measure of the integration of different activities (e.g. dwellings, workplaces, shops, schools, and health care services) in the same area. Diversity produces a more balanced demand for public transport over time (reducing differences between peak and off-peak periods) and space (in terms of direction of flow), while encourages walking (Ewing and Cervero, 2010);

Table 3. Built environment indicators.

Spatial factors	Indicators	Abbreviation
Density	Population density	Pop_D
	Jobs density	Job_D
	Activities density	Act_D
Diversity	Land use mix (entrophy)	LUM_1
	Land use mixed-ness	LUM_2
	Job-housing balance	JHB
Proximity	Population proximity	Pop_P
	Jobs proximity	Job_P
	Activities Proximity	Act_P
Connectivity	Average block lenght	ABL
	Density of street intersections	DSI
	Route directness	RD

- *Spatial proximity* refers to the spatial relationships between the location of urban activities and the location of the railway station. The significance of proximity is that the more people living and/or working near a railway station, the greater the likelihood the service will be used (Lindsey et al., 2010);
- *Connectivity* refers to the directness of links and the density of connections (i.e., intersections) in a street network. A station area with a highly-connected street network has streets with many short links, numerous intersections, and few dead-ends and cul-de-sac. Well-connected street patterns favour access to stations on foot and increase transit ridership (Ryan and Frank, 2011).

The aforementioned spatial factors have been subsequently transformed in twelve measurable indicators able to describe the urban environment of station catchment areas, as reported in Table 3. The proposed indicators aim to answer to the following questions: i) How dense and diverse is the land use surrounding a railway station? Which are the spatial relationships between the location of a railway station and the location of its surrounding urban activities? iii) Does the design of the urban environment around a railway station encourage walking?. We calculated these indicators for each station catchment area and then aggregated the results at the corridor level. A detailed description of each indicator and the aggregation method will be provided in a forthcoming paper. In this section, we briefly illustrate the GIS-based procedure we implemented to calculate such indicators at the station level (Figure 1).

The procedure consists of four steps as follow[1]:

1. *Estimating population and jobs for a regular grid cell.* The use of data from multiple sources and the variable size of census tracts might be problematic when estimating population and jobs within walking distance from a transit station (Papa and Bertolini 2015). Consequentially, we selected an exagonal grid as the basic spatial unit of our analysis (Fig. 1). Then we allocated jobs and population to the grid cells using the area-ratio method (Gutiérrez and García-Palomares 2008);
2. *Creating a walking network from OpenStreetMap (OSM) data.* OSM provides free vector geographic databases using contributions from Internet users. It has been extensively used as a source of spatial data for many researchers. However, the accuracy and the completeness of OSM data depends, among other things, on the number of contributors (Haklay, 2010). In order to accurately model walking routes to transit stations, some topological corrections were made to the original data, while some missing street segments where

1. In this project, the desktop GIS package ArcGIS 10.4 is selected as a platform for the implementation of the built environment indicators.

Figure 1. Defining station catchment areas.

manually added. Furthermore, street segments classified as "motorway" were removed from the network;

3. *Defining station catchment areas*. Station catchment areas are broadly based on an understanding of how far people are willing to walk to take transit. One key aspect to take into account when defining station catchment areas is the delimitation of the station radius of influence. Different radius have been used in the literature. In this research, a radius of 500 m was considered to reflect an 8-min walking time (Pagliara and Papa 2011). Using the Network Analysis tool, a catchment area around each station was created and overlapped with the grid cell layer, the street network layer and the buildings layer (Figure 1);
4. *Calculating built environment indicators*. Using the Spatial Join Analysis tool, the following information were associated to each station: population and jobs (by sector); building's footprints; street segment's lengths; Euclidean and network distances from cell centroids to the associated station. After these operations, built environment indicators were calculated for each station. Subsequently, these values were aggregated for each line.

5 RESULT

The results of our analysis are summarized in Figure 2 and Figures 3a, 3b, 3c and 3d. For sick of brevity, we report here the results for the Milan Line 3 and then a comparison between the four transit corridors.

In the city of Milan, results show a significant decline in built environment indicators, moving from the historical city centre to the outlying parts of the city (Figure 2). Built environment attributes of station areas are similar for historical neighbourhoods that, as expected, show a high degree of connectivity, a good functional mix, and relatively high densities. On the contrary, low densities, mono-functionality and non-pedestrian friendly environments characterize post-World War II urban expansions such as Rogoredo or Affori. Indeed, these areas present mainly residential buildings of great dimension and a great degree of functional separation.

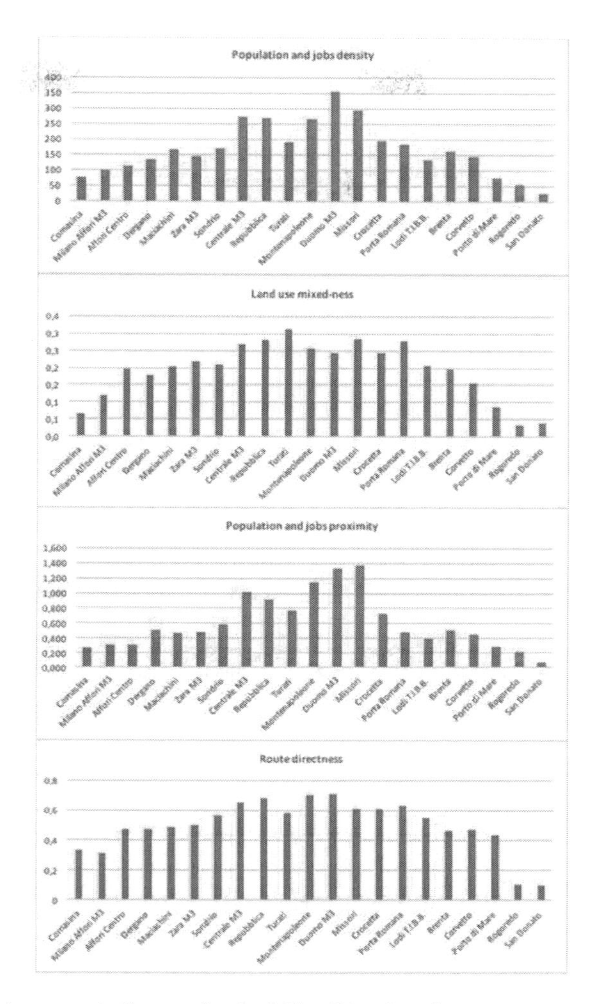

Figure 2. Built environment indicators for the Milan Line 3 station areas.

Activities density ranges from a maximum of 350 activities per hectare in the central station of Duomo to a minimum of 25 activities per hectare in the outlying station of San Donato. While activities density decries from the city centre to the periphery, also the number of jobs compared to the numbers of inhabitants declines. These is reflected in the pattern of land use mix. Finally, it is interesting to note that, in the central areas, economic activities tend to concentrate in close proximity to the railway stations, while in the periphery the opposite is true. This means that, for economic activities, the proximity to the railway station is more important in the central area then in the suburbs.

The results for the four transit corridors are presented trough means of radar charts (Figures 3a, 3b, 3c and 3d). Results show that the Milan Line 3 has the higher degree of transit orientation as railway stations are located within dense, mixed-use activity centres. For the Naples Line 1, results indicate a high level of population density, a high degree of street connectivity and a medium to low degree of functional mix. The Rome metro B presents the highest performances in terms of jobs and population density. However, especially in the periphery, station areas are characterized by mono-functionality and non-pedestrian friendly environments. Finally, the Turin Line 1 presents lower population and job density, a medium to low degree of land use diversity and a relatively high level of street connectivity.

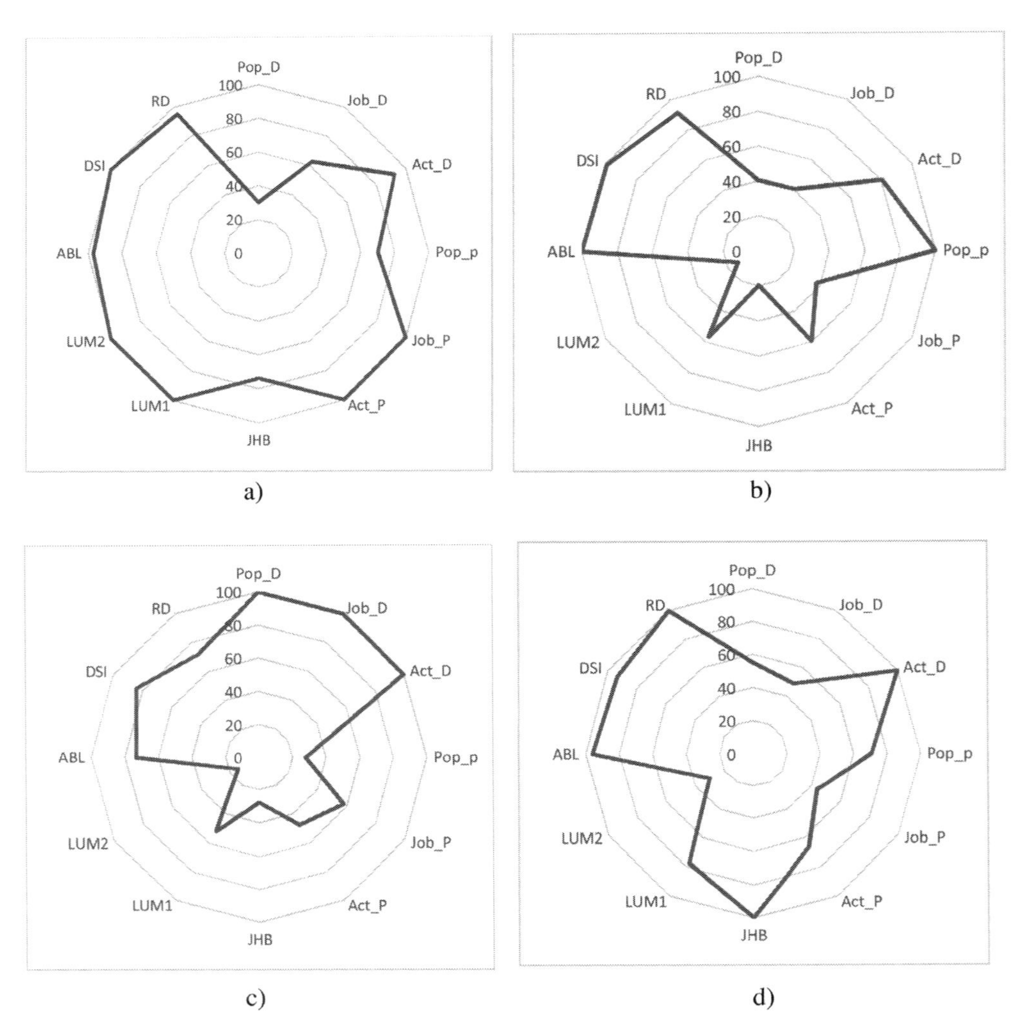

Figure 3. a. Built environment indicators for the Milan Line 3 corridor, b. Built environment indicators for the Turin Line 1 corridor, c. Built environment indicators for the Rome Line B corridor, d. Built environment indicators for the Naples Line 1 corridor.

6 CONCLUSIONS

A supportive built environment is considered a fundamental attribute for the successful implementation of any railway policy. In this study, we presented a GIS-based methodology to evaluate the degree of transit orientation of stations catchment areas and transit corridors. The proposed methodology is applied in four relevant Italian case studies, revealing, for the first time, density, diversity, proximity and connectivity patterns along the railway corridors of the four largest Italian cities.

The outputs of our analysis can be used for policy formulations aimed at increasing transit use trough transit corridors and station area re-development. The results of our investigation call for a more coordinated approach to land use and transport planning. They also call for site-specific solutions, especially for suburban areas, to alleviate accessibility problems and support transit use by crating dense, mixed, and pedestrian-friendly environments around transit stations.

Finally, we identified a number of further research developments. First, in our opinion, it would be useful to synthetize the characteristics of the built environment in a single metric. This could be done by means of multicriteria spatial analysis as in Singh et al. (2014). Second, we plan to extend our analysis to the whole urban railway network of the four cities, in order to make comparisons not only between lines, but also between cities.

REFERENCES

Ewing, R., & Cervero, R. (2010). Travel and the built environment: a meta-analysis. *Journal of the American planning association,* 76(3), 265–294.

Gutiérrez, J., & García-Palomares, J.C. (2008). Distance-measure impacts on the calculation of transport service areas using GIS. *Environment and Planning B: Planning and Design*, 35(3), 480–503.

Haklay, M. (2010). How good is volunteered geographical information? A comparative study of OpenStreetMap and Ordnance Survey datasets. *Environment and planning B: Planning and design*, 37(4), 682–703.

Lindsey, M., Schofer, J.L., Durango-Cohen, P., & Gray, K.A. (2010). Relationship between proximity to transit and ridership for journey-to-work trips in Chicago. *Transportation Research Part A: Policy and Practice*, 44(9), 697–709.

Papa, E., & Bertolini, L. (2015). Accessibility and transit-oriented development in European metropolitan areas. *Journal of Transport Geography, 47,* 70–83.

Pagliara, F., & Papa, E. (2011). Urban rail systems investments: an analysis of the impacts on property values and residents' location. *Journal of Transport Geography*, 19(2), 200–211.

Redman, L., Friman, M., Gärling, T., & Hartig, T. (2013). Quality attributes of public transport that attract car users: A research review. *Transport Policy*, 25, 119–127.

Ryan, S., & Frank, L.F. (2009). Pedestrian environments and transit ridership. *Journal of Public Transportation, 12*(1), 39–57.

Singh, Y.J., Fard, P., Zuidgeest, M., Brussel, M., & van Maarseveen, M. (2014). Measuring transit oriented development: a spatial multi criteria assessment approach for the City Region Arnhem and Nijmegen. *Journal of Transport Geography*, 35, 130–143.

Sung, H., & Oh, J.T. (2011). Transit-oriented development in a high-density city: Identifying its association with transit ridership in Seoul, Korea. *Cities*, 28(1), 70–82.

Suzuki, H., Cervero, R., & Iuchi, K. (2013). Transforming cities with transit: Transit and land-use integration for sustainable urban development. World Bank Publications.

Integrated tools for town and transport planning

Town and Infrastructure Planning for Safety and Urban Quality – Pezzagno & Tira (Eds)
© 2018 Taylor & Francis Group, London, ISBN 978-0-8153-8731-2

On building a common awareness of territory by a transport-system-integrated assessment

Valentina Colazzo
Università degli Studi di Roma La Sapienza, Roma, Italy

ABSTRACT: Nowadays the concerning about environmental issues are dramatically increasing.

The infrastructure is the backbone of the transport system and it affects the area in which it is, creating a set of macro-implications: Externalities, Impacts, Effects.

This research proposes an integrated approach which evaluates the infrastructures according to the three specific categories written above: economic, environmental and territorial; using different variables.

A "good assessment" uses the EIE cognitive analysis scheme (externalities, impact, effect) and adopts a EET vision (Economics, Environment, Territory). This method may represent the key-point for a good vision, and so a good evaluation, of the land transformation inevitably induced by the creation of a new infrastructure. Moreover, it can point out how creating a new infrastructure could be a positive venture in economic terms, but, at the contrary, could not be environmentally-friendly.

1 TOPIC ANALYSIS

Nowadays the concerning about environmental issues are dramatically increasing.

Speaking of this, the transport-system is the main field that can tangibly intervene through public policies.

The environmental consequences linked with the transport system derive from its features and from an increasing need for more and more mobility: the transport system which uses non-renewable energy source produces environmental consequences like air pollution, noise and hydrogeological instability, just a few of the climate change related consequences.

The infrastructure is the backbone of the transport system and it affects the area in which it is, creating a set of macro-implications. They might be classified into three categories:

- Externalities: positive and negative economical implications;
- Impacts: environmental implications;
- Effects: territorial-economic related implications.

This distinction is generated from an analysis of the cultural contexts of different transportation implications and it shows how a sectorial way of thinking prevail in the current assessment tools.

Furthermore, this is the direct consequence of a previous lack of tools able to evaluate different aspects of the transportation system. Eventually, this affects the evaluation outcome, restricted to a specific analysis area.

For these reasons, to take these different categories in consideration through an integrated assessment is pivotal in decision-making. In fact, this will draw up an overall scenario of every aspect (positive and negative) regarding the implementation of infrastructure in order to create awareness about induced transformations and to point the main issues out. Then, how can the integrated evaluation contribute to the reduction of climate change?[1]

Integrated assessment builds an information system related to the infrastructure itself (and to all the infrastructure-category in general) and to the local context in which it is inserted. Each infrastructure, in fact, had a different evaluation process depending on the subject and the tool which has been used.

For example, an infrastructure that, to be built, is required by law to take environmental issues in consideration will focus on those, because this tool exists to evaluate these aspects. But, if the same work is evaluated through other tools for specific assessments, it's impossible to get a comprehensive judgement which analyses the work in its entirety.

This research proposes guidelines of an integrated approach which can be applied in the initial analysis phase and it's functional in building a cognitive framework of the works to be realized.

Actually, some features are not considered just because it's not a habit to do so: for example, the data regarding the induced employment is often relegated exclusively to the construction phase; the use of land, the hydrogeological collapse, the landscape etc. These are only some environmental elements that have been modified due to the infrastructure realization. A "good assessment" uses the EIE cognitive analysis scheme (externalities, impact, effect) and adopts a EET vision (Economics, Environment, Territory). This method may represent the key-point for a good vision of the land transformation inevitably induced by the creation of a new infrastructure. Moreover, that points out how a work under a certain kind of evaluation is successful (for example, environmental analysis) may not receive the same positive judgment by a different analysis (for example, economical analysis).

In this case, the infrastructure would be positive to some elements, but negative to other elements: the evaluation would therefore be different from the evaluated component and therefore to its method.

With the Code of Conduct reform, it is possible to find positive support for the integrated approach as the technical and economic feasibility project (PDF) and the public debate are introduced in the first stage of evaluation:

The PDF will define the best cost-benefit solution following a series of in-depth technical analysis, while public debate can be an integrated collective sharing path.

2 INTEGRATED EVALUATION: COMPONENTS AND OBJECTIVES

Urban planning is a fundamental element in the present conformation of the territory, regardless of the reference scale.

Historically, urban transformations were an evolution of society and assumed the role of symbol of urban development.

A similar function is also the one that today lives the territorial transformation: change, beyond the purely physical transformation, becomes a decision-making tool, a political instrument, an opportunity for growth at every level.

The components of the urban system (environmental, economic, social, cultural, transport, etc.) have identified their dynamic equilibrium, adapting, or at least trying to adapt itself to the various changes induced by anthropic intervention.

Within these components, the transport system plays a central role, almost indispensable in territorial transformation.

This characteristic that the transport system carries with itself in every transport choice has clashed several times over the years with the natural opposite dimension, that is, the environmental one. The importance of environmental protection has arisen in economic culture through a series of world conferences from the 70 s until now.

---

1. Vallejo, L. and M. Mullan (2017), "Climate-resilient infrastructure: Getting the policies right", OECD Environment Working Papers, No. 121, OECD Publishing, Paris.

The main consequence of these actions for sustainable development led to major transformations in the transport sector and it changed the evaluating tools for infrastructure projects.

What arises is the need to identify new design and evaluation elements for infrastructure works, which are not based merely on economic efficiency, as in the past, or on minimizing the investment, but which relate to the social and environmental dimension in a global vision.

Transport is characterized by internal costs and external costs (externality), "those economic advantages or disadvantages that are not among the received (advantages) or endured (disadvantages) by the producer or the consumer (user) of the good or service corresponding to each activity".[2]

In world literature[3] and in the research results, other terms are used and replaced to the term "externalities": the term impact and the term effect.

By analysing the cultural origin of each of the terms, one can understand how each terminology has its precise chronology of formation and attribution of a specific meaning: the three terms if ideally exchangeable, are different spheres of interest.

Externality refers to an economic environment in which it represents an external factor outside the traditional market that causes a loss of well-being for the user.[4]

A loss of well being is generated when an external agent alters that balance.

This is the case of externalities, which modify the ideal conditions for the establishment of an efficient market.

The term "impact" refers to the environmental sphere: it is no coincidence that there is a tool called "environmental impact assessment", better known as EIA (VIA).[5]

The term impact is correlated to a number of repercussions directly and indirectly attributable to the environmental system.

The EIA is the main manifestation of an ethical and normative environmental consciousness, attributing value to a public good considered up to now marginal as the environment.

The term effect refers to a consequence, positive or negative, concerning the urban sphere.

By the term effect the territorial implications of an infrastructure investment are meant; infrastructures or interventions on the transport system can affect the economic development of a place: there may be, for example, a polarizing effect of the work that initiates or increases productivity of private investment, or urban transformation, like a superior accessibility.

In conclusion, as the three terms discussed above are conceptually similar, they belong to different cultural spheres, which describe a constructive path of terminology and what results from it.

Use of meanings is neither an alternative nor an exchange of words.

Competence and the affiliate sector influence the vision, use, and definition of tools or assessment methods.

This cultural distinction affects the specificity, lack of complementarity of the evaluations produced and its outcomes and creates a situation of sectionalism in the application and results obtained through current tools and assessment methods.

Existing tools and methods do not allow a global evaluation, but a sectoral one, which is a limit to the same achievement of the work.

An overall assessment allows the project to be inserted into a well-defined scale, as externalities, impacts and effects are not limited to the work and cannot be circumscribed in its area, but they indirectly affect larger spatial contexts.

It is necessary to develop an integrated approach to build an overall vision of the work through an economic, environmental and territorial assessment.

---

2. Musso A., Piccioni C. (2010). Teoria dei sistemi di trasporto, II Edizione, Edizioni Ingegneria 2000.
3. Commissione Europea, (2014). Update of the Handbook on External Costs of Transport.
4. Danielis R. (2001). La teoria economica e la stima dei costi esterni dei trasporti. ANFIA-ACI. I costi e i benefici esterni del trasporto
5. Commissione Europea (1985). Direttiva Comunitaria 85/337/CEE.

3 BRESCIA METRO AND TRAMWAY OF FLORENCE: SECTORAL EVALUATION EXAMPLES

The importance attached to the construction of infrastructure works has been confirmed by the analysis of the process of realization of some national interest urban transport operations (for their economic relevance and for the areas of realization): the Brescia Metro (Metropolitana Automatica di Brescia) and the Tramway of Florence.

The study of the documentation submitted has been structured according to a few key points: whether the documentation submitted was sufficient to make a decision, whether the issues examined were adequately dealt with and whether the evaluations carried out had considered externalities, impacts and effects. What matters in these projects is the path that led to the creation of the work.

The Brescia subway was subject to EIA procedure, according to art. 10 of the d.p.r. 12.04.1996 and art.1 of L.R. 20/1999; of substantial considerations of Brescia's Province, in particular it is noted that "[...] the documentation submitted is lacking in specific studies relating to the demand for transport in relation to the territorial area concerned from the intervention and the needs of local mobility; that from the design point of view the technical characteristics and operation of the transport structure identified should be specified; that various alternatives should be explored which allow, in different scenarios, the choice of the most satisfactory solution from a cost and benefit analysis, improvement of air quality and minimization of impacts on sensitive receptors identified; that from the environmental point of view, the ante-operam analysis and the impacts arising from direct and indirect interferences with the water environment, acoustic impact and vibration, impact with historical and archaeological heritage, and with the urban landscape".[6]

The required documentation, as imposed by the national legislation, was represented by the Environmental Impact Study (SIA), which however has been recognized a preventative function, without defining the role in detail.

SIA should be the cognitive tool for excellence and create the most comprehensive picture framework.

The study of such effects is absent, while the section on traffic studies appears to be very summary.

The absence of a global vision is highlighted, for example, in the absence of a proper study on traffic reduction, and on landscape impacts, which have then been verified after completion.

Concerning the Tramway of Florence, this work, according to the Province documentation, was not subject to EIA procedure; In fact, according to art.11 of the L.R. 79/98, the work had to be subject to EIA verification procedure and only later to the EIA procedure.

The verification has failed, because, according to the Exploratory Report presented by the proposing subject (ATAF Spa), "overall, the impacts generated by the realisation of the project under consideration can be considered known and measurable, those about the construction phase have to be considered largely reversible if it is handled correctly [...][7]".

The Province had therefore ordered that the Tramway of Florence should not be subjected to an environmental impact assessment procedure.

This decision was taken with respect to the documentation submitted by the client: the Investigation Report played a central role in whether or not the work was being submitted to VIA.

What has become glaring is that the documentation submitted, that is the General Report accompanying the EIA verification procedure, may be considered as deficient, as some issues have been handled insufficiently.

For example, the section on identifying possible impacts and envisaged mitigation is very superficial and generic, especially when dealing with traffic and air quality issues, without presenting an objective and specific territorial correlation.

6. Document Brescia's Province for Metro EIA.
7. Document Firenze's Province, Rapporto Istruttorio.

The document in general, in addition to presenting a large historical reconstruction that is unimportant to the project in question, appears to be inadequate because of the importance it attaches as the documentation used to decide on a possible EIA. The exclusion of the tramway from the EIA procedure and the entire cognitive pathway which then constituted the basis for making the decision are therefore rather poorly, related to the quality of the information provided.

4 INTEGRATED EVALUATION: OPPORTUNITIES FOR CHANGE

The two case studies outlined above have made it possible to identify shortcomings in the current evaluation procedures.

In both cases, in fact, the evaluation or the lack of it was conditioned by the procedure performed and by the information provided in support of the decision. Regional law is fundamental in this case to define object and information for evaluation.

Law 79/1999, Tuscany, considering the Exploratory Report provided by the proposer, stated that it did not submit the work to VIA: the decision has been taken on the submitted present information, which however, unfortunately, showed the deficiencies described above.

Law 20/1999, Lombardy, declined to submit to VIA the work, because of the shortage of documentation, such as transportation demand or the urban landscape impacts.

The consequence is a specific assessment of specific problems that may filter the judgment of the work itself: if an EIA is applied, the judgment will have an environmental focused path; for instance, if a cost-benefit analysis is applied, the judgment will have an economy focused path.

The consequence of this is a targeted and specific assessment of precise problems that may put a "filter" on the judgment of the work: if an EIA is applied, judgment will have an environmental impact; For example, if a cost-benefit analysis is applied, judgment will have an economic impact.

The central point is therefore that the application of a certain type of assessment (environmental, economic, ...) excludes another and cuts out the result presented a number of possible negative consequences.

Infrastructures carry with them economic externals, environmental impacts, and territorial effects: the evaluation should therefore become aware of these components as a fundamental part of the work and evaluate them in the most complementary way possible, as one influence the other.

The integrated evaluation resumes the concept of territorial impact assessment or VIT (Camagni, 2006) referring to an evaluation process divided in two parts: the first is knowledge analysis and the goal is to build an information system that includes the economic, environmental and territorial aspects of an infrastructure work; the second is methodological integrated approach for evaluate through the information provided what the work will bring in the territorial context in which it will fit and with respect to these information decide whether the work is doable or not.

The integrated assessment does not intend to replace the current assessment procedures but it acknowledges its limits.

The new Code of Conduct (50/2016) revises the design levels of the works and replaces the feasibility study and the preliminary project with the technical and economic feasibility project (PDF), which "identifies among the more solutions the one with the best ratio Cost-benefit to the community in relation to the specific needs to be met and the performance to be provided" (50/2016).

The importance of integrated and global information lies in the possibility of defining the PDF only after cognitive investigations, preliminary environmental studies and other territorial analysis.

Annex DEF 2017 identifies two elements as infrastructure policy cornerstones, namely strategic planning and ex-ante evaluation of works. Moreover it shows how the public debate is central to government action.

This method underlines the concept of an evaluation that provides the widest possible information and which interests every aspect related to the creation of a work on the territory.

This should not be seen as a new evaluation method: the integrated assessment does not replace the current procedures but it provides a supporting framework to the structure that should have an ex-ante evaluation.

The leader will no longer have a type of information and a sectoral, partial or spoiled outcome, due to the use of a certain evaluation procedure of the problem, but he can compare a general framework according to the following scheme:

1. Constructing information background: by taking the EIE definition, the information here refers to externalities (site costs, construction costs, construction costs, future occupation, variation in the value of areas and buildings), impacts (categories Environmental issues identified by 12/04/1996 Presidential Decree, considering hydrogeological disruption, soil consumption and resuming the new categories introduced by European Directive 2014/52/EU) and the effects (quality of life, integration and sustainability with existing transportation, potential territorial conflicts);
2. Building a global work scenario (EAT) according to the information provided in the analysis in point 1;
3. Build a cognitive and decision-sharing method with the community through the public debate tool.

The integrated approach is therefore articulated as a three-point schema, which sets the guidelines to make an ex-ante evaluation in line with the recent regulatory changes. Addition of new evaluation elements is also linked to the existence of new underrated issues, such as land use or territorial conflict.

REFERENCES

Atto di indirizzo e coordinamento per l'attuazione dell'art. 40, comma 1, della L. 22.02.1994, n. 146, concernente disposizioni in materia di valutazione di impatto ambientale.

Borgia E., (2000). Studi d'impatto ambientale nel settore dei Trasporti. Consiglio Nazionale delle ricerche.

Commissione Europea, (1985). Direttiva Comunitaria 85/337/CEE.

Danielis R. (2001). La teoria economica e la stima dei costi esterni dei trasporti. ANFIA-ACI. I costi e i benefici esterni del trasporto. Torino.

Musso A., Piccioni C., (2010). Teoria dei sistemi di trasporto. Edizioni Ingegneria 2000.

Provincia di Brescia, Documentazione Metropolitana.

Provincia di Firenze, Rapporto Istruttorio.

Town and Infrastructure Planning for Safety and Urban Quality – Pezzagno & Tira (Eds)
© 2018 Taylor & Francis Group, London, ISBN 978-0-8153-8731-2

Urban modifications and infrastructural system: A research to ease integrated and flexible approaches

R. De Lotto, G. Esopi & E. Venco
Università degli Studi di Pavia, Pavia, Italy

ABSTRACT: The paper aims to elaborate a useful tool for tracking and monitoring urban transformations and their impacts. The tool is based on layers' overlapping that combines the information related to settlement system and to mobility and technological networks: changing a parameter it is possible to evaluate the effects on the others.

Authors demonstrate that the overlap of urban modification and technology services allows a comprehensive and integrated vision of current and future situation and build the basis for a large-scale strategic planning.

The paper focuses on the identification of an 'observatory' of structural changes that allows mapping the urban modifications, to prevent possible conflicts and to manage infrastructure services at medium-long term. The tool allows defining rules and actions to be taken for the future of the city by simulating new scenarios. The urban development simulation helps to understand large-scale consequences of spatial planning decisions in a complex urban system.

1 INTRODUCTION

The definition of city as a complex system derives from the studies of different authors and it is widely debated within the scientific literature (i.e. Batty, 2009; Batty, Marshall, 2009; Portugali 2011). The complex urban system is subject to incessant modifications of different nature (socio-economic, political, geographic, etc.) that take place more and more rapidly. The modifications speed affects the demands by citizenship towards the city and its features: urban system must be able to adapt to temporary situations in a flexible way (De Lotto, 2011).

From the physical point of view, city consists of subsystems such as settlement and infrastructure system. The first is territory areas (built or not built) characterized by the presence of urban activities and functions (residential, productive, economic and managerial areas, services and public city places, etc.). The latter represent the backbone for the development of settlement processes. Infrastructures can be classified according to their function in mobility networks (intended for the people and goods transport) and technological one (serving the human activities). In spatial planning, mobility infrastructures imply big scale corridors, roads, interchange nodes, airports, logistics platforms and pedestrians and cycle paths. Technological services include among all energy networks, integrated water management systems, radio and television networks, waste management and district heating equipment.

During city planning history, the mutual connection between these two systems has been continuously emphasized because they both can be treated as cause or effect of urban phenomena depending on the study point of view (Mitchell et al., 1954; Oduwaye et al., 2011).

In particular, land use causes the traffic flow within the city: fundamental factors, which determine people and goods movement, are the quantity and nature of activities performed in daily life in a certain city (Mitchell e Rapkin, 1954). It is clear that existing cities modify functions but, often, infrastructural networks do not change consequently: this is one of the main inner problem for the good performance of urban systems.

At the local scale, planning process involves mobility and technology networks' adaptation according to location and functional choices through punctual actions, unstructured in an overall view. Such approach leads to a loss of information and, therefore, of the overall complexity of the dual system settlement-infrastructure.

In order to ensure an urban system able to respond to different internal and external stresses, it is necessary an integrate and flexible planning process. This process is able to study and to monitor together causes and effects of different urban modifications. This kind of approach can help to improve both efficiency than quality of urban environment.

2 OBJECTIVE OF THE PAPER

Starting from the recognition of urban systems modifications, the paper aims to elaborate a theoretical and methodological research in order to create a tool to facilitate integrated and flexible planning approaches. A 'Spatial and Temporal Map of Urban Modifications', that is defined in the following paragraphs, promotes the integration of information (through an Observatory of structural changes) about different city components: settlement system, mobility and technology networks (artificial elements), demographic system (urban agents).

The map, related to the analyzed territory, allows the detection of urban modifications, the analysis of the effects produced in terms of mobility demand change and the impacts monitoring in the short and long term.

This tool is useful for planning choices and for the consultation and diffusion of open access information related to urban transformation processes on the territory.

3 METHODOLOGICAL APPROACH

The methodology followed to achieve the objective can be summarized in these steps:

- Territorial (urban an local—neighbourhood and single block) analysis: present situation at time t = 0 regarding settlement system, mobility and technological networks, demographic system and people and goods flows, necessary aspects for weakness, strengths and needs identification.
- Mapping and detecting of urban transformations: at time t = 1 there are spontaneous modifications, implementing plans in place/adopted, planning hypothesis (scenarios).
- Evaluating and monitoring of impacts in the short-long term through the simulation of future scenarios.
- Definitions of optimal solution(s) that are established by the municipality within a participatory decision-making process.

The proposed tool is transversal to all steps.

4 DESCRIPTION

In order to pursue a planning model based on an integrated and flexible approach between settlement system and infrastructure and technology networks, it is crucial to analyze links and relationships that arise among the systems. The link between the two systems is complementary (completion and integration): in fact, both are needed for efficient operation of the city. Integration implies that actions carried out on a system are affected by various effects on the complementary system. As regards relationships, the analysis of urban modifications show that those relate with settlement system and mobility networks are linked by a two way relation: a change in one of these systems generates impacts on the other. Instead, between settlement system and technological networks there is a one way: the latter are subject to impacts (in terms of urban load) due to settlement changes. This is because technological networks depends strictly on settlement system and, consequently, it changes only as a result of its transformations. These comments come from a physical observation; while, from performance point of view, mobility and technological network modifications usually imply a better efficiency of the settlement system.

The main categories of urban modifications are shown below.

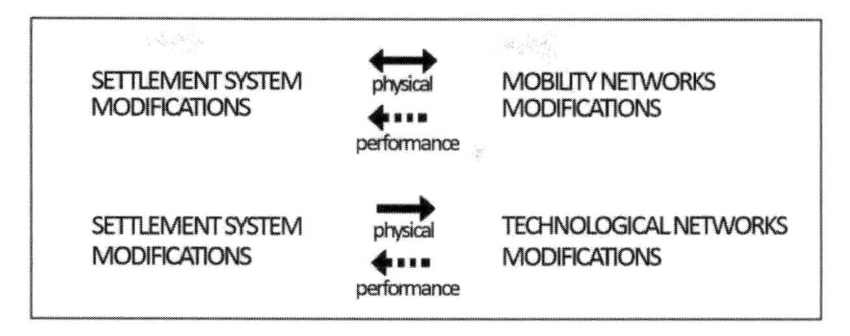

Figure 1. Relationships between urban modifications.

4.1 *Main settlement system modifications*

4.1.1 *Urban edges completion (urban scale)*

The urban edges correspond to transition situations between urbanized areas and undeveloped territory (also rural-urban edges). They can become opportunities for transformation through projects of green suburban bands to enhance environmental sustainability and mitigate negative impacts through coordinated and innovative uses.

4.1.2 *Brownfields regeneration and re-functionalization (urban scale)*

Brownfields are large or small-scale areas within the existing urban fabric characterized by the presence of neglected structures and spaces, which was abandoned due to deindustrialization phenomena. The interventions for the development of these areas are usually new construction works.

4.1.3 *Population density changes (urban scale)*

Population density changes are due to an increase or a decrease in residents. In the past centuries, cities industrialization was one of the main causes of urban expansion; today, several cities are subject to the decentralization phenomena. This relocation into peripheral and first-belt areas is due to a series of well-known reasons: it leads to an intensification of flows from the hinterland towards the city centers increasing the pressure on mobility network system.

4.1.4 *Buildings change of use; reuse; restoration (local scale)*

The process of city adaptation to contemporary users and uses implicates the change of building intended use: from apartments to offices, from religious buildings to public services, from handmade spaces to residence and so on. Reuse interventions of historical heritage imply the transition from an original function to a new compatible one: in some cases, the conversion is not only functional, but also typological-structural.

4.1.5 *Urban spaces regeneration (urban and local scale)*

Interventions on urban fabric concern built and not built space. In urban territory there are large amount of public and private spaces that are underutilized and abandoned. Such spaces, if appropriately redeveloped, could host new functions or increase existing ones and generate new places for citizenship. Open space interventions can be punctual (requalification of specific areas) or at urban scale (strategies for whole urban fabric regeneration).

4.2 *Categories of mobility networks modifications*

4.2.1 *Development actions (urban and local scale)*

The main actions include the extension of roads, railways, metro lines, tramway and local public transport network and, also, the creation of new urban stations and belt services.

4.2.2 *Urban environment quality actions (urban and local scale)*

The main actions include diffusion of soft mobility, development of cycle-pedestrian paths and resolution of specific and local problems to allow a better and safer use of public space.

4.2.3 *Management actions (urban and local scale)*

The main actions include parking lots regulation, traffic management and control, establishment of restricted zones to non-residential citizens and so on.

Apart from separations between the two subsystems, it is obvious that in a city many phenomena are based on the system considered as a whole and the investigation scale depends on modification typology.

4.3 *Urban modifications and flows*

Urban modifications involve a change of people and goods flows. The study of mobility demand evolution is crucial in order to estimate the loads on transport system (number of users that use an existing transport system or those who would use a future transport system). However, the estimation of mobility demand in an urban territory is a complex process because of its dynamism. The methods of mobility demand estimation refer to time analysis (trend evolution) or to spatial relationships through land use variables (Ortuzar, Willumsen, 2004). About people distribution, modification of flows follows the most variable parameter so, aspects such as demographic composition are less relevant than socio-economic factors. These influences are useful for understanding movement nature, type and mode. In example, the income level (issue related to economic and social status of the population) affects mobility demand especially in relation to the movement mode, and the high percentage of elderly population entails a greater demand of specific public transport.

4.4 *Monitoring and detection of urban transformations*

In recent years, researchers increase and develop the use of widespread sensory and immersive monitoring (a type of survey that allows to integrate the multidisciplinary knowledge for a territory proper management) to analyze and monitor the territory throughout sensors and big data (i.e. City Sensing, Senseable City Lab MIT). The main element is the use of technological devices that allow the data acquisition through specific and different monitoring points on the territory, keeping them connected to analyze the complex urban contexts from a within point of view and also (if necessary to the specific analysis) with real-time data collection and evaluation. Moreover, the monitoring activity (punctual or on large scales related to modification typology related to modification typology) allows to systemize all the information for a correct contextualization of phenomena occurring in space and time, in relation to different pressure factors.

Actually, urban space is an interconnected space: data deriving from the present situation analysis, data relate to planned modifications (municipality/public administration manages information and their restitution on, usually, a GIS platform) and data flows from monitoring devices are then aggregated instantly and organized within a database.

These integrated components provide the fundamental information for the decision-making process, for planning and for management.

The territory immersive monitoring projects is related to the concept of a participated and shared cognitive framework to support decision-making processes, and to create a better and widespread availability of information resources for decision makers.

The main elements monitored are:

- People flows, as users of different types of services (authors analyze mobility system);
- Physical consequences generated by urban transformations;
- Social and economic aspects related to urban transformation (evaluation of driven forces, changing of elements and flows).

4.5 *General modifications consequences*

4.5.1 *Demand for infrastructure services*

Each type of settlement includes private elements (dwellings, offices, shops) and public ones consisting of support structure (primary and secondary urbanization). The demand for infrastructure services is the effect produced by all the elements as demand of structures and collective works, depending on the number of people settled on a given territory. The amount and type of functional changes must take into account not only the maximum and minimum limit on the use of infrastructure and networks, but also the density generated after changes in land uses.

4.5.2 *Traffic management*

Infrastructure implementation, urban edges completion, abandoned areas regeneration, congestion and pollution cause negative externalities. Even the phenomenon of people decentralization, due to economic and social changes, leads to increased traffic flows towards city center along the main urban directions.

4.5.3 *Environmental and acoustic pollution*

Interventions related to new infrastructures construction involve negative impacts on air quality, water system, natural and anthropic environment and ecosystem. They cause also acoustic pollution especially for sensitive services (hospitals, schools and so on) (Zambrini, 1991).

4.5.4 *Physical degradation and urban quality reduction*

Interventions on the mobility network can lead to negative externalities such as physical degradation and reduction of urban quality that can also occur far from the specific area.

4.6 *Examples of Monitoring tools and strategies for mobility*

It is well known that data related to traffic flows (i.e. in real time or hour by hour) are easily reachable in a lot of different typology of infrastructure and urban context. In the following section, there are some examples of mobility-related data-monitoring devices and strategies currently used for monitoring the vehicles flows:

- Traffic trends on freeway (Sohn, Hwang, 2008), urban roads, traffic and origin-destination flow estimation (Sohn, Kim, 2008) are studied with fixed sensors, GPS receivers and specific real-time sensors (such as inductive turn, non-intrusive technologies, above ground devices as radar, microwave, laser);
- There are several projects regarding feasibility of mobile equipment location-based ITSs (intelligent transportation system) using ICT to monitor transportation system and to reduce congestion problems have been developed in Europe (Ygnace et al., 2000) and North America (Yim, Cayford, 2001; Herring, 2010; Fletcher 2000);
- Some devices allow the identification, tracking and monitoring of goods, vehicles and people. They are useful to manage access control services, fleet management, urban logistics, cash/cash-free access and control of disabled persons license;
- Some devices allow monitoring traffic flows by hours, on urban and/or suburban roads. They record the passage of vehicles distinguishing the various categories: the analysis of these data allow the rationalization of mobility system and the improvement of cities livability.

5 RESULT OF THE RESEARCH

Nowadays, in view of the wealth of mobility data, settlement system mapping is missing: its modifications are traceable in long period but not in short term. As first step it is therefore necessary that traffic data and those related to urban transformations must be comparable also from the temporal point of view.

So, to reach this goal, it is necessary a planning model based on an integrated and flexible approach that implies the use of innovative tools for managing the various components of complex urban system.

The presented methodology aims to the creation of the 'Spatial and Temporal Map of Urban Modifications', an interactive and integrated tool of the evolving territory and continuous monitoring (with a timetable suitable for each elements and modifications) of the spontaneous or planned modifications. It is an analytical and descriptive map for settlement system, mobility and technological networks and demographic data (relate to citizens and city users).

Several overlapping layers with technical specifications constitute the 'Spatial and Temporal Map': population with sex/age/social status/employment specifications; residents/city users and their different typologies; urban settlements and fabric information; urban functions/public services/attraction elements/facilities; infrastructure system (roads, railways, channels) pedestrian and cycle paths; technological services (networks and main nodes); non urbanized/green/agricultural areas. So information come from entire city scale and punctual buildings paper-works.

The definition of each objects in a territorial context and the ability to define spatial and not-spatial relations among them are the essential element for the creation of this kind of database.

Considering the defined cartography (current situation, modifications in progress, hypothesized scenarios), it is possible to:

- Define elements and points of sensitivity and territorial/socio-economic weakness;
- Highlight the conflicting points that arise from the interactions of the different sub-systems: built settlement-environment, environment-infrastructure, infrastructure-built settlement, population-built settlement, population-infrastructure;
- Develop an advanced and integrated planning process that can anticipate any risks/malfunctions arising from the critical interrelationships among sub-systems. Monitoring in a proper time-period (for both the subsystems) allows to reduce costs providing increased public safety;
- Develop different scenarios to study and assess future planning interventions and processes.

To analyze the spontaneous modifications, the implementing plans and the planning hypothesis and then to define the optimal solution(s), authors use the classical definition 'Scenario Planning': the wide scale development simulation (through dynamic and stochastic models) helps to understand large-scale consequences of spatial planning decisions in a complex urban system. Interactive control of proposed planning interventions and their impacts assist the public interactively engaging in planning processes and contributing to planning decisions.

At the same time, the database with an open (for almost all the information) access defines the Observatory of structural changes that allows the mapping of the urban modifications and the management infrastructure services at medium-long term. This research structure is based on a multidisciplinary approach and composed by human experts and technological resources. The information are always updated (in relation to their time-period of modification) and available. So, detailed and easy understandable information about planning decisions and the expected impacts support the negotiations with citizens during planning processes increase public participation and support to these decisions (http://www.urbanapi.eu).

The 'Spatial and Temporal Map' creates and gives back:

- Synthetic, interactive and easily understandable representation and descriptive analysis of collected information also with in-depth analysis on local and urban scale;
- Improved communication with decision-makers and citizens/city users;
- A way for testing transport technology innovations, investigating travel behavior, verifying the state of the different infrastructure systems, preventing potentially dangerous situations;
- Identification of areas requiring systematic interventions on infrastructure networks and their definition (time and operating methods).

6 CONCLUSIONS

The issues faced in the paper give a qualitative understanding of how the use of this tool can provide valuable services to municipality, citizens and city users.

The proposed method aims to elaborate from theoretical and methodological point of view an useful tool for planning decisions during the control of process related to urban transformation actions (from city planning to singular buildings) and the definition of intervention priorities in the territory. It allows studying and monitoring together causes and effects of urban transformations in order to define the best planning choices. Through the integration of spatial information, related to urban interventions and to spontaneous actions, the map allows integrated and multi-scalar management of different urban aspects. The strength of a tool able to ensure the monitoring of urban evolution and the spatial view of information lies in a greater understanding of current urban dynamics and above all of their territorial distribution. This kind of tool is completely feasible using actual digital technologies while it is requires a multidisciplinary or flexible approaches by involved technicians and managers.

REFERENCES

Batty M., Marshall S. (2009). The evolution of cities: Geddes, Abercrombie and the New physicalism. Town Planning, no. 80, vol. 6. 551–574.

Batty M. (2009). Cities as Complex Systems: Scaling, Interaction, Networks, Dynamics and Urban Morphologies. Encyclopedia of Complexity and Systems Science, Springer. 1041–1071.

Bologna Municipality (no date). Piano Strategico Metropolitano. Mobilità e Trasporti.

De Lotto R (2011). Flexibility principles for contemporary city. In Shiling Z., Bugatti A. (Eds.) Changing Shanghai—from Expo's after use to new green towns. Officina, Rome. 73–78.

Fletcher D.R. (2000). Geographic information systems for transportation: a look forward. Transportation in the new millennium. Washington, DC: Committee on Spatial Data and Information Science, Transportation Research Board.

Herring R., Hofleitner A., Amin S., Nasr T., Khalek A., Abbeel P., Bayen A.(2010). Using mobile phones to forecast arterial traffic through statistical learning. Proceeding 89th Transp. Res. Board Annu. Meeting, Washington, DC.

Milano Municipality, (2015). Piano Urbano Mobilità Sostenibile, Valutazione Ambientale Strategica. Sintesi non tecnica.

Mitchell R.B., Rapkin C. (1954). Urban traffic: a function of land use. Columbia University Press, New York.

Oduwaye L., Alade W., Adekunle S. (2011). Land Use and Traffic Pattern along Lagos—Badagry Corridor, Lagos, Nigeria. Proceedings REAL CORP 2011. 525–532.

Ortuzar J., Willumsen L. (1990). Modeling Transport.Wiley, Hoboken, NJ.

Portugali J. (2011). Complexity, Cognition and the City. Understanding Complex Systems. Springer-Verlag, Berlin.

Sohn K., Hwang K. (2008). Space-based passing time estimation on a freeway using cell phones as traffic probes. IEEE Trans. Intell. Transp. Syst., no. 3, vol. 9,.559–568.

Sohn K., Kim D. (2008). Dynamic origin—Destination flow estimation using cellular communication system. IEEE Trans. Veh. Technol. no. 5, vol. 57. 2703–2713.

Ygnace J.L.,. Remy J. G, Bosseboeuf J.L., Da Fonseca V. (2000). Travel Time Estimates on Rhone Corridor Network Using Cellular Phones as Probes: Phase 1 Technology Assessment and Preliminary Results. Arceuil, French Dept. Transp.

Yim Y.B. Y., Cayford R. (2001). Investigation of vehicles as probes using global positioning system and cellular phone tracking: Field operational test. California PATH Program, Univ. Calif., Berkeley, CA.

Zambrini M. (1991). Valutazione Impatto Ambientale. Infrastrutture stradali ed autostradali. Provincia Autonoma di Trento, Dipartimento Territorio, Ambiente e Foreste, Servizio protezione Ambiente, Ufficio per la valutazione di impatto ambientale.

SITOGRAPHY

Senseable City Lab. Available at: http://senseable.mit.edu.

Interactive Analysis, Simulation and Visualization Tools for Urban Agile Policy Implementation. Available at: http://www.urbanapi.eu.

Istituto Superiore per la Protezione e la Ricerca Ambientale (ISPRA). Soil Consuption. Available at: http://www.isprambiente.gov.it/it/temi/suolo-e-territorio/il-consumo-di-suolo.

Town and Infrastructure Planning for Safety and Urban Quality – Pezzagno & Tira (Eds)
© 2018 Taylor & Francis Group, London, ISBN 978-0-8153-8731-2

The contrast to the Urban Heat Island phenomenon to increase the urban comfort as an incentive to slow mobility. A study on the city of Parma

P. Rota
Comune di Parma, Parma, Italy
Università degli Studi di Parma, Parma, Italy

M. Zazzi
Università degli Studi di Parma, Parma, Italy

ABSTRACT: The paper focuses on adaptation to climate change and proposed, for the city of Parma, a risk map induced by heat islands on sensitive people, as a tool to support planning decisions for urban regeneration interventions.

The study, has confirmed the connection between the temperature change and land use and found that, when it crosses the 80% threshold, the heating rate increases considerably.

One of the most successful solutions is the equipment of vegetation increase. Continuity of pedestrian and cycle routes, and their overlap with ecological corridors, could be a solution to increase of green infrastructure, as they allow at the same time re-sewing between rural and urban areas at different scales.

Strategies and processing practices recommended for the heat island mitigation can have positive effects on several components of the territorial and urban system. Starting from environmental criticalities, the recommendations for action can have positively impact on different sectors.

1 INTRODUCTION

«Earth's 2016 surface temperatures were the warmest since modern recordkeeping began in 1880, according to independent analyses by NASA and the National Oceanic and Atmospheric Administration (NOAA). Globally-averaged temperatures in 2016 [...] 0.99 degrees Celsius warmer than the mid-20th century mean. This makes 2016 the third year in a row to set a new record for global average surface temperatures» (NASA's Goddard Institute for Space Studies, 2017).

In Emilia-Romagna, average regional temperatures have risen since 1961, by 1.1°C; in the province of Parma the average temperature has gone from 12.8°C in the thirty years 1961–1990, to 14°C in the next 1991–2015 (ARPAE, Emilia-Romagna Region, 2017).

Compared to the 1961–1990 reference period, the representative heat waves index, the warm spell duration index (WSDI) ranked the Italian summer of 2015 in the fourth place of the series, with an anomaly of +28 days a year compared with the standard of the period considered (ISPRA, 2016).[1]

---

1. «In particular, the average number of tropical nights, that is with a minimum temperature greater than 20°C, recorded in 2015 the second highest value in the whole series since 1961 (after 2003), with an anomaly of +26 nights compared to normal value» (ISPRA, 2016). The figure is particularly alarming as the poor circadian thermal excursion during summer periods prevents night-time recovery for the most fragile people (Morabito, et al., 2015).

Table 1. Monthly series of deaths related to residents in the city of Parma from 2011 to 2015. Source: Municipality of Parma, Statistical Service.

	2015	2014	2013	2012	2011
January	206	189	194	185	171
February	182	173	167	192	144
March	169	169	164	176	170
April	177	176	152	168	157
May	141	138	139	138	173
June	163	172	133	155	151
July	184	158	159	157	132
August	165	147	122	145	154
September	152	155	121	131	137
October	162	167	174	168	136
November	137	159	149	146	175
December	187	181	196	170	177
TOT	2.025	1.984	1.870	1.931	1.877

Although no unique heat wave definition exists, the most recent reading shows an increase in cardiorespiratory mortality with variations ranging from +4% to + 16% depending on the intensity and duration of the phenomenon (Xu, FitzGerald, Guo, Jalaludin, & Tong, 2016). Some studies on the impact of heat waves on mortality in European cities have highlighted the greatest occurrence of the phenomenon in Mediterranean cities, where there is a strong incidence on the population, especially females, belonging to the age group between 75 and 85 years old[2] (D'Ippoliti, et al., 2010).

Epidemiological studies have shown an inverse correlation between winter and summer mortality: the increase in winter deaths is followed by a decrease in the summer months, although denied by the data on deaths that occurred in 2015. The peak at national level, confirmed locally (Table 1), during the summer of 2015 – characterized by very intense and long term heat waves occurring in July, especially in the cities of the Center-North—was preceded by an excess of mortality occurred in the early months of the year. This circumstance may have made the impact of heat waves less dramatic, given the numerical reduction of the most fragile band of the population[3] (Michelozzi, et al., 2016).

The effects of heat waves on the health and well-being of the population are not only a major topic of public health, but in a context of climate change, they represent a challenge for adapting urban systems.

In the urban areas, the so-called Urban Heat Island (UHI) phenomenon is witnessed, characterized by the temperature difference between the urban and the rural areas (Oke, 1981; 1995).[4] In densely-built cities, characterized by low percentages of green areas and strong soil sealing, solar radiation is captured thanks to the high thermal and low albedo capacities of building materials in urban fabrics. The phenomenon is associated with other concurrent causes such as pollution or anthropogenic inputs.[5]

2. The phenomenon «may be attribuable to social conditions of elderly women living alone and to physiological differences, such as a reduced sweating.capacity that affect the ability to respond to heat stress» (D'Ippoliti, et al., 2010, p. 6).

3. Michelozzi et al. (2016) report the high mortality rate in the summer of 2015 in Rome even among 35–64 year olds.

4. Generally, the difference is around 3°C, but in some special circumstances it can reach 12°C, as it happened in Paris in the summer of 2003.

5. In this respect, it is not to be underestimated the contribution of housing cooling systems. It has been estimated that for each increase of 1°C in the intensity of UHI, the demand for energy increases from 2 to 4% (Shafaghat, et al., 2014; Taslima et al., 2015), contributing to feed, in a vicious circle, the UHI itself.

The repercussions on the use of the city by the inhabitants may have conflicting outcomes. Although from a social point of view the modal shift from the car to the bicycle has beneficial effects on the health of the population, due to the reduction of polluting emissions and greenhouse gases and through greater physical activity, at the individual level scenery it may have features of the opposite order.

The greater exposure to pollution and high temperatures, the perceived risk of road accidents contribute to the limitation of the comfort of pedestrians and cyclists and can be considered as disincentive factors to the use of soft mobility vehicles. In addition, the policies addressed to the urban environment and the public space rarely take into account the climatic pressures, especially the increase in temperatures, and the expected impacts in the medium and long term.

In planning and designing public space, it is therefore increasingly important to adopt devices to increase urban comfort levels.

While initially acting on individual aspects, the recommended transformation strategies and practices for mitigating the UHI may have positive effects and spin-off on several components of the territorial and urban system. Starting from environmental concerns, intervention recommendations, because of process characteristics and activated solutions, may indeed have a same positive impact on different sectors.

2 A MAP RISK TO SUPPORT PLANNING CHOISES

There are many variables acting in UHI formation, diversified depending on the considered urban context. The factors depend on the air temperature, the earth's surface temperature, and then on the physical characteristics of its components, the morphology of the territory, the morphology of the building fabric, the distribution of green areas and of course on the territorial distribution of the population. Calculating the risk to which the population's sensitive bands are subjected to at the single building level, and displaying it through dedicated maps can be useful in assessing the safety of existing and projected cycle paths, as well as in giving directions to the devices and materials to be used along trajectories. Spatialization of the population by age groups, interpolated with the presence of public and collective equipment, suggests in fact which are the preferential walkways, and above all reveals the most dangerous situations that occur during the summer.

The study presented is based on a research doctorate that has tackled the theme of adaptation to climate change and proposes a map of risk, caused by the UHI on sensitive people, for the city of Parma as a tool to support urban planning choices for urban regeneration interventions. The study, which has made use of a collaboration with IBIMET-CNR, has highlighted the link between thermal variation and soil consumption.

Before dealing with this topic, we consider useful to point to the Community Potential, a neighborhood-related indicator that measures the wealth of the community in relation to the location of services and their accessibility, used for drawing up the Master Plan of the Neighborhoods, which is preliminary to the drafting of the Municipal Structural Plan (PSC) in Parma, currently being completed. «The analysis of community potentials aims to highlight the territorial components that make up the community's connectivity network in the neighborhoods, thus identifying the cornerstones of the identity recognition of the inhabitants and the effective distribution of neighborhood services» (Burdett et al., 2010, pp. 43–44). The model shows the polarization of services in the historic center and in the districts built since the Second post-war on the basis of zoning plans and Peep districts and connections through the soft mobility.

3 THE RISK MAP. DATA AND METHODS

According to the definition provided by the European Commission, Risk is the likelihood of the damaging consequences or expected losses, coming from a particular hazard, to a given element at risk or in danger for a specified period of time (Schneiderbauer & Ehrlich, 2004).

Research emphasized risk dependence on the three components: Danger, Exposure and Vulnerability as Crichton (1999) showed with the so-called Risk Triangle.[6]

The map obtained correlates the climate data of the temperature variation with the resident population within each building and verifies the possible causal relationship with soil consumption.

In order to clarify the relationship between microclimate and geometric, optical and thermal characteristics, a series of related variables were selected and included in the analysis.

The spatialization of the resident population forced to link the censorship data with the building ones by utilizing the information obtained through the common factors of the two data classes, determining the amplitude of the sample on which to conduct the analyses. Although partial, data were however enough to demonstrate the effectiveness of the model.

In the present study, the "triangle formulation" of the algorithm contemplated the three fundamental aspects of Hazard, Exposure, Vulnerability so identified:

Hazard: is the climatic input; in this case, Brightness, air temperature and surfaces are used, as derived from NASA Landsat and Aster satellite surveys conducted in the summer of 2015;

Exposure: is the data of the spatially distributed population obtained from the Municipal Registry.

Vulnerability is divided into two factors:

Vulnerability 1: is the spatialization of the sensitive population, considered to the thresholds: over 75, over 65[7], or under 5.

Vulnerability 2: is the parameter of urban-environmental sensitivity, represented by soil consumption, referring to ISPRA's National Land Map. For the compilation of the risk map in the present study, the radius of the surrounding area was taken 200 m from the centroid of the considered building. Elaborations were carried out with rays of 57 and 100 m, with similar results. Vulnerability is no doubt the most complex component, as social and temporal factors are introduced. And it is also the most dynamic parameter that requires constant updates for monitoring. All data have been normalized. The Crichton methodology was then applied to obtain the Summer Heat Risk Index, calculated using a weighted sum of the four parameters illustrated in Fig. 1.

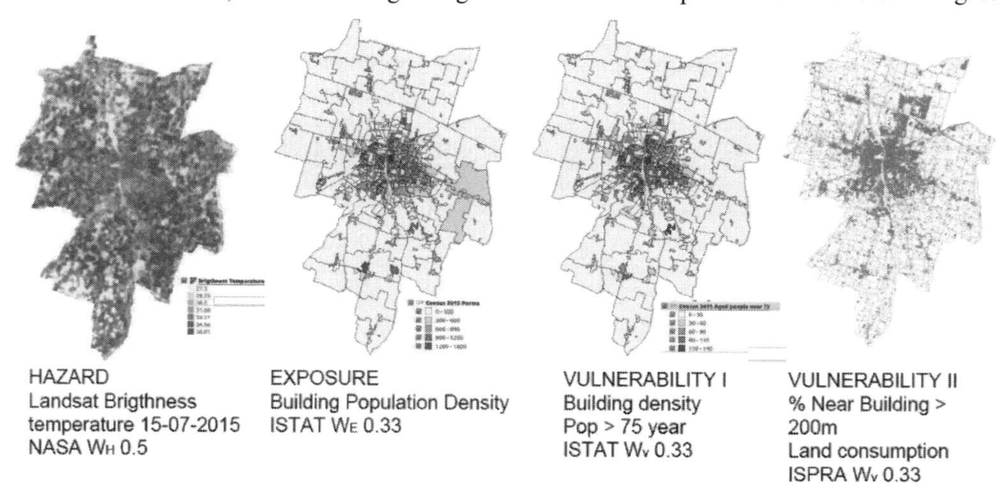

Figure 1. Maps of danger, exposure and of the two vulnerabilities with indication of the weights assigned to every factor. Source: (Crisci, et al., 2016).

6. In the triangle of Crichton Hazard, Vulnerability and Exposure are arranged along the sides. The risk is measured by the triangle area and can be reduced by directing actions to one side, in search of easier solutions. If you can eliminate one side, then you cancel the risk (Crichton, 2006, 2011).

7. Over sixty-five were considered as representatives of a numerically significant age group, exposed in the near future. From the municipality's data reported on December 31, 2015, they are 22% of the municipal population.

4 SOME CONSIDERATIONS ON USING THE RISK MAP

The maps show incontrovertibly the weight of soil consumption in determining the heat island risk on sensitive population bands. Where soil consumption in the surrounding of the considered building is over 80%, the risk rises suddenly as shown in Fig. 2. This means that temperatures are also exacerbated in public spaces, with serious limitations on their attendance.

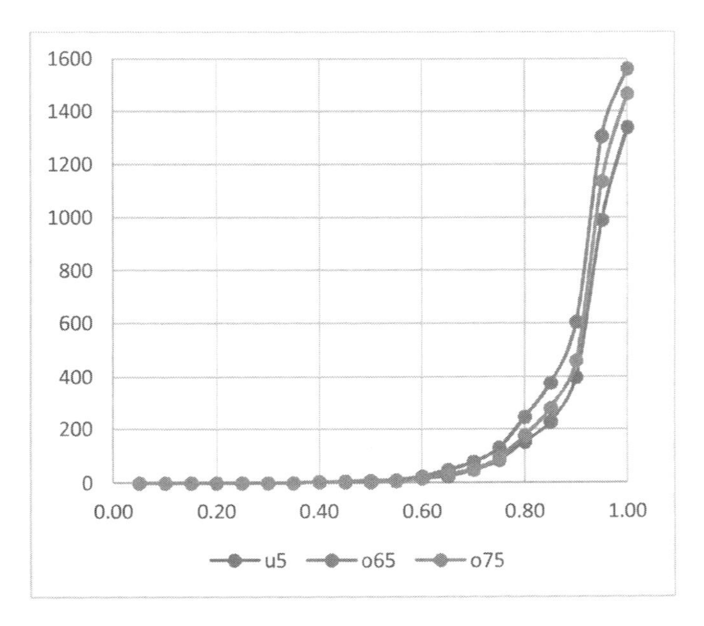

Figure 2. Risk distribution chart for fragile population classes and soil consumption.

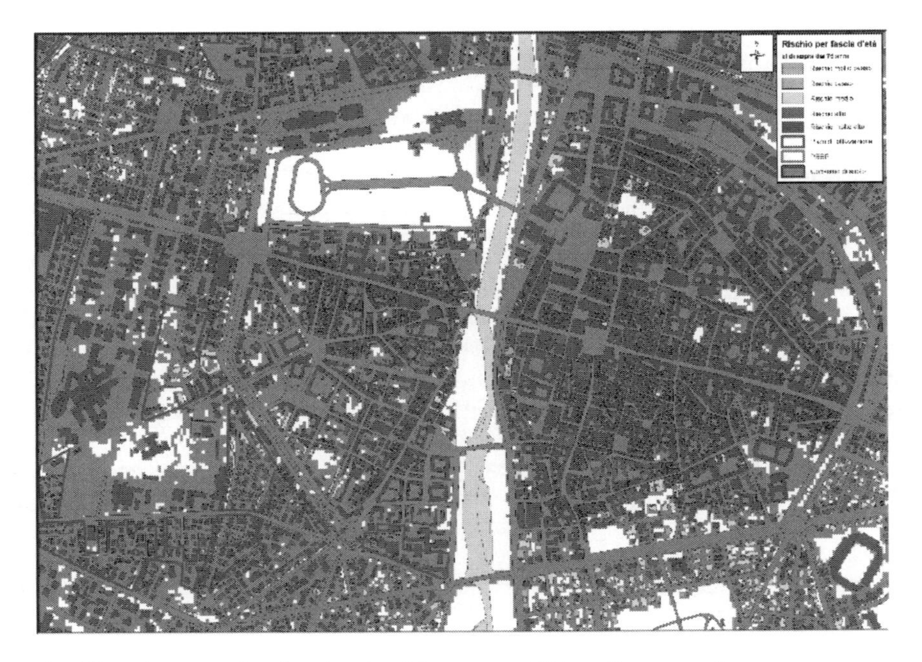

Figure 3. Historic centre. Overlapping the risk map over 75 to soil consumption map.

Figure 4. Overlapping the risk map over 75 to soil consumption map.

Despite the partial nature of the data, it is interesting to note how the distribution of risk within the city reflects urban history. To the historic, compact, sealed core, whose building fabric turns its back to the creek that crosses it, thus denying the opportunity to enjoy the breezes that blow from the riverbed—note that only a few buildings of the Oltretorrente on the left bank, facing directly on the water course, have a low risk (Fig. 3)—it follows the crown of the districts, arisen especially from the fifties of the last century, based on parcelling plans and plans for popular economic building (Peep) (Fig. 4). The first ones, the result of the expansion of the city in the decades immediately following the end of the Second post-war, whose realization was guided by criteria aimed at obtaining the maximum of the rent; the second ones programmed and designed also to balance the shortage of public services and equipment.

The public city that in Parma, as in the Emilia region, had a function of rebalancing mainly through the provision of public areas and services, if it was successful in the assimilation from the point of view of the functional endowments and consolidation of the Social networks, it has proved insufficient in adapting to climate change.

However from the comparison of different situations (parcelling vs Peep, parcelling vs. parcelling) it emerges that under conditions of similarity of the age datum, it is the second component of the vulnerability, that is the soil consumption, to adversely affect the risk situations.

The result highlights the caution in the direct adoption of one of the most important urban development strategies: urban densification which, while being functional in reducing energy consumption, if not controlled, triggers and intensifies the UHI effect, resulting in contrast with adaptation to climate change strategies.

Among the others, the wind is indeed an attenuating factor of UHI and friction against tall buildings slows down the eolic flows. In this case, the roughness of the surfaces becomes significant and influences the conditions of urban comfort.

In order to counteract the negative effects of building concentration, free spaces are needed, hopefully functionally interconnected and carefully designed so as to have minimal soil coverage.

The problem therefore is to re-internalize and partially redesign the open spaces, mainly public if any, and to intervene where sealing is high, through planning strategies that can mainly be attributed to the treatment of the four aspects albedo, vegetation, shadowing, ventilation.

Especially the continuity of cyclopedonal pathways, in particular those in their own place, and their potential overlap with existing or futur ecological corridors, make the adoption of solutions to increase green infrastructure particularly interesting, as they enable the connection between urban environments and rural to different scale.[8]

Green areas and open spaces on the urban, public and private scale have a positive impact on different areas: recovery of ecological value, improvement of the quality of life of residents, maintenance of services for adaptation to climate change. Green infrastructures are also one of the constituents of urban drainage systems.

5 CONCLUSIONS

The risk map is itself an address to develop multi-level actions designed at the specific situations of the sites examined, where possible adaptive solutions are compared with the physical-morphological characteristics of the sites, and are related to the identity and value even intangible of the places themselves.

Consider, for example, the delicacy of treatment required by the historic city where the morphological features are intangible and unmodifiable. Historic city, which for its peculiarities, is intrinsically one of the sites most susceptible to the phenomenon of the heat islands.

This, of course, does not mean neglecting the importance of preserving the identity of the city, but instead of reinforcing it, seeking in the history of the city the possibility of a natural evolution.

To go deeply on a resilient regeneration of the post-war city, a preliminary distinction should be made between those which are the consolidated practices to combat the island of heat and the use to which the risk card for the population Fragile city of Parma, the subject of this research work. The representation of each building risk has a dual role and in two different ways its functionality can be looked at. First, by linking the population data resident to the single building, the map can provide information on the type of equipment available to provide public space, lightweight, mobile, and flexible equipment. It can indicate which parts are the city's fabrics on which to prioritize, encouraging and promoting extraordinary maintenance and building renovation.

The role that public administration in this process assumes is multifaceted: providing resources through the provision of positive incentives, targeted primarily at areas that are most vulnerable; interventions on public space maintenance and infrastructure planning; as a facilitator of processes in raising awareness of citizenship.

In a context of lack of public resources, becomes increasingly important the sharing of objectives and management models of territory, environment and cultural public resources, with attention to the opportunities offered by collaborative economies.

8. In the enrichment of vegetation equipment, low or zero tree species of volatile organic compounds (VOCs) should be selected. A class of chemical compounds that extends the life of hydrocarbons increases the formation of ozone. (Dessì, et al., 2016).

However, the risk map obtained needs further insights. It would be worthwhile to know the physic consistence of the city, made of different materials: minerals and plants. In order to estimate, for example, the collective benefits resulting from the stimulation of specific mitigation measures.

Despite the limits described, due to the lack of cognitive data, the methodology used for mapping can be replicated to other urban realities, if appropriately validated.

REFERENCES

Akbari, H. & Levinson, R., 2008. Evolution of Cool-Roof Standards in US. Advances in Building Energy Research, 2(32), pp. 1–32.

Conticelli, E., Proli, S., Santangelo, A. & Tondelli, S., 2015. Salvaguardia e promozione del verde nella città compatta attraverso le politiche di pianificazione: il caso del RUE di Faenza. Urbanistica Informazioni —special issue, Volume 263, pp. 44–47.

Crichton, D., 1999. The Risk Triangle. In: J. Ingleton, a cura di Natural Disaster Management. London: Tudor Rose, p. 38–47.

Crisci, A., Congedo, L., Morabito, M. & Munafò, M., 2016. Summer Heat Risk Index: how to integrate recent climatic changes and soil consumption component. PeerJ Preprints.

Dessì, V., Farnè, E., Ravanello, L. & Salomoni, M. T., 2016. Rigenerare la città con la natura. Strumenti per la progettazione tra mitigazione e adattamento ai cambiamenti climatici. Santarcangelo di Romagna: Maggioli Editore.

D'Ippoliti, D. et al., 2010. The impact of heat waves on mortality in 9 European cities: results from the EuroHEAT project. Environ Health, 16 07. pp. 1–9.

EEA, 2012. Climate change, impacts and vulnerability in Europe 2012. An indicator-based report. Copenhagen: EEA.

Georgiadis, T., 2015. Cambiamenti climatici ed effetti sulle città, Bologna: Regione Emilia-Romagna.

Gherri, B., 2012. Il comfort outdoor per gli spazi urbani. In: M. Maretto, ed., Ecocities. Il progetto urbano tra morfologia e sostenibilità. Milano: Franco Angeli.

ISPRA, 2015. Il consumo di suolo in Italia, Rapporto: 218.

ISPRA, 2016. Consumo di suolo, dinamiche territoriali e servizi ecosistemici., Rapporto: 248.

ISPRA, 2016. Gli indicatori del CLIMA in Italia nel 2015. Stato dell'ambiente 65/2016. s.l.: ISPRA.

Michelozzi, P. et al., 2016. Sull'incremento della mortalità in Italia nel 2015: analisi della mortalità stagionale nelle 32 città del Sistema di sorveglianza della mortalità giornaliera. e&p Epidemiol Prev, 40(1), pp. 22–28.

Morabito, M. et al., 2017. Summer surface temperatures of residential buildings in the city of Parma: the effect of imperviousness in different urban areas, Orvieto: CNR.

Morabito, M. et al., 2015. Urban-Hazard Risk Analysis: Mapping of Heat-Related Risks in the Elderly in Major Italian Cities. PLoS ONE, 10(5).

Morabito, M. et al., 2016. The impact of build-up surfaces on land surface temperatures in Italian urban areas. Science of the Total Environment, Volume 551–552, pp. 317–326.

Munafò, M., Luti, T. & Marinosci, I., 2016. Il consumo di suolo in Italia. In: A. Arcidiacono, et al., ed. Nuove sfide per il suolo. Rapporto 2016. INU Edizioni: Roma, pp. 57–61.

Musco, F. & Zanchini, E., 2014. Il Clima cambia le città. Strategie di adattamento e mitigazione nella pianificazione urbanistica. Milano: Franco Angeli.

Oke, T. R., 1981. Canyon geometry and the nocturnal urban heat island: comparison of scale model and field observations. International Journal of Climatology, Vol. 10, pp. 237–245.

Oke, T. R., 1995. The heat island characteristics of the urban boundary layer: Characteristics, causes and effects. Wind Climate in Cities, p. 81–107.

Xu, Z. et al., 2016. Impact of heatwaves on mortality undere different heatwaves definitions: A systematic review and meta-analysis. Environmental International, Volume 89–90, pp. 193–203.

Town and Infrastructure Planning for Safety and Urban Quality – Pezzagno & Tira (Eds)
 ISBN 978-0-8153-8731-2

Toward a sustainable mobility through. A dynamic real-time traffic monitoring, estimation and forecasting system: The RE.S.E.T. project

V. Torrisi, M. Ignaccolo & G. Inturri
Università degli Studi di Catania, Catania, Italy

ABSTRACT: The Department of Civil and Architecture Engineering of Catania, within the project RE.S.E.T., has implemented an "ITS Laboratory". Specifically, it is a traffic monitoring, estimation and short-term forecasting system, equipped with radar sensor and a central control station for traffic data elaborations.

The main focus of this study concerns the model development and the implementation of this system and a systematic investigation of its functionalities is made in order to identify opportunities to enable the next generation of ITS technologies to contribute to lower energy usage and GHG emissions and to optimize the existing infrastructure and transportation systems.

The final aim of this work is to increase the knowledge of the actual dynamic of road traffic in urban areas and to obtain useful data and information that can contribute to the implementation of an optimal control of mobility system.

1 ABOUT THE PROJECT

The realization of the "ITS laboratory" for the urban area of Catania was financed by the EU from the European Regional Development Fund under Regional Operational Program FESR Sicily 2007–2013 within the project "Network of laboratories for the safety, the sustainability and the efficiency of transports in the Sicilian Region" – RE.S.E.T. Beneficiary of the project is a partnership consisting of the four Sicialian universities[1] and three public authorities. In particular, the partnership aspired to an investment campaign that, through the acquisition of highly technological equipments, could make the research center a benchmark for the transport sector, the logistics and the sustainable mobility throughout the Mediterranean basin.

The University of Catania, partner of the project, has collaborated with the Department of Civil Engineering and Architecture (DICAR) through the involvement of four scientific-disciplinary groups: ICAR 05 Transport, ICAR 04 Railways and Airports, ICAR 06 Topography and Cartography, ICAR 20 Engineering and Urban Planning, and still with the contribution of the MoMact, a university internal structure with the mobility management function.

In the design reference framework above mentioned, the contact information of the University of Catania are as follows:

Prof. Ing. Matteo Ignaccolo,
Scientific director of the project RE.S.E.T.
Department of Civil Engineering and Architecture, University of Catania
Via Santa Sofia 64, Catania 95123, Italy
http://www.dicar.unict.it/Personale/Docenti/Docenti/Ignaccolo.html

1. Universities of Palermo, Enna, Messina and Catania.

2 INTRODUCTION

Transport is key to growth and competitiveness, providing the physical networks and services for the movement of people and goods. Nevertheless, with the high-speed development of city construction, waste of energy caused by traffic congestion and inefficient traffic has become a common problem concerned (Xiao et al., 2009).

Intelligent Transport Systems (ITS), with the integration of information and communication technologies, are crucial for sustainable development and they play an important role in shaping the future ways of mobility and the transport sector.

ITS technologies for promoting sustainable mobility consist in real-time, accurate and efficient integrated systems for transport, traffic monitoring and management, which take advanced information technology, data communication transmission technology, electronic sensor technology, electronics and computer control technology as a whole into ground transportation management system and they play an important role in modern transportation and traffic management (Ning et al., 2006).

According to this, many cities are going to develop ITS as a major way to solve contradiction between the roads and the vehicles, therefore, traffic flow forecast and traffic congestion control are the precondition to establish intelligent urban traffic.

Starting from these remarks, the main focus of this study concerns the model development and the implementation of a traffic monitoring, estimation and short-term forecasting system, with a systematic investigation of its functionalities.

The paper is structured as follows: after the introduction about the capability of ITS in the context of urban sustainable mobility presented in this section, the research approach for the model development used to implement the traffic control and supervisor system and the related infomobility services will be dealt in section two; the case study will be presented in section three by a description of the territorial framework and some related applications of the system; the final section will summarize the major conclusions and directions for further development of the research.

3 RESEARCH APPROACH FOR THE MODEL DEVELOPMENT

The proposed research approach allows the identification of the conceptual model of a traffic control and supervisor system, that represents the basis to define the system's architecture, in order to identify a set of required services and functional relationships.

The model development is based on the typical cyclical structure of the control systems. It consists of *sensors* for monitoring the real system (vehicle-driver-infrastructure-environment); *elaboration system* for data processing and the identification of strategies to be implemented; and *actuators* for real-time regulation and sharing user information.

The general operating scheme of this ITS system is shown in Figure 1 and it is characterized by three phases: input, elaboration and output.

The input systems (radar pad sensors, cameras, inductive loops, floating car data and bluetooth data, etc.) detect traffic data in real time. These data are collected, filtered and processed by the elaboration system, to be eventually transformed into operator information and subsequently available to users.

This phase of elaboration is realized by using software that simulate actual and future traffic conditions, based on input data and historical databases.

Finally, by using this simulation data, output systems provide decision-making support, both in planning and management transport process. In addition, they provide information to vehicles and drivers and/or actively intervene on traffic regulation and transport network operation.

This schematic diagram, even though it is being proposed for the broader application level, i.e. network transport system, it remains valid even for individual components of the system, up to the level of a single vehicle's control. Therefore, the proposed model is quite general: in fact, in some applications, some components or some interactions may be absent.

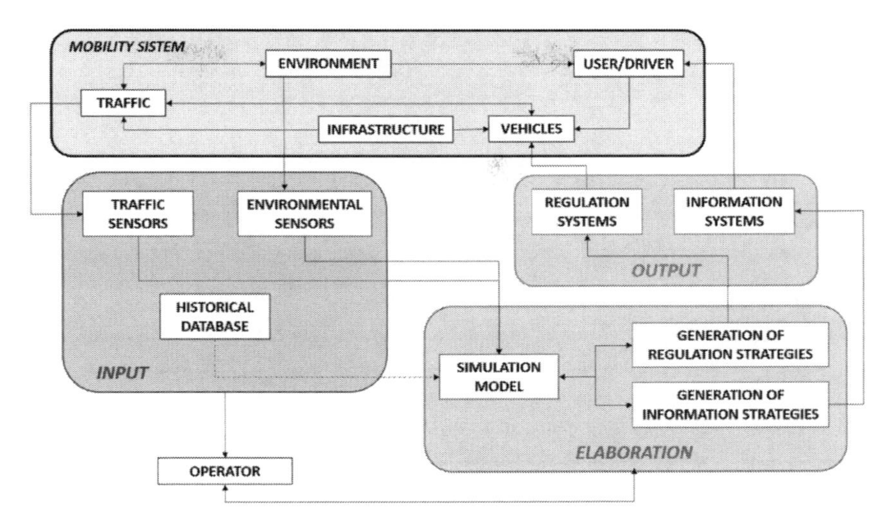

Figure 1. Model development and components' interactions of the ITS system.

The next section describes the components of the implemented system.

4 CASE STUDY

4.1 *Territorial framework and set-up of the system*

Catania is a city of about 300.000 inhabitants and it is located in the eastern part of Sicily; it has an area of about 183 km^2 and a population density of 1.754,54 inhabitants/km^2 (Ignaccolo et al., 2017). It's part of a greater Metropolitan Area (750.000 inhabitants), which includes the main municipality and 26 surrounding urban centers, some of which constitute a whole urban fabric with Catania. The main city contains most of the working activities, mixed with residential areas (Ignaccolo et al., 2016) characterized by a high commuting phenomenon which has led to a heightened inclination to private mobility and the direct consequence of traffic congestion.

The study area is broadly represented in Figure 2, by the white coloured portion of the territory, whereas the grey portion identifies the neighboring municipalities, which are not included (Torrisi et al., 2016). The analyzed transport network coincides with the territory of the urban area of Catania, which is of municipal and partly provincial competence, as well as state-owned network sections managed by ANAS[2] company.

Basically, the network transport model embodies the following network objects: 89 zones, 304 connectors, 7871 nodes and 17528 links. The purpose is to identify shortcomings and bottlenecks for each modelled link in the current and forecast traffic conditions.

Figure 3 illustrates the architecture of the implemented system in Catania. As stated in previous section, this architecture consisting of three levels: a peripheral and a central level, connected by a communication level.

The initial basis of traffic data is essentially represented by the continuous records from floating car data and by the detected flows by traffic sensors installed on transport infrastructure (peripheral level), which have been configured by using the MobilTraf MANAGER software and remotely managed via the MMobility interface.

These information, together with those related to any anomalous events, are transmitted through a GPRS cellular network (communication level) in real-time to Optima software, that

2. Azienda Nazionale Autonoma delle Strade.

Figure 2. Study area and simulated road network (Torrisi et al., 2016).

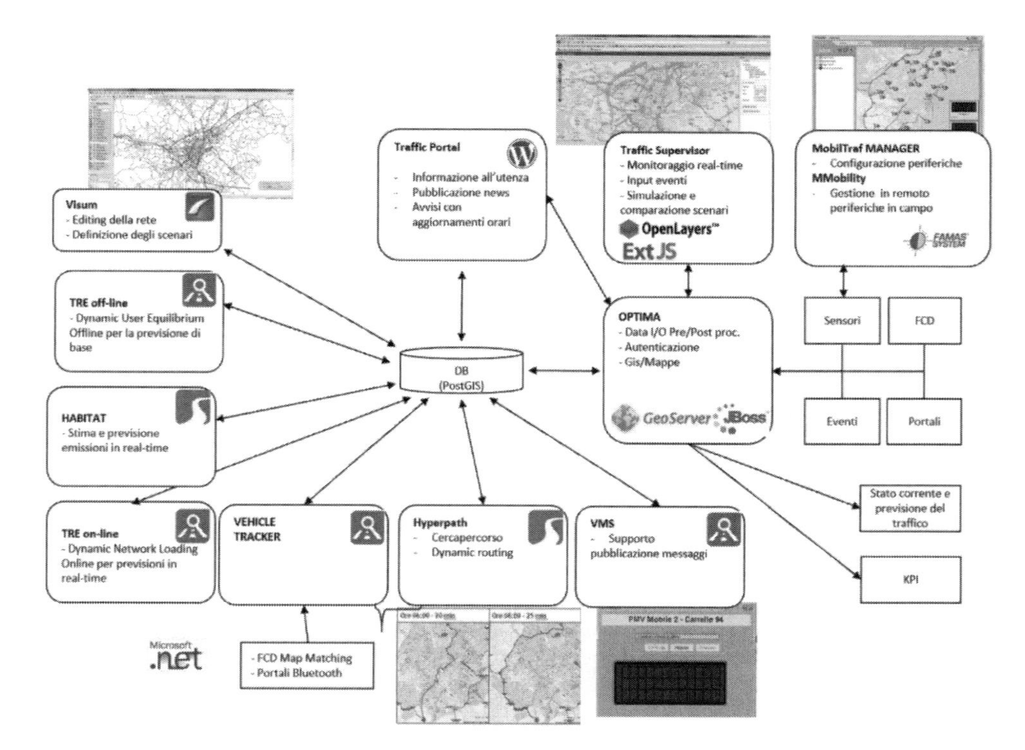

Figure 3. Architecture of Catania's ITS system.

represents the "brain" of the system, through which traffic estimates and forecasts are simulated (central level). As far as the traffic flow estimating and forecasting method, it is based on the integration of a dynamic assignment simulation model (Meschini and Gentile, 2011) and the application of periodic update with movable horizon called rolling horizon (Mahmassani, 2001), in order to reproduce the dynamic interaction between the level of congestion and users route choices of a congested road network, that is significant non-linear. It propagates the spatial and temporal traffic flow measured in less than 1% of the network links and reaching a high level of precision in short term traffic predictions over the whole graph.

In more detail, this system supplies traffic data including flow, speed, density and queue from simulations integrating the available radar detectors' data and FCD. More in detail, it is able to provide real time traffic estimations and forecasts on the main streets of Catania with 15 minutes' time intervals.

Optima has two interfaces: the Operator Interface Traffic Supervisor and the User Interface Traffic Portal, through which it is possible to immediately view on a map the current and future traffic condition of the network, i.e. speeds, presence of congestion and unexpected events.

These elaborations are stored in real-time within a database and they are continuously transmitted to the additional modules. Thereby, Hyperpath—the dynamic intermodal and multimodal journey planner—is able to determine the optimal path, taking into account real-time and future conditions of the network, by using different transport modes (foot, bicycle, public transport, possibly combined with private vehicle). As long as, Habitat—the environmental module—calculates the vehicular emissions based on COPERT4 model, starting from the estimated traffic flows.

This part of the architecture, is the on-line side of Optima's functioning. Instead, the off-line side is represented by the use of Visum software for the network editing and the static assignment procedure required for calculating basic traffic data.

The complex set-up of the system has comprised the following phases: the definition of the study area and the simulated road network by means the network graph and the demand segments. Subsequently, the identification of the optimal location of traffic sensors and the integration of FCD in the data model are followed. Then, the operations of installation, calibration and configuration of the system have been performed. Empirical data from external sources are represented by traffic counts and they were used to calibrate and to validate the model. Moreover, it has been carried out a statistical analysis of the results to verify the reliability of the traffic reliefs and estimations respectively obtained by the traffic sensors and the simulation model. Finally, the implementation of the additional modules have been executed.

4.2 *Application and use of the ITS system*

After installation, calibration and validation of the traffic monitoring and forecasting system, it can be used to assess current and future conditions also in the presence of anomalous events such as accidents or congestion (Figure 4), to identify potential strategies and

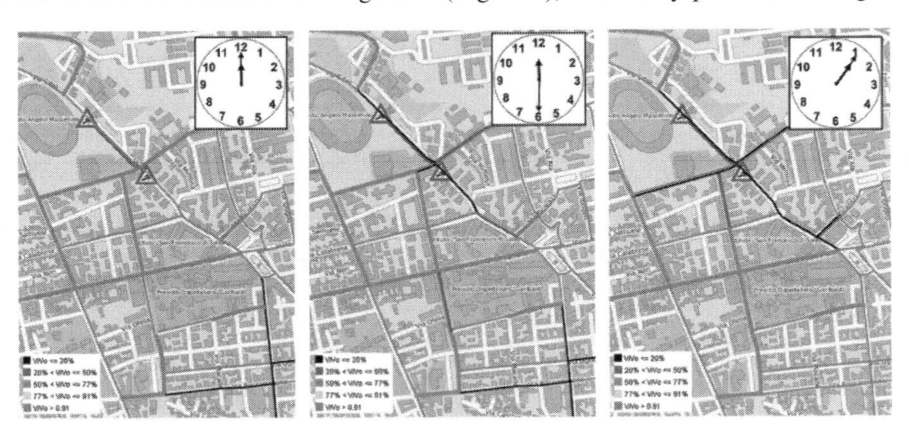

Figure 4. OPTIMA TrafficSupervisor: Displaying the current and forecast traffic condition.

Figure 5. OPTIMA TrafficSupervisor: Simulation of a Limited Traffic Zone—ZTL.

measures to improve sustainable mobility, i.e. Limited Traffic Zone—ZTL (Figure 5), and to assess the impacts of these strategies and measures on the transport system itself and on influenced processes, like the environmental impacts, the congestion reduction and the achieved modal shift.

The system, conceived in this way, is able to provide real time information on traffic congestion on the main streets of Catania, with 15 minutes' time intervals and traffic forecasts for a period of up to 60 minutes—even after unexpected situations. Although, this information is useful for just-in-time delivery operation, or for private cars movements inside the city, in order to provide a description of how congestion evolves, on each arc of the road network, during the day.

This traffic flow analysis is also the basis to determine external effects like environmental impacts emission of pollutants, through the environmental module.

Taking into account this transportation negative externalities, the journey planner is able to able to determine the optimal path on the transport network, based on different transportation and environmental point of views.

5 CONCLUSIONS AND FUTURE RESEARCHES

As stated in the introduction, information and communication technologies are crucial for sustainable mobility because their use accelerate the efficiency, the safety and the "greening" of transportation.

In this context, analyzes of current and predicted future traffic conditions and transport impacts are an important input to develop strategies and measures to alleviate the identified shortcomings and to transform the transport system in accordance with more sustainable transport visions and objectives.

The paper proposes a research approach to develop a traffic monitoring, estimation and forecasting system, through the integration of real-time traffic data in a dynamic simulation model.

This system is capable of scientifically and correctly estimating and forecasting future developments and changes in traffic conditions. Accordingly, quantitative results can be obtained, allowing to determine impacts of the strategy alternatives on traffic conditions and environment, in order to promote sustainable mobility.

The procedure adopted to implement this system, conceived in this way, is scalable and exportable in various metropolitan contexts where the phenomenon of congestion is high and the variation of travel times and traffic flows is not negligible throughout the day.

To conclude, further research will be conducted through the spatial extension of the system and its evolutionary maintenance for the improvement of its current capabilities and adding any new functions.

REFERENCES

Ignaccolo M., Inturri G., Giuffrida N., Torrisi V. (2016). Public transport accessibility and social exclusion: making the connections. Proceedings ICTTE – 3rd International Conference on Traffic and Transport Engineering Belgrade 2016, pp. 785.

Ignaccolo, M., Inturri, G., Giuffrida, N., Le Pira, M., Torrisi, V., 2017. Structuring transport decision-making problems through stakeholder engagement: the case of Catania metro accessibility. In: Ed(s) Dell'Acqua, G. and Wegman, F. (Eds.) "Transport Infrastructure and Systems: Proceedings of the AIIT International Congress on Transport Infrastructure and Systems (Rome, Italy, 10–12 April 2017)". CRC Press.

Mahmassani, H.S. 2001. Dynamic network traffic assignment and simulation methodology for advanced system management applications. Networks and Spatial Economics, 1 (3–4): 267–292.

Meschini, L.; Gentile G. 2011. Real-time traffic monitoring and forecast through OPTIMA—Optimal Path Travel Information for Mobility Action. 2nd International Conference on Models and Technologies for Intelligent Transportation Systems. Leuven, Belgium, 22–24 June 2011.

Ning, S., Minggui, T., Yaqing, W., Zhongfa, C., Yan, C., Yong, Y., (2006). A Bp Neural Network Method for Short-term Traffic Flow Forecasting on Crossroads. Computer Applications and Software, Vol. 23, No. 2, Feb. 2006, pp.32–33.

Torrisi V., Ignaccolo M., Inturri G., Giuffrida N. (2016). Combining sensor traffic simulation data to measure urban road network reliability. Proceedings ICTTE – 3rd International Conference on Traffic and Transport Engineering Belgrade 2016, pp. 1004.

Xiao, L., X., Wang, Z., Xu, B., Hong, P., (2009). Research on Traffic Monitoring Network and its Traffic Flow Forecast and Congestion Control Model Based on Wireless Sensor Network. International Conference on Measuring Technology and Mechatronics Automation, Zhangjiajie, Hunan, China, China, 11–12 April 2009. IEEE. DOI: 10.1109/ICMTMA.2009.405.

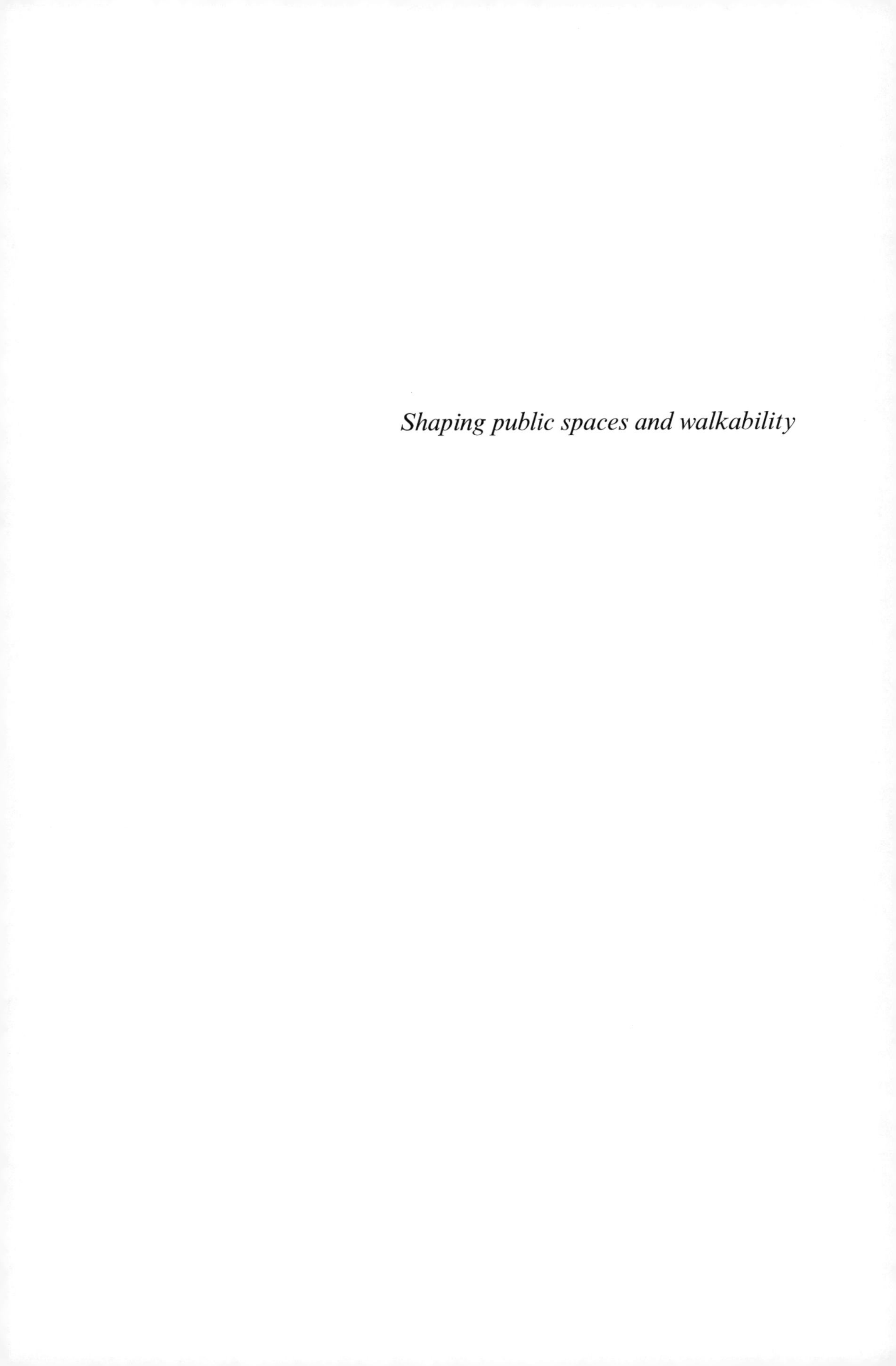

Shaping public spaces and walkability

Town and Infrastructure Planning for Safety and Urban Quality – Pezzagno & Tira (Eds)
© 2018 Taylor & Francis Group, London, ISBN 978-0-8153-8731-2

The influence of the built environment on walking experience and walking intentions. A case-study from Bristol, UK

A. Bornioli, G. Parkhurst & P. Morgan
University of the West of England, Bristol, UK

ABSTRACT: This study compares affective walking experiences (AWE) and walking intentions (WI) in five locations of the city center of Bristol, UK and tests the relationship between AWE and WI with inferential analysis and qualitative research.

An online experiment was conducted with 269 adults who work and/or study in Bristol.

Findings confirmed that WI varied between settings and that WI were influenced by affective walking experiences.

Policy implications are also discussed. First, the development of traffic-free areas in urban settings is recommended. Second, improvements on the walking infrastructure can contribute to WI. Third, cleanliness and maintenance of streets and public areas should be enhanced. Fourth, results strengthen rationales for preservation and conservation of historic areas, buildings, and elements such as lampposts. Fifth, incorporating green and blue elements in cities is a successful strategy to improve psychological wellbeing and increase WI. Finally, spaces to socialise should be a priority for city designers.

1 INTRODUCTION

Increasing walking levels is a priority for the urban sustainability agenda. This is due to the fact that walking entails psychological wellbeing benefits[1]. Shifting from vehicular to walking modes would also reduce car dominance in cities, thus reducing air and noise pollution. However, settings vary in the extent to which they support particular activities, and the benefits of walking are moderated by the environments in which walking is performed[2]. It is, therefore, increasingly important to explore which characteristics of current built environments support psychological wellbeing.

According to environmental psychology theories, positive affective experiences encourage behaviours and intentions, while negative ones elicit avoidance[3]. These ideas also apply to travelling and walking behaviours. The affective traveling experience also influences future behaviours and intentions, with individuals also likely to choose the travel mode that provides a positive affective experience[4]. Therefore, as noted by Gatersleben and Uzzell[5], affective appraisals of the travel experience produce important implications for the promotion of sustainable transport, as they provide "insight into the reasons that people prefer certain travel modes to others" (p. 417).

However, the role of affect in travel behaviours has received limited attention from scholars, despite it having 'considerable explanatory potential' in the walkability debate[6]. Gatersleben and Uzzell[7] have attested that positive affective appraisals—relaxation and excitement—

1. Robertson et al., 2012; Gatrell, 2013.
2. Johansson, Hartig, and Staats, 2011.
3. Russell and Lanius, 1984.
4. Gatersleben and Uzzell, 2007; Páez and Whalen, 2010.
5. 2007.
6. Andrews et al., 2012, p. 1930.
7. 2007.

are associated with positive attitudes—e.g. individuals' likes or dislikes about, for example, walking—towards travel options. Johansson, Sternudd and Kärrholm[8] showed that affective valence predicted intentions to avoid or to choose specific routes. In addition, valence mediated perceived urban qualities—presence of greenery, level of upkeep and maintenance, and perceived complexity and aesthetic quality, with high valence encouraging walking intentions.

This research assessed the affective benefits of walking in several built settings[9] and whether and how walking intentions are influenced by such affective experiences. While the former topic is discussed in a different work[10], this paper illustrates results related to the latter.

2 OBJECTIVE OF THE PAPER

The current study compares affective walking experiences (AWE) and walking intentions (WI) in five locations of the city center of Bristol, UK and tests the relationship between AWE and WI with inferential analysis and qualitative research.

3 METHODOLOGICAL APPROACH AND/OR RESEARCH APPROACH

An online experiment was conducted with adults who work and/or study in Bristol, UK. Participants were 269 individuals (69.1% females) ranging in age from 18 to 67 years old ($M = 31.69$, $SD = 13.63$) and mainly White British (82%). One-hundred and twenty nine were undergraduate psychology students and 140 were employees of public and private organizations based in Bristol city center. Five environmental simulations involving different videos, each of a walk in a single location in the city center of Bristol (UK) were used:

1. Pedestrianized Historic Environment, PedHist (Figure 1). A cobbled, historic street with no greenery located in Bristol's Old Town.
2. Pedestrianized Modern Environment, PedMod (Figure 2). A modern street with no evident greenery.
3. Pedestrianized mixed environment, PedMixed (Figure 3). A stone-paved pedestrian/cycle route framed on one side by Bristol Cathedral, and on the other side by a semi-open area with grass, trees, and lampposts.
4. A commercial area with traffic, CommTraf (Figure 4). It has many high street retail outlets and cafes. Traffic is moderate.
5. An inner city urban park, Park (Figure 5).

A one-minute video of a simulated walk was filmed for each environment. Measures of AWE were taken before and after the video; these included Hedonic tone (hedtone: happy, sad, content, sorry) and Tense arousal (stress: relaxed, nervous, calm, edgy) from the UWIST MACL Scale[11], both on 16-point scales. Measures of attractiveness (ugly–beautiful, unpleasant–pleasant, unfriendly–friendly, unenjoyable–enjoyable, repulsive—inviting) and interestingness (uninteresting–interesting, average–exceptional, dull–exciting) measured on a 5-point Likert scale, were also included. Walking intentions (*If this kind of environment was on your way to work, would you be more likely to walk to work more often?*) was measured on a 5-point scale.

The analysis was supported by qualitative research. 14 photo-elicited semi-structured interviews[12] were conducted. Participants were asked to take a walk in the city centre and photograph elements of the surroundings that made them feel good or bad during the walk. Photographs were then discussed during the interview.

---

8. 2016.
9. Bornioli, Parkhurts, and Morgan, 2017.
10. Bornioli, Parkhurts, and Morgan, 2017.
11. Matthews et al., 1990.
12. Guell and Ogilve, 2013: Belon et al., 2014

Figure 1. PedHist.

Figure 2. PedMod.

Figure 3. PedMixed.

Figure 4. CommTraf.

Figure 5. Park.

4 DESCRIPTION

4.1 *Innovative aspects*

This research measures AWE with a innovative experimental methodology and supports quantitative findings with qualitative research.

Results will show that the impact of environmental perceptions on WI is mediated by AWE. In other words, physical features have an impact on walking intentions because they influence affective variables, which in turn influence WI. It is the AWE that guides WI, with positive experiences—e.g. relaxing and/or pleasant journey—encouraging future walking, and negative experiences—e.g. stressful and/or unpleasant—discouraging future walking.

4.2 *Research and/or technical developments*

Barriers and enablers of WI are identified. It is shown that traffic, busyness, and poor aesthetics have a negative effect on walking intentions by making participants stressed, annoyed, and concerned. Conversely, natural elements and interesting features elicit positive affective responses—relaxation, happiness, sense of safety, etc. – and this increases WI.

5 RESULT OF THE RESEARCH

The mixed ANOVAs revealed that the walk in PedMixed increased hedtone and decreased stress, the waks in PedHist and Park increased hedtone, and the walk CommTraf increased stress and hedtone[13].

Descriptive statistical analysis of the variable *walking intentions* showed that PedHist, PedMod, PedMixed, and Park were associated with positive WI, while CommTraf was associated with low WI (Table 1).

A one-way analysis of variance (ANOVA) identified a significant main effect with a large effect size of setting, $F(4, 262) = 21.587$, $p = .000$, $\eta_p^2 = .210$. Post hoc analyses using the Scheffe post hoc criterion for significance indicated that WI in the Park were significantly higher than in CommTraf ($p = .000$), PedHist ($p = .009$), and PedMod ($p = .014$), but not different from PedMixed. WI in CommTraf were significantly lower than in PedHist ($p = .002$), PedMod ($p = .000$), PedMixed ($p = .000$) and Park ($p = .000$).

Multiple linear regression analysis was carried out to explore the association between *WI* and a series of independent variables: affective appraisals (AWE), environmental perceptions, walking habits, and socio-demographics. The model was significant, $F(8,206) = 11.113$, $MSE = 88.906$, $p = .000$, $R^2_{adj} = .350$ (Table 2). The Sobel test indicated mediation of attractiveness by Δstress, $t = 3.22$, $p = .001$, and by Δhedtone, $t = 6.27$, $p = .000$.

Interviews revealed the elements of the built environments that encourage or deter walking intentions. These are summarised below.

Table 1. Walking intentions descriptive statistics.

Walking intentions (5-point scale)		
Setting	N	M
PedHist	70	3.41
PedMod	73	3.43
CommTraf	76	2.55
PedMixed	75	3.88

13. For extended results see Bornioli, Parkhurst, and Morgan, 2017.

Table 2. Walking intentions regression model.

Model 1: Walking intentions					
	Unstandardized Coefficients		Standardized Coefficients		
Model	B	Std. Error	Beta	t	Sig.
(Constant)	1.633	.382		4.276	.000
Δstress	.061	.025	.172	2.407	.017
Δhedtone	.070	.031	.171	2.254	.025
Interestingness	.267	.109	.229	2.441	.016
Attractiveness	.195	.103	.185	1.894	.060
Age	.003	.005	.043	.711	.478
Female1	–.061	.146	–.025	–.421	.675
Walks to work2	.112	.123	.052	.908	.365
Heavy Walker3	.134	.152	.051	.883	.378

Comparison groups:
1 = Male
2 = Other modes (cycling, public transit, car)
3 = Non-walker (e.g. walks less than 4 days a week)

6 BARRIERS OF THE AWE

CommTraf was associated with low WI. The qualitative phase highlighted that this is related to motor-traffic, city busyness, and lack of aesthetic value. Interviews showed that participants "deliberately chose a route that avoided the main road' and avoid busy routes. This confirms previous research on the low walking levels in large and busy roads[14] and the negative effects on walking intentions of the presence of motor traffic[15].

City busyness—crowds, high buildings, scaffolding and construction sites—and poor aesthetics are deterrents for walking intentions. participants reported that they prefer to avoid crowds. Scaffolding and construction sites that obstruct the pavement represent another feature that makes participants avoid certain routes. This is in line with previous research that attested the poor desirability of low-quality pedestrian infrastructures during walking[16]. Poor aesthetics deters WI due to safety concerns.

7 ENABLERS OF THE AWE

The quantitative analysis highlighted that walking intentions were highest in the settings with greenery and/or blue elements—the park and PedMixed. During the interviews, the presence of natural areas in the city was discussed in detail, and nature emerged as a motivator for people to walk. The presence of nature encouraged participants to walk, and when they had the option of choosing between a built and green path, they tend to opt for the latter because it is less busy and crowded. Hence, the current results extend previous research on the positive effects of presence of natural elements on WI[17].

Moreover, participants reported that particular architectural features attract their attention during walking. Interesting architectures are related to higher WI compared to static designs of tower blocks and housing estates, and this emerged to be especially true for older

14. Adkins et al., 2012, Cain et al., 2016.
15. Belon et al., 2014; Mindell, 2017.
16. Gatersleben and Uzzell, 2005, Belon et al., 2014, Cain et al., 2016.
17. Cain et al., 2016, Johansson, Sternudd and Kärrholm, 2016.

architectures, which evoke a sense of history and encourage walking. This confirm Ewing and Handy's[18] ideas on the importance of *imageability* on walking intentions; the current results extend previous empirical research that has assessed quantitatively that the presence of 'historic buildings', 'iconic or distinctive architecture' and 'articulated, non-rectangular buildings' contribute to walkability[19].

Also, social liveliness contributes to WI, thus confirming previous research[20].

8 CONCLUSIONS

Findings confirmed that WI varied between settings and that WI were influenced by affective walking experiences (Δstress and Δhedtone). By examining the AWE with qualitative research, it was possible to identify aspects of built settings that influence walking experiences and intentions. The relevance of the AWE echoes findings by Guell and Ogilve[21] that showed that 'wellbeing' is among 'the most prominent narratives' of traveling. Their research attested that enjoying natural scenery or interesting built features, and experiencing relaxation and 'me-time' were very important aspects of commuting practices, and on the other hand, the 'lack of wellbeing' was a major issue in participants' accounts.

Within the current research, traffic was among the most important barrier of the AWE and WI, and this is related to noise and air pollution related to traffic. These represent a source of stress and attentional fatigue. Traffic also causes interruptions to the walking flow and forcing pedestrians to wait at crossing points, two elements which bear further negative effects on psychological wellbeing. Finally, safety concerns and power dynamics also contribute to a negative affective walking experience, with perceptions that cars represent a danger and that drivers "always win". Conversely, walking in pedestrianised areas had a neutral or positive affective response, and was associated with positive WI. This was due to the absence of the barriers related to motor-traffic. In fact, pedestrianised areas are less noisy and air-polluted, there is more freedom of movement, and there are fewer safety concerns.

Other than motor-traffic, city busyness, which refers to the feeling that "the city never seems to rest", was firstly accentuated by the lack of space in pavements due to pedestrian crowds, narrow or low-quality pavements, or objects that obstruct walking (e.g. scaffolding on the pavement). Such lack of space tends to make walking uncomfortable and frustrating. Second, the multitude of stimuli in the urban setting—signs, shops, and tall buildings—enhances perceptions of crowding, thus increasing stress levels. Finally, poor aesthetics—litter, fly tipping, overflowing bins, and tagging—make walking unpleasant and triggers safety concerns.

Several enablers of WI emerged. The pedestrianised historic street and the pedestrianised setting with green and built elements were associated with positive WI. These outcomes appeared to be partially related to the sense of history that some individuals experience during walking in historic areas. Experiencing older buildings, historic elements such as lampposts or cobble floor offers a distraction during walking and triggers imagination and curiosity. On a deeper level, the sensation of "feeling the ground" also contributes to a sense of safety and comfort. In some cases, being exposed to historic architectures triggers place attachment and pride in individuals. These outcomes are related to the details of older architectures, their unique "character", and the association with Bristol's history and culture. Therefore, the current research confirmed ideas from urban planning on the role of old buildings and historic elements in the city[22].

Variety in the built environment is an additional element that contributes to the affective walking experience. interviews revealed that the positive affective walking experience in the Ped-Mixed setting was partially related to its urban design variety, that included natural and historic elements but also a mix of people (see following section: Social liveliness). The interviews also

18. 2009.
19. Ameli et al., 2015.
20. Brown et al., 2009.
21. 2015.
22. Lynch, 1984.

highlighted that architectural variety, the architectural mix of old and new, the mix of natural elements and built features, panoramic and hidden views are generally stimulating and inspiring. These features make the walking journey more interesting and make it go quicker.

Walking in natural and semi-natural settings triggers affective and restorative benefits[23]. Within the experimental phase of research, the two simulated walks in in the settings with green elements were associated with high WI. Interviews also revealed that the piece and quietness of natural spaces and the solitude offered by such spaces are among the crucial features that encourage walking. In addition, nature details, such as flowerpots, falling leaves, or glimpses of sky all contributed to the AWE and WI.

9 POLICY IMPLICATIONS

The development of traffic-free areas in urban settings is recommended. When this is not possible, further measures on car traffic could be considered by policy makers. First, noise and pollution from vehicles, especially buses and trucks, could be reduced in order to decrease the negative impact of motor-traffic on pedestrians[24]. It seems that electric vehicles could bear a smaller negative impact on walkers than fuel vehicles, thanks to the reduced auditory and environmental impact. Second, results imply that improvements on the walking infrastructure can contribute to improve the affective walking experience. For instance, better separations between pavements and roads could be introduced in order to increase pedestrians' perceived safety. In addition, safety could be improved at crossing points, for example with raised platforms; waiting times for pedestrians could also be reduced in order to aid the walking flow.

Urban design improvements could also contribute to the AWE. These include enlargement and improvement of pavements and prioritisation of pedestrians during road works and constructions, interventions which would minimise the discomfort for pedestrians during walking. Finally, results indicate that general cleanliness and maintenance of streets and public areas is crucial to support affect during walking, hence these should be enhanced.

Results strengthen rationales for preservation and conservation of historic areas, buildings, and elements such as lampposts. In addition, policy makers could improve interpretation and awareness of such elements, for example by introducing interpretation plaques in the urban setting, or by introducing education programmes that promote the value of historic elements. Also, walking itineraries in historic areas could be recommended by policy makers.

Results confirmed that incorporating green and blue elements in cities is a successful strategy to improve psychological wellbeing and increase WI. These elements include parks, small parks, canals and bodies of water, but also micro natural elements such as flowerpots, bushes, and street trees.

Spaces to socialise should be a priority for city designers. These include active frontages, outside seating, and public squares results highlighted that sense of community triggers positive feelings, hence place making, community events, exhibitions, busking etc. could be promoted.

REFERENCES

Ameli, S.H., Hamidi, S., Garfinkel-Castro, A. and Ewing, R., 2015. Do better urban design qualities lead to more walking in Salt Lake City, Utah?. *Journal of Urban Design*, *20*(3), pp. 393–410.

Andrews, G.J., Hall, E., Evans, B. and Colls, R., 2012. Moving beyond walkability: On the potential of health geography. *Social Science & Medicine*, *75*(11), pp. 1925–1932.

Belon, A.P., Nieuwendyk, L.M., Vallianatos, H. and Nykiforuk, C.I., 2014. How community environment shapes physical activity: perceptions revealed through the PhotoVoice method. *Social Science & Medicine*, *116*, pp. 10–21.

23. Ulrich, 1983; Kaplan and Kaplan, 1989.

24. Taylor, 2003.

Bornioli, A, Parkhurst, G., and Morgan, P., 2017. The psychological wellebeing benefits of walking in attractive urban settings: an experimental design. Under review.
Brown, B.B., Werner, C.M., Amburgey, J.W. and Szalay, C., 2007. Walkable route perceptions and physical features: Converging evidence for en route walking experiences. *Environment and behavior*, *39*(1), pp. 34–61.
Cain, K.L., Millstein, R.A., Sallis, J.F., Conway, T.L., Gavand, K.A., Frank, L.D., King, A.C., 2014. Contribution of streetscape audits to explanation of physical activity in four age groups based on the Microscale Audit of Pedestrian Streetscapes (MAPS). Soc. Sci. Med. 116, 82–92.
Ewing, R. And Handy, S., 2009. Measuring the Unmeasurable: Urban Design Qualities Related to Walkability. *Journal of Urban Design*, 02/01; 2015/03, vol. 14, no. 1, pp. 65–84.
Gatersleben, B. And Uzzell, D., 2007. Affective Appraisals of the Daily Commute: Comparing Perceptions of Drivers, Cyclists, Walkers, and Users of Public Transport. *Environment and Behavior*, May 01, vol. 39, no. 3, pp. 416–431.
Gatrell, A.C., 2013. Therapeutic mobilities: walking and 'steps' to wellbeing and health. *Health & Place*, 7, vol. 22, pp. 98–106.
Guell, C. and Ogilvie, D., 2015. Picturing commuting: photovoice and seeking well-being.
Johansson, M., Hartig, T. And Staats, H., 2011. Psychological Benefits of Walking: Moderation by Company and Outdoor Environment. *Applied Psychology: Health and Well-being*, vol. 3, no. 3, pp. 261–280.
Johansson, M., Sternudd, C. And Kärrholm, M., 2016. Perceived urban design qualities and affective experiences of walking. *Journal of Urban Design*, 03/03, vol. 21, no. 2, pp. 256–275.
Kaplan, R. and Kaplan, S., 1989. *The experience of nature: A psychological perspective*. CUP Archive.
Karmanov, D. And Hamel, R., 2008. Assessing the restorative potential of contemporary urban environment(s): Beyond the nature versus urban dichotomy. *Landscape and Urban Planning*, 5/26, vol. 86, no. 2, pp. 115–125.
Lynch, K. 1984. *Good city form*. MIT press.
Matthews, G., Jones, D.M. and Chamberlain, A.G., 1990. Refining the measurement of mood: The UWIST mood adjective checklist. *British journal of psychology*, *81*(1), pp. 17–42.
Páez, A. And Whalen, K., 2010. Enjoyment of commute: a comparison of different transportation modes. Transportation Research Part A: Policy and Practice, vol. 44, no. 7, pp. 537–549.
Robertson, R., Robertson, A., Jepson, R., & Maxwell, M. 2012. Walking for depression or depressive symptoms: a systematic review and meta-analysis. Mental Health and Physical Activity, 5(1), 66–75.
Russell, J.A., & Lanius, U. F. 1984. Adaptation level and the affective appraisal of environments. Journal of Environmental Psychology, 4(2), 119–135.
Russell, J.A., 1980. A circumplex model of affect. Journal of Personality and Social Psychology, vol. 39, no. 6, pp. 1161.
Taylor, N., 2003. The aesthetic experience of traffic in the modern city. Urban Studies, 40(8), pp. 1609–1625.
Ulrich, R.S., 1983. Aesthetic and affective response to natural environment. In Behavior and the natural environment (pp. 85–125). Springer US.

Town and Infrastructure Planning for Safety and Urban Quality – Pezzagno & Tira (Eds)
© 2018 Taylor & Francis Group, London, ISBN 978-0-8153-8731-2

Tram-train line as infrastructural corridor for the requalification of Pontina Bioregion

V. Mazzeschi
Università degli Studi di Roma La Sapienza, Roma, Italy

ABSTRACT: Many cities across the world are implementing plans to reduce the private road mobility, encouraging the modal shift to public transport. Many authors make reference to the well-established American practice of Transit Oriented Development (T.O.D.), where the strategy for the government of urban transformation is developed within the areas of influence of the stations. Value capture, is the tool by which is possible to finance part of the costs of the infrastructure by intercepting the added value of real estate that are located at a certain distance from the railway station.

This paper focuses on the project of a railway system, served by tram-train, in Pontina Bioregion. The project, thought like a new infrastructural corridor in this territory, introduces also actions directed to the revaluation of the whole area, through the requalification of the hill old towns, the transition to low carbon mobility and the safeguard of the natural environment.

1 INTRODUCTION

In the last decade for the first time in the history of the world, the urban population became bigger than the rural population. The new urban areas are land settlements that are diffused on vast territories and they are characterized by high mobility of inhabitants, as an effect of the urban sprawl phenomenon.

Indeed, in order to contrast the negative externalities, it is necessary to introduce the Bioregion concept. Different authors such as Patrick Geddes, Lewis Mumford and Peter Berg developed this idea. In this paper, we consider the definition given by Alberto Magnaghi. Magnaghi defines the Bioregion as a place to find new relationships of coexistence between inhabitants-producers and regional territory; local self-reliance communities, ensuring that the natural environment can sustain them permanently; local closure of cycles (water, food, energy, waste, ecological networks to reduce the ecological footprint); development of local economic systems for the exploitation of the territorial and landscape heritage; food sovereignty and enhancement of multifunctional agriculture (Magnaghi, 2011).

Moreover, the Bioregion concept could contrast the inhabitants' cultural preference of cars for the everyday journeys. This is also an urban sprawl effect, that causes traffic congestion, increase of the air pollution, and other negative externalities. To contrast these externalities many cities are implementing policies to reduce the car use in favour of public transport. An example is the "London congestion charge", a fee that motor vehicles, which want to operate within the CCZ (Congestion Charge Zone), in London city centre, have to pay.

Nevertheless, these policies have their limitations because their focus is only on the modal shift from the private means of transport to public transport (buses, subway, trams) and not on the theme of urban transformation and the densification of residential buildings, with the possible integration between the transport planning and the urban and regional planning.

2 OBJECTIVE OF THE PAPER

For these reasons, the objective of this paper is to study infrastructural corridors, not only as a line that link two cities, but as an axis which is able to redevelop urban and rural areas. Moreover, the infrastructure chosen is the tram-train, an innovative means of transport, employed for the first time in Germany in 1992. The main focus that will be described in the next paragraph is the analysis aimed to project a tram-train line in Pontina Bioregion, between the cities of Sabaudia and Ceccano.

3 METHODOLOGICAL APPROACH

After a literary review on well-known practices such as Transit Oriented Development and value capture with their integration in the Copenhagen case of study, the paper describes the guidelines and the analysis for a new infrastructural corridor, served by a tram-train system, in Pontina Bioregion.

4 DESCRIPTION

The Transit Oriented Development (T.O.D.) is a particular urban development that create liveable, walkable and pedestrian-oriented communities centred on railway systems (www.tod.org). The transit stop is the pivot area, surrounded by a high density mixed land use, with the density that gradually becomes lower in relation to the distance from the centre. Indeed, the densest areas are located within a radius of half mile around the transit stop, because this distance is considered suitable for pedestrians. In this way, also the last mile problem is solved because the citizens will use their own cars only for reach their final destinations, located not so far than one mile from the transit stop.

The main characteristics of the Transit Oriented Development are:

- Mixed Land use: there is the presence of shops, houses, offices. This mix permit to do different activities without take any means of transport.
- Value Capture: the possibility to recover part of the amount used for building the infrastructure.
- Transit stop: it is the point where the requalification of the whole area starts.

Moreover, Transit Oriented Development is an interesting application of the integration between urban and transport planning. In conclusion, the potential of T.O.D. are several (Cervero, 2004):

- Environmental: the car use reduction, more walkability areas, the obstacle to urban sprawl,
- Social: pedestrian-oriented neighbourhood,
- Financial: the use of land value capture tools.

Starting from this last point, we can consider that land and property values increase due to the improved access to the area. If we want to catch this gain, for example, in order to recover part of amount for the construction of a railway system, we have to introduce the theme of land value capture, which is "a way to capture the increase in the value of land and development generated by the improved accessibility of transportation" (National Bank, 2014). In other words, areas near urban rail transit stations can be valuable because they are more accessible (Cervero, 1994), making these areas more attractive than other places. Indeed, it is very important to make people aware of the benefits, so they are willing to pay for them. Moreover, there are different tools to capture the value gain, tools that depend on the local circumstances, such as the urban density and the tax regime. Nevertheless, value capture cannot completely replace public financial support (Budoni, 2014), but they are important to establish a relation between the infrastructure and the citizens which use it.

An interesting case study of T.O.D. application strictly connected to the use of a land value capture tool is the Ørestad line in Copenhagen, Denmark. The main characteristic of the Ørestad model is the integration between the project of a new neighbourhood and the construction of a metro line (Book et al., 2010). In the early '90, the city council decided to develop the Ørestad area, at that time it was an abandoned area, as a new "rail-based" district (Book et al., 2010), in order to governing the land use development. Although the choice of metro line was the most expensive alternative, the Ørestad Development Corporation (ODC), the agency formed with the aim to plan this project, managed to get funding by two different ways: the first one was the traditional state budget for the infrastructure, the second one was the land banking, a land value capture tool. Through it, the ODC recovered part of the amount of money spent to realize the whole project, by selling the land, whose value increased after the construction of the infrastructure, near the metro line stop (Sumiraschi).

Other examples are Metro Gran Paris (Sumiraschi, 2013), Warsaw Metro System (Medda et al., 2010), New Castle Tyne and Wear Metro (Milotti et al., 2008). After this short literary review, we can notice that there is a new way to consider the infrastructure corridors role, especially the railways. They are not only a line linking two different places, but they are also a catalyst to the redevelopment of the bioregional territory. Moreover, the entire urban and rural structure is reoriented towards the transit stop with high density building near the station, in order to reduce the use of cars and also, in this way, making a modal shift to the public transport.

The public means of transport chosen in this paper is the tram-train. The tram-train system is a transport system that integrates railway lines and tramways using tram vehicles. These vehicles can circulate without any intermediate stop on both types of infrastructure (railway and tramway), in order to connect the suburban territory with the city (Molinaro, 2013). The key word of the tram-train is the interoperability (Spinosa, 2010), that is the possibility of the vehicle of being able to circulate both on railway and tramway tracks. The first tram-train system was built in the city of Karlsruhe, in Germany in 1992 and it was a very successfully experience. The tram-train therefore offers a high accessibility service, with a high capillarity in the urban areas and able to guarantee a high frequency of service. Furthermore, this means of transport can become a requalification tool for urban and rural areas affected by urban sprawl externalities.

5 RESULT OF THE RESEARCH

An example of the construction of an infrastructural corridor and its positive externalities is the project of the tram-train line Sabaudia-Ceccano in Pontina Bioregion.

The identification of Pontina Bioregion was carried out using the seminar work developed during the attendance of "Territorial design[1]". The students, including the author of this paper, analyzed the territory divided into four groups. Each group worked on a different theme: the analysis of the phases of territorialisation, the analysis of socio-economic characteristics, the study of the land settlements and the focus on the environmental characteristics of the area. According to the investigations conducted by the four group, each of them identified its own Bioregion boundaries, which were obviously different from each other. The last phase was the most intense and complicated one, since those different boundaries were to be synthesized in a single boundary, which could be the expression of all analyzes carried out. After many meetings, where several debates took place, the boundary of Pontina Bioregion was defined. The boundaries are: to east and southeast there are the Lepini-Ausoni Mountains, on westbound there is the sea and to the north there is the volcanic area of Colli Albani. The Pontina Bioregion includes the cities of:

Anzio, Aprilia, Nettuno, Cisterna di Latina, Latina, Cori, Rocca Massima, Segni, Montelanico, Norma, Bassiano, Sermoneta, Sezze, Maenza, Roccagorga, Carpineto Romano,

1. Master degree in "Environmental Engineering for sustainable development", Sapienza University of Rome, Latina University Campus. The Territorial design course was held by Prof. Alberto Budoni.

Figure 1. Pontina Bioregion.

Giuliano di Roma, Villa Santo Stefano, Vallecorsa, Amaseno, Ceccano, Terracina, Pontinia, San Felice Circeo, Sabaudia, Monte San Biagio, Fondi (Fig. 1).

The problems present in the area are mainly due to the strong impact the anthropic matrix has on the environmental system. The impact has surely become more pressing since the years of integral reclamation, the 30s of the last century, when an entire ecosystem was destroyed to "redeem" the swampy areas, transforming those wetlands into lands that have not always been as fertile as it had been planned. Particularly undeniable are the urban sprawl effects, such as the massive use of the car for inhabitants' journeys. Indeed, this area is sadly known for the high number of deadly car accidents.

For these reasons, in this paper the author introduces the project for the construction of the tram-train line between Sabaudia, a city closer to the sea, and Ceccano, a city in the inland[2]. In this way, different objectives could be reach:

- Enhance the accessibility of the cities on the Lepini-Ausoni mountains and in some cases, it could encourage the repopulation of these areas, with the aim of avoiding the excessive urbanization of the plain municipalities,
- Create a connection, through a railway, between the seacoast and the mountains, with the goal of making the modal shift from private vehicles to public transport,
- Make a connection between the Rome-Formia-Naples railway and the Rome-Cassino-Naples railway, that it could be a rapidly line of communication between Latina and Frosinone, the two main cities in the South Lazio.

The cities affected by this line, both those physically crossed by the line and those that fall into the line basin are: Sabaudia, San Felice Circeo, Terracina, Priverno, Pontinia, Roccasecca dei Volsci, Prossedi, Maenza, Roccagorga, Sonnino, Amaseno, Villa Santo Stefano, Giuliano of Rome, Ceccano, Patrica, Frosinone and Carpineto Romano. In particular, the cities belonging to the line basin are included considering a journey time of less than 20 min-

2. This project was the main focus of author Master degree thesis. The supervisor was Prof. Alberto Budoni and the assistant supervisor was Prof. Stefano Ricci. In particular, Prof. Stefano Ricci, who is a transport engineer, oversaw the technical feasibility of the project and he suggested the different rail services.

utes by car to reach the nearest station of the project tracks. Moreover, these cities are characterized by an average commuting population rate of 20%.

In the following table, we can find some information about the population, the commuting rate and the number of the everyday journeys (to the outside).

The line object of the present study and has an extension of 85 km, divided into three main sections:

Table 1. Commuting rate.

City	Population	Total journeys	Commuting rate (%)
Sabaudia	20331	3461	18
Terracina	45850	4972	11
San Felice Circeo	10018	1493	15
Pontinia	14871	3342	23
Priverno	14495	2875	20
Roccasecca dei Volsci	1153	302	27
Prossedi	1212	271	23
Maenza	3077	785	26
Roccagorga	4587	1353	30
Sonnino	7483	1739	24
Amaseno	4314	927	22
Villa Santo Stefano	1750	392	23
Giuliano di Roma	2373	701	30
Ceccano	23504	5136	22
Frosinone	46529	5236	12
Patrica	3147	633	29
Carpineto Romano	4581	1274	28

Table 2. Tram-train's stations and stop.

Station and stop	City
Sabaudia Carlo Alberto	Sabaudia
Sabaudia Molella	Sabaudia
San Felice Circeo	San Felice Circeo
Terracina Sisto (stop)	Terracina
Terracina Badino	Terracina
Terracina (already exist)	Terracina
Priverno Fossanova (already exist)	Priverno
Priverno Mezzagosto	Priverno
Giuliano di Roma	Giuliano di Roma
Ceccano (already exist)	Ceccano

Table 3. Types of services on Circeo Express line.

Section	Type of service	Description
Rome-Latina-Terracina-Sabaudia	"fast" regional train	No stop between the different stations
Latina-Terracina-Sabaudia	Regional train	The train stop in every station
Terracina-Sabaudia	Tram	High frequency

Table 4. Types of services on Rome-Frosinone via Latina.

Section	Type of service	Description
Rome-Priverno Fossanova	"fast" regional train	One stop in Latina
Priverno Fossanova-Frosinone	Regional train	The train stop in every station

Table 5. New population and number of "touch and go" tourists.

Cities crossed by the line				Cities in the line basin			
City	Population	New population	N° of tourists (touch and go)	City	Population	New population	N° of tourists (touch and go)
Sabaudia	20331	519	71159	Pontinia	14871	371	2082
San Felice Circeo	10018	459	35063	Maenza	3077	7	431
Terracina	45850	487	160475	Roccagorga	4587	6	643
Sonnino	7483	69	1048	Amaseno	4314	23	604
Priverno	14495	212	2030	Villa Santo Stefano	1750	2	245
Prossedi	1212	–6	170	Roccasecca dei Volsci	1159	11	163
Giuliano di Roma	2373	18	333	Carpineto Romano	4581	–24	642
Ceccano	23504	139	3291	Patrica	3147	39	441
				Frosinone	46529	–33	32571

- 28.5 Km to be constructed, between Sabaudia and Terracina, where there is a tram service, with frequent and capillary stops,
- 20 Km, already present, between Priverno Fossanova and Terracina, using the existing railway linking the two stations; there are no stops or stations in this section because the user basin is very small,
- 36,5 Km, to be constructed, between Priverno Fossanova and Ceccano, developing through the whole Amaseno valley and ending at the station of Ceccano, providing a railway service.

On the tram-train line, nine stations have been hypothesized, plus one stop, the terminus are Sabaudia and Ceccano. Of these stations, three already exist as they are stations served by railways. The following table shows the names of stations and stop. The names derive from the locations where they are.

In particular, for each line there are different services. Let see them separately.

The first one is the Rome-Sabaudia via Priverno Line. This line uses the Roma-Formia-Napoli line to the Priverno Fossanova station and then takes the branch to Terracina and from that station continues until the Sabaudia terminal on tram tracks. The route would allow a quick connection between Rome and the coast, especially for San Felice Circeo and Sabaudia, cities where many Roma residents have a vacation home. Moreover, the two municipalities are part of the Circeo National Park, which is the destination of thousands of tourists from all over Italy every year. For these reasons the line could be named "Circeo Express".

In the following table, we can find different services for the Circeo Express line.

The second one is the Rome-Frosinone via Latina line: the aim is the connection between Latina and Frosinone, the two main cities in South Lazio. Indeed, there is also a concern between the Faculty of Engineering of Latina university campus and the two Chambers of Commerce, that are interested in the possibility of a connection through train between the two territories.

The service includes a fast-regional service between Rome and Priverno Fossanova, with the only stop in Latina, and then the trains are inserted in the new tram line, in Priverno Fossanova station. After crossing the Amaseno valley they arrive in Ceccano where they enter in the regional line "Rome-Cassino" and finally arrive in Frosinone.

As for the Circeo Express line, now we can see the different services.

Then, in conjunction with the hypothesis of tram-train line, there is also a regional planning of the area crossed by the line. The actions designed for the final set-up concern (Fig. 2):

- The environmental system, such as the creation of new protected areas (Monti Lepini Regional Park, Rural Park) and the incentive for multifunctional agriculture,

Figure 2. Pontina Bioregion asset.

- The socio-economic system, with the opening of commercial points where local products could be bought;
- The settlement system, which would undergo a strong change through forms of densification-concentration in the cities on the hill.

Particularly noticeable is the reorganization, also, of the road system that sees the presence of three parkways:

- Litoranea (SP 39), parkway of the Circeo National Park, located along the coast line,
- Appia (SS7), parkway of the Rural Park of the Appia, located halfway between the coast and the mountains,
- Carpinetana (SR609), parkway of the Monti Lepini Regional Park, connecting the mountain centres each other.

A territorial set-up of this kind can obviously generate attraction phenomena. For examples:

- People who decide to move into the territory as new residents,
- Occasional tourists, that, given the increased accessibility, can choose the area as their journey destination,
- Residents in the "second homes", who can take advantage of the tram-train system to reach their homes during the holiday periods.

For these reasons, an abacus of cities has been drawn up, divided into cities crossed by the tram-train line and the cities that fall into the basin of the line. In this abacus, some data are reported to understand what are the potentials, in terms of settlement of new inhabitants or number of "touch and go" tourists that could be attracted by the reorganization of urban and rural spaces linked to the construction of the tram-train system.

Let us analyse the columns.

- Residents: the data released by ISTAT on June 1, 2015.
- New Residents: are the percentage of population that, according to the estimates of the last five years, will be located within each municipality. As can be seen, some municipalities have negative values since in these realities there are currently phenomena of regression of the population.
- Number of tourists: LUISS's research report highlights that 34% of tourists in San Felice Circeo are "touch and go" tourists (Caroli, 2005). The first step was to calculate the 34% of all the tourists in San Felice Circeo. After that, the second step was to create a ratio between tourists and population. This ratio was 3.5. It was multiplied to the population of Sabaudia and Terracina. Indeed, for the city of Frosinone, the population was multiplied to the ratio and then divided to 5. In conclusion, for the other cities, their populations were multiplied to the ratio and then divided to 25. These numbers, 5 and 25, derives from the assumption, extrapolated by the data, that the tourist's presence on the coast is respectively 5 times higher respect Frosinone and 25 times higher respect the hillside cities.

Furthermore, the construction of railway requires a large amount of money, which is not present in the local government funds, deciding in most cases the non-economic feasibility of the intervention.

As we have already seen, land value capture can be a mechanism for recovering some of the funds invested in the construction of a railway and in general is a redistribution of positive externalities that are generated by the realization of an infrastructure. For the tram-train project, different land value capture mechanism has been thought: linkage capture, for the new inhabitants, tax on shops and enterprises, tax increment financing, for the inhabitants.

5 CONCLUSIONS

The tram-train line "Sabaudia-Ceccano" is the realization of the objective of creating an infrastructural corridor, connecting mountain areas to the coast. There are different research themes that will be explored by the author during the next years. The themes are: the change of cultural preference to public transport instead of cars, the impact on urban patterns, that could be redeveloped by the application of T.O.D., the problem to introduce a new taxation, such as land value capture tools, due to the global economic crisis. In conclusion, it can be said that infrastructural corridors are powerful "tools" to requalify a bioregional territory, but it is also necessary rethinking methods and actions in order to integrate the regional and urban planning with the transportation planning, pushing the people to public transport, especially train, tram and tram-train, that contribute to a low carbon mobility.

REFERENCES

Book et al., (2010), Governing the balance between sustainability and competitiveness in urban planning: the case of the Ørestad model, Environmental policy and governance n°20, 382–296.

Budoni A. (2014), Capturing land value to develop local rail networks, Ingegneria Ferroviaria n°5, Anno LXIX.

Cervero R. (1994), Rail Transit and Joint Development: Land Market impacts in Washington D.C. and Atlanta, Journal of the American Planning Association, 60:1, 83–94.

Cervero R. (2004), Transit Oriented Development in the United States: Experiences, Challenges and prospects, Transportation Research Board of the National Academics.

Magnaghi A. (2011), Bioregione urbana e sostenibilità: applicazioni progettuali nella Toscana centrale." Comunicazione ai corsi di formazione alla efficienza energetica e alla sostenibilità. Ordine degli Architetti della Provincia di Prato.

Medda et al. (2010), Land value capture as a funding source for urban investment: the Warsaw Metro System, Ernest and Young Better Government Program.

Milotti et al. (2008), La cattura del valore come metodo di finanziamento per le infrastrutture di trasporto: tra casi a confronto, S.N., S.L.

Molinaro E. (2013), Le linee guida per il sistema tram-treno, 5° convegno nazionale sistema tram, Roma 31/01/2013–01/02/2013.

National Bank (2014), Land value capture as a source of funding of public transit for Greater Montreal, National Bank Report.

Spinosa A. (2010), Progetto tram-treno: dall'infrastruttura al progetto urbano, vol.1, Roma, 2010.

Sumiraschi C. (2013), Catturare il valore. Politiche innovative per finanziare le infrastrutture, EGEA, Milano.

SITOGRAPGHY

www.tod.org last access 22/04/2017.

Town and Infrastructure Planning for Safety and Urban Quality – Pezzagno & Tira (Eds)
© 2018 Taylor & Francis Group, London, ISBN 978-0-8153-8731-2

Services and commerce within a walkable distance from home

M. Olitsky, Y. Lerman & E. Avineri
Afeka Academic College of Engineering, Israel

ABSTRACT: In the context of the crisis condition of the Israeli housing market in recent years, it is of interest to examine the housing preferences for walking distance to land uses in Israeli neighborhoods. The main purpose of this study is to assess the impact of walking distance to shopping centers, public service facilities, open spaces and recreational sites, on the housing decision. A stated-preference analysis was conducted using a choice-based conjoint model, based on data collected from 184 respondents of a cross-sectional, online survey. The results indicate an overall preference toward apartments that are in a walkable distance from a commercial street, rather than a shopping mall. This finding alone is sufficient to indicate a demand for walkability, as shopping strips are generally pedestrian friendly. Based on the findings, the paper provides some policy recommendations with regard to efficient implementation, and discusses generalization of these findings to other countries.

1 INTRODUCTION

The concept of walkability integrates transportation and land use in a holistic manner, incorporating various social, economic and environmental aspects. Walkability describes areas ability to accommodate walking for functional or recreational purposes. For pedestrians, walkable neighborhood is an interesting, convenient and safe place, allowing a wide range of opportunities for social encounters and various activities through improved accessibility (Speck 2012). Accordingly, walkable neighborhoods are characterized by a relatively high residential density, accessible mixed land uses, street connectivity, suitable infrastructure, aesthetics, road safety and low crime rates (Leslie et al. 2005; Saelens et al. 2003).

Increasing and encouraging personal, non-motorized transportation, such as walking and cycling, is vastly supported by academic literature on walkability advantages (Talen and Koschinsky 2013; Frank et al. 2006; Jaskewicz and Besta 2014). Pedestrian oriented, mixed-use neighborhoods complement social and community engagement and improve public health by encouraging physical activity (Sallis et al. 2009; Sallis et al. 2015; Doyle et al. 2006; Leyden 2003). Compared to car-oriented neighborhoods, real-estate value, within walkable neighborhoods, is higher, and businesses enjoy up to 80% sale increase in such areas (Leinberger and Alfonzo 2012; Hack 2013). Moreover, as a result of reducing car usage and energy consumption, walkable neighborhoods decrease the ecological footprint and limit urban sprawl (Ewing et al. 2010; Van der Ryn and Calthrope 1986).

The relationship between transportation and the built form is complex, for being multidirectional and affected by other non-physical elements, such as individual preferences. However, it remains that walking behavior is in fact influenced by diversity of land use, and the number of destinations within a walkable distance (Ewing & Cervero 2010). In addition, Cao, Handy and Mokhtarian (2005) showed that not only the distance factor is significant in predicting frequency of walking, but preferences toward residential location and neighborhood characteristics also influence walking behavior.

2 OBJECTIVE OF THE PAPER

The purpose of this paper is to examine wether walkability, in terms of walking distance to specific land uses, affects the residential choice. It is of interest to compare results of previous walkability preference studies, usually conducted in Europe or the United States, and test their performance. The research was carried out in Israel, one of the most densely populated countries in the world with a considerably high population growth rate, however, similar to many other countries in terms of government land use policy and trends of urban spatial development (Carmon 2001; Orenstein & Hamburg 2010; Omer & Zafrir-Reuven 2015).

3 METHODOLOGICAL APPROACH AND/OR RESEARCH APPROACH

Discrete choice models, such as revealed and stated preference, have been widely applied in the context of travel decisions (Ben-Akiva & Bielaire 1999). However, the larger part of research on transportation and land use, practice revealed preference methods. That is, gathering and analyzing data that illustrates decisions already made in the past. Yet, it can be argued that the yielded results are lacking, in light of studies showing that desired travel behavior itself, influences the housing choice (Boarnet & Sarmiento 1998). Comparisons of revealed preference versus stated preference methods were conducted in both scopes, that of transportation and of housing separately. Revealed preferences in housing research normally involve aspects such as market condition and availability, thus, obscuring the isolated effect of preference on the actual choice, while in transportation research, stated preferences reasonably account for actual choices in reality (Wardman 1988; Timmermans, Molin & Noortwijk 1994).

4 DESCRIPTION

A stated preference choice based Conjoint (CBC) model was selected for the purpose of this study. Mainly due to the ability of the CBC analysis to simulate a real-life decision process (Bergen 2011). Additionally, CBC model is among the most popular methods in market research and consumer preference analysis, and specifically beneficial for estimation of housing location preference (Earnhart 2001; Orme 2009). Lastly, although more recent CBC method developments, such as Hierarchical Bayes and Latent Class, are gaining recognition, empirical evidence show that they are similar to the original aggregate model, in terms of real-world performance (Natter & Feurstein 2002).

A CBC study usually begins with attribute and attribute-levels selection and definition. The attributes and levels should be distinct, and relevant to the hypothesis. Next, different profiles, consisting of the abovementioned attribute levels, are constructed. The minimal amount of profiles and comparisons required for the experiment depends on the expected sample size and total number of attribute levels. Then a survey is administered. Respondents choose one profile, the most preferred, of the profiles presented in each choice task. The collected data is analyzed using a conditional logit model, providing results in form of part-worth utilities for every attribute level, which can also be calculated to assess overall importance of each attribute.

Attribute levels were defined as amenities within a walkable distance from the apartment profiles, presented in the survey. In order to maintain a framework, applicable for future consideration by policy makers and planners (Jacobs and Manzi 2000), amenity attributes and attribute levels were derived from official government planning manuals and later on refined with respect to literature on walkability (Ewing & Cervero 2001; Talen & Koschinsky 2013; Leslie et al. 2005). Table 1 presents the selected attributes and their associated levels.

The only attribute that is not related to walkability and whose levels consist of nominal values "Price" is crucial for comprehending the other attributes true relative importance. This divergent attribute plays a significant role for a proportion, according to which a comparison between built environment and inherent housing characteristics could be made. Moreover,

Table 1. Apartment attributes and levels included in Conjoint analysis.

Attributes	Levels
Commerce (within a walkable distance from residence)	Shopping mall
	Shopping strip
Services (within a walkable distance from residence)	Health clinic
	Educational institution
Leisure amenities (within a walkable distance from residence)	Sport center
	Cultural center
	Restaurant/café/pub
Open space (within walkable distance from residence)	Park
	Promenade/Boulevard
	Public garden
Price	1,200,000 NIS
	1,400,000 NIS
	1,600,000 NIS

weighing price is a natural and common component of value perception and the purchase decision process (Zeithaml 1988).

Systematically different apartment profiles, in terms of price and walkable distance to amenities, were constructed via an optimized fractional factorial design. This approach reduces the initial amount of profiles that a full factorial design would require (2*2*3*3*3 = 108). Due to the direct link between number of profiles and sample size, Orme (2010) suggested a general thumb rule, enabling to determine a minimal sample size for conjoint experiments. According to this rule, the minimal sample size for the current study results in n = 63.

An online cross-sectional survey was administered through email and social networks. The first questions addressed demographic transportation usage and current housing features. Next, the CBC survey presented 12 sets of 2 apartment profiles, from which respondents were asked to choose one they prefer. The apartments were described as completely identical with regard to characteristics such as condition, size and number of rooms and so forth. Thus, Intrinsic physical apartment, were determined as constants, therefore, allowing a categorical identification of preferences related to the surrounding environment.

Respondents' stated preferences were analyzed in order to assess the importance of specific amenities within a walkable distance from place of residence. Results of the CBC analysis stand for coefficient values of the utility function for the examined attribute levels. Part-worth utilities signify the preference towards each attribute level. Computation of the part-worth utilities estimates and statistics was carried out using a Multinomial Logit regression. In addition, based on the ratio between the Log-Likelihood of the resulting model and the null model, goodness of fit was evaluated by Mcfadden's Rho square (1974).

Note that values of part-worth utilities measure relative preference, and cannot be directly compared across different attributes or models (Hauber et al. 2016). Therefore, assessing an attribute relative importance requires calculation of the difference between the most preferred level and the least preferred level of the same attribute across all the attributes. Then, the sum of the all differences provides a proportion for the differences within each attribute, thus importance of each attribute is presented in form of percent.

4 RESULT OF THE RESEARCH

The online survey reached an initial amount of 231 participants, yielding 184 complete and valid samples in total. General demographic data of the respondents is presented in Table 2. Most of the respondents (40.76%) travel to work by private cars, 26.63% by public transportation, 12.50% by bicycle, 10.33% walk, the rest either work at home or other. 67.39% reported that work trips take under 30 minutes. Almost half of the respondents reported that

Table 2. Demographics of respondents.

Characteristic	Value	Percent	Total (n = 184)
Gender	Female	44.56%	82
	Male	55.44%	102
Age	20–29	21.73%	40
	30–39	50.00%	92
	40–49	16.85%	31
	50–59	5.98%	11
	60+	5.43%	10
Income	Significantly below average	28.26%	52
	Slightly below average	16.30%	30
	Average	11.95%	22
	Slightly above average	27.71%	51
	Significantly above average	15.76%	29
Children in household	Yes	42.40%	78
	No	57.60%	106

Table 3. Results of conjoint analysis.

Attribute	Relative Importance	Level	Estimate	Standard err.	Wald χ^2	$\|P\|>\chi^2$
Commerce	23.05%	Shopping mall	–0.405	0.033	148.966***	<0.0001
		Shopping strip	0.405	0.033	148.966***	<0.0001
Services	11.57%	Health clinic	–0.203	0.040	25.272***	<0.0001
		Educational institution	0.203	0.040	25.272***	<0.0001
Leisure	16.00%	Sport center	–0.228	0.064	12.56***	0.000
		Cultural center	–0.106	0.047	5.088**	0.024
		Restaurant/café/pub	0.334	0.053	39.737***	<0.0001
Open-Space	6.14%	Park	–0.124	0.072	2.917*	0.088
		Promenade/Boulevard	0.031	0.080	0.156	0.693
		Public garden	0.092	0.068	1.840	0.175
Price	43.23%	1,200,000 NIS	0.725	0.072	10.0555***	<0.0001
		1,400,000 NIS	0.068	0.058	1.370	0.242
		1,600,000 NIS	–0.794	0.056	–11.2239***	<0.0001
–2 Log likelihood		2508.132				
McFadden's R^2		0.181				

*Significant at 90%; **Significant at 95%; ***Significant at 99%.

they are planning to move to a new residence in the next 1–3 years. 42.93% of the respondents have purchased an apartment at least once in the past.

Results of the CBC analysis including the relative importance of attributes are illustrated in Table 3. Not surprisingly, compared to other attributes, the most important attribute affecting residential choice is price (43.23%). Commerce within a walkable distance from place of residence is the most important attribute among the neighbourhood facilities (23.05%). All the results are statistically significant at a 95% level, except for "open space" attribute levels: "park" being statistically significant at 90% level, "public garden" and "promenade/boulevard" not significant.

5 CONCLUSIONS

Although accessible and diverse land uses are a principal aspect of walkable neighborhoods, the relationship between walking behavior and the built environment, to include actual preferences, has yet been firmly established. This study examines residential choice via stated preference experiment to assess importance of walkable distance to various land uses. The data was collected and analyzed using a choice based conjoint method. The results of this study generally support previous residential stated preference studies, suggesting that in fact, accessible land use characteristics have a significant impact on housing choice (Molin & Timmermans 2003). Furthermore, the assessed importance of commerce and services is in accordance with findings of recent research on sustainable urbanism in Israel (Rofè, Pashtan, Hornik 2017).

Shopping centers within walking distance of residence play a major role in housing choice. As Newmark, Plaut and Garb (2004) showed that shopping malls cause a shift from pedestrian to motor vehicle modes, the strong preference of a shopping strip over a shopping mall, reported in the current study, provides another significant and independent evidence on peoples demand for walkable neighborhoods. Likewise, educational institutions are preferred over health clinics, assumedly, because they are visited more regularly and frequently.

Walkable distance to land uses is as important as price; both affect residential choice in a similar magnitude. This finding advocates promotion of policies and planning toward walkable and sustainable neighborhoods, both by the public and the private sector. The study could also be implemented in future examination of preferences regarding sustainable urban planning, walkability and residence location choice. The applied choice based conjoint methodology is relevant for researchers in fields of transportation and could assist in assessing desired travel behavior. For in some cases, personal attitudes affect travel behavior even more than land use characteristics (Kitamura, Mokhtarian & Laidet 1997; Bagley & Mokhtarian 2002; De Vos et al. 2012).

REFERENCES

Bagley MN, Mokhtarian PL. The impact of residential neighborhood type on travel behavior: a structural equations modeling approach. The Annals of regional science. 2002 Aug 18;36(2):279–97.

Ben-Akiva M, Bierlaire M. Discrete choice methods and their applications to short term travel decisions. InHandbook of transportation science 1999 (pp. 5–33). Springer US.

Bergen van, J., An Exploration of the Buying Decision Process of Residential Consumers: The Application of a Choice-Based Conjoint Experiment to Reveal Housing Preferences, Eindhoven University of Technology, 2011.

Bierlaire M, Robin T. Pedestrians choices. InPedestrian Behavior: Models, Data Collection and Applications 2009 Nov 19 (pp. 1–26). Emerald Group Publishing Limited.

Boarnet MG, Sarmiento S. Can land-use policy really affect travel behaviour? A study of the link between non-work travel and land-use characteristics. Urban Studies. 1998 Jun 1;35(7):1155–69.

Cao X, Handy SL, Mokhtarian PL. The influences of the built environment and residential self-selection on pedestrian behavior: evidence from Austin, TX. Transportation. 2006 Jan 1;33(1):1–20.

Carmon N. Housing policy in Israel: Review, evaluation and lessons. Israel Affairs. 2001 Jun 1;7(4):181–208.

Caspi I. Testing for a housing bubble at the national and regional level: the case of Israel. Empirical Economics. 2016 Sep 1;51(2):483–516.

De Vos J, Derudder B, Van Acker V, Witlox F. Reducing car use: changing attitudes or relocating? The influence of residential dissonance on travel behavior. Journal of Transport Geography. 2012 May 31;22:1–9.

Doyle S, Kelly-Schwartz A, Schlossberg M, Stockard J. Active community environments and health: the relationship of walkable and safe communities to individual health. Journal of the American Planning Association. 2006 Mar 31;72(1):19–31.

Earnhart D. Combining revealed and stated preference methods to value environmental amenities at residential locations. Land economics. 2001 Feb 1;77(1):12–29.

Earnhart D. Combining revealed and stated data to examine housing decisions using discrete choice analysis. Journal of Urban Economics. 2002 Jan 31;51(1):143–69.

Ewing R, Cervero R. Travel and the built environment: a synthesis. Transportation Research Record: Journal of the Transportation Research Board. 2001 Jan 1(1780):87–114.

Ewing R, Cervero R. Travel and the built environment: a meta-analysis. Journal of the American planning association. 2010 Jun 21;76(3):265–94.

Frank LD, Sallis JF, Conway TL, Chapman JE, Saelens BE, Bachman W. Many pathways from land use to health: associations between neighborhood walkability and active transportation, body mass index, and air quality. Journal of the American Planning Association. 2006 Mar 31;72(1):75–87.

Hack, G. et al. Business Performance in Walkable Shopping Areas, Robert Wood Johnson Foundation, 2013.
Handy S. Critical assessment of the literature on the relationships among transportation, land use, and physical activity. Transportation Research Board and the Institute of Medicine Committee on Physical Activity, Health, Transportation, and Land Use. Resource paper for TRB Special Report. 2005 Jan;282.
Hauber AB, González JM, Groothuis-Oudshoorn CG, Prior T, Marshall DA, Cunningham C, IJzerman MJ, Bridges JF. Statistical methods for the analysis of discrete choice experiments: a report of the ISPOR Conjoint Analysis Good Research Practices Task Force. Value in health. 2016 Jun 30;19(4):300–15.
JACOBS, Keith; MANZI, Tony. Evaluating the social constructionist paradigm in housing research. Housing, Theory and Society, 2000, 17.1: 35–42.
Jaśkiewicz M, Besta T. Is easy access related to better life? Walkability and overlapping of personal and communal identity as predictors of quality of life. Applied research in quality of life. 2014 Sep 1;9(3):505–16.
Kitamura R, Mokhtarian PL, Laidet L. A micro-analysis of land use and travel in five neighborhoods in the San Francisco Bay Area. Transportation. 1997 May 1;24(2):125–58.
Kevin M. Leyden. Social Capital and the Built Environment: The Importance of Walkable Neighborhoods. American Journal of Public Health, September 2003, Vol. 93, No. 9, pp. 1546–1551.
Leinberger, C.B. and Alfonzo, M. Walk This Way: The Economic Promise of Walkable Places in Metropolitan Washington, D.C. Washington, DC: Metropolitan Policy Program at the Brookings Institution, 2012.
Leslie E, Saelens B, Frank L, Owen N, Bauman A, Coffee N, Hugo G. Residents' perceptions of walkability attributes in objectively different neighbourhoods: a pilot study. Health & place. 2005 Sep 30;11(3):227–36.
Molin E, Timmermans HJ. Accessibility considerations in residential choice decisions: accumulated evidence from the Benelux. In Präsentiert auf dem "82nd Annual Meeting of the Transportation Research Board" (TRB), Washington, DC 2003 Jan 12.
McFadden D. Conditional logit analysis of qualitative choice behavior. In: Zarembka P, ed. Frontiers in Econometrics. New York, NY: Academic Press, 1974. p. 105–42.
Natter M, Feurstein M. Real world performance of choice-based conjoint models. European Journal of Operational Research. 2002 Mar 1;137(2):448–58.
Newmark G, Plaut P, Garb Y. Shopping travel behaviors in an era of rapid economic transition: evidence from newly built malls in Prague, Czech Republic. Transportation Research Record: Journal of the Transportation Research Board. 2004 Jan 1(1898):165–74.
Omer I, Zafrir-Reuven O. The Development of Street Patterns in Israeli Cities. Journal of Urban and Regional Analysis. 2015 Jul 1;7(2):113.
Orenstein DE, Hamburg SP. Population and pavement: population growth and land development in Israel. Population and Environment. 2010 Mar 1;31(4):223–54.
Orme B.K., Which conjoint method should I use?, Sawtooth Software Research Paper Series, Sequiem, WA: Sawtooth Software, Inc. 2009.
Orme B.K. Getting started with conjoint analysis: strategies for product design and pricing research. Research Publishers; 2010.
Rofè Y, Pashtan T, Hornik J. Is there a market for sustainable urbanism? A conjoint analysis of potential homebuyers in Israel. Sustainable Cities and Society. 2017 Apr 30;30:162–70.
Sallis JF, Saelens BE, Frank LD, Conway TL, Slymen DJ, Cain KL, Chapman JE, Kerr J. Neighborhood built environment and income: examining multiple health outcomes. Social science & medicine. 2009 Apr 30;68(7):1285–93.
Sallis JF, Spoon C, Cavill N, Engelberg JK, Gebel K, Parker M, Thornton CM, Lou D, Wilson AL, Cutter CL, Ding D. Co-benefits of designing communities for active living: an exploration of literature. International Journal of Behavioral Nutrition and Physical Activity. 2015 Feb 28;12(1):30.
Saelens BE, Sallis JF, Black JB, Chen D. Neighborhood-based differences in physical activity: an environment scale evaluation. American journal of public health. 2003 Sep;93(9):1552–8.
Speck J, Farrar S, Like G. Walkable City. Farrar, Straus and Giroux, New York, NY. 2012.
Talen, E. and Koschinsky, J. The Walkable Neighbourhood: A Literature Review. International Journal of Sustainable Land Use and Urban Planning, Vol. 1 No. 1, pp. 42–63, 2013.
Timmermans H, Molin E, Van Noortwijk L. Housing choice processes: Stated versus revealed modelling approaches. Journal of Housing and the Built Environment. 1994 Sep 1;9(3):215–27.
Van der Ryn S, Calthorpe P. Sustainable communities. Sierra Club: San Francisco. 1986.
Wardman M. A comparison of revealed preference and stated preference models of travel behaviour. Journal of transport economics and policy. 1988 Jan 1:71–91.
Zeithaml VA. Consumer perceptions of price, quality, and value: a means-end model and synthesis of evidence. The Journal of marketing. 1988 Jul 1:2–2.

Town and Infrastructure Planning for Safety and Urban Quality – Pezzagno & Tira (Eds)
 ISBN 978-0-8153-8731-2

When transport infrastructure shapes the public space: Public transport stops as urban places

E. Vitale Brovarone
Politecnico di Torino, Turin, Italy

ABSTRACT: The paper will analyse the interest over public transport stops (PTS) in the scientific debate, examining the approaches to this issue. Three main spheres of the scientific debate that express some interest on this issue will be explored: debate on modal choice (transportation engineering, territorial planning); debate on public spaces, urban design, landscape design; manuals and guidelines on the design of public transport stops. Furthermore, the paper will focus on some realised projects: some interesting realisations show that conceiving PTS as urban places is possible. Combining the scientific debate with the design proposals and realised projects, the paper will define some design criteria and directions for the development of an approach to the design of PTS based on liveability, on quality and on the harmony of PTS with the urban context in which they are inserted.

1 INTRODUCTION

Albeit the design of transport systems and infrastructures have been usually ascribed to the technical-engineering domain, the interest over it is recently growing in different fields of studies, among which urban planning and design.

From an urban design perspective, transportation infrastructures are something more than functional elements that guarantee an efficient mobility: infrastructures can be considered as active and important components of a place, being it urban or suburban.

Within the field of public transportation, the idea of transport infrastructure as an urban place in all respects is chiefly linked to major railway, metro or intermodal stations. Several authors have studied this aspect, analysing the role that major railway, metro or intermodal stations play into the urban landscape and their impact on travellers' behaviour (Cascetta & Cartenì, 2014; Kido, 2005; Thorne, 2001).

Ordinary public transport stops are widely spread throughout the city and physically integrated into the urban context, but their important role in shaping urban places is neglected.

Public transport stops (PTS) are considered as part of the transport system, and their design must comply with technical criteria that are often predefined without any specific adaptation to the urban context where they will be inserted.

On the contrary, the synergistic union of functional aspects with the harmonic insertion of PTS into the urban context is a key element for their liveability. In its turn, their liveability produces an influence both on the quality of the urban landscape and on modal choice.

Even if public transport stops are usually considered with respect to their functionality, they are part of an urban place, being active and distinguishing elements in its definition process. Therefore, they deserve more attention by the disciplinary debate.

2 OBJECTIVE OF THE PAPER

Starting from the assumption that PTS can play a key role in the process of definition of a public space, the main goal of the paper is to explore the validity of this assumption, fostering the debate. This goal is linked to some sub-objectives aimed at its fulfilment:

- To detect the theoretical bases within the scientific debate;
- To identify some realised projects that can testify this assumption;
- To identify some possible criteria to define how PTS can actually become shaping features of the urban environment.

3 METHODOLOGICAL APPROACH

The research started from an analysis of the debate on the role of PTS in shaping the urban landscape.

Since the issue is quite narrowed, the literature review started from a broader analysis of the debate over the design of transport systems and infrastructures in the urban environment. Subsequently, the analysis focused on the insertion of PTS in the urban landscape and on their liveability.

The basic assumption of the paper is that the liveability of transit stops is as an expression of the synergistic union of functional aspects with a balanced insertion into the urban context, that in its turn produces an influence on the quality of the urban landscape and on modal choice.

A search for realised projects has also been conducted, to identify cases in which PTS can be considered as shaping features of the urban environment.

Since the literature review lead to quite scarce results in terms of case studies, this phase has been developed mainly through a web search.

The criterium for the selection of realised projects has been to prioritise the ones that were expression of an explicit aim to create a place, and-or that were inserted into a process devoted to this issue instead of sporadic design initiatives.

A third phase has been the definition of a set of design criteria oriented towards the liveability and the quality of PTS so that they can definitely become shaping features of the urban environment.

4 DESCRIPTION

The interest over PTS can be divided into three main spheres of the scientific debate:

- *Debate on modal choice* and on the possibility to influence travel behaviour through the design of PTS. The interest on stops is part of a wider set that includes several different elements, considered by different points of view and with different methods (Beirão & Cabral 2007; Buys & Miller 2011; Carse, 2011; Clark, Chatterjee, & Meli, 2016; Frank & Pivo, 1994; Taylor et al., 2002; Wardman, 2004). Inside this debate, the role of PTS as urban places is often considered as secondary to other factors, such as population density, cost of parking, public transport system efficiency, socio-economic conditions, etc.
- *Manuals and guidelines on the design of public transport stops*. Here public transport stops get their specific attention: rules, parameters and suggestions are defined, to ensure their efficiency and integration in the urban transport system. Nevertheless, also in this case the functional component dominates considerations on aspects regarding PTS as urban places. The integration with the urban context is often considered as ancillary aspect, while technical and functional aspects are much more important (see e.g. APTA, 2012; Ceder, 2007; SEPTA, 2012).
- *Debate on public spaces, urban design, landscape design*: here the interest over the role that PTS play in the urban space is much more considerable. Albeit the literature on this issue is not so rich and mostly focused on major public transport nodes, here they are acknowledged as real and authentic urban places, being able to generate values and identity and to foster dialogue and exchange (Caroselli 2011; Cascetta & Cartenì, 2014; Imbrighi, 1999; OTREC, 2013; Zhang 2012). Within this debate, functional aspects do not prevail over the aesthetic ones, allowing attempts of integration and openness towards a constructive

dialogue, aimed at the improvement of quality and liveability of the urban environment as well as to the usability of public transport.

Beyond the scientific debate over PTS as urban places, the interest over this issue is attested also in the professional field, through initiatives specifically devoted to PTS, such as the "Thinking Beyond the Station" project in the UK, "Subart" in the San Francisco Bay, "Bus:Stop" in Austria.

Albeit a specific theoretical debate on this issue is hard to be detected, some affinities can be recognised among the different approaches to it: the PTS is seen as a determinant for the definition of a place. This way of considering the PTS is recognisable also in several realised projects; the following paragraph, shows some projects where the PTS is conceived as something more than a functional element of the transport system.

5 REALISED PROJECTS

1. Portland: The Portland Transit Mall

The aim is to combine the infrastructural renovation with the opportunity to realise a wider urban redevelopment project being immediately recognisable.

Stemming from the specific features of the urban context, the project identifies seven different "urban rooms", characterised by their own attributes, opportunities and limits.

The leading idea is to overcome the view of public transport exclusively with respect to mobility and accessibility, acknowledging its role for the definition of public urban places.

The renewal included the introduction of a light rail line with seven stations, new shelters, repaving and renewal of street furniture, and artistic installations.

Each PTS is rooted in its urban context, with a careful design of features, materials and typologies, sometimes privileging very light structures, transparencies and slight elements, others opting for more robust and recognisable marks.

2. Providence: The TransArt project

The goal is twofold: to improve the comfort of users waiting for the bus, and to realise unique and recognisable stops, expressly linked to their urban context.

A selection of 12 pilot PTS has been made, based on several factors, such as the volume of users, the number of lines, population density around the stops.

Each of them is strongly related with its urban context, becoming integral and representative parts of it.

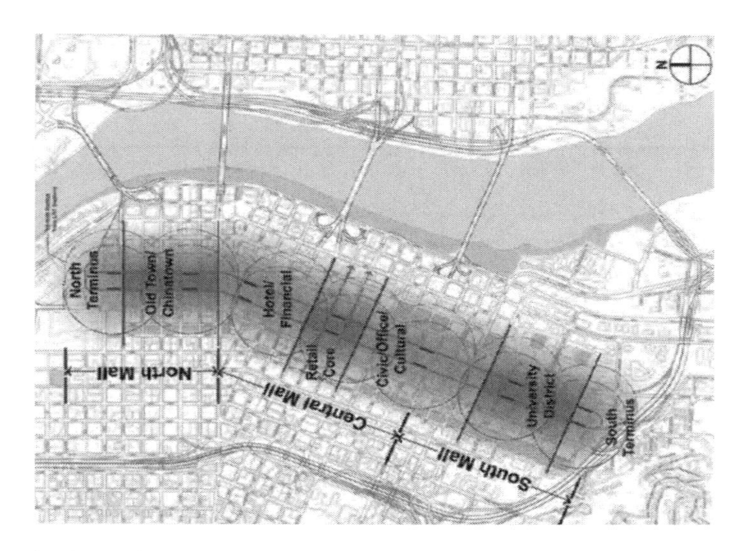

Figure 1. Portland Transit Mall—The "urban rooms". Source: Portland Bureau of Planning.

Figure 2. Office district: transparency and lightness are the main features of the stop. Here form and function are strictly intertwined, and physical, visual and interpersonal relationships are maximised. Ph: Landperspectives.

Figure 3. Bank district: the PTS is the expression of the urban place and foster relations; the structure, realised in the '70s, is preserved and transformed in a little coffee cart. Ph: Jean Senechal Biggs.

3. London: The Edible Bus Stop Project

The Edible Bus Stop (EBS) project is aimed to transform disused sites situated on a London bus route into design-led public growing spaces, directly involving local inhabitants and fostering the creation of neighbourhood hubs.

A 3-stage process lead the development of each stop: engage the community, animate the existing space and transform the site through design.

Each EBS is developed according to its specific context, utilising or referencing available materials or existing landmarks.

This approach encourages the sense of place by the community and prevents from acts of vandalism. It also promotes social inclusion: the growing spaces are open to all, volunteers take on the responsibility of gardens upkeep and can enjoy its harvests. Albeit this approach is nowadays limited to a single bus route, it can be easily extended to other routes, creating

Figure 4. CommTraf.

Figure 5. TransArt—Wickenden Street Stop, Providence. Ph. RIPTA.

Figure 6. The Original Edible Bus Stop before the intervention. © 2016 The Edible Bus Stop® All Rights Reserved.

Figure 7. The Original Edible Bus Stop after the intervention. © 2016 The Edible Bus Stop® All Rights Reserved.

a network of growing spaces and green corridors into the urban context, as it doesn't imply huge economic or operating efforts.

4. Highland Town, Baltimora:

B-U-S Here the PTS is the place, becoming the main element of the urban renewal project, that meets the needs and constraints laid down by the context, creating an iconic urban meeting point.

The project, realised in Baltimore by the Spanish collective "mmmm..." jointly with two local sculptors, is part of the TRANSIT—Creative Placemaking with Europe in Baltimore initiative, that includes three permanent installation located in different zones of Baltimore and is aimed to join public art with public transportation.

To represent the vitality, dynamism and plurality of that neighbourhood, the intention was to create a place for the community to enjoy, interact and meet while waiting for the bus.

Figure 8. BUS. The PTS *is* the place. Ph. Mmmm...

Figure 9. BUS. The PTS *is* the place. Ph. Mmmm...

The project combines a strong imageability, immediateness of perception, practicality and plurality of uses.

6 RESULT OF THE RESEARCH

The theoretical debate and the realised projects shown in the previous chapters let emerge some criteria that can be defined for the development of an approach to the design of PTS based on liveability, on quality and on the harmony of PTS with the urban context.

The following design criteria are not meant to be alternative to the functional parameters, but they can be considered as part of a system in which form and function are constituting part of an integrated approach.

- **Figurability**: The PTS deserves to become an active element for the definition of the public space. According to the specific features of the context, the PTS will develop its own, influencing to a different extent the definition of the space.
- **Belonging and integration**: The PTS is part of the urban space, sharing the same tangible and intangible features. As opposite to the widespread practice, that sets a structural and morphologic homologation of PTS, the idea itself of a place identity implies a diversification.
- **Usability and comfort**: To make the PTS fully perform its function of place for staying—an intrinsic aspect of public transport—all the recommended functional elements (seats, shelters, racks, …) must be designed as expressions of sensitiveness and attention to people's comfort.
- **Transparency and visibility**: The PTS is part of the urban environment and sets up a dialogue with it, combining transparency and visibility to and from the surroundings, also contributing to the real and perceived safety of the PTS.
- **Accessibility**: The ease of access to the PTS, especially with non-motorised modes, is a crucial factor for its liveability. Besides the physical accessibility, virtual accessibility is equally important (e.g. wi-fi, etc..), as well as granting access to additional services and contents.

A cross-cutting element resulting from the interaction of the former ones, is the **vitality** of PTS. The role of people, that are at the same time percipients and active elements for the definition of the place is essential for the vitality and liveability of the PTS.

7 CONCLUSIONS

If on the one side the analysis of the debate and the realised projects confirmed the marginal attention devoted to this issue, on the other hand some elements and approaches attesting the interest over the role of PTS in shaping the urban environment can be detected. Yet, where considered, the percective aspects are often dealed with in a separate way and as ancillary elements, in deep contrast with their intrinsic meaning.

The results are still scarce in terms of observational studies, data gathering, empirical evidence on outcomes, making difficult a quantitative evaluation of this issue.

Nevertheless, some rare but valuable studies and realised projects stemming from the acknowledgment of the role of PTS as shaping features of the urban environment can be detected.

The PTS can and should be considered as a shaping feature of the urban environment, concurring to the definition of a *place*.

Albeit the interest over this issue is still fragmentary and chiefly oriented towards the aesthetic aspects, the realised projects and design criteria presented in this paper are proposed as elements for the development of a broader debate.

REFERENCES

APTA (2012). Design of On-street Transit Stops and Access from Surrounding Areas - APTA Recommended Practice. American Public Transportation Association.

Beirão G., Cabral S. (2007). "Understanding attitudes towards public transport and private car: A qualitative study", Transport Policy, vol. 14, pp. 478–489.

Buys L., Miller E. (2011). "Conceptualising convenience: Transportation practices and perceptions of inner-urban high density residents in Brisbane, Australia", Transport Policy, vol. 18, n. 1, pp. 289–297.

Caroselli M. (2011). Architettura delle fermate del trasporto collettivo, Maggioli.

Carse, A. (2011). Assessment of transport quality of life as an alternative transport appraisal technique. Journal of Transport Geography, 19(5), 1037–1045.

Cascetta, E., & Cartenì, A. (2014). The hedonic value of railways terminals. A quantitative analysis of the impact of stations quality on travellers behaviour. Transportation Research Part A: Policy and Practice, 41–52.

Ceder A. (2007). Public Transport Planning and Operation. Theory, modeling and practice, Elsevier.

Clark, B., Chatterjee, K., & Meli, S. (2016). Changes to commute mode: The role of life events, spatial context and environmental attitude. Transportation Research Part A, 89, 89–105.

Frank, L. D., & Pivo, G. (1994). Impacts of mixed use and density on utilization of three modes of travel: single-occupant vehicle, transit, and walking. Transportation research record, 1466, 44–52.

Imbrighi G.,1999. L'architettura delle pensiline. Mobilità e sperimentazione a Roma, Edizioni Kappa.

Kido, E. (2005). Aesthetic aspects of railway stations in Japan and Europe, as a part of "context sensitive design for railways". Journal of the Eastern Asia Society for Transportation Studies, 6, 4381–4396.

OTREC (2013). From Transit Stop to Urbanity Node: Field Audit for Measuring Livability at the Transit Stop, OTREC, Portland.

SEPTA (2012), Bus Stop Design Guidelines, Delaware Valley Regional Planning Commission, Philadelphia.

Taylor B., Iseki H., Miller M., Smart M. (2009). Thinking Outside the Bus: Understanding User Perceptions of Waiting and Transferring in Order to Increase Transit Use, Institute of Transportation Studies, Berkeley.

Thorne, M. (2001). Modern Trains and Splendid Stations,Merrel.

Wardman, M. (2004). Public transport values of time. Transport Policy, 11, 363–377.

Zhang J.K. (2012). Bus Stop Urban Design. Techniques for Enhancing Bus Stops and Neighborhoods.

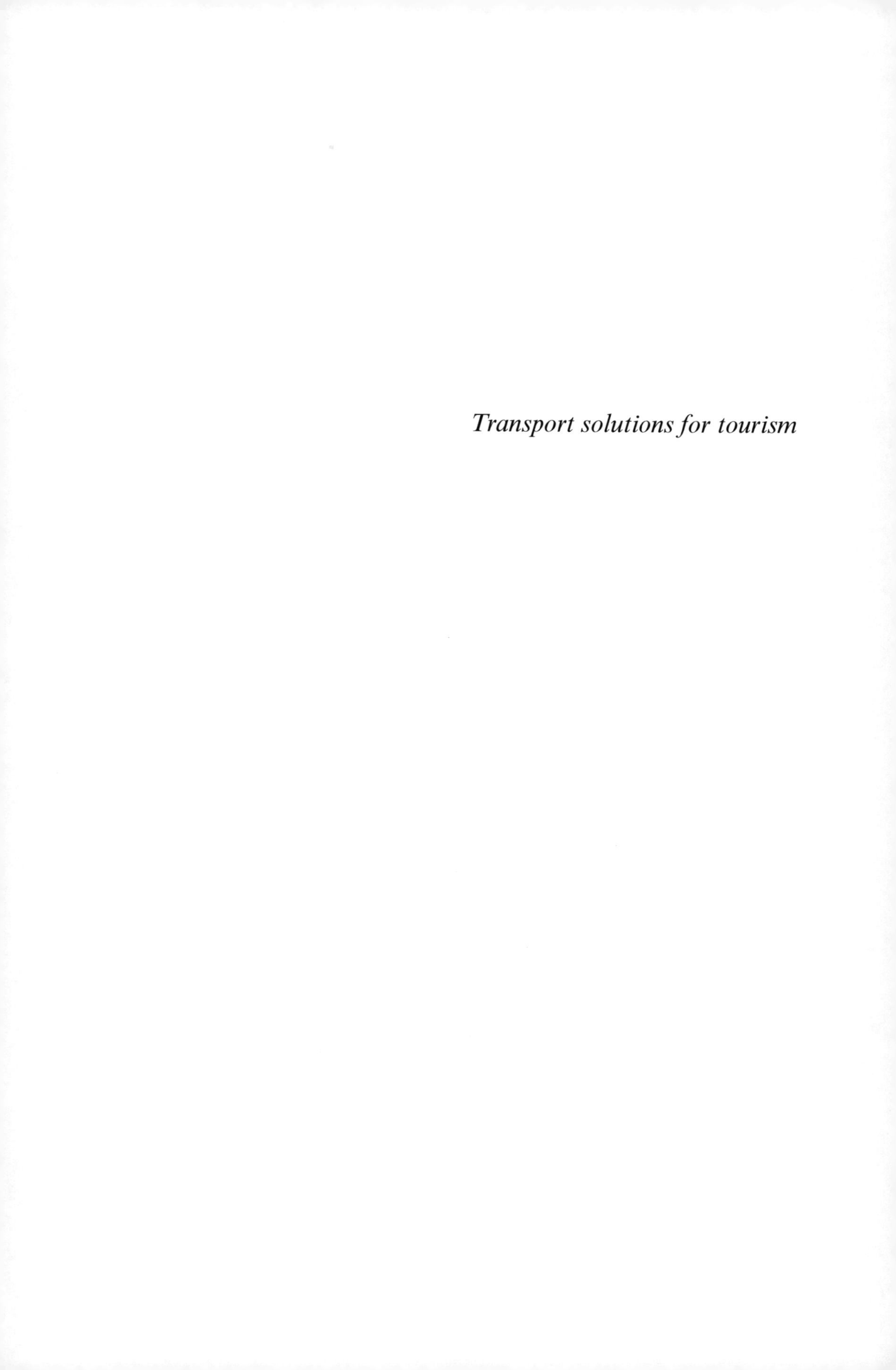

Transport solutions for tourism

Town and Infrastructure Planning for Safety and Urban Quality – Pezzagno & Tira (Eds)
 ISBN 978-0-8153-8731-2

The role of greenways in the sustainable mobility development: The study case of western municipalities of Sicily

F. Cutaia
Università degli Studi di Palermo, Palermo, Italy

ABSTRACT: The case-study presented in this paper explores the opportunities originated by the cooperation of five municipalities which, at this time, are jointly working towards the implementation of policies and measures for the Urban Agenda of western Sicilian cities, during the 2014–2020 programming period. Having examined the territorial potentials, the infrastructural system and the local and inter-municipal demand of mobility (systematic and occasional), the paper presents the model of a green infrastructure that will allow alternate paths to be made by bicycle with transfers by train, frequently matching bicycle path and railway track. This plan, aiming to increase the efficiency of the current railway service, offers two opportunities: its conversion to "tram-train", giving to the city a new public service; the making use, where possible, of areas belonging to the railway line, for cyclists, people in wheelchairs and pedestrians, being possible to take advantage of the linear tendency and the moderate slopes of the railway infrastructure.

1 INTRODUCTION

Since the end of the 80's, the term greenway has entered into the dictionary of regional and landscape planners (Fabos, 1995). Close to the evolution of the concept of "open space", we need to observe the changing of vehicular traffic composition (Uzzell et al., 2000), which determined some changes to the prerequisites of the greenway's plan, and the need to enjoy open spaces, everyday more distant from our residential places.

2 OBJECTIVE OF THE PAPER

The case study presented proposes to consider greenways from a double point of view: on the one hand as an alternative to motorized traffic, on the other hand as an occasion for access to the historical-cultural heritage diffused throughout the countryside. These two profiles are associated by the concept of multifunctionality.

The rural space constitutes an important field for three reasons: its ecological value (Turner, 2001); the economic activities that take place in it (Mastronardi, 2000); the cultural heritage contained within it (Valentini, 2005). Considering these premises, greenways allow us to consider a territory in its entirety and to gather its several qualities, tracing them back to a unique system.

The role of the greenway we have proposed—about 90 km—goes beyond the "mending" of relationships between landscape-cultural heritage and people, and that of reactivating ecosystem connections. In fact, on the administrative hand, "mending" implicates identifying and choosing communal strategies for territorial promotion. In so doing, the greenway becomes the "thread" able to gather together the policies to be undertaken: territorial renewal and promotion, sustainable mobility, environmental protection, etc.

3 RESEARCH APPROACH

The scientific-disciplinary frame of this paper is that delineated by urban and landscape planning, by the sustainable mobility policies, by the transport disciplines and by urban and peri-urban renewal.

The experimental area of this research is that of the district system which pivots on four principal urban areas of medium size: the conurbation of Trapani-Erice, the area of diffused urbanization of Marsala, the urban poles of Mazara del Vallo and Castelvetrano, which are jointly working for the implementation of policies and measures for the Urban Agenda of Sicilian western cities, during the 2014–2020 programming period (Municipality of Marsala et al., 2017). We want to propose a method for integrating the rail service with a cycling infrastructure, in order to promote sustainable mobility and to collect—in the same system—all the assets of the same sector.

4 DESCRIPTION

4.1 *Innovative aspects*

Nowadays, many municipalities are involved with the definition and promotion of policies, procedures and interventions aimed at improving "slow traffic". The Urban Agenda of Western Sicily municipalities is also focused on this topic and addresses its attention to heterogeneous spatial configurations, but associated by solid relationships of synergy and complementarity because of their different economic and functional profiles. Furthermore, at the boundaries and inside that district, there are several cultural and environmental resources—some of international interest—that define communal potential, also in terms of tourist attractions (Fig. 1). Alongside this heritage, we also have to mention the fine food and wine production: wine, salt, oil, cheese etc. For this reason, in our opinion, the greenway planning could be enriched by new contents, like these typical of "taste" itineraries. In fact, together with production places, this kind of itinerary allows the enjoyment of artistic-monumental heritage, archaeological areas, highly natural areas and, alongside them, of the reception and hospitality services. In order to discipline the itineraries and to fix the minimum quality standards, two important legal devices have been introduced in Italy: the L. 268/1999 and the D.M. of the 27th July 2000[1]. This represents an important advantage for greenway planning, because they can find a solid support for their definitive institutionalization, regulation and diffusion in the territory.

The new scenario needs an infrastructural net different from the traditional one, because it should be based on multifunctional criterion, but with one restriction-bind-bond: the network's part of higher environmental quality has to be restricted by motorized mobility (Socco et al., 2007). In urban environments, the creation of a greenway represents an opportunity for the improvement of the open space systems; in non-urban environments they can facilitate the access to areas of high historic-landscape value. Moreover, combining the movement by bicycle with that by personal car, by bus or train, it is possible to set up multimodal connections able to cover movement of broad reach, also at an inter-communal scale (Cantarella, 1997).

In many European countries, transporting bicycles on suitable rail coaches is possible on most trains (La Rocca, 2008). The benefits from the integration of the two means of transport, essentially, are:

- the creation of an ideal interchange for daily displacement;
- the increasing of cycle-tourism for long, medium and short distances.

With the first point we refer to the opportunity of alternating tracks to do by cycling and by placing a bike on a train. This is pretty realizable in those cities crossed by the railway—as in our case—since trains could be used like a tram. Regarding the second point, we think that

1. Respectively, the Law on "Wine Roads" and the Ministerial Decree for the minimum quality standards for Wine Roads.

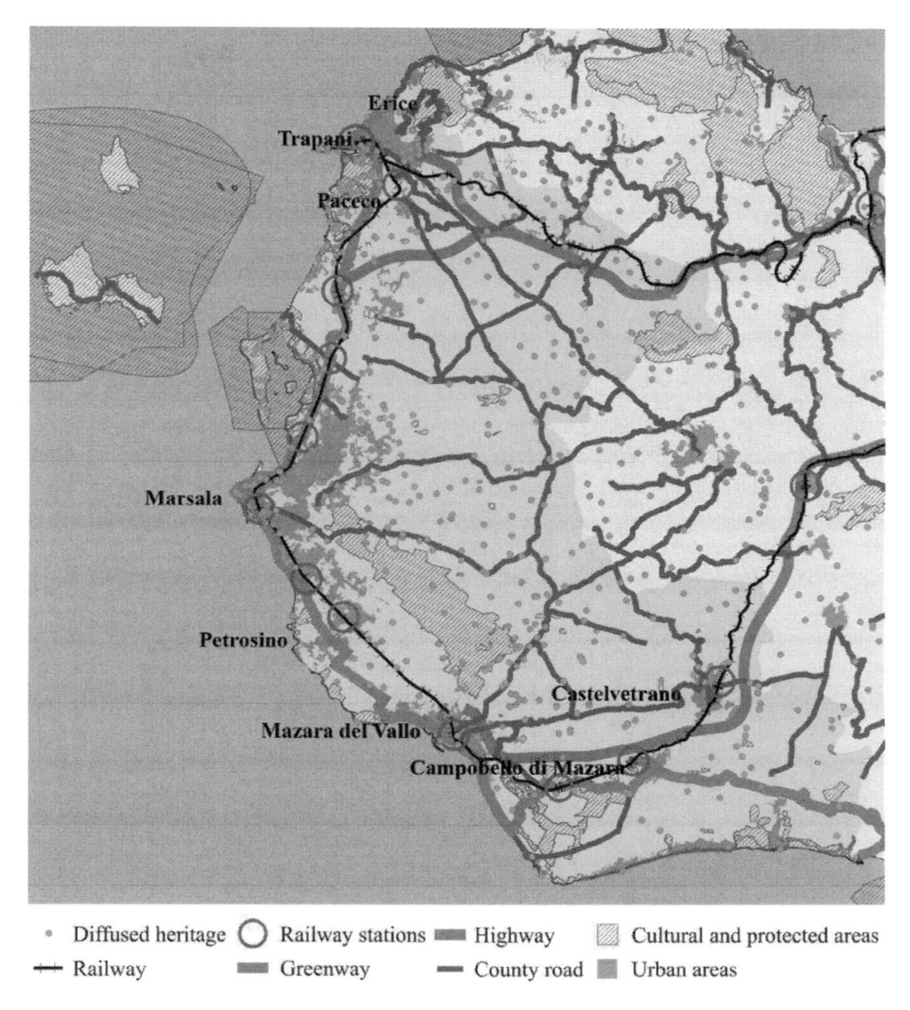

Figure 1. Territorial framework and definition of the "greenway's layout".

the intermodality of train-bicycle is indispensable in order to easily reach the heritage spread throughout the countryside and to satisfy the important demand today known as "cycle tourism" and "sustainable tourism" (Amoroso, Migliore, 2002). In general terms, the intermodality between bicycle and public transport has a multiplicative effect, both on touristic potentialities, and on daily transport (Pileri, 2014).

4.2 *Mobility in the study area: Supply and demand*

The main axe of travel communication in the urban system of the Western Sicilian area leans on two different infrastructure networks:

- the highway A29, which has its terminals in the urban centres of Trapani and Mazara del Vallo, and in the airport of Trapani-Birgi;
- the main road 115, which, going along the coast, connects the urban centres of the five municipalities of the Urban Agenda and the airport and the main road 188 (at the neighbourhood of Marsala) too.

Moreover, this urban system enjoys a railway service extended for 75 km, going through the main urban areas; students and commuters mostly use this service (Municipality of Marsala

Table 1. OD Matrix for study trips.

Municipality of Origin \ Municipality of Destination	Castelvetrano	Erice	Marsala	Mazara del Vallo	Trapani	Grand Total-Origin
Castelvetrano	4.889,55	3,03	1,00	56,23	31,15	5.031,42
Erice	0,00	2.943,39	2,00	0,00	1.623,33	4.576,06
Marsala	2,30	25,50	11.949,13	45,99	205,45	12.327,25
Mazara del Vallo	342,76	0,00	89,44	8.477,48	43,84	9.011,11
Trapani	0,00	1.212,20	21,73	2,59	9.940,33	11.237,91
Grand Total—Destination	5.234,61	4.184,12	12.063,30	8.582,29	11.844,10	41.908,42

Table 2. OD Matrix for work trips.

Municipality of Origin \ Municipality of Destination	Castelvetrano	Erice	Marsala	Mazara del Vallo	Trapani	Grand Total-Origin
Castelvetrano	6.163,80	2,00	82,49	181,34	206,16	6.788,94
Erice	14,26	2.689,98	121,27	12,47	3.539,44	6.503,16
Marsala	105,24	73,97	14.554,45	438,91	862,31	16.519,60
Mazara del Vallo	317,39	5,78	323,56	8.047,19	266,90	9.172,19
Trapani	27,29	1.723,44	317,72	53,04	12.935,74	15.384,74
Grand Total—Destination	6.628,48	4.495,17	15.399,49	8.732,95	17.810,55	53.066,64

et al., 2017). These cities are also served by local public transport systems different for dimensions and capillarity. The already mentioned airport "V. Florio" is one of the most important doors of this territorial system, which has received about 1,5 million passengers in 2015, 75% from national departures. Regarding the strategic axis of the PO Fesr Sicilia 2014–2020[2] dedicated to the definition of the Regional Urban Agenda, interventions on mobility must respond to the need of reinforcing transport systems in order to guarantee sustainability in the different ways of movement, also by the creation of a bicycle path network.

In urban areas, making use of bicycles can become an alternative mode for conventional transport; in the countryside, it lends accessibility to environmental resources from the cycle point of view (Bergamaschini, 2014).

In order to estimate the composition of systematic mobility in the territory, we analysed the Origin-Destination Matrix (OD) of daily home to work trips and home to study trips, based on the 15th Italian population census (2011). The cross-section is constituted by 94.975 individuals, about 14% of the district's population, of which 41.908 move for study reasons and 53.066 for work. Regarding the inter-municipality study trips (Table 1), the main flows are registered from Erice to Trapani and vice versa.

Observing the other OD pairs, Trapani looks like the bigger pole of attraction, followed by Marsala. Regarding movements inside the nodes, Marsala shows visible differences, compared to nearby cities, arriving at a figure of 16,8% more respect to the inside flows of Trapani, and 29% more with respect to the figures for Mazara del Vallo.

2. PO Fesr Sicilia 2014–2020 establishes the Regional Strategy to help pursue the European Cohesion Policy objectives and defines, for each Thematic Objective, the type of operations that can be financed.

Table 3. Tourist flow.

	Arrivals			Nights spent			Average stay		
City	2014	2015	Variation	2014	2015	Variation	2014	2015	Variation
Castelvetrano	88.660	93.461	0,8%	387.047	353.662	−8,6%	4,37	3,78	−0,58%
Erice	31.395	30.304	−3,5%	90.022	84.526	−6,1%	2,87	2,79	−0,08%
Marsala	63.088	61.770	−2,1%	145.850	170.301	16,8%	2,31	2,76	0,45%
Mazara del Vallo	34.543	25.133	−27,2%	79.793	65.077	−18,4%	2,31	2,59	0,28%
Trapani	77.623	85.705	10,4%	222.228	208.294	−6,3%	2,86	2,43	−0,43%

The work flow dynamics are very similar (Table 2). In fact, most of them are recorded from Erice to Trapani and vice versa, but the phenomena proportions change because of the bigger displacement demand. In fact the trips from Erice to Trapani are 54,2% more than these for study reasons, and from Trapani to Erice 29,6% more. With regard to the data of city internal displacements, the relationship is confirmed once again. In fact, Marsala occupies the highest level and is followed by Trapani and Mazara del Vallo.

This area is one of the most affirmed touristic districts in the Region, in which we can notice a touristic index (10,4) second only to that of Messina (16,8) – the unique area which records values close to the national average. Trapani represents the best affirmation of the last ten years, considering a touristic demand increase of about 660.000 units (+67,1%), against the general demand contraction in other Mediterranean areas. The effect of the "America's cup", even before the Ryanair establishment at the Birgi airport, introduced an important fly-wheel for the growth of the local tourism industry, in a period in which the international competition has exponentially increased and Italian wealth has decreased (Municipality of Marsala et al., 2017).

5 A GREENWAY FOR THE SICILIAN WESTERN CITIES

The feasibility study of a greenway to service the western Sicilian district identifies a layout differentiated in a main road and a secondary one. The blueprint has been planned according to three criterion: to connect in only one system the area's resources, to offer a "sustainable" alternative for systematic movements, and to promote the bicycle-train intermodality, all reasons for which the greenway frequently intercepts stations. The latter aspect has a double value: on the one hand it allows the taking advantage of the linear trend and the low slopes of the railway infrastructure, on the other hand it reinforces its efficiency, up to the point of its final conversion into a "tram-train". In fact, this could allow for the running of an urban tract with many stops (tram), uninterruptedly, followed by an interurban one with less stops and at higher speed (train).

We analysed the potential path taking into account:

- general context—urban, peripheral and rural;
- largeness of the road section;
- functional characteristics;
- average travel speed time;
- vehicles admitted on the carriageway;
- geometric characteristics and technical parameters;
- frequency of intersections;
- regulation of the pit stop;
- regulation of the pedestrian flow.

Based on these criterion, we have identified eight "model sections" and equally "model solutions" of intervention. Specifically, two typologies include the development of the greenway layout next to the railway. Usually, greenways are planned on already existing and neglected railway. In this case, the railway is still in use and, in spite of its inefficiency, represents for the local community an important alternative to private means of mobility (Fig. 2).

The rectilinear trend and the minimal inclines suggest, where it is possible, to use the adjacent spaces for the realization of a greenway with exclusive lanes for cyclists, pedestrians and, in some cases, for equestrians. The first typology (Fig. 3) is called "free areas adjacent to railways" and describes a road section in which the global largeness varies from 10 and 20 meters. This is the case of secondary non-urban roads, flanked by the train tracks, that go

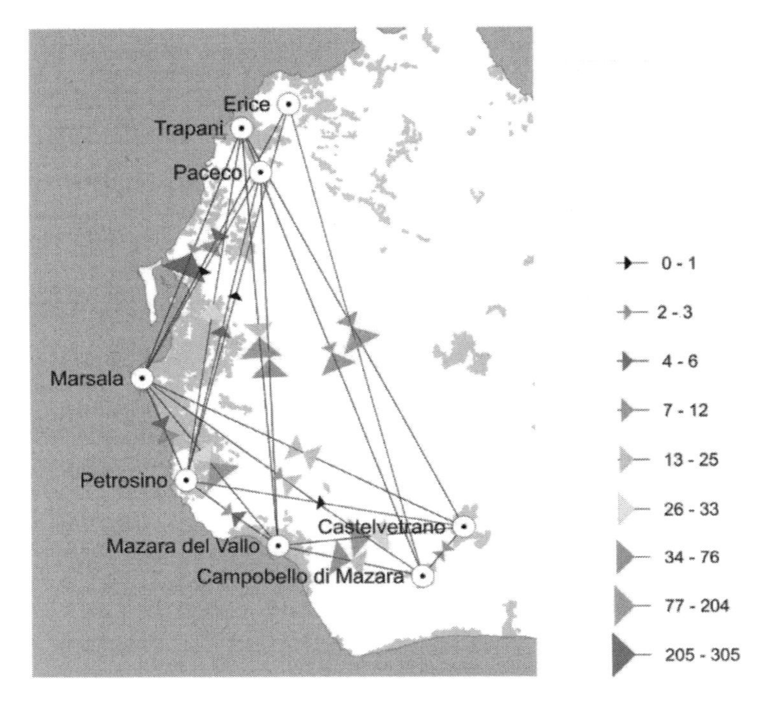

Figure 2. OD graph for displacement by train.

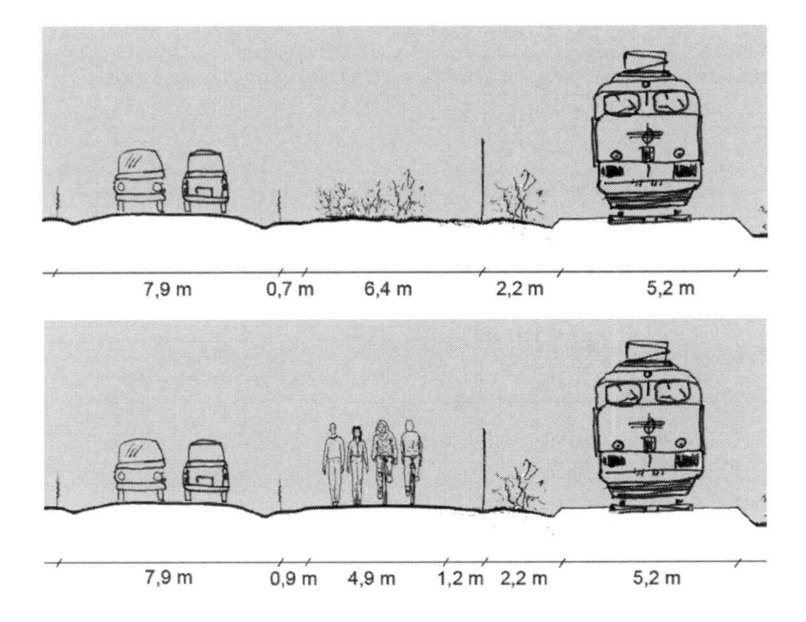

Figure 3. Free areas adjacent to railways.

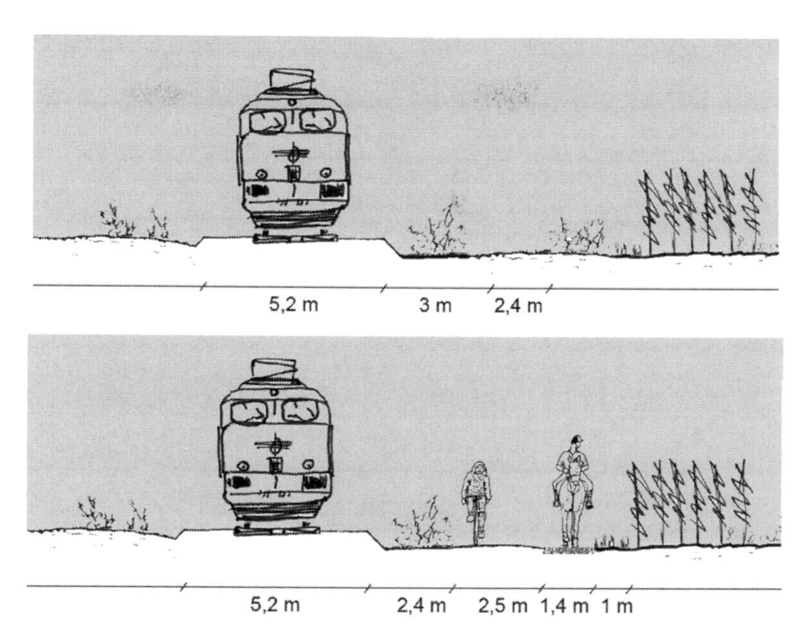

Figure 4. Rural roads along the railways.

through areas with different levels of urbanization. Moreover, the presence of the railway, reducing the frequency of grade-level crossing, facilitates the installation of platforms for non-motorized mobility between the pavement and the same railway.

The second typology (Fig. 4) differs from the first because its platform for non-motorized mobility is situated on rural roads, next to the railway, hence the appellation "rural road along the railways". For its realization, it could be necessary to acquire a tight strip of land beside the road, in order to allow for better comfort. The new platform will be located 3 meters from the railway tracks, ensuring security for users.

In the perspective presented, the Sicilian western greenway represents an opportunity for the continuity of the ecological system and, in general, for a network of cultural resources, due to the organic net—corridors and parks—developed not only on the geomorphologic support, but also on anthropogenic linear infrastructures (Angrilli, 1999).

6 CONCLUSIONS

The real difference with other greenway experiences is in the role of the railway. Usually, we note its conversion into a light infrastructure; in this case, we suggested the use of the railway for a double objective: on the one hand, in order to take advantage of an existing lane suitable for the realization of platforms for non-motorized mobility; and on the other hand to increase the combined usage of bicycle and train for different travel reasons.

The analysis about movement flow showed that bicycle use is already well diffused in the area, in daily home-work and home-study trips too, beyond the residential municipalities. In spite of the minimum percentage thresholds, these data suggest the improvement of the infrastructural supply. In fact, together with the promotion of intermodality, it could facilitate greenway usage by several types of user. This is especially true in the case of the Marsala-Trapani itinerary, both for the advantageous morphologic profile, and for the presence of a railway which already includes eight stations and twelve daily trips in both directions.

Regarding the everyday growth of touristic demand, the current mobility infrastructures, together with the realization of a greenway on the western Sicilian waterfront, could finally give an important impulse to the development of the area.

REFERENCES

Ahern J., (1995). Greenways as a Planning Strategy. Landscape and urban Planning, n. 33. 131–155.
Amoroso, S., Migliore, M. (2002). Bike & railway: an integrated approach for an urban sustainable mobility. Proc. IX International Conference Living and Walking in Cities, Brescia e Piacenza, 13–14 giugno 2002.
Angrilli M., (1999). Greenways. Urbanistica, n. 113. 92–97.
Bergamaschini I., (2014). Grandi ciclovie tra turismo e mobilità sostenibile. Ecoscienza, n. 3. SDA Bocconi.
Cantarella G.E., (1997). A general fixed-point approach to multimodal multi-user equilibrium assignment with elastic demand. Transportation Science, n. 31. 107–128.
Fabos J., (1995). Introduction and overview: the greenway movement, uses and potentials of greenways. Landscape and urban planning, vol. 33. 1–13.
La Rocca R.A., (2008). Mobilità dolce e trasformazioni del territorio: esempi europei. TeMA, n. 3. 57–64.
Mastronardi N., (2000). Il recupero della viabilità rurale storica. Quaderni, vol. 1. Accademia dei Georgofili.
Mennella V.G.G., (2004) (ed. by). Greenway per lo sviluppo sostenibile del territorio, Il Verde Editoriale.
Municipalities of Marsala, Castelvetrano, Erice, Mazara Del Vallo, Trapani, (2017). Agenda Urbana Delle Città Della Sicilia Occidentale: Preliminare di Strategia.
Pileri P., (2013). Una strada leggera come Vento. Tekneco, n. 18. 64–79.
Socco C., Cavaliere A., Guarini S., (2007). L'infrastruttura verde urbana. Osservatorio Città Sostenibili, Dipartimento Interateneo Territorio—Politecnico e Università di Torino.
Turner T., (2001). Greenways: theory and history. Lecture at Agricultural and Food Sciences Faculty of Milan.
Uzzell D., Groger J., Leach R., Wright A., Ravenscroft N., Parker G., (2000). User Interaction on non-motorized shared use routes. Final Report to the Contryside Agency, University of Surrey.
Valentini A., (2005). Mettere in rete le risorse: le greenway quali strumenti per il progetto del paesaggio periurbano. Quaderni della Ri-Vista Ricerche per la progettazione del paesaggio. Firenze University Press. 15–26.

Town and Infrastructure Planning for Safety and Urban Quality – Pezzagno & Tira (Eds)
© 2018 Taylor & Francis Group, London, ISBN 978-0-8153-8731-2

Mobility as a service for the last mile—a Bulgarian case

L. Ilieva & P. Rizova
Club "Sustainable Development of Civil Society", Bulgaria

ABSTRACT: The project LAST MILE aims to find sustainable flexible solutions for regional mobility systems. It makes sure that visitors travel the "last mile" in their travel chain sustainably and provide alternatives to car use for residents and their daily trips as well. LAST MILE supports regions in creating tailor made solutions for multiple users, interacing with main public transport lines and hubs and introducing low emission and energy efficient vehicles, that shall lead to sustainable and financeable regional mobility options. The project is funded by INTERREG EUROPE. The LP is the Austrian Environmental Agency. There are 7 project partners from 6 countries (Austria, Bulgaria, Luxemburg, Poland, Slovakia and Spain). The Bulgarian partner is CSDCS (www.csdcs.org).

1 INTRODUCTION

Most people agree nowadays that we need to change the way we travel. One of the main causes of climate change is the emissions of carbon dioxide from the transport sector. Many municipalities are investing in public transport and better cycling and walking infrastructure, but at the same time the sale of new cars continues to increase. The intense car traffic has many other negative impacts on cities and environment—congestion, noise and air pollution and the transport facilities—roads and parkings occupy a substantial amount of public space, that could otherwise be used for housing, green spots, parks etc.

Mobility as a Service (MaaS) is increasingly promoted as an opportunity to reduce demand for travel by private car, and hence as a way to reduce the car traffic. There are many definitions of MaaS because various actors define the concept in different ways, depending on their starting point of view.

The definition of Kamargianni is:[1]

> *"The term 'Mobility as a Service' stands for buying mobility services as packages based on consumers' needs instead of buying the means of transport. Via 'Mobility as a Service' systems consumers can buy mobility services that are provided by the same or different operators by using just one platform and a single payment."*

Following Heikkilä[2]:

> *"Mobility as a Service (MaaS) - a system, in which a comprehensive range of mobility services are provided to customers by mobility operators."*

The definition of the Maas Alliance says:[3]

> *"Mobility as a Service (MaaS) puts users, both travellers and goods, at the core of transport services, offering them tailor-made mobility solutions based on their individual needs. This means that, for the first time, easy access to the most appropriate*

1. Kamargianni et al. (2016). 'A critical review of new mobility services for urban transport'. Transportation Research Procedia 14 (2016).
2. Heikkilä (2014). Mobility as a Service—A Proposal for Action for the Public Administration. Case Helsinki.
3. MaaS-Alliance: http://maas-alliance.eu/.

transport mode or service will be included in a bundle of flexible travel service options for end users."

To create a complete MaaS solution, many different competences are needed. In addition to all mobility services that are to be integrated, an IT platform is necessary to handle information, booking, payment etc., and a service provider who sells the integrated service to the end customer.

2 OBJECTIVE OF THE PAPER

The main objective of this paper is to show the importance and the results of the LAST MILE Project for Bulgaria as a first initiative introducing MAAS in the country.

LAST MILE aims to find sustainable flexible solutions for regional mobility systems. It makes sure that visitors travel the 'last mile' in their travel chain sustainably and provide alternatives to car use for residents and their daily trips as well. LAST MILE supports regions in creating tailor made solutions for multiple users, interacing with main public transport lines and hubs and introducing low emission and energy efficient vehicles, that shall lead to sustainable and financeable regional mobility options. The project is funded by INTERREG EUROPE. The LP is the Austrian Environmental Agency. There are 7 project partners from 6 countries (Austria, Bulgaria, Luxemburg, Poland, Slovakia and Spain). The Bulgarian partner is CSDCS (www.csdcs.org).

For Bulgaria, the target region is the famous Black sea resort area in Varna district where policies and concrete measures are elaborated in line with the regional structural funds programme in regard to potential funding of new projects.

3 METHODOLOGICAL APPROACH

The project focuses on user oriented services for the travel chain's last segment in remote destinations offering and promoting door-to-door accessibility. Still, in terms of the full distance to cover between origin and destination, there is often a bottleneck on the last link of the journey, i.e. the distance between the regional railway station and accommodations. This missing link is crucial for deciding what kind of transport to use. Experiences have shown that a demand-responsive transport system combined with regular public transport is a thankful enhancement in many cases. The last-mile problem can be solved by introducing a variety of flexible transport services (FTS) making the transportation multimodal, on-demand, seasonal, shared, and increasing passengers' choice and convenience. The overall transportation system step by step becomes more digital and therefore more efficient by better matching demand and supply.

4 DESCRIPTION

The project is developing on 2 phases. The Phase 1 (2016–2019) starts with a thorough research including 3 different joint analyses, which build the backbone of the interreg exchange:

- Analysis of the national legal and institutional frameworks and economic aspects related to sustainable demand-responsive/flexible transport systems and the identification of the barriers that are hindering the implementation of especially small scale systems in remote areas/hinterland
- Analysis of the technical state-of-the-art of sustainable transport, in particular of flexible systems in the different regions
- Analysis and evaluation of existing practices in regional flexible transport policies. This joint research evaluates good practices of and beyond the regions, taking into account former best practice collections of other projects as well. During the study visits the

regional approach including the specific framework conditions, financing structures and the concrete mobility systems is discussed and evaluated using two specific questionnaires prepared by CSDCS. In this regard, potentials for optimization or innovation are identified, and if applicable concrete solution approaches are elaborated.

The analyses prepared by each partner region are consolidated and summarized thus allowing the preparation of a **Synthesis and policy recommendations.** Based on the results, it derives recommendations for policy makers at different levels. They will be used for elaborating **Regional action plans for implementing flexible transport services** that prepare actions and investments to improve the door-to-door accessibility of peripheral tourist/recreational destinations benefitting also the inhabitants. These plans will make sure that lessons learned from the research and interregional exchange are integrated in the regional policies. Actions defined will be put in practice during the Phase 2 of the project (2019–2021).

5 RESULT OF THE RESEARCH

For Bulgaria, the project is a real challenge because of the lack of the MAAS-concept in the regional transport schemes. The target region is the Black sea resort region of Varna. CSDCS already analyzed the national legal and institutional frameworks and the economic aspects related to sustainable mobility services and the identification of the barriers hindering their implementation in remote areas. The goal is to elaborate Regional Action Plan for implementing flexible transport measures that prepares actions and investments to improve the door-to-door accessibility of Varna peripheral recreational destinations benefitting also the inhabitants.

The main novelty is in the large collaboration for providing last mile FTS. Establish contact with and sell the last mile idea to players in the fields of tourism, transport, and environment (tourist entrepreneurs, transport authorities, protected area management, local politicians etc.) who might have a vested interest in the policy or project, and whose involvement could have a positive financial or political impact is crucial. For the first time the tourism and transport sectors in Bulgaria meet and discuss their common problems and interests having the possibility to exchange experience with more advanced European regions. The project made it possible to bring together all the important subjects involved in people's mobility to try and find points in common and to encourage partnerships both locally and across borders.

6 CONCLUSIONS

The transferable element of the project is the Learning from Best Practice. Its collection is focused on regional solutions of FTS thus allowing the pilot regions to become familiar with case studies of similar and/or dissimilar conditions and learn about core success factors for regional tourism and mobility concepts. Though 100% transferability is rare, best practice examples provide future project ideas and knowhow about coperations or technologies. During the study visits each regional approach is evaluated. The best practices are collected, thoroughly described and evaluated. In this regard, potentials for optimization or innovation can be identified and some concrete solution approaches elaborated.

The expected impact of the project is that thanks to the MAAS measures introduced the socioeconomic benefits of tourism increase and are more equitably distributed in the target regions. The outcome will be increased tourism receipts benefitting people living in underdeveloped segments of the target regions. Project outcomes would be: (i) improved urban-rural connectivity; (ii) improved regional environmental conditions; (iii) strengthened institutional capacity for tourism destination management; and (iv) effective project implementation and knowledge management. The project is very important especially for Bulgaria representing one more step to the appeal of a seamless, technologically-facilitated transportation ecosystem, universally accessible yet designed for maximum efficiency and site-specificity.

REFERENCES

ELTIS portal: www.eltis.org.

Heikkilä (2014). Mobility as a Service—A Proposal for Action for the Public Administration. Case Helsinki.

Kamargianni et al. (2016). 'A critical review of new mobility services for urban transport'. Transportation Research Procedia 14 (2016).

MaaS-Alliance: http://maas-alliance.eu/.

Sochor, Strömberg och Karlsson (2015). An innovative mobility service to facilitate changes in travel behaviour and mode choice. 22nd ITS World Congress, Bordeaux, France, 5–9 October 2015.

Town and Infrastructure Planning for Safety and Urban Quality – Pezzagno & Tira (Eds)
© 2018 Taylor & Francis Group, London, ISBN 978-0-8153-8731-2

The project of infrastructures for sustainable mobility as driver to join urban public areas, the production landscape and areas of high environmental and artistic value

G. Marinelli, M. Bedini & F. Rossi
Università Politecnica Delle Marche, Ancona, Italy

ABSTRACT: A virtuous process has been started in many European cities to construct sustainable cycling-pedestrian mobility networks, in order to establish a reciprocal contamination of uses and functions between public urban spaces, the production landscape and the many diffuse excellences of the territory (nature reserves, historical-architectural buildings and monuments, diffuse tourist facilities).

The new light infrastructural frames may constitute a territorial reinforcement, around which projects may be created to regenerate urban peripheral areas, agricultural production areas for social use, new urban services and new social-economic drivers through programmes for tourist-cultural improvement, activating contamination and unprecedented territorial relations between the production landscape and territorial excellences.

This paper will define the methodological and project guidelines to plan light cycling-pedestrian infrastructures for the city and territory by describing one project currently underway in Marche Region.

1 INTRODUCTION. GREEN MULTIFUNCTIONAL INFRASTRUCTURES: THE GREEN COMET IN THE NEW URBAN PLAN OF ANCONA

Green urban areas are very important not only in terms of their recreational and aesthetic value (Barton, Pretty, 2010), but because they may also provide important ecosystemic services, improving, for example, the quality of air and water, generating and restoring soil fertility (Dybas, 2001), controlling temperature fluctuations and housing a host of natural habitats. Each of these valuable functions also plays an important role in adapting to climate changes (Rosenzweigh, Solecki, Hammer, 2011).

The new light infrastructural frames may constitute a territorial reinforcement, around which projects may be created to regenerate urban peripheral areas, agricultural production areas for social use, new urban services and new social-economic drivers through programmes for tourist-cultural improvement or wine and food routes and routes of cultural excellences such as the Wine and Art Routes, activating contamination and unprecedented territorial relations between the production landscape and territorial excellences (nature reserves, diffuse historical and architectural buildings and monuments, diffuse tourist facilities).

The "Cometa Verde" ("*Green Comet*") is one of the priority urban redevelopment projects proposed by the Program Document for the new Urban Plan of Ancona (Gasparrini, 2010). The Concept Plan provides creation of environmental infrastructures within a context of urban green areas assuming the role of the structure of a sequence of timely interventions for the upgrading and enhancement of degraded and marginal areas. The main objectives are three, which are the synthesis of the environmental project:

1. Ecological and functional continuity;
2. Transversal mobility systems within the city;
3. Rehabilitation of public and private building heritage and degraded urban areas.

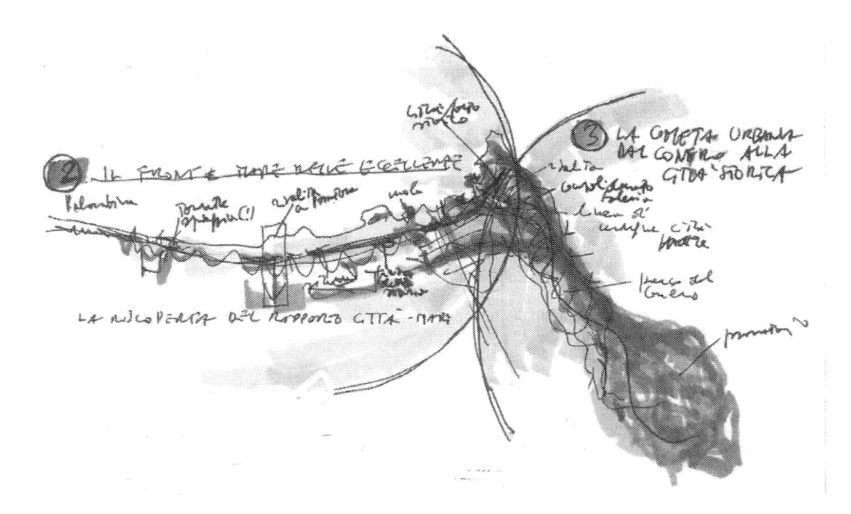

Figure 1. “Cometa Verde”, Ancona Concept Plan (*Source: C. Gasparrini, 2010*).

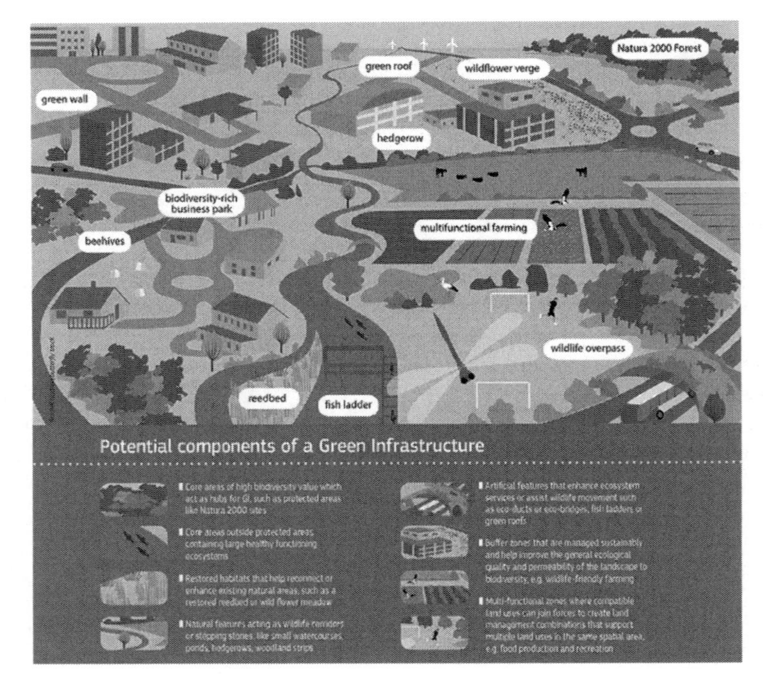

Figure 2. Potential components of a Green Infrastructure (*Source: UE—Life 2020*).

The Urban Planning Division of SIMAU Department started the study and planning work on the basis of the most important urban regeneration projects of the Programmatic Document for the New Urban Plan of the City of Ancona.

The objective of the project is to redefine four extensive linear park systems that may act as public territorial reinforcement in the city for sustainable cycling-pedestrian mobility at the service of the city and the nearby Conero Natural Park and production landscape of excellence.

The sustainable cycling-pedestrian mobility network may act as a frame for the city and territory to define a system of environmentally-friendly cultural corridors in order to:

- stop degradation of the territory and fragmentation of natural environments;
- relaunch, improve and network resources of the historical territory;
- improve cultural, beneficial and environmental relations between the parties;
- protect the biodiversity and eco-compatibility of farming systems;
- improve agricultural landscapes;
- *encourage land defence.*

2 METHODOLOGICAL APPROACH: "FROM FRAGMENTS TO FORM", FROM QUANTITY TO QUALITY

The starting point of the methodology is the definition of the strategies used to join urban public areas by intertwining different cognitive systems: thematic mapping of public property; analysis of the implementation level of the Urban Plan; interpretation of the morphological and territorial aspects. The scope of this paper is to define a systemic project for light infrastructures, planned according to an "adaptive logic", strongly integrated with natural values and the production landscape (Poinsot, 2008). The methodological approach is based on the following three steps:

(A) The first step is the implementation of the vision of the urban public areas (urban standard D.M.1444/68) according to qualitative-performance models, whose raison d'être is that of offering services for the quality of urban life. In many cases, the rationale behind the standards ended up with producing merely "quantitative" effects (an increase of green areas that is often only tabular but not substantial), but from a qualitative point of view, green areas did not produce efficient values and functions (Sanesi, Lafortezza, 2002), while parks, open spaces and other forms of urban green may provide essential services for both carriers (sustainable mobility and accessibility). In this way, green urban areas guarantee a further social function, that of public order and urban "spatial justice" (Allmendinger, Haughton, 2014), since in many cities, districts and communities with different ethnic groups, that very often are in the low-income bracket, the challenges of public health tend to be the most critical node in terms of overcoming inequalities (Bedimo-Rung, Mowen, Cohen, 2005) and the main brake to social innovation in cities (Barnett, 2001).

After completing this preliminary but fundamental step, a number of measures (B) will be necessary together with an "inter-institutional *roadmap*" and a complementary time schedule, in order to define an "organic figure" of continuous and morphologically recognisable public space. This is a synthesis of proprietary systems and coordination between (public and private) bodies for the construction of linear urban space networks integrated with green infrastructures. Besides safeguarding nature in order to preserve its eco-systemic services, the

Table 1. Municipality of Ancona: "urban standard" (public areas) for residential settlements as referred to Italian town-planning regulations D.M. 1444/68. The per-capita (33 sqm/ab) is in terms of areas much higher than the minimum requirements (12 sqm/ab).

		23	24	25	26
Zone terrioriali omogenee	Fipologiadi servizi (D.M. 1444/68)	Superfici convenzionali servizi residenziali (mq.)	Totale abitanti (insediati + insediabili)	Standard abitante (col. 23/col. 24) (mq/ab)	Riferimenti di legge Standard minimi (mq/ab)
	Interesse comune	629.330		5,44	2,00
A + B + C	istruzione	540.055		4,67	4,50
	parcheggi	605.478	115.758	5,23	2,50
	verde	3.823.604		33,03	9 + 3 = 12
Totalegenerale		5.598.467		48,36	18 + 3 = 21

hope is that these may be extended and inserted in urbanised areas where they are more useful in direct contact with the population (Barton *et al.*, 2012). A wide range of possibilities is offered through the adaptive use of obsolete and/or underused urban infrastructures, such as railway corridors, abandoned areas, or areas awaiting reclamation.

In dozens of medium-sized European cities (Eurocities, 2011), many dilapidated architectural structures which have been transformed into green infrastructures with walking and cycling routes, playgrounds, gym areas and areas for social interaction, have become famous, and also provide a strategy for the balanced distribution of urban runoff and the supply of eco-systemic services (Newell et al., 2013; Wolch *et al.*, 2011). The infrastructure that the digital city has accustomed the large modern metropolis to use in a different way has created the root of what is now called the *smart city*, in which many different technologies are at the service of the city in order to improve the quality of life of users and ensure its sustainability.

It is to ensure the sustainability of these steps that Step (C) is designed, and includes the planning of new infrastructures, in order to create a smart grid (Scaglione, 2015), in which the services indicated by step one are offered through a multi-layered complementation, where public priorities and functions are ranked and prioritised.

This creates an "urban equation" capable of encompassing complexity both in terms of design and in terms of services offered.

Ecosystem services	Smart Mobility	Social innovation

The sum of these three systems provides a service delivery to the citizen and at the same time constitutes a set of independent layers so defined:

1	Ecosystem services + Smart Mobility + Social innovation	Services delivery
2	Redefining urban form and Shrinkage	Reconnection of urban spaces
3	Reconnection of the urban, periurban and neighboring rural areas	City-country connection

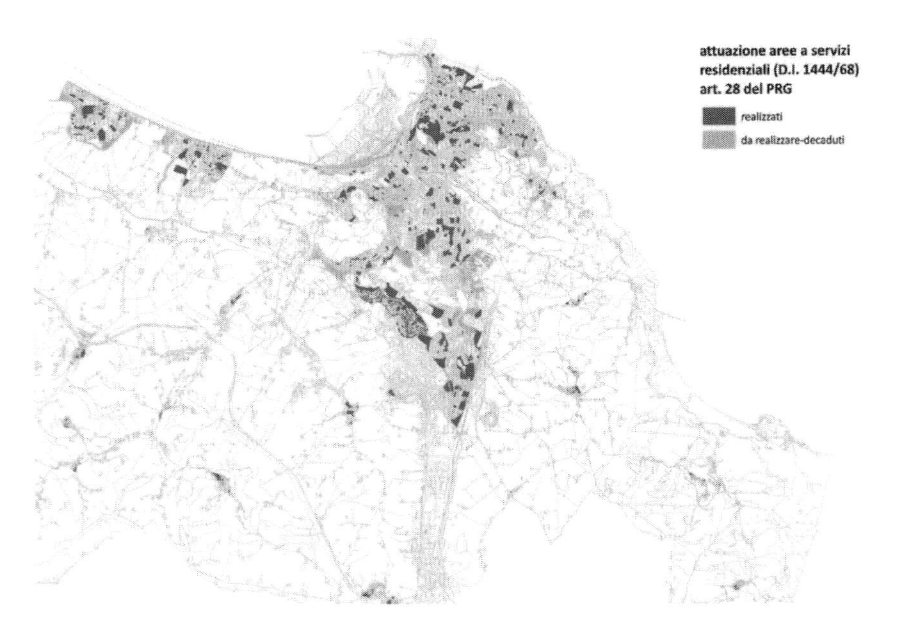

Figure 3. City of Ancona: Public green areas in the city center (Italian town-planning regulations D.M. 1444/68). Picture focus on the high level of fragmentation of green and service public areas (Source: Municipality of Ancona).

The "Cometa Verde" masterplan is divided into four interdependent environmental systems, each of which relate directly to elements of functional excellence linked to culture, tourism and the provision of quality services with multiscalar potential both urban and Metropolitan rank. In this paper, a single sub-project is considered: "Vallemiano Park", characterized by significant environmental potential, urban regeneration and tourist reconnection of the periphery with the natural areas of the Conero Regional Park.

3 PILOT RESEARCH PROJECT: THE GREEN INFRASTRUCTURE OF VALLEMIANO

Vallemiano is made up of a large green lung with great extension and great potential. Elements to be considered to re-evaluate it as a "gateway" from Ancona to the Conero Natural Park, especially by light mobility, are:

- urban accessibility (ease of parking and access to the public area of the headland below the well-equipped axis, the main entry point to the south of the city);
- landscape characterized by a slight uphill of the two parallel valley roads, not particularly abundant morphology);
- The public property fragments: rappresent an opportunity to reconstruct unified path to the Conero Natural Park and transversal traces linking to exsistent urban routes.

The amplitude of the green belt along the route will vary depending on the opportunities that will be created for the realization of sports facilities, in addition to the communal pool exits, but also for urban gardens, easily irrigable. Once the perennial agricultural area has been reached, for km0 of foodstuffs (Carter, *et al.*, 2003), there are intertwined possible touristic attractions: to Portonovo and the Natural Parks on one side, towards the beaches of the Riviera and the city On the other, always using cyclopedonian paths dedicated to integrate progressively with multi-level smart-grid (Scaglione, 2015), bike-sharing and public transport.

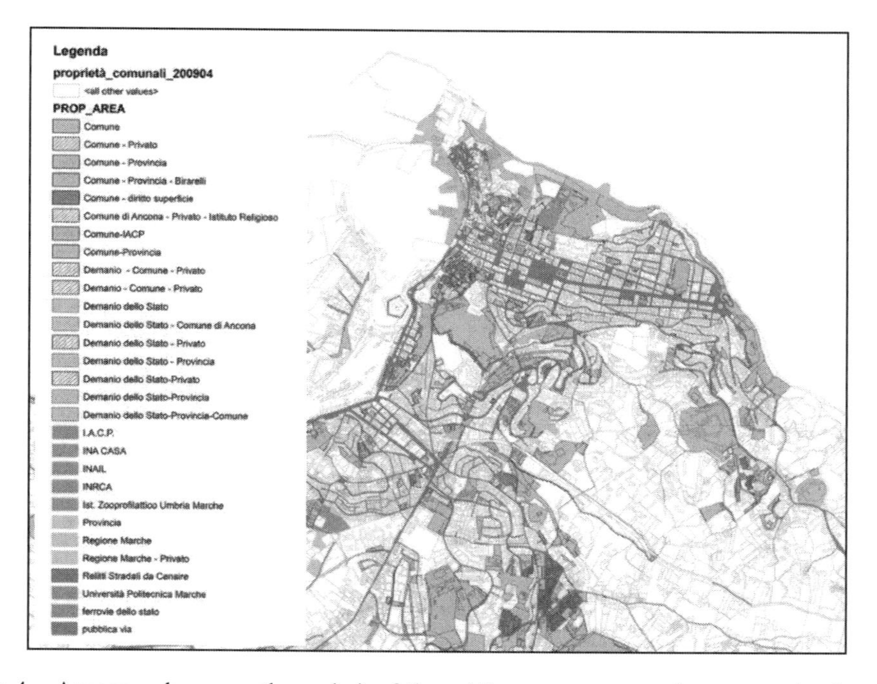

Figure 4. Ancona, urban area: the analysis of the public property system shows a supply of continuous public areas to define new linear urban parks.

Figure 5. Ancona, potential urban parks (green infrastructures) between city center and periurban countryside (graphic design).

Figure 6. Vallemiano Park, Ancona Pilot Area/Case study (Graphic Design).

Vallemiano Park	
Ecosystem services	Hydrogeological control and agricultural production
Smart Mobility	Touristic routes from the consolidated city to the receptive and bathing facilities. – Relationship between city and rural environment extend to food supply width Logistical and transport connections to and from the agricultural land and the city;
Social innovation	Urban gardens; Centers for new citizenship and educational and recreational functions
Redefine urban form	Landscape: metropolitan agriculture alone can not solve the problem of the degradation of the urban fringe landscape, but at at the same time it can be a good assumption for this to happen.

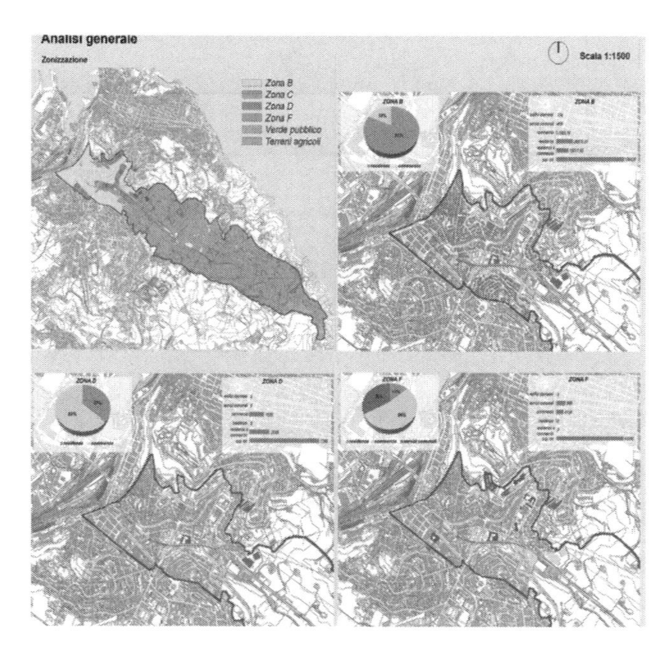

Figure 7. The park of Vallemiano, Ancona. Urban Components, layers: residential area, urban public functions and productive infrastructure (Graphic Design).

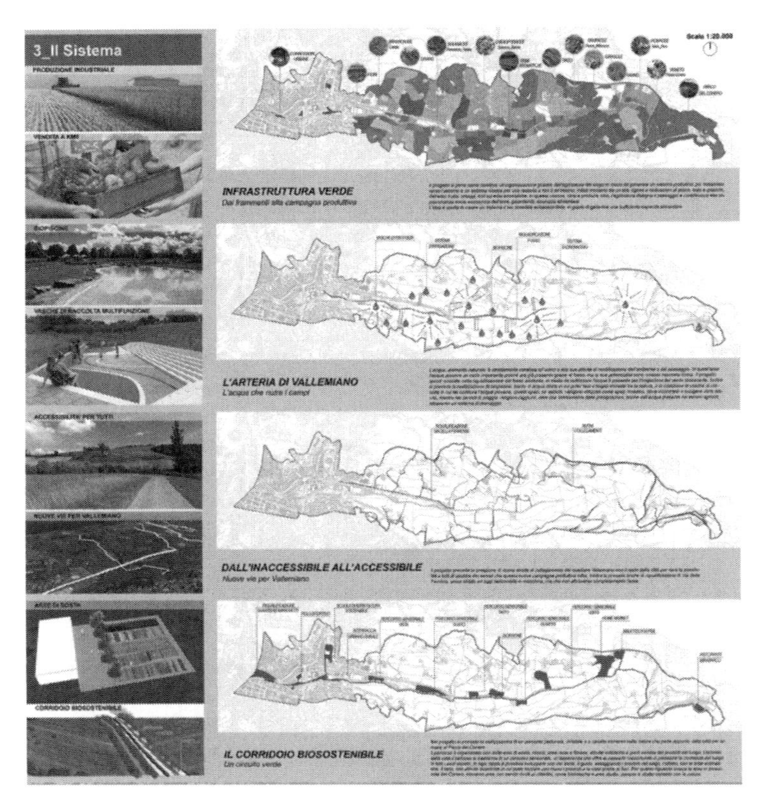

Figure 8. The park of Vallemiano, Ancona. Layers of integrated Masterplan (Graphic design by A. Caporossi, A. Santoni).

Figure 9. The Park of Vallemiano, Ancona: "La Porta del Conero", conections between urban area and rural territory (Graphic design by A. Caporossi, A. Santoni).

4 RESULT OF THE RESEARCH

More than ever today, the environmental question in Europe seems to acquire a broader sense, and the experimental work developed leads to a number of considerations for the next generation of urban plans:

1. Growing contamination between the urban plan and project (Scaglione, 2008);
2. Greater complexity outside the size of the urban plan required to rediscover the sectorial strategic-management dimension (green plan, adaptation plan, and urban landscape);
3. Integration of urban green and biodiverse urban areas, with the slow and cycling mobility system as the "vocation" of the green system in order to construction new multifunctional complex urban sections;
4. A new demand for wildness beyond the urban plan as an expression of spontaneous experiences for the re-appropriation of biodiverse urban areas on a national and European scale (such as the creation of self-managed urban gardens) (Van Leeuwen, 2010).

To do all this, economic significance must be attributed to social and environmental values, defining new types of public spaces, in order to mitigate the distance between centrality and marginality, compact cities and the "urban countryside" (Donadieu, 2005), restructured parts and parts awaiting redevelopment. With this change of approach, both in terms of method and strategy, behaviours and lifestyles and the financial and environmental costs associated with the current urban conditions may be changed, and new economies, job opportunities, and experimental practices of social solidarity are created. Evaluating open spaces in terms of fragmentation and the ability to provide ecosystemic services (such as evapotranspiration) is a crucial to territorial planning, because these indicators provide a decisive approach on the use that is to be made of a particular site: its transformation into a nature park, green belt, farming plot or playground may depend on the relationship between fragmentation of natural matrices and the levels of evapotranspiration (La Greca *et al.*, 2010). Three major project dimensions were developed in the pilot project of the case of Ancona, that may be used as methodological guidelines for the construction of green urban infrastructures:

1. *the strategic-programming dimension:* urban green as reinforcement that steers parts of the project towards policies of an urban vision, adaptation and resilience.

   For example, like that already successfully developed in the general urban plans of Milan, Bologna, Bergamo, Ferrara and Sesto Fiorentino;

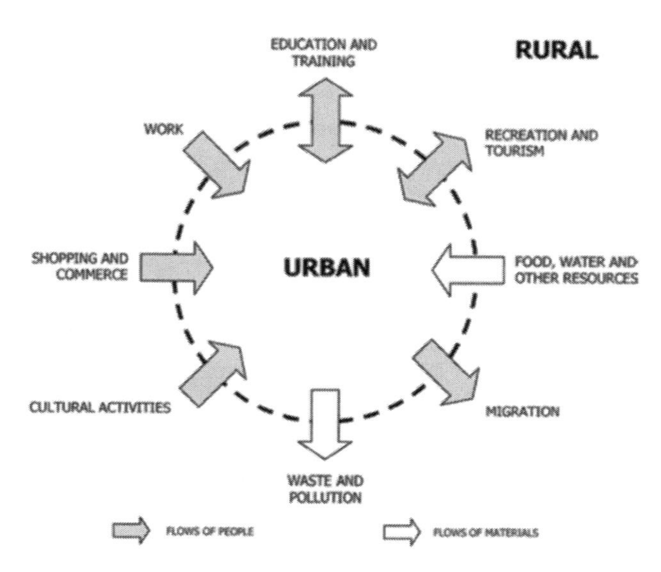

Figure 10. The flows of people and goods between urban and rural areas (Source: V. Nadin and D. Stead, 2000).

2. *the ecological-performance dimension:* public and private green as a strategic reference for urban regeneration and eco-systemic performance.

   For example, as in the general Italian urban plans of Florence, Ivrea, La Spezia, Rome and Todi;
3. *the systemic dimension*: "biodiverse urban areas":
   - as a quality element, that is able to converge the actions of the plan into a transcalar vision (vast area) of continuity/proximity between the city and the periurban territory,
   - strongly linked to the principles of the European Landscape Convention.

5 WORKING SCENARIOS/CONCLUSIONS

New regional planning for tourism and recent investment policies introduced in Marche region, based on an innovative concept of "Advanced Cultural District", may represent interesting opportunities for experimentation in planning.

Growing attention to contamination of production landscape values and the infrastructure project opens interesting perspectives on the role of urban and territorial networks for sustainable mobility and cycle touring in the urban planning project and greater area to improve the quality of public spaces in peri-urban fringe settlements and the production landscapes that characterise the territories in central Italy in a multi-faceted manner. The environmental capacity and intervention opportunities of the new ecological Marche Network, R.e.m. to reach a definition of the design of the different light infrastructures planned according to an "adaptive logic" strongly oriented to integratation with natural values and the productive landscape.

Three macro-themes are examined:

- urban equalization/compensation for the public/private exchange of land;
- ecosystemic services in landscape design with respect to use of the cycling-pedestrian mobility network;
- financial instruments from EU programming for Green and Blue Infrastructures integrated with sustainable mobility systems;
- relations with food production at KM0 for the city and culturally orientated tourist networks.

The urban planning culture does not seem to be ready, in this stage of territorial-institutional reorganisation, to give strong multi-sectoral project responses for these hybrid border urban-rural territories.

Figure 11. Vallemiano "#Direzione parco": temporary set-up for citizen participation in park construction (Source: Casa delle Culture, Ancona, April 2017).

The territories of the "urbanized countryside" contain «phenomena of urbanisation that still cannot be classed under the definition of city» (Gibelli, Salzano, 2006), but urgently require new approaches of integrated planning on the relevant scale, as well as mutations of paradigms and innovative territorial visions to relaunch the production landscape of the intermediate territories in a proactive manner (Bronzini, 2012).

REFERENCES

Allmendinger P., Haughton G., (2014). Revisiting Spatial Planning, Devolution, and New Planning Spaces. Environment and Planning C: Government and Policy, 31, 6: 953–957, DOI: 10.1068/C3106.

Barnett H., (2001). The Chinatown cornfields: Including environmental benefits in environmental justice struggles. Critical Planning, 8, 50–60.

Barton, D.N., Lindhjem, H., Magnussen, K., Norge S., Holen, S., (2012). Valuation of Ecosystem services from Nordic Watersheds. From awareness rising to policy support? TemaNord. Nordic Council of Ministers, Copenhagen, p. 506.

Barton J., Pretty J., (2010). What is the best dose of nature and green exercise for improving mental health? A multi-study analysis. Environmental Science and Technology, 44(10), 3947–3955.

Bedimo-Rung A.L., Mowen A.J., Cohen D.A., (2005). The significance of parks to physical activity and public health. American Journal of Preventative Medicine, 28, 159–168.

Bronzini F., (2012). The Secret Thread that Binds Cities and Territories. In Bronzini F., Bedini M.A., Marinelli G. (eds.), Marche. The Heartbeat of My Land. Il Lavoro Editoriale, Ancona.

Carter A., Mann P., Smit, J., (eds.) (2003). Urban Agriculture and Community Food Security in the United States: Farming from the City Center to the Urban Fringe. A Primer Prepared by the Community Food Security Coalition's North American Urban Agriculture Committee. Published by the Community Food Security Coalition, Venice California.

Dybas L., (2001). From biodiversity to biocomplexity: a multisciplinary step toward understanding our environment. BioScience, 51.

Donadieu P., (2005). Dall'utopia alla realtà delle campagne urbane. Urbanistica, 128.

Eurocities (2011), Cities cooperating beyond their boundaries: evidence through experience in European cities, Brussels.

Gasparrini C., (2010). Policy Document for the New Plan of Ancona, Comune di Ancona, Ancona.

Gibelli M., Salzano E., (eds.) (2006), No Sprawl. Perché è necessario controllare la dispersione urbana e il consumo di suolo. Alinea Editrice, Firenze, ISBN 88-6055-063-7.

La Greca P., La Rosa S.D., Martinico F., Privitera R., (2010). From land use to land cover: evapotraspiration assessment in a metropolitan region. In: Las Casas G., Pontrandolfi P., Murgante B. (eds.), Atti della Sesta Conferenza Nazionale Informatica e Pianificazione Urbana e Territoriale INPUT 2010, 367–377. Libria Editore, Melfi, ISBN: 978-88-96067-45-1.

Nadin V., Stead, D. (2000), Interdependence between urban and rural areas in the West of England, Centre for Environment and Planning, Working Paper 59, University of the West of England, Bristol.

Newell J.P., Seymour M., Yee T., Renteria J., Longcore T., Wolch J.R., Shishkovsky A., (2013). Green Alley Programs: Planning for a sustainable urban infrastructure? Cities, 31, 144–155.

Poinsot Y., (2008). Comment l'agriculture fabrique ses paysages. Un regard géographique sur l'évolution des campagnes d'Europe, des Andes et d'Afrique noire. Karthala édition, Paris.

Rosenzweigh, C., Solecki, W., Hammer, S., (2011). Climate Change and Cities: First Assessment Report of the Urban Climate Change Research Network (ARC3). Cambridge University Press.

Sanesi, G., Lafortezza, R, (2002). Verde urbano e sostenibilità: identificazione di un modello e di un set di indicatori. Estratto da "Genio rurale ed estimo. Il Sole 24 Ore, Edagricole, n. 9.

Scaglione G., (2008), SlowCity Landscape, Ecourbanism, Strategy, Project. List Laboratorio Internazionale Editoriale, Barcelona-Trento 156 p, ISBN: 9788895623092.

Scaglione P., (2015). Spostamenti intelligenti verso nuovi paesaggi ecologici. List, Rovereto, 95 p., ISBN: 9788898774685.

Van Leeuwen E., (2010), Urban-Rural Interactions. Towns as Focus Points in Rural Development. Springer, Berlin.

Wolch J., Jerrett M., Reynolds K., McConnell R., Chang R., Dahmann N., Brady K., Gilliland F., Su J.G., Berhane K., (2011). *Childhood obesity and proximity to urban parks and recreational resources: a longitudinal cohort study. Health Place, 17(1), 207–214.*

Town and Infrastructure Planning for Safety and Urban Quality – Pezzagno & Tira (Eds)
© 2018 Taylor & Francis Group, London, ISBN 978-0-8153-8731-2

The railway Noto-Pachino as a transport solution for a better tourism

M. Ronzoni
Università degli Studi di Bergamo, Bergamo, Italy

ABSTRACT: The paper is focused on an area rich in environmental and cultural emergencies, with a wide range of local products and an important infrastructure, but cumbersome: a 27,5 kilometers long railway line, abandoned in 1986, the Noto-Pachino line, in the province of Syracuse, Sicily.

It' easy to combine these aspects to develop a restoration project of this railway, a project that aims to systematize all polarities along the route. Polarities that the urbanistic investigation has highlighted. The project of restoration seeks to introduce elements of research and experimentation and suggests the use of rolling stock and energies, both innovative. The project relies on inter-modality between the train and the bicycle. This one enables to penetrate the territory, starting from the different train stops and reach its splendid riches such as the beaches with unique characterizations, the marshes of Vendicari, the villa of Tellaro, the Eloro ruins, the museum of sea of Calabernardo, the Pizzuta column, the citadel of Maccari, the town of Noto and Pachino.

1 INTRODUCTION

Quality of life means, among other things, to fully enjoy the opportunities that a territory can offer. The mobility system can do much to help improve the quality of life of a place. From the physical point of view, as social, environmental and economic. In the case dealt with and shown here we are in the presence of an area rich in environmental and cultural emergencies, with a wide range of local products and an important infrastructure, but cumbersome: a 27,5 kilometers long railway line, abandoned in 1986, the Noto-Pachino line, in the province of Syracuse, Sicily[1]. (Fig. 1).

The VI Italian Report on sustainable tourism and ecotourism, dated February 2016, edited by the Univerde Foundation, points out that nature, landscapes, historical, artistic and cultural events represent the most popular topics in the choice of a tourist destination, all aspects present in the territory of Noto that the railway infrastructure could further put in value.

The same report, when it has been asked if it would be willing to give up the car finds a 67% (multiple answers) open minded to the waiver if it will be possible to reach the goal by rail, and even a 75% interested in planning a tourist itinerary using an old railway.

2 OBJECTIVE

It' easy to combine these aspects to develop a restoration project of this railway, a project that aims to systematize all polarities along the route. Polarities that the urbanistic investigation has highlighted. The project of restoration seeks to introduce elements of research and experimentation and suggests the use of rolling stock and energies, both innovative. The project relies

1. Noto is a municipality of about 23.913 inhabitants (at 2017), characterized by a wide territorial extension (551,12 km^2) with a density of 43,4 inh/km^2; Pachino is a municipality of about 22.205 inhabitants (at 2017), with a territorial extension of 50,47 km^2 and a density of 440 inh/km^2.

Figure 1. The Noto-Pachino railway line, currently abandoned.

Figure 2. Cathedral of Noto.

on inter-modality between the train and the bicycle. This one enables to penetrate the territory, starting from the different train stops and reach its splendid riches: the numerous beaches with unique characterizations, the marshes of Vendicari, the villa of Tellaro, the Eloro ruins, the museum of sea of Calabernardo, the Pizzuta column, the citadel of Maccari, the village of Marzamemi without neglecting the departure station, the town of Noto, with its precious Baroque and scenic strength of its spaces (Fig. 2) and the arrival city, Pachino, with its rich offer of D.O.C. and D.O.P. to IGP brand. On a territory that presents a catchment basin of 46,100 inhabitants, that in the months of highest tourist numbers, can get to reach 200,000 presences or more, a speech thus conceived goes in the direction of reducing the CO_2 amount attributable to the field of mobility and makes easier to shift within this beautiful area to tourists that populate the coast or the historical centers of Noto and Pachino.

3 METHODOLOGICAL APPROACH

The methodological approach is based on the school of Vincenzo Columbo, which attributes to urban investigation and data collection a fundamental and indispensable role in the construction of the project. As I have repeatedly stated, a careful reading of the territory points out to the planner the right choice, in harmony with the character of the place, able to profit the resources present and mitigate the effects of weaknesses and threats investigated.

Reading the territory at the different scales, since, as it is well known, a project of this dimension can not only to be analyzed about the railroad track, but it must be considered within a wider context.

The survey therefore dealt with the study of historical factors, not just the history of the railway as an artifact in itself, built in 1935[2] as a link to bring foodstuffs, mainly fish industry products and wine from Pachino to the south of the France,[3] but also the history of the places that this crosses, to extract elements of characterization and enrichment of the proposal being elaborated.

Physical survey is focalized to collect data on environmental emergencies, but also gives attention to the energy resources, that are on the site and to the aspects related to landscape significance.

Social Survey, understood as a collection of demographic and human factors, but also of kinematic, building and technological factors. Finally, last but not least, the economic survey, which has paid particular attention to the tourist offer, as the hotel offer, but also that of second homes, hotels, farmhouses, bed and breakfasts and places bed in campsites.

Another important methodological element to get directions and expectations is the consultation of the planning tools that operate on the study area. In this way we fall into line with a previously territorial design. In particular, the PTCP of the Province of Siracusa and the landscaping plan of Areas 14 and 18 of Siracusa territory have been analyzed.

The research has prompted several insights and has also led to the organization of two workshops that have allowed to integrate the aspects investigated, working not so much on the path of the old line as on the outline elements.

4 DESCRIPTION

The project, as mentioned, concerns the territory of the municipalities of Noto and Pachino in its coastal part, rich in historical, environmental and landscape emergencies. If we want to condense here the most relevant elements extracted from the survey, which can be traced back to the project we can remember the following. With regard to the environmental aspects, it allows access to areas of high natural and landscape value. Just remember how the trace of our linearity goes through the Oasis of Vendicari, (Fig. 3; Fig. 4) one of the richest territories of biodiversity in Italy.

Here, we can find the Caretta Caretta turtle nest, and find numerous habitats of ornithological species including the heron and the flamingo.

Among the essences present the myrtle, the thymus of the Iblei, the caper, the dwarf palm, the lentiscus, the autumnal mandrake. Research has made it possible to capture visitors data of the Oasi of Vendicari, which in 2015 have reached 127,491 units. In the year 2012 the peak of visitors with 139,957 presences. Among the new hypothesized stops, some fall within the Oasis: Calamosche, Tonnara of Vendicari and Maccari village. Calamosche is one of Italy's most beautiful, sheltered, cozy coves, full of contrasts between the blue sky, the white/golden beach and the greenery of the surrounding Mediterranean vegetation (Fig. 5).

Few years ago it was the set for Vodafone advertising, since then it is almost always crowded and this crowding is not good. The Tonnara stop allows access to the beach strip on the Stagnone of Vendicari, a marshy area marked by birdwatching cabins, where you can also visit the museum of the Ente Fauna. The Maccari village stop allows you to access the Oasis from the southernmost part, walking in the middle of the figs of India, perceiving in the breath the scent of history. But this is a feature that often occurs in these lands, even on the north-east side of the oasis, where the ancient settlement of Eloro is located, with the ruins of its forum,

2. It was inaugurated in 1935, but the first projects date back to the early 1900s. The exercise was suspended in 1986, shortly after the inauguration of the electrification of the rail crossings. The casting off was in 2002 and it is 2016 that a document transfers the use of the rail infrastructures to the municipalities of Noto and Pachino.

3. We have also documented the presence of weighhouse near the station of Noto, where in the past there were numerous factories for the food processing of tomatoes, which arrived through train from Pachino.

Figure 3. The ancient Via Elorina, which connected the settlement of Eloro with Syracuse.

Figure 4. Shore stretch in the Oasis of Vendicari.

Figure 5. Beach of Calamosche in the Oasis of Vendicari.

the traces of his Greek theater and his Temple. At sunset the air is filled with the odors of a past far away in time, but close thanks to the intensity of its scents. With regard to social aspects, in particular demographic flows and data, these relate to important tourist flows, with significant presences of foreigners. For the vast majority, however, they are local tourism, from the isle. There is a lot to work to grow and increase the share of international tourism. Guaranteeing safe, comfortable connections and connecting the most important local polarities, would surely provide an incentive for this tourist offer. As far as mobility is concerned, the information collected and the direct observation tell about traffic on rubber entrusted mainly to the private vehicle (Fig. 6; Fig. 7). In the last few years, car and caravan rental agencies have flourished alongside buses that, on the territory analyzed, are often characterized by uncertain timetables, they are not cadenced and make the service only in the highest season.

While these areas are also enjoyable in intermediate seasons, during which one of the disadvantages is the lack of links and the lack of mobility services. Certainly there is always the possibility of hiring a car, but this makes the holiday much more expensive and probably forces the tourists to turn to other destinations more competitive than ours. Sometimes it takes courage to unlock consolidated situations and even a certain amount of risk predisposition. Mobility is understood as a service that in absolute value is in loss, but with its numerous relapses it is a driving sector and it is able to recover the costs of the expenses for its maintenance, subtrahend it from the proceeds of the market induced. As for the economic aspects, especially the tourist sector, as this is said, this has significant potentialities, certainly not in the possibility of building other settlements or widening existing ones or new tourist structures:

Figure 6. The car now seems to be the only solution to reach the beaches and it forces pedestrians to walk in the middle of the street.

Figure 7. The nature defaced by man.

you must always be aware that this is a beautiful territory, but fragile. It should be safeguarded and we should begin to reflect on the need to block soil consumption. There is, however, a lot of disuesed, there are a lot of empty houses, particularly at Piano Alto in Noto. In recent decades, social life in Noto is mainly in the lower part of the city, with direction east west, while the upper part is largely abandoned. In one of the two workshops that I have conducted on this topic, we have tried to understand how the reactivation of the Noto Pachino Railway could attract in these places, and specifically on Noto, more tourist flows. These tourist flows could find proper placement in Piano Alto, in the buildings of the ancient nucleus, now abandoned, but easily recoverable in a proposal of widespread hotel. Overall, the analysis of the collected data makes it clear that the reactivation of the dismantled railway line could be useful support to the the tourist offer. In fact, the data on tourist flows, tell us of an important tourist presence, which could further Increase if there was an adequate supply of sustainable mobility at affordable prices. Official data probably are inferior to the actual ones, because the presences in the tourist facilities must be added to all those represented by the submerged, guests, friends, occupying the second homes. See the railway abandoned, run along a line that laps on historic, cultural, environmental, and landscaping polarities raises spontaneously the question: why not reactivate it? The territorial planning tools consulted seems to direct towards this choice. In particular the PTCP of the regional province of Siracusa. The landscaping plan too... The administration's choices are unfortunately directed towards transforming the line into a greenway. The municipality of Noto, together with the municipality of Pachino, acquired in free use thanks the company Ferservizi the stretches of the disused railway line Noto Pachino in order to establish structures in support of sustainable mobility and pedestrian (cycle tracks). There are a number of consulted documents that give testimony of this intention. Recently FS and RFI have presented an interesting document titled Atlante delle linee ferroviarie dismesse, where are mapped the different railway sections divided into dismantled definitively and discontinued for activation of variant of the trajectory, divided by North, Central Southern and islands. The introductory pages explain how the work is aimed at converting these railways into greenway so as not to disperse the advantages of a railway linearity: continuity of the route, absence of conflicting points. There are also several law proposals on the abandonment of old railways, but all proposals analyzed focus mainly on the opportunity to convert the old disused tracks into cycling tracks that are very appetite by tourists. In general, I agree with this position and I find it reasonable when there is no demand for mobility through train or there is a depopulation of a territory, either because the timing and needs have changed. In this case may be appropriate, also to protect the line and make it accessible to a question that has evolved over time, turn it into a greenway. Although sometimes I think the demand is missing not because it is absent absolutely, but simply because the quality of the service offered is so bad that it will keep far from this territory any traveler hypothetical. However, I do not agree with this kind of conversion with regard to the railway in question. These areas are characterized by a turbulent climate where in summer the thermometer reaches 47° and at nine in the evening it sometimes stagnates steadily at 34°. Surely it would be indispensable the pedal-assisted bicycle. This, however, with respect to the traditional bicycle, is more delicate, it needs more care and perhaps it is necessary to remember how it can be more easily subjected to vandalism. The costs would be significant if you wanted to keep bicycles in efficiency. It is also important to note that the track is sometimes protruding, but also for long stretches in trench. Walking or cycling it in the evening, in the dark, but also during the day, in the low season, could surely create insecurity[4]. In addition, the track has a total difference of level of about 60 meters.

The bicycle is not for everyone, while the train it is. I think that there are conditions for this abandoned railway to safeguard the line and reconsider the choice of disposal. In essence, I find myself in the Charter of Rome, signed on 23 February 2009, as a unique document expressly proposing *to encourage the recovery of disused railways by assessing the possibility of reactivating the service or, alternatively, of their possible reactivation consider their immediate transformation into a greenways that can be used with environmentally friendly means and from the widest categories of users.* In any case, the role of the bicycle is not marginal in this proposal. In fact, the project envisages its fundamental presence as a means of ensuring intermodal mobility: from each station there is a path, often as a ring, that penetrates the territory in the direction of the coast to reach the beaches. The train should therefore be set up for bicycle transport, however, at every stop the project always provides a parking area reserved for the service of bike sharing.

Compared to the original stations, which were 8 (Noto, Falconara Iblea, Noto Marina, Noto Bagni, Roveto Bimmisca, San Lorenzo Lo Vecchio, Marzamemi e Pachino) the proposal foresees 16 stops, in addition to those already listed, Villa Favorita, Cala Bernardo, Noto Lido, Colonna Pizzuta, Eloro, Corte del Sole, Calamosche, Vendicari.

5 INNOVATIVE ASPECTS

These mainly concern the technical aspects of the choice of rolling stock.

Among the objectives of the research there is that to find a transport solution that looks at the use of renewable energies, making the most of the resources present in situ. I also think that the approach that looks to the profitability of the project considering its outcomes in the long run, can be seen as innovative in comparison with the approach that consider its outcomes in the short time. In this last case, of course, the train line would been surely seen losing against a greenway proposal.

4. A journey to be attractive must ensure to be continuous, direct, safe, as short as possible.

6 RESULT OF THE RESEARCH

The research has managed to highlight the many aspects that characterize this territory, it has brought them to a system and it has developed a route that reaches everyone of this and helps the visitor to become aware of its.

It also has awaked the debate on this issue.

After the outcome of the first workshop on this topic was presented at Noto, there were numerous meetings, initiatives, articles, thesis work focused on this subject. Other workshops have been generated by this research, which is not and should not be a point of arrival, but a stage that contributes to the transformation of a territory.

7 CONCLUSIONS

Being able to let leave again this train can indeed bring many benefits.

Environmental. Where it reduces the impact of mobility on the environment by eliminating cars from the road, but also helping to put in value a territory that has significant potential. If you are moving on the road yourself, it becomes more difficult to grasp the peculiarities of the territory you are going through. If you use the train, this in its path and its stops, should help you to focus on the aspects that characterize the place and help you to become aware of what is around you.

Economics, considering that on this territory so rich in cultural and physical-environmental emergencies now the shortage of public transport and the absence of rail links to the touristic centers on the coast is extremely damaging to the development of the tourism sector.

Social, because it would contribute to improving the quality of life not only of tourists but also of residents. It would also surely create jobs, some at the infrastructure itself, but many in the tourist sector.

8 BARRIERS AND DRIVERS

Surely the biggest limit is the costs. Costs of installation as well as the costs of management and maintenance of such an infrastructure. However, in the opinion of those who write, the restoration of the rail service on this line could at the same time help to set in motion the economy and recall a large number of tourists, currently get out of the way because of the high costs of the trips.

At present, with the exception of local tourism, tourists arrive by plane, staying in the hotels of the coast, where they are conveyed by coaches taking them from the airport. Then for the duration of their stay, they are locked in paradisiac places, but which could offer much more in cultural terms and even as generic knowledge of the territory if a greater supply of sustainable mobility was guaranteed. The cycle track does not offer the same opportunities for movement, either for the climate, too hot in the summer months, insecure in some stretches at night, unsuitable for some class of age. A train link would ensure during the day the possibility to reach autonomously the various beaches and marine areas that characterize this stretch of coast, approaching Noto to its sea. The train link would make the city competitive with respect to the tourist structures of the coast. During the evening, it would guarantee to the many guests staying in hotels and structures on the coast, the possibility to reach the cities of Noto and Pachino where, especially with regard to Noto, there is always a rich cultural offerings.

REFERENCES

Canale C.G., Noto—la struttura continua della città tardo-barocca, Flaccovio Editore Palermo 1976.
Carta di Roma, 23 febbraio 2009.
FS con RFI (a cura di), Atlante delle linee ferroviarie dismesse.
Tobriner Stephen, La genesi di Noto, Edizioni Dedalo, Bari 1989.
VI Rapporto italiani, turismo sostenibile e ecoturismo del febbraio 2016 a cura della Fondazione Univerde.

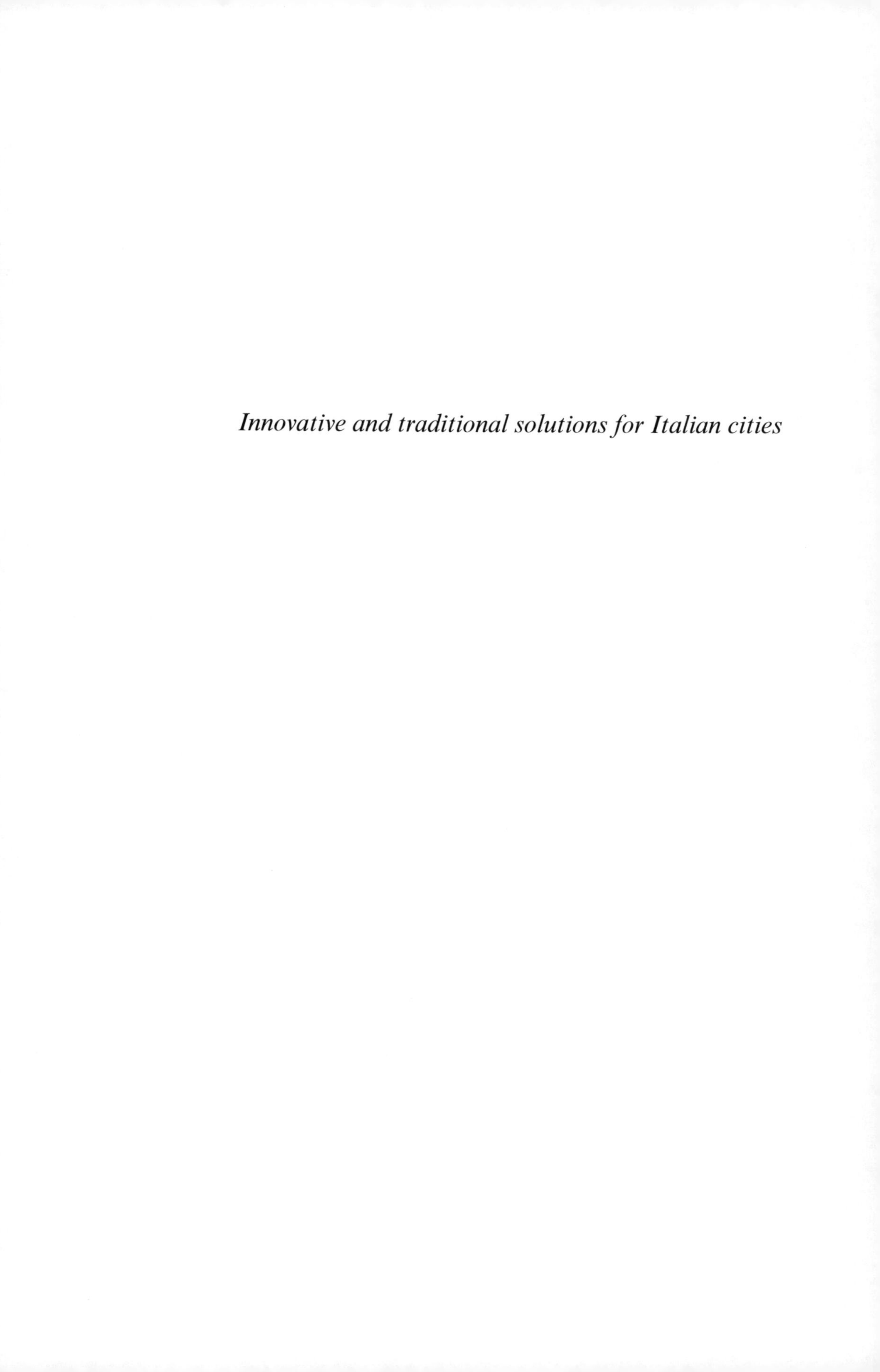

Innovative and traditional solutions for Italian cities

Town and Infrastructure Planning for Safety and Urban Quality – Pezzagno & Tira (Eds)
© 2018 Taylor & Francis Group, London, ISBN 978-0-8153-8731-2

Soft mobility in the "oblique city"

R. Fistola & M. Gallo
Università degli Studi del Sannio, Benevento, Italy

R. La Rocca
Università degli Studi del Sannio, Benevento, Italy
Università degli Studi di Napoli Federico II, Naples, Italy

M. Raimondo
Università degli Studi del Sannio, Benevento, Italy

ABSTRACT: This study aims to define a methodology for the planning of a network of urban connections inside the "oblique city" particularly in reference to urban historical centers characterized by steep differences in height. The methodology proposed refers to the individuation of a network of urban paths for soft mobility starting from specific "urban poles" (open places, squares, monuments, etc.). The network itself can be an attractive element for improving a sustainable way to visit the city. The city of Benevento, in the south of Italy, represents a meaningful case for testing the methodology proposed. The project proposes the definition of a network for soft mobility that, on one side, is aimed at increasing accessibility to the historical center, and on the other side, considers the possibility of interconnecting Benevento to other historic centers nearby (Pietralcina and Foglianise) that serve as additional poles in the network.

1 THE SMART OBLIQUE CITY: MAKING VERTICAL CONNECTIONS

Urban smartness can nowadays be related to many features that the modern city has to include in order to be considered "intelligent." As mentioned in other studies (Fistola, 2013), this characteristic is not merely the result of the quantity/quality of the technology that urban systems have within their physical space and its proper use in order to monitor the city in real time, but also the ability to support and implement sustainability, compatibility and energy-saving transformational actions. Through Intelligent Technology Adoption, the urban functional system can be led to a state of efficiency and effectiveness that is beneficial to the promotion of social capital and, in general, to raising the levels of urban living.

Without a doubt, mobility is the urban function that forms the essence of a city. The movement of goods, people and information determines the vital streams of the urban system. It should also be considered that the modern city was born from Ford's economic conception and was based on vehicle movement (automobile city), with much of its own growth and territorial spreading process being supported by fossil propellants. Polluting emissions resulting from this type of mobility, coupled with the generation of pollution as the result of other activities (production, residence, etc.), have produced the current state of crisis in modern urban systems that today seek for alternative forms of physical city mobility (Fistola, Gallo, La Rocca, 2012). Modern urban and metropolitan systems are putting policies into place supporting soft mobility, which will allow for urban shifting through the use of muscle strength (feet, bikes, skates, skateboards, etc.) or, alternatively, electric vehicles. This last category includes all mobility devices that allow you to move, in part or in whole, with the assistance of an electric motor: e-bikes and personal mobility devices (PMD) such as electric scooters, hoverboards, segways, golf-carts, electric cars, etc. (Fistola,

Gallo, La Rocca, 2013). Such devices are generally linked to green mobility, although traction electricity is, in any case, produced in power stations, located in non-urban areas, and generally powered by fossil fuels. It is evident that these devices are of particular utility, especially in hilly urban contexts, within which it is often necessary to overcome slopes of some relevance. Assistance in overcoming slopes makes urban fruition of the city easier. In Italy, there are numerous hilly cities located in the Apennines in particular and in subalpine areas. The idea is to develop a network of bicycle paths (pedestrian and cycling) capable of supporting sustainable mobility and of overcoming the slopes that constitute a very restrictive obstacle to urban movement, especially for specific categories of user (the disabled, the elderly, children, pregnant women, etc.).

Referring to these typologies of urban systems, we intend to propose a meta-project methodology based on the creation of networks dedicated to "soft" mobility. The methodology is based on the identification of a set of strategic nodes, within the context of a hilly urban site, that become fulcrums for the construction of a network of "urban connections," capable of linking the different areas of the city horizontally and vertically. In particular, the methodology considers an initial phase of analysis of the structure of the city articulated in:

- recognizing the characteristics of urban patterns and roadways;
- identifying the perimeters of a number of functional areas (areas featuring widespread or concentrated activities that structure the city);
- locating a number of POIs to connect to the network (paths).

In a second step, the "oblique city" is divided into a series of main horizontal circuits, drawn by connecting the functional areas and POIs at the same altitude. Consequently, a number of parallel sections of the city (planes) are created, located at different levels (Fig. 1).

The circuits placed on these planes themselves develop along flat surfaces (at a constant height) and allow easy access to the different urban areas where the activities are located.

The third step concerns the identification, within the development of the paths, of nodes that can be connected to each other via a system of oblique and/or vertical connections. In the first case, they can be represented by other paths, characterized by a certain slope, which can be overcome through the use of e-bikes, PMDs or with appropriately studied technical aids. In the second case, vertical lifts or elevators must be designed in order to carry up users and PMDs.

The node system is therefore characterized by the following typologies:

- single vertical interchange node;
- double vertical node;
- vertical interchange node (horizontal/vertical and ground/slope)

At last, the network that is designed in the territory draws a series of paths in an extremely schematic way (with concentric rings or open paths), located at different heights and interconnected by either oblique or vertical connections. As mentioned before, such connections can be expected through the predisposition of mechanized devices due to their physical

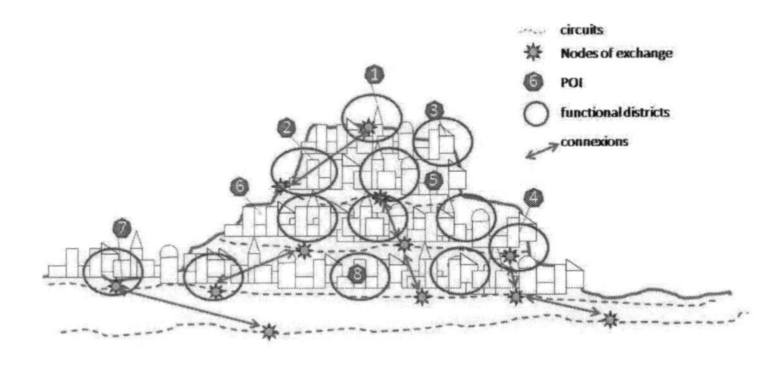

Figure 1. A scheme of the circuits, nodes and connections.

peculiarities (steepness of the path, severe jumping, etc.) and functional peculiarities as well (localization of major urban functions, concentration of tourist-cultural attractions, etc.).

As will be seen later, the proposed methodology has been tested in the city of Benevento, which represents a significant case of experimentation both for the urban dimension and for the orographic characteristics of a primarily hilly area.

2 TECHNOLOGICAL SOLUTIONS FOR SOFT MOBILITY IN OBLIQUE CITIES

Promoting soft mobility in urban areas is an effective strategy for improving urban sustainability, especially in small-medium size cities. Indeed, in these cities, a large percentage of trips have distances compatible with pedestrian and cycling mobility.

The provision and design of pedestrian areas and cycle lanes, sometimes coupled with bike-sharing systems, are the most common transport policy interventions for encouraging soft mobility. For oblique cities, in which there are significant slopes to overcome, some other specific solutions have to be considered; with the exception of assisted bike users, potential soft mobility users (inhabitants and tourists) can be discouraged by steep slopes. This issue is particularly important for the many Italian cities that have their historical centres located in hilly areas; in these cases, some vertical or oblique connections have to be identified and equipped with specific facilities (escalators, funiculars, etc.).

Another crucial point is the intermodality between transit systems and bicycles: all transit vehicles (buses, funiculars, metro trains, etc.) should allow users to bring personal bicycles on board easily, and the main terminals (railway and bus stations) should offer safe bike parking.

From the angle of a systemic design, the horizontal circuits should offer cycle lanes, cycle-friend transit systems (preferably with low or zero emission vehicles), and pedestrian paths; these circuits have to be linked at several points by equipped connections. Some connections can be dedicated to pedestrians (e.g., escalators), others to cyclists (e.g., bike escalators) and others to both users (e.g., funiculars). In every case, two near horizontal circuits have to be linked by at least one connection useful for pedestrians and one connection useful for cyclists; at the same time, mobility along the horizontal circuit, if the distances are sensible, should be served by (bike-friendly) transit systems.

The choice of more suitable technology for equipping oblique connections should be driven by a feasibility analysis from the technical, economic and financial points of view. For instance, escalators may be an appropriate solution if the distances to cover are short (no more than a few hundred meters) and, if the gradients are not too high, they can be replaced with moving walkways (also useful for cyclists). Short vertical connections may be ensured by lifts that may also be designed for carrying bicycles. For longer distances, funiculars and cableways are the more suitable solutions, even if costs rise significantly.

Figure 2 schematically summarises some possible technological solutions in terms of distance and gradient functions.

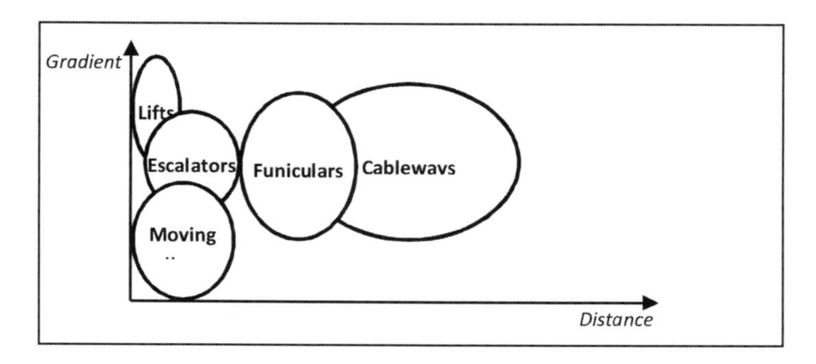

Figure 2. Technological solutions for oblique cities.

With this conceptual vision, the historical centre should be closed to traffic and the trips inside it should be supported by PMDs, which should be available to users through a system analogous to bike-sharing (limited time rental); specific and dedicated systems should also be available for disabled people.

3 OBLIQUE CONNECTIONS, TOURIST FRUITION AND TERRITORIAL PROMOTION

In this part of the paper, tourist mobility in urban areas has been understood as a specific demand within use of the city that is particularly appropriate for the promotion of soft mobility and sustainable urban policies. This aptitude can be related to certain aspects that characterize this kind of urban mobility. We can refer to a) the time spent using the city; b) the perception of the urban space; c) the aptitude to be framed within dedicated circuits; d) the possibility of being a part of the global supply system of the city. In general, we can state that these kinds of urban movement can refer to a *leisure mobility*, meaning the mobility generated by motivations different from those of work, and in this regard, it can be associated with the flow of tourism within the city. In this case, movement itself is an occasion for discovery and knowledge, and represents the main means for visitors to live their own tourist experience.

Urban mobility (for residents and users of the city) is considerably different from tourism mobility, in regard to both spatial perception and the time spent to move. Urban mobility fluxes, in fact, are characterized by rapidity, and do not concentrate on the quality of the path they travel to reach their destination (La Rocca, 2015). On one side, if, in the case of cities in which tourist activity is prevalent and is central for the urban economies, these differences can generate conflict between tourists and residents, on the other side, in the case of middle-sized cities, tourism can act as a driving function for improving urban requalification, to promote the system of resources (territorial, environmental, urban, social, etc.) of the city and to stimulate alternative forms of visiting the city (modifying behaviors and fostering sustainable tourism).

Our opinion is that this is possible only if the planning of tourist activity is integrated into urban planning for a common objective of improving the global urban supply of services and spaces, with specific attention to mobility issues. Movement is, in fact, the essence of tourism,[1] and an efficient mobility system can improve the competitiveness of a city, becoming a factor of attractiveness itself, able to guarantee, on the one side, accessibility to urban resources (places and POIs) that could not be achieved otherwise, and on the other side, representing the uncommon fruition of the city.

It should be considered that the realization of mobility infrastructures may have an excessive cost compared to the number of real users. Hence, it becomes essential that the oblique connections are well integrated into the transport system after it is already in effect, or that they are mainstreamed into the planning of the urban mobility system.[2]

In Italy, there are some example of medium-sized[3] hilly cities (Assisi, Perugia, Reggio Calabria, Siena, Spoleto, Urbino) in which the building of automated oblique connections (hectometer systems) triggered an urban requalification process of their historical centers (Naldini and Ambrosi, 2010). Most of these examples refers to an improvement in the acces-

1. Tourism is a social, cultural and economic phenomenon, which entails the movement of people to countries or places outside of their usual environments for personal or professional purposes (UNWTO2008).

2. We refer to the mobility system as the integration between the demand of movement and the supply of transport in an urban context. An Urban Mobility System (UMS) is, in fact, a concept that goes beyond the provision of public transport, referring to services and infrastructures that enable citizens and city-users to satisfy their mobility requirements (Macàrio, 2011).

3. We refer not only to the number of population, but also to the role that these cities play inside the region, being the intermediate pole between large metropolitan conurbation and small urban centers, as it concerns the Italian territorial context.

sibility of historical centers, sometimes disregarding the possibility that a larger design based on the presence of a network, according to a systemic vision of the city, could offer.

Considering these cases, the present study proposes integration of the objectives strictly connected to accessibility enhancement, for the purpose of promoting new forms of fruition of the city using urban paths connecting several POIs and dedicated to soft mobility. In other words, within the network of urban connections, the tourist points of interest (T-POI) assume a hierarchical role by identifying the places in which it is possible to integrate several innovative types of service that can support tourists during their urban visit (information, bike sharing, wi-fi guides, electric vehicles, etc.). In this respect, tourism plays the role of a driving function of positive processes of urban requalification, also considering the economic contribution of tourism to the local economies[4].

The methodology presented in this study thus proposes a planning procedure that is based on the definition of a network of internal connections that can also improve tourist attractiveness of the urban system involved. The possibility of combining public transport with cycling and/or walking is one of the most important elements of the planning methodology that could represent a useful tool for the definition of integrated action of urban planning and mobility management.

The case of Benevento is a first challenge for testing the proposed methodology, as it represents both a hilly, medium-sized urban system and a potential pole of excellence in promoting a sustainable form of tourism within the Campania regional tourism system.

4 SOFT MOBILITY AND THE OBLIQUE CITY: A PROPOSAL FOR BENEVENTO

The city of Benevento is a medium-sized city (about 61,000 inhabitants), characterized both by the presence of urban functions of a metropolitan level (University, Hospital, Courthouse, etc.) and by a satisfactory level of tourism attractiveness (about 18,000 total arrivals in 2016). It also lies in a strategical position close to the religious pole of Pietrelcina (1,500,000 million peregrines in 2016) and to the environmental pole of Monte Taburno (Regional Natural Park, 12,370 ha).

The main objective of the proposal consists in the definition of a network for soft mobility capable of both improving accessibility to the historical center of Benevento and promoting territorial resources for the development of sustainable tourism (Fig. 3).

According to the methodological premises, the network is articulated in levelled circuits and equipped poles linked by oblique or vertical connections along the rise of the slopes.

Equipped poles coincide with existing large parking areas (Pomerio and Porta Rufina), where bike-sharing stations for electric bicycles have been located so as to allow users (citizens, commuters, tourists, visitors) to access the historical center.

Another important equipped pole (Rotonda dei Peltri) allows users both to reach the historical center and to travel along existing routes to reach the external poles of Pietrelcina (to the east) and Foglianise (to the west), thus promoting a larger vision that considers the potentialities connected with a collaborative and strategical policy of territorial promotion.

Alternatively, from the equipped poles, which can be thought of as new city doors, users can also choose to travel (by cycle or on foot) along the "blue paths" by the rivers (Sabato and Calore) in the lower part of the city. The "blue paths" have a dual role within the network for urban soft mobility. On one side, they are dedicated paths providing safe conditions for the flow of pedestrians and bicycles. On the other side, they represent smart urban infrastructure, equipped to monitor the level of the river's water, to collect rainwater, and to contain the river's course (Fig. 4).

Along the "blue paths" two main poles join two existing cycle-lanes ("Ciclovia Spina Decumano" and "Raccordo Teatro Romano") that drive users to the main POIs within the historical center. Among these points, particular attention has been paid to the individuation

4. In Italy, the tourism tax is a local tax charged to tourists staying in accommodational structures. Local administrations decide on adoption of the charge. The tax revenues are intended to finance accommodation facilities, as well as interventions in the field of local public services. Even though it has been very controversial, the tax has been adopted in the most Italian tourist cities.

Figure 3. Map of the project.

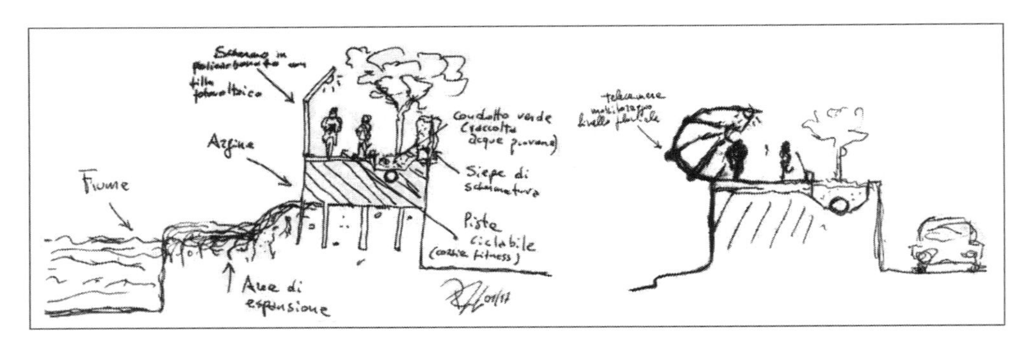

Figure 4. Sketches of the "blue paths".

of T-POIs, whose individuation was carried out by TripAdvisor using the web scraping technique. This allows us to map the usable T-POIs with an acceptable level of accuracy (Fig. 5).

Within the network, the levelled circuit that runs along the perimeter of the historical center also serves as a lane for electric buses dedicated to the movement of persons with reduced mobility, elder people, families with children, pregnant women, etc.

Oblique and vertical connections are in a specific node (vertical interchange nodes), corresponding to relevant slopes or differences in height. In the first case the connection occurs through the use of a cyclocable (Fig. 6); in the second case, lifts or escalators are disposed as connections inside the network (Fig. 7).

The project considers other integrative interventions aimed at improving the supply of sustainable mobility services. In particular:

- localization of charging stations for electric vehicles at strategic points within the city (the main access road, the train stations, the equipped poles, the interchange nodes);

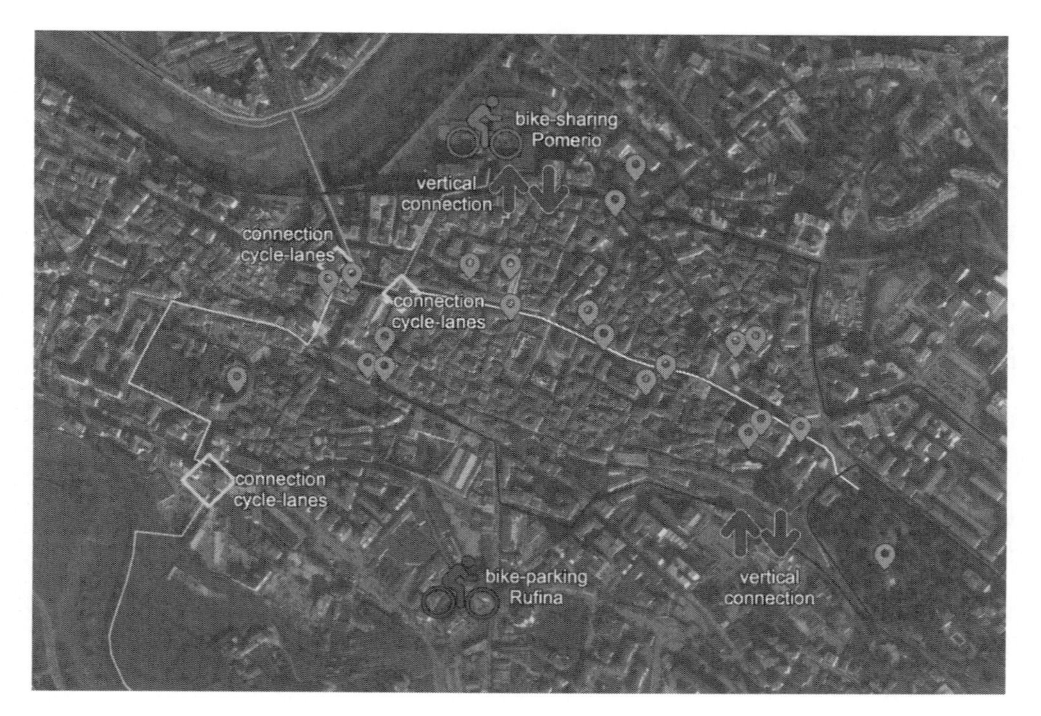

Figure 5. Map of T-POIs: most of T-POIs can be reached through the levelled circuit that runs along the perimeter of the historical center.

Figure 6. Image of the cyclocable.

- design of a circuit for electric buses connecting the historical center with other parts of the city;
- improvement of the [AU: Improvement of the what?]

The planning of an efficient network dedicated to urban sustainable mobility that also ties into the quality of the local public transport. In this respect, actions must provide:

- a shuttle service connecting the central station with the equipped pole in Pentri;
- the delocalization of a terminal bus from Colonna Square also to the equipped pole in Pentri.

The proposed project aims to improve the potential role that Benevento can play for the promotion of sustainable forms of visiting and knowledge of territorial resources.

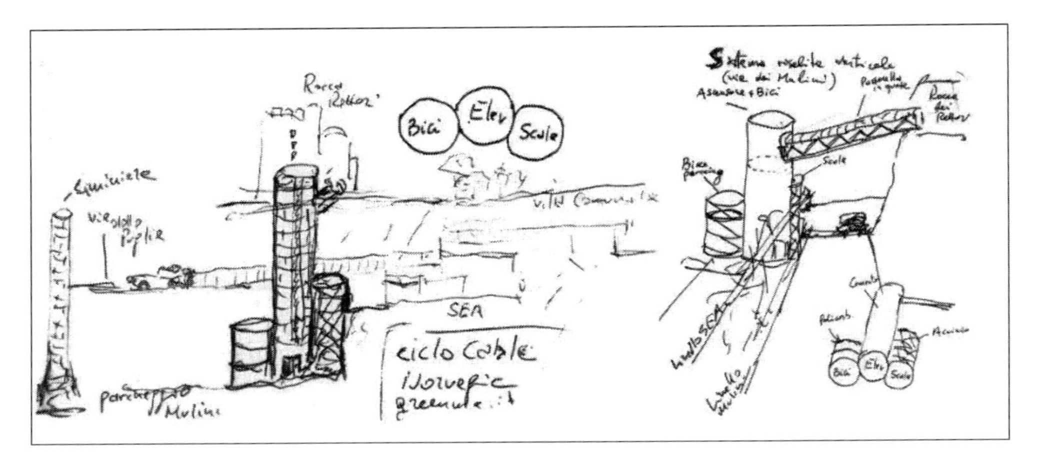

Figure 7. Sketches of vertical connection: the image of a vertical interchange node in which three towers have been designed in order to overcome 20 mt in height. The towers contain parking for bicycles, lifts for people and bicycles, and stairs.

5 CONCLUSIONS

The methodology of the proposed meta-project is articulated in a series of phases that allow for the development of an interconnected network of soft mobility, especially within hilly urban contexts. Connections between the network's nodes ensure widespread use of the oblique city, along with adoption of innovative technologies for overcoming the slopes and the forecasting of electric devices run along the ground circuit that play a useful role for TPL, tourist mobility and, in general, for connection among the horizontal/vertical interchange nodes. The procedure could also provide theoretical support for the development of territorial plans and instruments for the purpose of integrating town planning and goals of urban mobility. Furthermore, the proposed methodology can represent a useful way to develop technical tools, particularly in regard to the fruition of tourism in oblique historical cities.

REFERENCES

Fistola, R., (2013). "Smart city: riflessioni sull'intelligenza urbana", in TeMA. Journal of Land Use, Mobility and environment, 06 (1). University of Naples Federico II. 47–60. doi:http://dx.doi.org/10.6092/1970-9870/1460.

Fistola, R., Gallo, M., La Rocca, R.A., (2012). "La mobilità insostenibile: impatti dannosi del traffico veicolare sulla salute umana", in Moccia F.D. (a cura di), Città senza petrolio. Edizioni Scientifiche Italiane. 75–87.

Fistola R., Gallo M, La Rocca R.A., (2013). Nuovi approcci per la gestione della "mobilità dolce", in Moccia F.D. (ed), La Città Sobria Collana: Governo del territorio e progetto urbano—Studi e Ricerche, 7. Edizioni Scientifiche Italiane, Napoli. 301–310.

La Rocca, R.A., (2015). "Tourism and mobility. Best practices and conditions to improve urban livability" in TeMA. Journal of Land Use, Mobility and Environment, 8 (3). University of Naples Federico II. 311–330. doi: http://dx.doi.org/10.6092/1970-9870/3645.

Macàrio, R., (2011). Managing Urban Mobility Systems. Emerald. UK.

Naldini, L., Ambrosi, M., (2010). "La gestione dei flussi turistici tramite sistemi di trasporto innovativi: l'esempio di Perugia" in TeMA. Journal of Land Use, Mobility and Environment, 3 (3). University of Naples Federico II. 85–96. doi: http://dx.doi.org/10.6092/1970-9870/151.

UNWTO (2008) Understanding tourism, basic glossary. http://statistics.unwto.org/sites/all/files/docpdf/glossaryen.pdf. Last.

Town and Infrastructure Planning for Safety and Urban Quality – Pezzagno & Tira (Eds)
© 2018 Taylor & Francis Group, London, ISBN 978-0-8153-8731-2

The environmental conditions of Italian cities: A spatial analysis of the smartness components

F. Balducci & A. Ferrara
ISTAT (Italian National Institute of Statistics), Roma, Italy

ABSTRACT: Despite the research efforts and the growing number of articles the concept of smart city has not yet found a common qualification nor a testable definition. In this application the domains of smartness are obtained from the official data of the Italian National Institute of Statistics (Istat) describing the adoption of smart policies in the urban environment (ICT use, smart mobility, green innovation, sustainable governance etc.). The information, referred to all the Italian chief towns (116 cities), has been normalized and synthetized with statistical techniques in order to derive a taxonomy for the smartness' domains. Then the performances have been analysed for each domain, with regards to the geographic location of cities. The spatial interactions are examined through localized spatial autocorrelation. The results show a significant territorial discrepancy which go beyond the well-known Italian North-South polarization. There is evidence of localized clustering and interactions among neighbouring cities.

1 BACKGROUND

1.1 *The concept of smart city in the scientific debate*

The idea of "smart city" had proliferated since the late eighties, following the rapid technological advances and the continuing process of urbanization, in both developed and developing countries. Those paths of innovations were associated to positive outcomes (such as the increase in productivity and the dissemination of knowledge) as well as to problematic conditions in the environment of cities getting bigger, concerns towards sustainability and inequality in the distribution of wealth. Such mix of issues gave input for "smart" policies to be designed, with the aim of guaranteeing a sustainable economic development, paying attention to the environment, societal welfare, and people's health. ICT and new technologies were intended as a tool to support people and to make those targets effective (European Commission, 2012).

Following those inputs, the research interest in the "smart" area increased rapidly (Cocchia, 2014). However, in spite of research efforts the concept of smart city maintained a certain vacuity: a shared definition and a common taxonomy of its constituent items have yet to emerge in the literature. The definitions ranged from a narrow view, typically related to ICT and technological aspects (Harrison et al./IBM, 2010; Northstream; 2010), to broader ones, encompassing well-being, health and social relations (Caragliu et al. 2011; Giffinger et al. 2007).

Many contributions focused on one specific aspect of smartness. The use of ICT was widely explored, with regards to increasing urban capacity (Lee et al. 2014), energy consumption (Kim et al. 2012; Yamagata and Seya 2013), urban planning (Arribas-Bel, 2014), living improvements and Internet-of-Things (IoT) (see Anthoupulos et al. 2015 for a full review). Lee et al (2014) introduced a economic-oriented classification, while others focused sociological aspects, classifying the smart cities with regard to their resilience (Desouza and Flanery; 2013).

The environmental aspects, however, were less studied and not fully exploited even if they represent a constituent part of the smart city concept (Cocchia, 2014).

1.2 *Measurement issues*

A number of attempts to measure and rank the cities according to their degree of "smartness" have been proposed (Ojo et al. 2016). According to a widely cited and re-proposed framework the domains of smartness could be classified into: Governance, Economy, Mobility, Environment, People, Living (Giffinger et al. 2007). The classifications are sometimes wider (including health care, buildings and urban planning; Piro et al. 2014), or narrower (restricted to a *hard domain* – infrastructural, and a *soft domain* – referred to people, Neirotti et al. 2014).

Considering the heterogeneity in definitions and applications, the quantification of the degree of cities' smartness had been a contentious task. From a statistical point of view, the issues pertained to an unclear definition of the territorial unit (boundaries of the cities, e.g. Functional Areas rather than administrative limits etc.), to an arbitrary and heterogeneous selection of the indicators and to the absence of a weighting or aggregation procedure (De Santis et al. 2010 and 2014). The reliability of data coming from varied and non verified sources is also a issue. Sometimes, the scarce availability of information at urban level caused the aggregation of data pertaining to different time periods or territorial scales. Also, the results presented in form of rankings are partly non–transparent (Taylor, 2011; OECD, 2008).

As a result, the complex interrelations among sub-dimensions, and their relation to smartness, were treated only partially in the literature. The correlation between some of the aspects of smartness (urban wealth, creative professionals, transportation networks, ICT, human capital) and GDP had been analysed descriptively (Caragliu et al. 2011). Others checked the consistency of Giffinger's (2007) framework with statistical techniques (De Santis et al. 2014, Lombardi, 2013). Finally, Lazariou and Roscia (2012) proposed a robust methodology to assess and compare smart city models using fuzzy logic criteria.

2 OBJECTIVE OF THE PAPER

In recent years, Italian (and worldwide) cities have adopted a series of policies targeted as "smart", towards a innovative management of the urban environment and society.

However, there is little knowledge of the kind and extent of those policies, which are heterogeneous and sometimes fragmented. This study firstly aims to synthesize and structure the information derived from the data on the innovative policies and environmental governance, in order to understand which are the main components. After having identified the indicators which are mostly correlated, a taxonomy of smartness is derived from the data, rather than imposing a pre-determined classification.

Then, rather than producing an overall ranking, the interactions among nearby cities in the various domains have been studied by means of a geographic-spatial analysis. The analysis of the scores obtained in the various domains highlights the different performances of the Italian cities.

In lights of the limitations of previous studies, this work uses consolidated multivariate statistical techniques and a official source of data. Covering all the 116 Italian chief towns, the empirical application contributes factually to the existing literature.

3 METHODS

3.1 *Data merging and normalization*

Data had to be cleaned and uniformed in order to create an integrated database out of original Istat tables (referred to 2013–2014). Variables had been recoded and reclassified when necessary, omitting missing values and outliers. The merged database contains more than one hundred indicators. Given the heterogeneity in ranges and units of the data, normalization have been made by means of the *min-max* method (OECD, 2008)[1].

3.2 *Extraction of components*

The normalized indicators have been processed through the Principal Component Analysis (PCA)[2], which is useful to synthesize into a compact structure the information coming from a large number of indicators (Jollands et al. 2004). It has been thus possible to extract a smaller number of variables that contain much of the information of the original dataset, and to understand the main determinants of the overall variability.

The selection of the principal components to be retained was based on several criteria (eigenvalue, screeplot and percentage of explained variance). As commonly done, the loadings have been rotated for an easier interpretation (using VARIMAX rotation), and prior to the PCA an analysis of correlations among the variables has been conducted (Rabe-Hesketh, 2007).

The extracted components were further aggregated into smartness domains (see Table 4). The components pertaining to the same thematic area (e.g. mobility) were aggregated via weighted average, using the proportion of explained variance as weight.

3.3 *Spatial analysis*

The scores extracted for each smartness domain were linked to cartographic data available from the ISTAT, using administrative boundaries.

After producing ESDA (*Exploratory Spatial Descriptive Analysis*), an analysis of spatial autocorrelation has been performed. A statistic known as Local Moran's I (pertaining to the LISA – *Local indicators of Spatial Autocorrelations*) allows to calculate such a correlation for each area of the study region and to assess whether a statistically significant difference exists, comparing the results with a random distribution (with 999 permutations). Global Moran's I statistics, related to the overall study region, were also computed (O' Sullivan and Unwin, 2003; Anselin, 2005). Since the area objects are non-contiguous, two different measures of distance (spatial weights) have been tested: a nearest-neighbour (NN) relation (considering a fixed number of 5NN for each municipality) and distance-based relations (testing various thresholds[3]). Results do not differ significantly.

4 DATA DESCRIPTION

Data come from the annual enquiry of the Italian National Institute of Statistics (Istat) about the environmental condition of cities. They are collected for all the 116 Chief Towns. Original data contain information on: *water*, *air quality*, *energy*, *waste*, *noise pollution*, *urban mobility*, *urban green*, *eco-management* (urban planning, reporting, governance). The items pertains to three *macro-areas*: a. new technologies adoption and use (*ICT*); b. innovative practices for the environment and society (*eco-social innovation*), c. Sustainable and participative governance (*governance*).

A total number of 33 indicators have been retained (11 for the ICT macro-area, 13 for eco-social innovation, 9 for governance) (Tables 1–3).

1. *Min-Max* normalises indicators to have an identical range [0, 1] by subtracting the minimum value and dividing by the range of the indicator values.
2. The aim of *Principal Component Analysis* is to describe the set of multivariate data as parsimoniously as possible using a set of derived uncorrelated variables, each of which is a linear combination of those in the original data (for formalization see Rabe Hesketh, Rabe-Hesketh, 2007).
3. Defining neighbourhood with a 100 km threshold (reasonable for Italian regions' size) the overall mean number of NN is seven. 45% of municipalities have a number of neighbours comprised between 5 and 9.

Table 1. Description of variables: *ICT* macro-area.

Variable	Italy		Centre		North		South	
	Mean or %	SD	Mean or %	SD	Mean or %	SD	Mean or %	SD
LED street lighting (% on total street lights)	4.83	14.02	1.34	2.32	3.97	6.64	7.33	20.86
Intelligent traffic lights[1] (% on total traffic lights)	32.18	34.25	35.62	37.43	41.95	34.27	19.75	29.03
Electric cars (% on total number of cars in circulation)	0.21	0.18	0.17	0.08	0.34	0.20	0.08	0.04
Online services for residents[2] (Nr. of available typologies 0–10)	3.97	2.60	4.23	2.83	5.02	2.37	2.79	2.25
– *No availability (%)*	*5.17*		*0.00*		*0.00*		*12.77*	
Info-mobility services[3] (nr. of available typologies 0–9)	2.66	2.79	2.73	2.39	4.15	2.99	1.15	1.83
– *No availability (%)*	*39.66*		*27.27*		*21.28*		*63.83*	
Plants for production of alternative energy[4] (nr. of available typologies 0–4)	1.01	0.47	1.09	0.43	1.17	0.43	0.81	0.45
– *No availability (%)*	*9.48*		*4.55*		*0.00*		*21.28*	
Use of alternative energy[5] (nr. of available typologies 0–3)	1.34	1.09	1.45	1.01	2.02	0.99	0.62	0.71
– *No availability (%)*	*25.00*		*9.09*		*6.38*		*51.06*	
Teleheating[6] (m^3 per person)	8.17	25.74	0.18	0.47	20.07	37.58	0.00	0.00
Availability of solar panels on public buildings (% of municipalities with panels)	67.24	47.14	77.27	42.89	89.36	31.17	40.43	49.61
Total power of solar panels installed (kW per 1000 people)	8.23	22.37	8.94	15.26	6.00	10.34	10.73	33.87
Electric cars charging stations availability (% of municipalities with charging stations)	29.31	45.72	27.27	45.58	46.81	50.44	12.77	33.73
Electric cars charging stations (% on total nr. of electric cars)	3.09	16.57	3.60	10.04	1.02	1.71	5.14	25.93

Notes:
1 – Controlled and coordinated or adaptive control traffic lights (automatically controlled in response to traffic conditions or upon request of pedestrians; e.g. see https://en.wikipedia.org/wiki/Traffic_light_control_and_coordination).
2 – Administrative practices that residents can do online: residence or address change, status of the submitted requests, self-certifications, certificates (birth, death, marriage, citizenship etc.), payment of school fees and childcare, payment of school-meals, payment of fines, booking appointments with municipal representatives, other online applications, other online payments.
3 – Automatic road signs (variable message), SMS traffic alerts, electronic payment of parking, dedicated info-applications for mobile devices, real-time public transport information, electronic ticketing on-board, interactive information on routes, timetables and waiting times, travel planner for optimal routes, dedicated website for travel tickets.
4 – Municipality owned plants for production of alternative energy: photovoltaic, hydroelectric, geothermal, wind.
5 – Efficient use of energy by the municipality from alternative sources of energy: solar thermal panels, biomass/biogas, high efficiency heat pumps.
6 – Remote or District heating.

As a general result, cities located in the South of Italy have lower means (worse performances) in the environmental indicators, with a higher standard deviation. The differences with respect to the Northern cities are most evident in the energy-related indicators (solar-powered equipment or teleheating/district heating) and in differentiated waste disposal. An exception is the diffusion of LED street lighting, which is more enhanced in the South.

Northern cities have higher means in the mobility area: car-sharing and bike-sharing, diffusion of electric cars and charging stations, info-mobility services etc. Also, Northern cities perform better than Central or Southern ones (50% against around 30%) according to the share of recycling.

Table 2. Description of variables: *eco-social innovation* macro-area.

	Italy		Centre		North		South	
Variable	Mean or %	SD	Mean or %	SD	Mean or %	SD	Mean or %	SD
Car sharing: Vehicles per 100 thousand people	1.9	5.0	0.6	1.6	4.2	7.2	0.2	1.0
Car sharing: Share of electric vehicles	5.5	18.1	4.9	21.3	9.2	19.3	2.1	14.6
Bike sharing: Bicycles per 10 thousand people	4.7	7.2	3.9	5.4	8.0	7.1	1.7	6.6
Bike sharing: Stalls per 100 km^2	8.2	16.7	2.8	3.4	17.4	22.9	1.5	4.8
Park & Ride stalls (Per 1.000 cars in circulation)	22.2	28.9	23.9	20.0	32.9	37.9	10.6	14.8
Pedestrian areas (m^2 per 100 people)	31.8	53.8	29.9	22.9	44.3	78.2	20.1	23.8
Limited traffic zones (ZTL) (km^2 per 100 km^2 land surface)	0.7	1.6	0.5	0.9	1.1	2.1	0.4	0.9
Recycling (selective collection) (% on total waste collected)	40.1	19.1	37.6	13.0	50.1	12.6	31.6	22.2
Recycling incentives[1] (nr. of available typologies 0–8)	5.6	1.3	6.0	1.0	5.9	1.2	5.0	1.4
– *Only one typology available (%)*	*0.9*		*0.0*		*0.0*		*2.1*	
Incentives for home composting[2] (nr. of available typologies 0–4)	1.7	1,4	2.1	1.6	1.9	1.2	1.3	1.4
– *No typology available (%)*	*36.2*		*31.8*		*23.4*		*51.1*	
Density of green areas (% on total surface)	16.4	14.9	17.5	13.3	17.6	14.0	14.8	16.5
Incentive to urban green areas[3] (nr. of available typologies 0–3)	1.2	0.9	1.5	0.9	1.7	0.7	0.6	0.7
– *No typology available (%)*	*28.4*		*13.6*		*6.4*		*57.4*	
Incentives towards lighting efficiency(% of municipalities that incentive)	79.3	40.6	72.7	45.5	91.5	28.3	70.2	46.3

Notes:

1 – Initiatives aimed at favouring recycling: presence of ecological islands; activation of pick-up mobile stations; planned collection of abandoned waste; information campaigns; collection "door to door"; collection of bulky waste on call; collection of green waste on call (pruning waste etc.), Other initiatives

2 – Initiatives aimed at favouring home composting of food waste: reduction of waste management fee; free composter; free courses; other incentives.

3 – Initiatives aimed at favouring the development of urban green: urban forestry, urban gardens, presence of botanical gardens.

4 – Activated policies for increasing the street light efficiency (timing etc.) or reducing the light pollution.

Table 3. Description of variables: *governance* macro-area.

Variable	Italy		Centre		North		South	
	Mean or %	SD	Mean or %	SD	Mean or %	SD	Mean or %	SD
Electric or hybrid vehicles owned by the municipality (% on total vehicles owned by municipality)	14.64	18.14	14.81	12.76	21.19	16.58	7.55	19.53
Typologies of recycled waste in buildings owned by the municip.[1] (nr. of typologies 0–7)	4.49	1.74	4.64	1.40	5.09	1.50	3.83	1.89
– No typology available (%)	*1.72*		*0.00*		*0.00*		*4.26*	
Recycled paper (% on total paper purchased by the municipality)	58.80	43.33	61.70	39.98	75.96	37.26	39.03	43.57
Eco-purchasing[2] (nr. of available typologies 0–6)	2.58	2.10	3.27	1.96	2.85	1.90	1.98	2.24
– No typology available (%)	*28.45*		*9.09*		*19.15*		*46.81*	
Use of biological or organic food on municipal canteen (% of municipalities)	66.38	47.45	77.27	42.89	85.11	35.99	42.55	49.98
Presence of noise-regulated areas (% of municipalities)	73.3	44.4	95.5	21.3	89.4	31.1	46.8	50.4
ISO and EMAS certifications (nr. of typologies 0–2)	0.46	0.65	0.64	0.79	0.64	0.67	0.19	0.45
– No certification	*62.93*		*54.55*		*46.81*		*82.98*	
– Only one certification	*28.45*		*27.27*		*42.55*		*14.89*	
– Both certifications (% of municipalities)	*8.62*		*18.18*		*10.64*		*2.13*	
Planning & governance tools[3] (nr. of available typologies 0–4)	2.84	1.34	2.95	0.89	3.42	1.21	2.21	1.39
– No typology available (%)	*4.31*		*0*		*0*		*10.63*	
Participative governance and reporting[4] (nr. of typologies 0–3)	0.72	0.86	0.50	0.60	0.85	0.98	0.68	0.84
– No typology available (%)	*51.72*		*54.55*		*48.94*		*53.19*	

Notes:
1 – Types of separated waste in municipal structures: paper, plastic, toner, glass, metals (including aluminium), batteries, WEEE.
2 – Types of purchases made according to at least one Environmental Policy Minimum criteria (CAM), in the following categories: electrical equipment, office furniture, stationery, cleaning products, energy and air-conditioning, construction materials.
3 – Policies adopted for efficient planning and governance: approval of the General Urban Tool (SUG) over the past 10 years, approval or preparation of the SEAP (action plan for sustainable energy), presence of the Unified Mobility Plan (PUM), urban green census with geo-referenced data.
4 – Presence of social report (social balance sheet), environmental report or participatory planning.

5 RESULTS

5.1 *The domains of smartness*

Twelve components have been extracted from the indicators by means of the PCA, accounting for 70% of the total variability (Table 4). The first components are related to mobility (diffusion of bike-sharing), the availability and use of new technology and IT, the production and use of green energy (solar panels etc.), and waste disposal.

The scores of the principal components can be aggregated in broader categories when pertaining to the same thematic area (e.g., bike-sharing, car-sharing, pedestrian zones, Park & Ride spaces, intelligent traffic lights within *Smart Mobility*; urban planning, citizen's participation, eco-purchasing within *Governance*, etc.). By doing so, six domains of smartness are obtained: *Diffusion of IT*, *Green energy*, *Waste management*, *Smart mobility*, *Governance* and *Environmental policies*. Those domains pair with the smartness domains already existent on the scientific literature, and they are easy to represent and analyse. The domains of smartness can also be re-categorized within the three macro-areas (compare Table 4 with Tables 1–3).

The taxonomy of Table 4 shows how it is possible to derive the domains of smartness from a data-based approach, avoiding a priori classifications. The structure is flexible and decomposable since each classification can be used depending on the scope of the analysis (see OECD, 2008).

The obtained scores can be graphed to visualize the positioning of cities according to the various dimensions of smartness (Figure 1). Northern cities perform better (Milano, Bologna, Genova, Trento, Bergamo, Venezia, Verona etc.), while southern cities tend to concentrate in the low-left quadrant. Some exceptional cases are worth noting, such as the high score of Milan in the mobility component or the low score of Venezia in the waste management domain. The environmental policy domain highlights some unpredictable results, such as the high score of Messina and the low score of Bolzano.

5.2 *Proximity and spatial interactions*

Since the smartness domains are inevitably influenced by regional governance, local policies and imitation effect, there could be an interaction among nearby cities, which could influence the scores in their smartness domains (according to Tobler's 1970 first law of geography).

The following maps highlights the clusters of cities characterized by a statistically significant spatial correlation among high or low scores, for each of the smartness domains.

The global spatial autocorrelation (*Global Moran's I*) is generally low, especially for the environmental policy domain (I = 0.06). However, when computing the same correlation at

Table 4. From components of the PCA to domains and macro-areas.

Nr. of component	Proportion of var.	Extracted component	Domain	Macro-area
Comp 2	8.14%	Diffusion of IT	Diffusion of IT	ICT
Comp 3	7.77%	Green Energy	Green Energy	
Comp 4	5.97%	Waste management	Waste management	Eco-social innovation
Comp 1	8.80%	Bike-sharing	Smart Mobility	
Comp 5	5.66%	Car-sharing		
Comp 8	5.10%	Park & Ride and pedestrian zones		
Comp12	4.12%	Intelligent Traffic lights		
Comp 6	5.44%	Eco-purchasing	Governance	Governance
Comp 7	5.43%	Urban Planning		
Comp 11	4.26%	Reporting and participative gov.		
Comp 9	4.74%	Noise policies	Environmental policies	
Comp 10	4.60%	Green urban areas		
Total	70.03%			

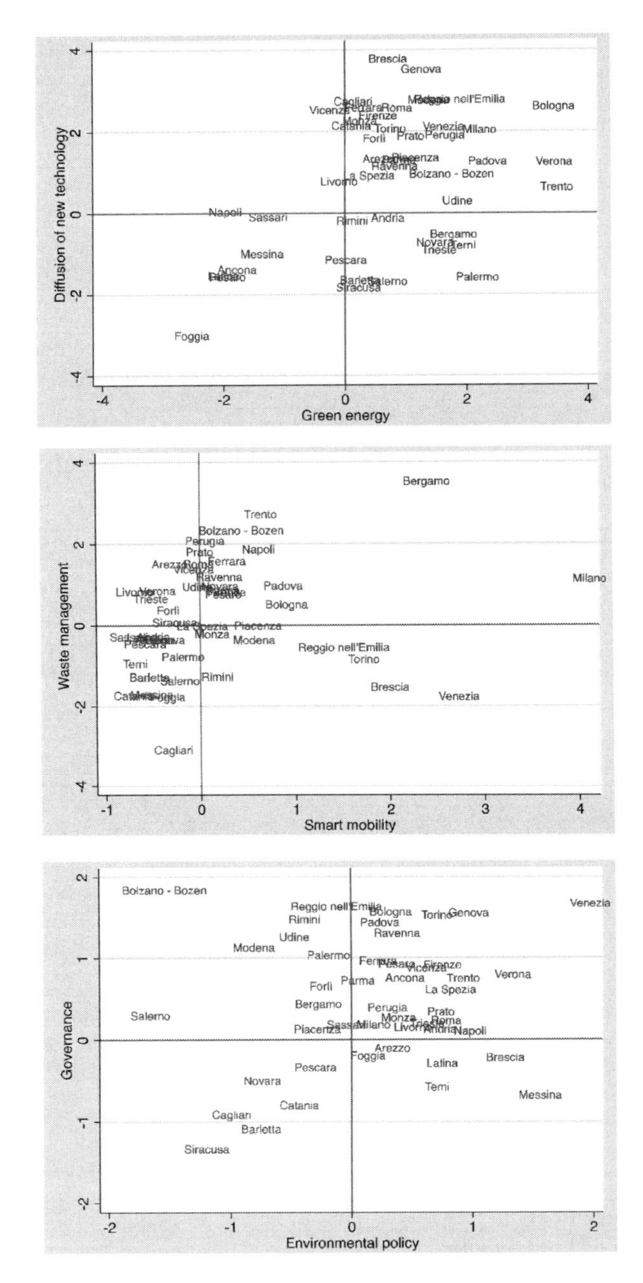

Figure 1. Scores for the smartness domains.
Note: Only the 50 biggest cities (population higher than 100 thousand inhabitants) are depicted in each graph.

a local scale (*Local Moran's I*) a significant clustering arises. The new technology and green energy domains show similar patterns. A large clustering of cities with high scores is evident in the Northern area of Italy (Bologna, Ferrara, Modena, Brescia, Verona, Milano etc.). On the contrary, the low-low clustering pertains to cities in Campania, Molise and Basilicata with the addition of Foggia in Puglia. Roma and Napoli are two outliers for the new technology domain: they have high scores but they are surrounded by municipalities with low values.

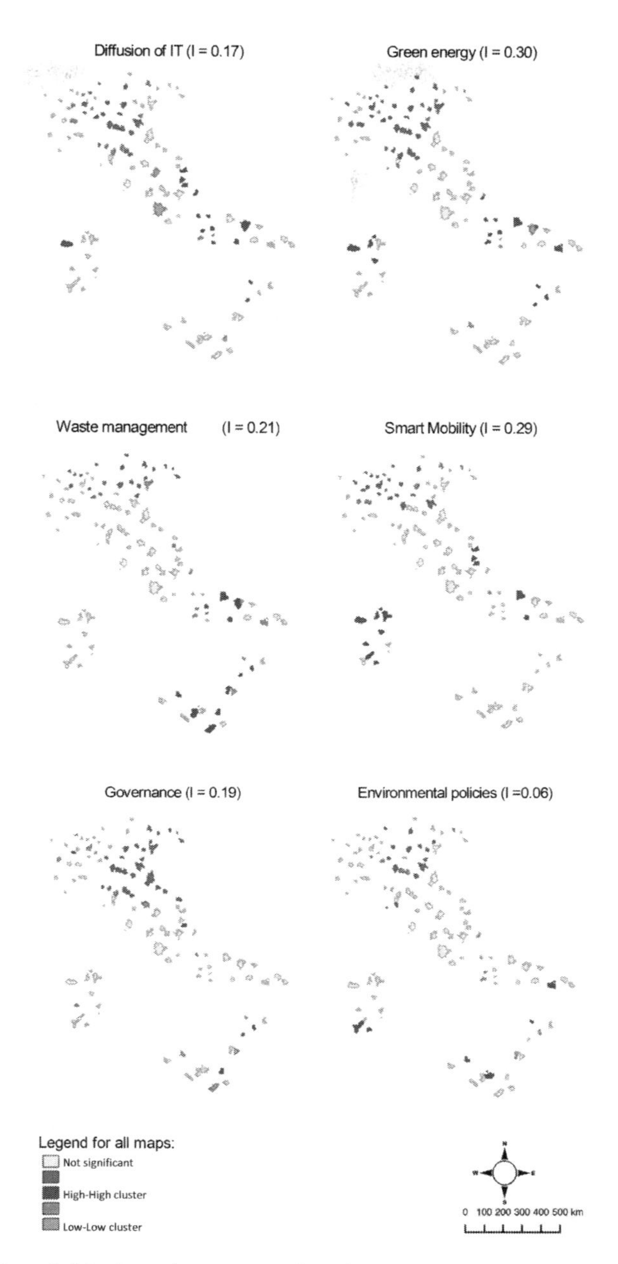

Figure 2. Clustering of cities in each smartness domain.
Note: LISA cluster map, signif. 5% (999 permutations). E.g. an high-high cluster indicates a city with high scores surrounded by cities with high scores. Global Moran's I (spatial autocorrelation) are also reported. Scale (approx.): 1:18.000.000. City names are reported in Appendix, Figure A.1.

No significant clustering is evident for southern regions such as Sicily and Calabria. Results are partially different for waste management and mobility, where a stronger polarization arise. In that case the low-low cluster concerns the Islands, Calabria and Puglia. The clustering of cities with high scores is instead restricted to the North-Eastern cities for waste management (Trento, Bolzano, Belluno, Udine, Treviso, Vicenza etc.; Venezia is an outlier). The area surrounding Milano have exceptional high scores for the mobility domain. Results for governance and environmental policy show similar patterns but are less significant from a statistical point of view.

6 CONCLUSIONS

The definition of a smart city, its determinants and measurement and – even more – its spatial implications, still miss a solid and shared framework of analysis. The methodologies adopted are also diversified and sometimes non rigorous.

This contribution attempts to classify the smartness moving from a comprehensive and official set of data, covering the whole Italy. The full national coverage improves Giffinger's (2007) framework, which is referred to a set of "mediu... cities", non univocally defined, non-comprehensive nor comparable.

The methodologies adopted are easy to apply and replicate in other contexts. Some limitations are innerly present in the PCA, however (such as being sensitive to outliers and updates; spurious variability).

The taxonomy of the smartness domains is derived directly from the data, with statistical criteria. Even with its limitations, this approach avoid arbitrary pre-determined classifications and ad-hoc selection of indicators.

The results clearly show that cities located in the Northern regions have higher scores in the smartness domains than Central or Southern ones. This result follows a well-known Italian evidence, confirmed by many other macro-economic indicators.

But this study demonstrates that, within the north-south polarization, there are localized effect related to groups of cities influencing one another. The fact of being part of the same administrative region, to which are transferred many environmental policies and governance practices, could be a possible explanation of the observed clustering.

Deliberately we do not super-impose the concept of smartness with the one of well being, which is widely treated elsewhere (e.g. see equitable and sustainable well-being, BES, Istat, 2016). A promising extension of the current study, however, is to relate the scores obtained with the PCA to one or more indicators of well-being, in order to understand if a statistical relation exists. In fact, in our vision a smart management of the urban environment should be the starting point for effectively enhancing people's well-being and quality of life. The underlying theoretical assumption is that technological advancement, environmental and societal innovation and intelligent governance should pursue the overall goal of promoting the well-being of the community who decides to adopt these tools.

REFERENCES

Anslein L (2005) Exploring Spatial Data with GeoDa, Center for Spatially Integrated Social Science. Available at: http://www.csiss.org.

Anthopoulos LG (2015) Understanding the Smart City Domain: A Literature Review in Rodríguez-Bolívar MP (ed.), Transforming City Governments for Successful Smart Cities, Public Administration and Information Technology 8. Springer Switzerland.

Arribas-Bel D (2014) Accidental, open and everywhere: Emerging data sources for the understanding of cities. Applied Geography, 49, 45–53.

Borruso G and Murgante B (2015) Smart Cities in a Smart World, in Rassia e Pardalos (Eds.), Future City Architecture for Optimal Living.

Caragliu A, Del Bo C and Nijkamp P (2011) Smart cities in Europe. Journal of Urban Technology, 18(2), 65–82.

Cocchia A (2014) Smart and Digital City: A Systematic Literature Review in R.P. Dameri and C. Rosenthal-Sabroux (eds.), Smart City. How to create Public and Economic Value with High Technology in Urban Space, Progress in IS. Springer Switzerland.

De Santis R, Fasano A, Mignolli N and Villa A (2010) Dealing with smartness at local level: experiments and lessons learned. Working Paper n.8, Fondazione Brodolini, Roma.

De Santis R, Fasano A, Mignolli N and Villa A, (2014) Il fenomeno Smart City, Rivista Italiana di Economia, Demografia e Statistica, LXVIII, 1, 143–151.

Desouza KC and Flanery T H (2013) Designing, planning, and managing resilient cities: A conceptual framework. Cities, 35, 88–89.

European Commission (2012) 'Smart Cities and Communities – European Innovation Partnership'. Report available at: http://ec.europa.eu/energy/technology/initiatives/ doc/2012_4701_smart_cities_en.pdf.

Giffinger R, Fertner H, Kramar Meijers E and Pichler-Milanovic N (2007) Smart cities: Ranking of European medium-sized cities. Vienna: Centre of Regional Science. Report available at: http://www.smart-cities.eu/download/smart_cities_final_report.pdf.
Gigliarano C, Balducci F, Ciommi M, and Chelli F (2014) Going regional: An index of sustainable economic welfare for Italy. Computers, Environment and Urban Systems.
Harrison C, Eckman B, Hamilton R, Hartswick P, Kalagnanam J, Paraszczak J and Williams P, Foundations for Smarter Cities, IBM Journal of Research and Development 54: 4 (2010) 1–16.
ISTAT (2015), Rapporto BES e Rapporto Ur-BES. Report available at: http://www.istat.it/it/misure-del-benessere/le-diffusioni/rapporti-bes.
Jolland N, Lermit J, Patterson M (2004), Aggregate eco-efficiency indices for New Zealand – A Principal Components Analysis, In: 2004 NZARES Conference, New Zealand Centre for Ecological Economics, Massey University and Landcare Research, Blenheim Country Hotel, Blenheim, New Zealand, June 25–26.
Kim SA, Shin D, Choe Y, Seibert T and Walz SP (2012) Integrated energy monitoring and visualization system for smart green city development designing a spatial information integrated energy monitoring model in the context of massive data management on a web based platform. Automation in Construction, 22, 55–59.
Lazaroiu GC and Roscia M (2012) Definition Methodology for the Smart Cities Model, Energy 47: 1 (2012), 326–332.
Lee J and Lee H (2014) Developing and validating a citizen-centric typology for smart city services. Government Information Quarterly, 31, 93–105.
Lombardi P (2013), Modelling the smart city performance. The European Journal of Social Science Research, Politecnico di Torino, Department of Housing and Cities.
Neirotti P, De Marco A, Cagliano AC and Mangano G (2014). Current trends in smart city initiatives: Some stylised facts. Cities, 38, 25–36.
Northstream (2010) White paper on revenue opportunities, from http://northstream.se/ white-paper/archive.
OECD (2008), Handbook On Constructing Composite Indicators: Methodology And User Guide. Report, available at: http://composite-indicators.jrc.ec.europa.eu/.
Ojo A, Dzhusupova Z and Curry E (2016). Exploring the Nature of the Smart Cities Research Landscape J.R. in Gil-Garcia et al. (eds), Smarter as the New Urban Agenda. A Comprehensive View of the 21st Century City, Public Administration and Information Technology, Springer Switzerland.
O'Sullivan D. e Unwin D. (2003), Geographic Information Analysis, John Wiley and Sons, New Jersey.
Piro G, Cianci I, Grieco LA, Boggia G and Camarda P (2014) Information centric services in smart cities. The Journal of Systems and Software, 88, 169–188.
Rabe-Hesketh S. and Everitt B. (2007), A Handbook of statistical analysis using STATA, 4th Ed., Chapman & Hall/CRC.
Repko e DeBroux (2012), Smart city Literature Review and Analysis, Emerging Trends in Information Technology, IMT 598 Spring 2012.
Stiglitz J, Sen A and Fitoussi JP (2009). Report of the Commission on the Measurement of Economic Performance and Social Progress. Paris. Report available at: http://www.stiglitz-sen-fitoussi.fr/en/index.htm.
Taylor Z (2011), 'Lies, Damned Lies, and Statistics' A Critical Examination of City Ranking Studies, report for the Intergovernmental Committee for Economic and Labour Force Development, Toronto, Canda.
Van Bastelaer B (1998). Digital cities and transferability of results. In: Proceedings of the 4th EDC Conference on Digital Cities.
Yamagata Y and Seya H (2013) Simulating a future smart city: An integrated land use-energy model. Applied Energy, 112, 1466–1474.

APPENDIX—SUPPLEMENTARY MATERIAL

Table A.1. Principal components of the PCA (loadings for each component C1–C12).

	C1	C2	C3	C4	C5	C6	C7	C8	C9	C10	C11	C12
LED street lighting									–0.42			
Intelligent traffic lights												0.76
Electric cars												
Online services for residents		0.49										
Info-mobility services												
Plants for production of alternative energy			0.48									
Use of alternative energy			0.41									
Teleheating												
Availability of solar panels on public buildings			0.59									
Availability of charging stations for e-cars		0.43										
Car sharing: Nr. of Vehicles												
Car sharing: Share of electric vehicles					0.65							
Bike sharing: Nr. of Bicycles	0.53											
Bike sharing: Stalls	0.54											
Park & Ride stalls								0.66				
Pedestrian areas								0.59				
Limited traffic zones												
Recycling (selective collection)												
Recycling incentives												
Incentives for home composting												
Density of green areas										0.56		
Incentive to urban green areas												
Presence of noise-regulated areas									0.73			
Electric or hybrid vehicles owned by the municipality										–0.47		
Typologies of recycled waste in buildings owned by the municipality				0.66								
Incentives towards lighting efficiency												
Recycled paper												
Eco-purchasing						0.64						
Use of biological or organic food on municipal canteen						0.50						
ISO and EMAS certifications												
Planning & governance tools							0.70					
Participative governance and reporting											0.75	

Note: Rotated components (VARIMAX rotation). Only loadings >0.4 are reported.

Figure A.1. Italian chief towns: names and population size.
Note: author computation on ISTAT data, scale 1:8,000,000.

Town and Infrastructure Planning for Safety and Urban Quality – Pezzagno & Tira (Eds)
 ISBN 978-0-8153-8731-2

Strategies and measures for sustainable mobility in Italian metropolitan cities

R. Battarra
Institute of Studies on Mediterranean Societies, Italy

M. Tremiterra & F. Zucaro
Università degli Studi di Napoli Federico II, Naples, Italy

ABSTRACT: The paper aims at evaluating whether and how the Smart Mobility concept is contributing to improve the efficiency, the sustainability and the liveability of 11 Italian Metropolitan Cities, through a set of parameters selected from some databases (e.g. ISTAT) and the collection of the major smart initiatives related to the mobility sector. The expected results can be summarized as follows: (i) an analysis of the adoption of the Smart Mobility concept within Italian cities; (ii) an evaluation of the status of Smart Mobility in the Italian cities by the comparison of parameters and initiatives. The paper is divided into four parts: the first provides a review of the scientific literature about the relation between Smart Mobility and sustainable mobility; the second explains the research methods; the third describes the main results of the analysis; finally, the last section provides some causes of reflections about the efforts that the Italian cities need to carry out in order to improve the "smartness" and the sustainability of urban mobility.

1 INTRODUCTION

Because of the numerous and rapid economic, social and territorial dynamics that affect cities in the last years, the "urban" dimension seems not to be longer adequate to cope with the complex challenges that cites have to face (i.e. climate change, energy saving). Some authors (Salet et al., 2003; Hamilton et al., 2004; Mazzeo, 2017) suggest referring to the metropolitan size that can be defined as multi-centered territorial system «which develops mainly along functional networks, cutting across institutionally defined territorial boundaries» (Kubler & Heinelt, 2002). According to some EU documents (EC, 2007; COM(2008) 616 final; BBSR, 2011), Metropolitan cities can really encourage Europe's economic development, as they provide with operational and strategic capabilities, in addition to catch particular attention at the international level. Furthermore, Metropolitan cities constitute important transport hubs both at local and transnational level, supplying (i) urban mobility within the cities; (ii) regional transport between cities and their regions; (iii) national and European long-distance transport (EMTA, 2000). These services require innovative measures to tackle the challenges and changes set off by metropolisation process and that do not relate only to transport planning, but also to the spatial one, as land use and transport development dynamics affect reciprocally. Within the wide range of initiatives aimed at improving transport system efficiency, reducing environmental impact and changing users' mobility behaviour, ICTs have been spreading easily by enhancing the flexibility of transport systems, providing for istance apps for real-time information on parking availability, traffic conditions and bike sharing.

Thanks to the huge benefits (and maybe also profits) provided by ICTs' use, Smart Mobility has more and more auspicated as the main option to seek more sustainable transport systems (Staricco, 2013; Benevolo et al., 2016). Even if there are several meanings and interpretations of Smart Mobility concept, such as for Smart City one (Papa et al., 2015), it can be defined as a network system mainly characterized by: connections, digital and physical ones, in order to satisfy people's fulfilment; use of appropriate technologies, to enhance performance and attractiveness of mobility system; sustainability, to reduce the need of travel and so energy consumption and carbon emissions, according to the previous studies on this issue (Lam & Head, 2012; Papa et al., 2016).

Nevertheless it is not only the massive use of innovative technologies that through a technocentric approach (Papa & Lauwers, 2015) makes the mobility system more appropriate to dela with the challenges of inclusiveness, accessibility and sustainability. In other words, if technology allows to improve mobility's efficiency and reduce its impact on the environment, it is also true that only a harmonious and integrated combination of multiple components such as Accessibility, Sustainability and ICT, make a mobility system suitable to support the development of urban activities taking into account the needs of its users. Therefore, it can be assumed that these three components, Accessibility, Sustainability and ICT, resume the main characteristics that Smart Mobility should have. Overcoming the plethora of labels and efforts of defining and explaining Smart Mobility, we can assume that it cannot be separated from the sustainable mobility approach, in order to guarantee that the best use of technology that we should make, can support a sustainable future of cities from the economic, social and environmental point of view. From the brief scientific framework provided, it is evident that mobility is a key sector on which cities have been investing resources and experiencing innovations and that it has a crucial role within Metropolitan cities' development, especially to increase their competitiveness level.

In Italy Metropolitan cities were formally instituted by Law 56/2014 "Provisions on Metropolitan Cities, on Provinces, on unions and mergers of Municipalities" which commits them three main objectives: to support the strategic development of the metropolitan territory; to promote and manage in an integrated way the services, infrastructures and networks of communication of metropolitan competence; to handle institutional relations with European cities and metropolitan areas. Italian Metropolitan cities collect 36% of national population, 40% of the value added and 35% of businesses. These few data are enough to figure out the economic and social perspective importance "of the new long-awaited administrative subjects from the best part of the country" (Papa et al., 2016) to enhance accessibility to urban services and increase sustainability, moreover as concern the transport system.

2 OBJECTIVE OF THE PAPER

In this context, through the analysis of the Smart Mobility initiatives and the definition of synthetic indicators of urban mobility, the paper aims at evaluating whether and how the Smart Mobility concept is contributing to improve the efficiency, the sustainability and the liveability of 11 Italian Metropolitan Cities.

3 METHODOLOGICAL APPROACH

The analysis about urban mobility measures has concerned 11 Italian Metropolitan cities—Turin, Milan, Genoa, Bologna, Florence, Rome, Bari, Naples, Reggio Calabria, Palermo and Catania (Fig. 1), according to an empiric study developed by Department of Civil, Architectural and Environmental Engineering (University of Naples Federico II) between 2014 and 2016.

The procedure for such analysis is divided into the following phases:

1. Selection of Smart Mobility initiatives for each Metropolitan capital city;
2. Classification of selected initiatives based on types of actions;
3. Identification of categories for each type of actions for analyzing initiatives;
4. Definition of a set of parameters for each category for auditing the status of Smart Mobility in the Italian cities;

Figure 1. Map of the Italian Metropolitan cities.

Table 1. Types of smart mobility measures and corresponding categories.

Types of smart mobility measures	Categories
A. Realization of new mobility infrastructures	Accessibility
B. Improvement of public transportation	
D. Strengthening of the car park system	
C. Promotion of soft mobility	Sustainability
F. Promotion of sharing mobility	
G. Promotion of e-mobility	
E. Logistics innovations	ICT
H. Implementation of info-mobility services	
J. Mobility platforms	
K. Mobile apps and other technology products	

5. Comparison among capitals of the Italian Metropolitan cities considering initiatives and parameters.

In the first phase, a screening of the most significant Smart Mobility ongoing initiatives in each city was made. The criteria[1] used for their choice was their level of technological innovation and "smartness".

In the second phase, such initiatives have been selected by the examination of several indirect sources (i.e. instruments for urban and territorial government, web sites, publications). Then, such initiatives have been classified into ten types of measures, which are defined in Table 1.

In order to have a synthetic overview of Smart Mobility, in the third phase the related measures have been collected in three categories: Sustainability, Accessibility and ICT. In

1. To clarify this criteria, it can refer to Papa et al., 2016.

particular, for each category a definition has been conceived for assigning each measure to a category (Table 1):

- Accessibility: measures that aim at enhancing the ability of places to be reached and guarantee safe and affordable transportations to urban community;
- Sustainability: measures that preserve the natural environment and promote the use of renewable energy resources and advocate for conservation of not-renewable ones;
- ICT: measures that can be entitled as ITS and have both the capability to improve the efficiency of the urban system and to impact on users' behavior.

In the fourth step, a set of parameters has been defined for each above-mentioned category. In particular, 28 parameters have been chosen and they have been articulated into the three categories (Table 2). Their selection has been aimed at measuring the status of urban mobility in the Metropolitan cities. The related data have been collected by consulting ISTAT database, referring to 2014. This year, according to the availability data, allowed to consider the implementation period of the initiatives selected. In order to evaluate the urban mobility condition in the Italian cities, for each parameters has been identified a benchmarking, that is the average value, calculated as follows:

$$\overline{x_i} = \frac{\sum_{j=1}^{n} x_{ij}}{n}$$

Table 2. Parameters selected for the three categories.

Category	ID	Parameter	Unit
Accessibility	A1	Public tranpsort demand	No. passengers/inh.
	A2	Public tranpsort supply	No. seats*km/inh.
	A3	Public transport lanes	km/100 km^2
	A4	Bus stops density	No. stops/km^2
	A5	Rail network	km/km^2
	A6	Rail network stops	No. stops/km^2
	A7	Toll parking	No. stalls/1000 vehicles
Sustainability	S1	Ecological buses (electric, natural-gas, GPL)	No.
	S2	Pedestrian zones	m^2/100 inh.
	S3	Restricted traffic zones	km^2/100 km^2
	S4	Cycle lanes	km/100 km^2
	S5	Ecological cars (electric, natural-gas)	No.
	S6	Car sharing demand	No. users/1000 inh.
	S7	Car sharing supply	No. available vehicles/100.000 inh.
	S8	Bike sharing supply	No. bikes/10.000 inh.
	S9	Bike sharing density	No. stations/100 km^2
ICT	ICT1	Road traffic signal systems	No./km^2
	ICT2	Variable message sign	1 or 0
	ICT3	SMS for traffic alerts	1 or 0
	ICT4	Electronic payment park systems	1 or 0
	ICT5	Applications for mobile devices	1 or 0
	ICT6	SMS for public transport information	1 or 0
	ICT7	Electronic bus stop signs	1 or 0
	ICT8	Electronic travel tickets	1 or 0
	ICT9	Electronic purchase of travel ticket by mobile devices	1 or 0
	ICT10	Information on routes, schedules and waiting times	1 or 0
	ICT11	LPT travel planner	1 or 0
	ICT12	Travel tickets online	1 or 0

Thanks to this value, it has been possible to make a comparison between the different cities also using a suitable graphic representations of the parameters. The analysis has been performed only for the Accessibility and Sustainability categories, as the ICT parameters are defined by dichotomous ones and do not provide more information than the analysis of the initiatives. Furthermore, a synthetic indicator for each categories was defined with a geometric mean value of corresponding parameters standardized[2], using the following formula:

$$I_X = \frac{\sum_{i=1}^{n} Xi}{n}$$

where:

X category

Xi standardized indicator of category

In the last step, through the comparison of results obtained by initiatives and synthetic indicators' analysis, the Smart Mobility status of Italian cities has been identified. Such analysis has highlighted weaknesses of each city, but also gaps among them.

3 RESULT OF THE RESEACH

Although the classification of the initiatives in the three categories identified (Fig. 2) may be partially conditioned by the specific "cut" adopted for the selection of the measures[3], results allow to obtain interesting insights on how the Italian cities are facing the challenge of smart and sustainable mobility.

As far as concerns Accessibility, some cities (Milan, Naples, Turin, Rome) are focusing on strengthening public transport through expensive infrastructural solutions, such as the realization of new urban rail lines with the integration of ICT devices. Other ongoing projects are related to parking lots (not only for interchange), the modernization of the traffic light network or the improvement of local public transport supply by low environmental impact transport modes (i.e. trams in Florence and trolleybus in Milan and Bologna). The 11 Metropolitan capital cities have been developing the equal number of measures related to Sustainability and ICT, although by interpreting these aspects in different ways. In fact, some cities have been realizing more Sustainability initiatives through the promotion of cycling mobility (Milan, Genoa, Turin), the introduction of congestion charger zones (Milan) and the sharing mobility as a suitable alternative to the private car use (i.e. Milan and Turin). In other cases, ICT plays a predominant role probably due to the interests of the manufacturing companies. For instance, Milan has adopted a relevant number of technologies and services both to provide users with information and electronic ticket and to manage traffic and logistic transports. Finally, one last consideration on lack of the territorial balance characterizing the distribution of the initiatives. About 90% of measures are concentrated in the Northern and the Central cities, while in the Southern cities the interventions are sporadic and represent pilot projects rather than broad-scale actions. Moreover, such few initiatives are mainly related to Accessibility rather than Sustainability or ICT and this highlights the still-present infrastructure gap characterizing Italy.

Fig. 3 shows that the Central Italian cities have a balanced distribution of the three Smart Mobility components, while the Northern cities, which have already achieved a higher level of efficiency and sustainability of transport system, are more aimed at ICT spread.

After analyzing the framework of Smart Mobility initiatives, the use of parameters and synthetic indicators (step 4 of the methodology) allowed to audit the Italian Smart Mobil-

2. Standardization was necessary to compare different indicator's parameters, which are expressed in different units. Therefore, it was used the Z-score standardization.

3. For the aim of this research, we have selected those initiatives that contribute to the adoption of a smart approach in the Italian Metropolitan cities, and so the ones that are explicitly relating to the use of ICT.

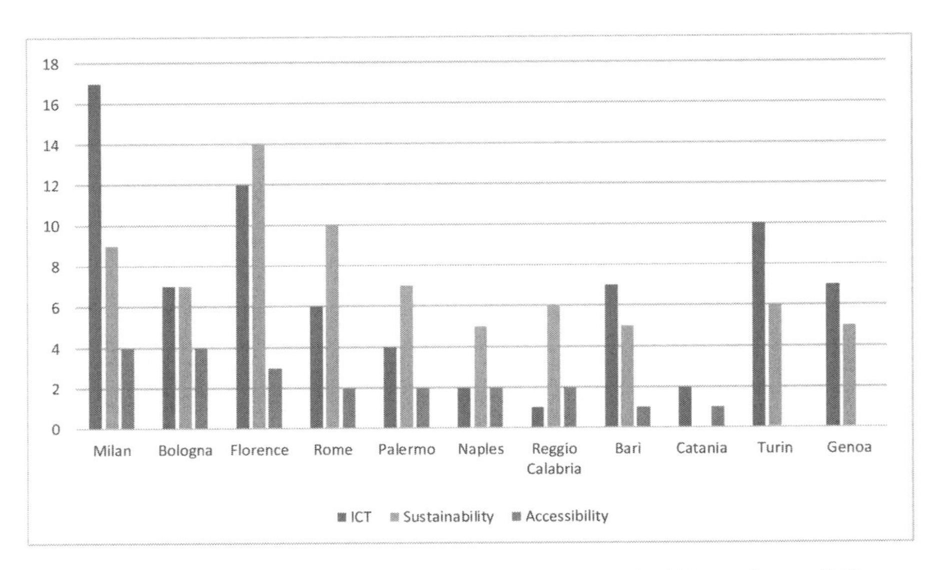

Figure 2. Percentage distribution of initiatives related to ICT, sustainability and accessibility.

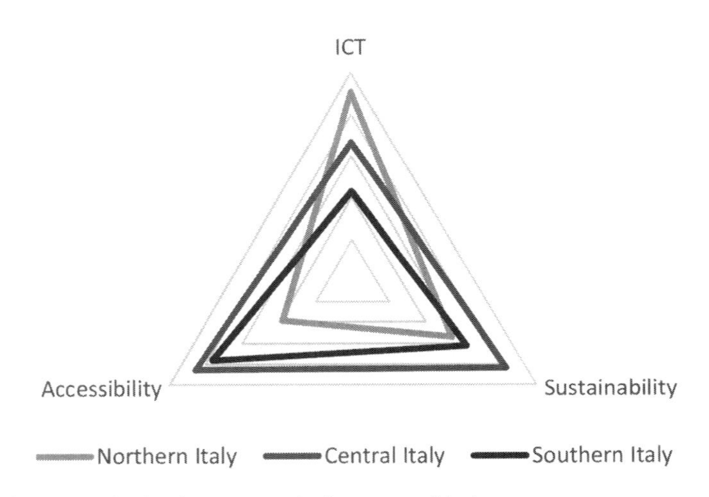

Figure 3. Performances in the three categories by geographical area.

ity status. With regard to Accessibility and considering the single parameters (i.e. local PT supply and demand, availability of parking lots) (Fig. 4) there is a significant gap between the Northern and the Southern cities.

This last ones have values' parameters below the average, with the exception of Naples as regards the rail network and the bus stops density (slightly above the average also for Bari) data. Milan and Turin are characterized by the best performances and, especially with regard to the local PT supply, they have far higher values than the average ones. The Central cities record higher average values except for local PT lines.

Moving to Sustainability (Fig. 5), the results are similar to those found for the Accessibility category, with the exceptions of Milan and Genoa. These two cities, indeed, show weaker performances, and in particular, Milan poorly performs with respect to Ecological buses and cars, while Genoa shows values below the average for all the parameters, with the exception of the two Car sharing ones. Finally, some cities are characterized by the high presence of pedestrian zones and restricted access zones (first of all Florence and Milan due to the traffic restriction for the entire historic center, as well as Naples and Palermo).

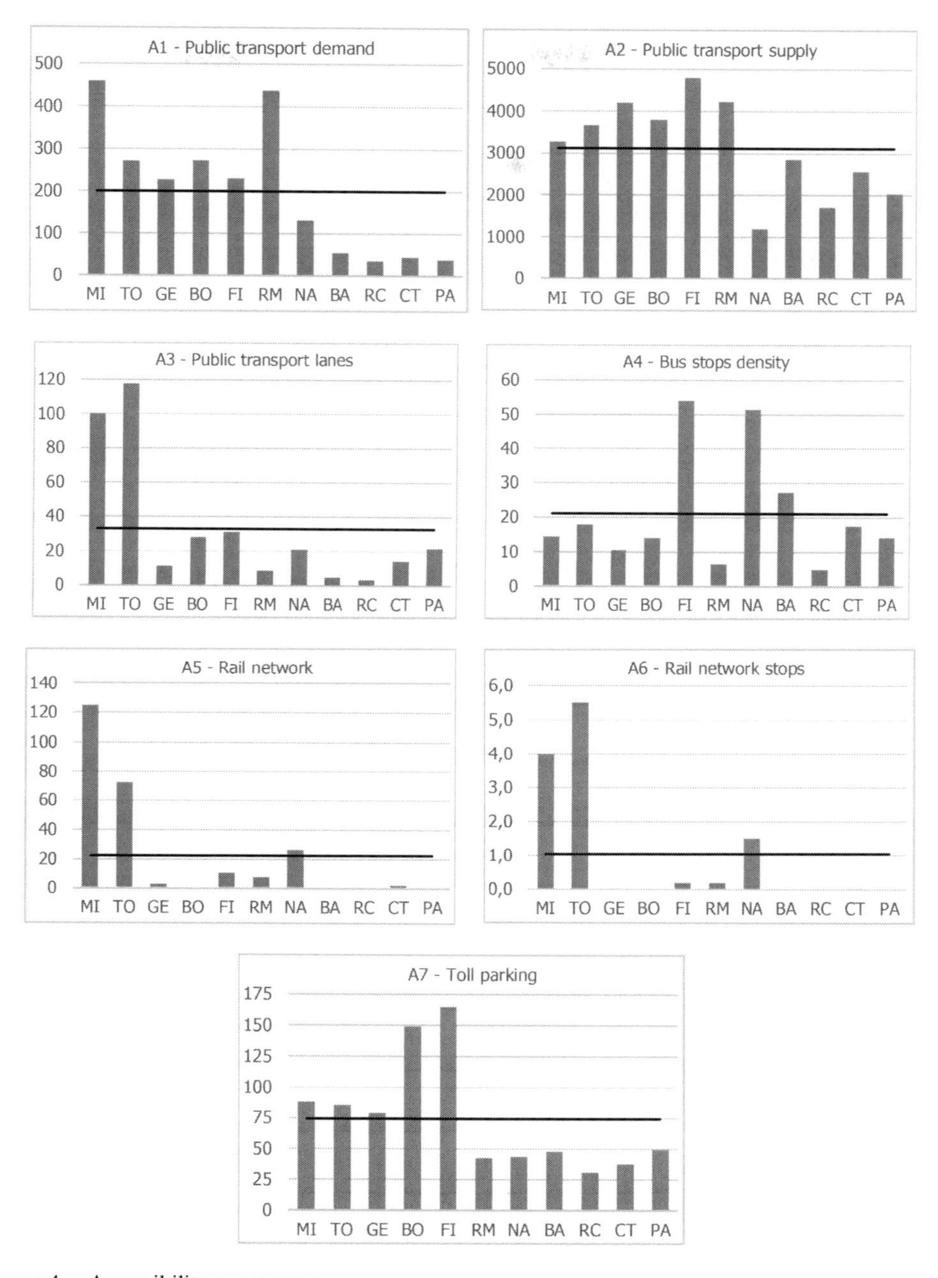

Figure 4. Accessibility parameters.

As regards Accessibility and Sustainability indicators (Fig. 6) Milan, Turin, Florence and Bologna have values above the average, while all other cities deviate from the mean value, obviously with different ranges.

The comparison between the Smart Mobility indicators and the framework of the initiatives described above, allows to assume that the Italian Metropolitan cities are not developing transport policies oriented to improve urban mobility smartness. For instance, although the Southern cities are characterized by a negative performance, the local decision makers seem no to pay particular attention to implement Smart Mobility measures.

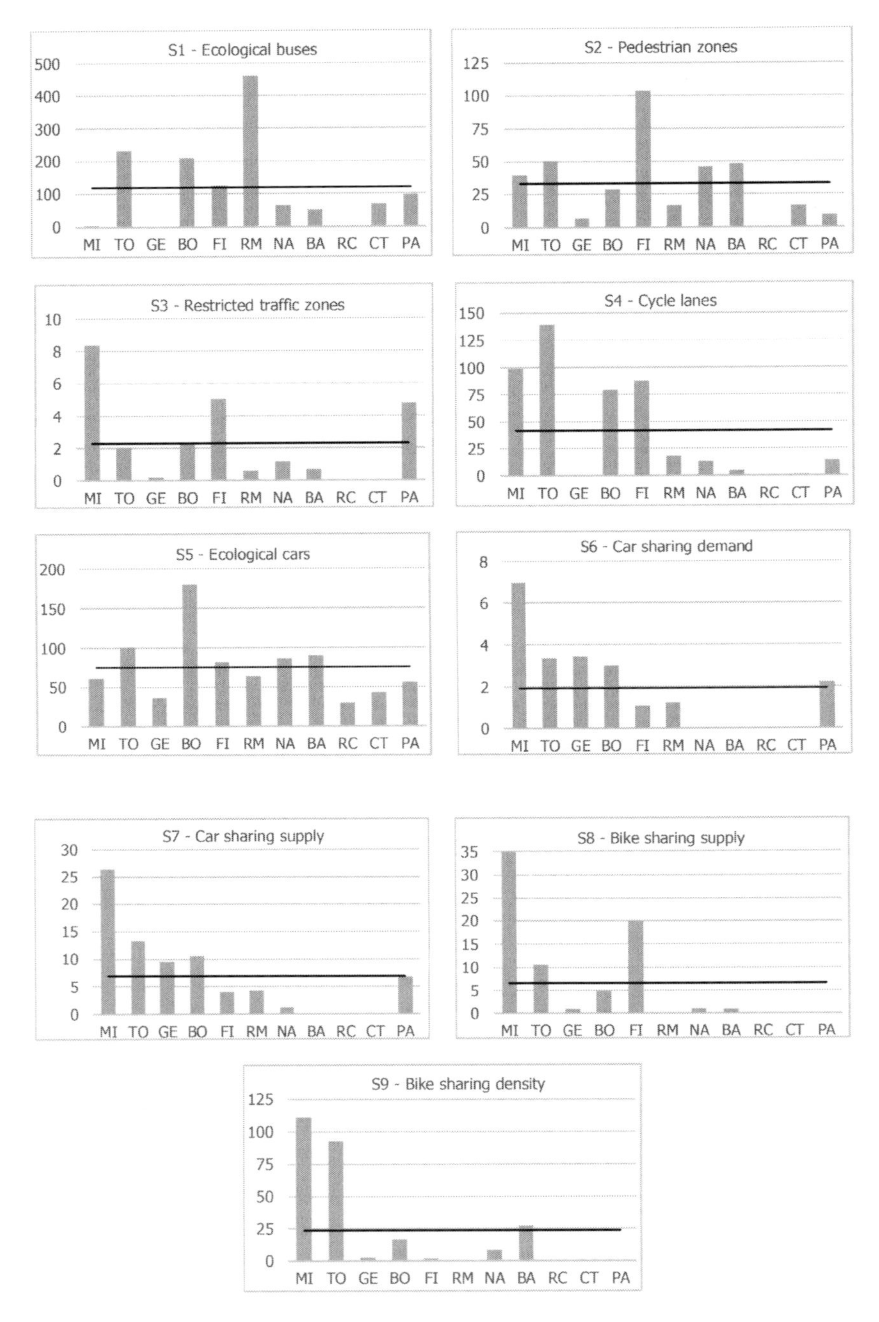

Figure 5. Sustainability parameters.

Cities that are investing more in Sustainability, ICT and Accessibility are those that already have the best performance, confirming the key role of an efficient transport supply in the most advanced urban systems.

Nevertheless, some critical issues emerge even for the "smartest" cities. For instance, Milan seems to invest more in ICT rather than on Sustainability issues, nevertheless its relevant air pollution levels; also Genoa has triggered sporadic initiatives in the Sustainability category, maybe because of the particular morphological conformation of its territory.

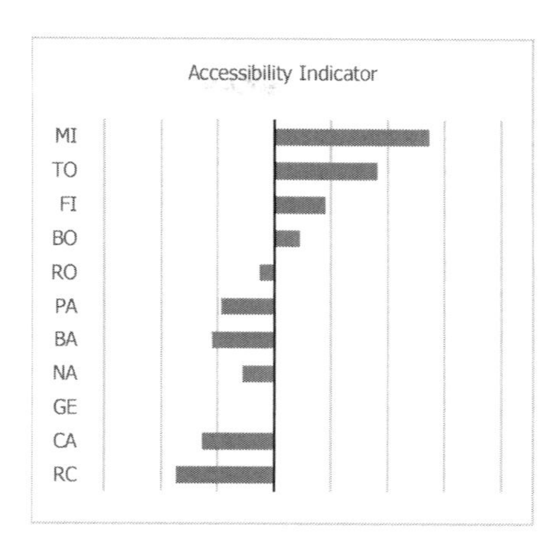

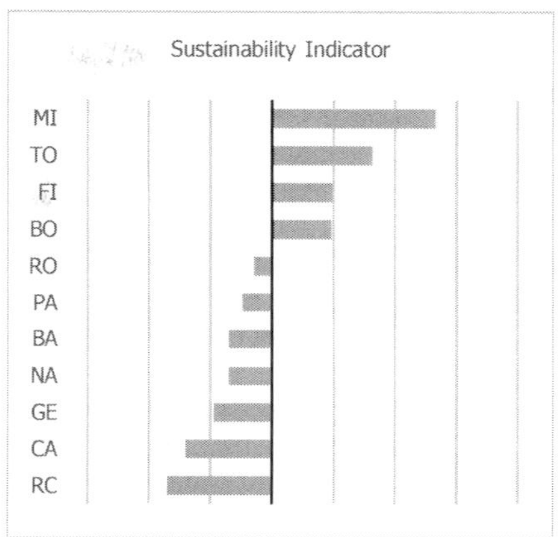

Figure 6. Accessibility and Sustainability indicators.

4 CONCLUSIONS

From the comparison between indicators and initiatives, several critical issues emerged about whether and how Italian Metropolitan cities are declining the Smart Mobility theme. The first, and (perhaps) predictable element is the huge gap between the different geographical areas. In fact, the Northern cities are equipped with efficient and sustainable transport systems and have reached a mature stage of ICT application, although still far from the standards of some European cities (for instance, Amsterdam according to http://www.smart-cities.eu). Despite in some cases initiatives are driven by interests of important business groups (e.g. Milan) and in others by EU funding (e.g. Genoa), the Northern cities record a number of measures and values of indicators above the average of the other cities considered. In the Southern cities indicators record values below the average ones and also the share of initiatives is reduced compared to other cities analysed. Furthermore, such initiatives are sporadic events and are not integrated into a policy framework aimed at increasing sustainable mobility.

For instance, why promoting soft mobility initiatives if the related infrastructure networks are not available and can be characterized by low levels of road safety? Moreover, in the Southern cities many interventions have been developed as European pilot projects, and it is not still clear whether and how they can be "scaled up" to Metropolitan areas. At the same time, the use of ICTs can result useful and effective in cities provided by an efficient transport system, while can be a catchy label if blindly "placed in" a backward context. Then, considering the new institution of Metropolitan cities, the topic of how ICT's use in mobility sector can support cities in dealing with the tasks assigned to them by the Law 56/2014 (e.g. the integrated management of public transport system) has not been sufficiently explored yet.

To conclude, it can be stated that the most successful cities have included ICTs into a transport system developed through an integrated and coordinated urban planning system. In fact, a well-defined vision of Smart City and the related role of mobility allow to implement initiatives and projects that, beyond the evocative slogan, are able to respond to community needs and effectively support human activities.

REFERENCES

Benevolo C., Dameri R.P. D'Auria B., (2016). Action Taxonomy, ICT Intensity and Public Benefits. Springer. In T. Torre et al. (Eds.), Springer, 13–28.

BBSR, (2011). Metropolitan Areas in Europe. BBSR-Online-Publikation, Nr. 01/2011. Federal Institute for Research on Building, Urban Affairs and Spatial Development.

European Commission, (2007). Growing Regions, growing Europe. Fourth report on economic and social cohesion.

Giffinger R. Fertner C., Kramar H., Kalasek R., Pichler-Milanović N., Meijers E., (2007). Smart Cities: Ranking of European Medium Sized Cities. Vienna, AU: Centre of Regional Science, Vienna University of Technology. Available at: http://www.smartcities.eu/download/smart_cities_final_report.pdf.
Hamilton D.K., Miller D.Y., Paytas J. (2004). Exploring the horizontal and vertical dimensions of the governance of metropolitan regions. Urban Affairs Review. 40(2), 147–182.
Kübler D., Heinelt H. (2002). An analytical framework for democratic metropolitan governance. 30th ECPR Joint Sessions of Workshop. Turin.
Lam D., Head P., (2012). Sustainable urban mobility. In: O. Inderwildi, D. King (Eds.), Energy, Transport, [illegible] the Environment. Springer-Verlag, 359–371.
Mazzeo, G. (2017). Planning assignments of the Italian metropolitan cities. Early trends. Tema. Journal of Land Use, Mobility and Environment, 10(1), 57–76.
Papa R., Gargiulo C., Cristiano M., Di Francesco I., Tulisi A., (2015). Less smart more city. Tema. Journal of Land Use, Mobility and Environment, 8(2),157–182.
Papa R., Gargiulo C., and Battarra R., (2016). Città Metropolitane e Smart Governance. Iniziative di successo e nodi critici verso la Smart City, FedOA Press.
Papa E., Lauwers D. (2015). Smart Mobility: Opportunity or Threat to Innovate Places and Cities. In: 20th International Conference on Urban Planning and Regional Development in the Information Society, Proceedings. Available at: http://repository.corp.at/36/.
Salet W., Thomley A., Kreukels A. (2003), Metropolitan Governance and Spatial Planning. Comparative Case Studies of European City-Regions. Spon Press.
Staricco L. (2013). Smart Mobility, opportunità e condizioni. Tema. Journal of Land Use, Mobility and Environment, 3, 289–354.

Town and Infrastructure Planning for Safety and Urban Quality – Pezzagno & Tira (Eds)
© 2018 Taylor & Francis Group, London, ISBN 978-0-8153-8731-2

Transportation system planning and town center: Case study of Bologna

S. Sperati & V. Colazzo
Università degli Studi di Roma La Sapienza, Roma, Italy

ABSTRACT: Transportation systems play a big role as an engine for urban transformation, significantly contributing to the shaping of the current and future city structure.

The importance of an exact public transportation system planning derives from the necessity to provide an efficient and effective service in response to the demand of mobility. However, the historical and cultural tradition conflicted with the transportation planning numerous times.

Such urban configurations often represent concrete issues for the Public Administration in the management of the necessary territory transformations.

The article focuses on such themes and on the city fo Bologna, analyzing its historic evolution through a brief history of its urban and transport plans to this day.

Finally, a specific in depth-analysis about a recent projet on creation of intermodal hubs linked with the city center, with the goal of providing an alternative to private transportation is presented.

1 INTRODUCTION

The infrastructural system has led Bologna to a unique position in the regional and national context.

The suburban explosion did not provide an adequate offer of infrastructure nor services and the historic center of the town lacked of free spaces.

Bologna was stifling: growing population, commuting, large poles of attraction in the historic centre, inadequate public transport and consequent paralysis of urban roads set the traits for intervention.

The PSC, innovative plan in urban planning identified the directions in order to transform the city thanks to a civic participation process.

Our idea, integrating pedestrian mobility project aims to restore the identity of the old town centre through the return of public spaces to the citizens.

The already existing and new rail infrastructure represents the starting point for the city implosion. The same infrastructural system that characterized Bologna in the 1900's now represents an invaluable obstacle for the city.

1.1 *Bologna city: The relationship between history and urbanism*

Since 1825, the railway development has led to a growth of the city, resulting in the depopulation of countryside and the abandonment of traditional urban model.

The influx of migrants and the rapid industrialization caused a crisis in the industrial city.

The new city structure got stuck in the same problems: undefined dimension, expansion of service industries, causing frequent congestions itself.

In the second half of the 1800s, Bologna became a communication key point through the realization of different rail connections.

Towards the end of the 19th century, Bologna embarked on a major transformation with the 1889 PRG, an expansion plan that included the demolition of the historic walls to allow the city to expand.

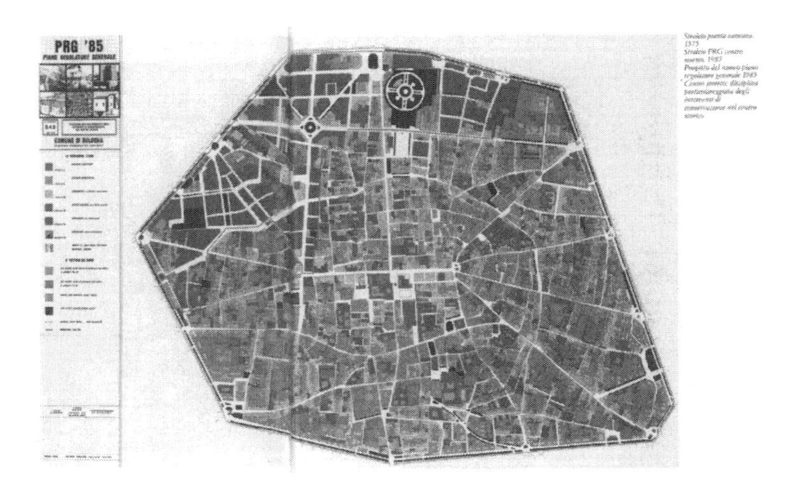

Figure 1. PRG 1985.

The demolition of the city walls started a modernation phase of the city, characterized by the first shifts from the outskirts to the historic center.

These migratory phenomena, in addition to the need to facilitate periphery-center communications, lead to an excessive agglomeration in the city center.

Bologna was subjected to a high demographic pressure and an unsatisfied housing demand.

In 1938 a Particular Plan, never approved for the absence of a General Regulatory Plan, was developed for several areas of the city center.

In 1940 a recovery plan was developed to resume the previous plan interventions.

In 1949 was developed New Regulatory Plan: a reconstruction post-war plan.

In 1958, the new General Governance Plan was adopted, only able to initiate speculative processes in the Old Town, extending the existing ones.

The need to rebuild as a result of war damage was witnessed by the PEEP Plan of 1964 as a response to the enactment of Law 167/1963.

The first addresses were identified in 1969, with the "Cervellati Plan": implemented over long periods due to the lack of economic resources, emerged as a variant of the PRG of 1958 in line with Ministerial Decree 1444/1968 and foresee a gradual recovery of the urban area of the Old Town.

In 1970 was issued a variant in order to regulate the entire municipal territory: the shortcomings in urban development envisaged by the 1958 PRG were introduced.

In 1978 the variant for the recovery of the building's heritage was adopted and it was not limited only to the historical center.

This strategy valorized the center of the city and was restored by the 1985 regulation plan, which aimed at recovering degraded buildings and areas (Figure 1).

The last urban plan goes back to 2008, Municipal Strategic Plan (PSC).

The PSC describes the urban transformation directives and, altogether with the Municipal Operational Plan (POC) and the Urban Buildings Regulation (RUE), schedules individual interventions and establishes the urban transformation boundaries respecting the land and the pre-existing buildings.

The existing urban plan has been a fundamental change in the history of Bologna's urban planning because, defining itself as a dynamic process rather than as a static tool, adapted to the diverse urban needs involving citizens in decision-making.

1.2 *Bologna: Evolving mobility*

Bologna has a firm urban structure: in the Middle Age the porticoes of the historic center were the main city streets.

Nowadays, the buildings arrangement does not fit an adequate road structure: the proximity of buildings results in limited mobility.

The road system was not structured to withstand the traffic of the city-center (200,000 daily inputs).

The city development was stressing the historic center the expansion process of service industries became pressing and the traffic focused on the radial roads.

It was attempted to intervene through a broader vision, where public transport took on greater importance.

The metropolitan intervention strategy was lead by W. Husler.

Husler focused on upgrading existing infrastructures: the 8-axis and the rail network were the key infrastructures to connect the major concentration centers and the centralities.

At the suburban level, the bus service covered the city borders needs, also integrating with the SFM (Metropolitan Rail Service) stops: from the station, buses would connect the outer areas of the city.

At the urban level, the bus network was supposed to distribute middle and long-distance travelers, who had to move around the city.

The car parkings with direct access to the SFM were essential to allow the link between the suburbs and the highway system.

A decisive change in the city urbanism occurred with Bernhard Winkler's 1985 traffic plan, which changed in the concept of the old town, intended as a space to be free from the congested traffic of vehicles.

Bologna portrays the start of a new phase in the urban development of the historic center.

For Winkler, the medieval city was not suitable for cars due to the lack of adequate space to fully exploit vehicle features.

Winkler's idea of a private car-free historic center has been the cornerstone of all the public administrations in recent years in Bologna.

From the second half of the 20th century, the dominance of both passengers and freight transport was represented by this mobility-system, efficient for its speed and flexibility.

The new development of the infrastructural system is represented by the occur of High Speed lines.

Bologna implemented new public transportation reforms, maily by enhancing the pre-existent SFM project.

The need for an efficient public transport service was not only on a local level: the purpose of the new project-plan was to provide fast and frequent connections between the suburbs and the old town.

All levels of planning recognized the need to restructure the rail system.

The activation of a metropolitan public transport service managed the flow of commuters, and provided an alternative to the private vehicle.

The SFM service represents the main mobility tool of the city. Its development started in the 90's, when the Naples-Milan High Speed Line project was being outlined.

This investment was the basis of the SFM project. The completion of the underground lines (High Speed Lines) indeed cleared the surface ones, employing them for the local (SFM) and regional services.

The SFM service consisted of a metropolitan rail transport in the center of Bologna distributed over 8 previously existing railway radial lines (service network of 280 km, radius of 30/35 km).

The integration of the different transport systems involved various points of exchange, 22 total, of which 6 in urban areas like as points of exchange in the urban area.

2 DESCRIPTION

2.1 *Bologna today: Evolution from ZTL to ZAP*

Bologna, following the guidelines of the European Community White Paper Transport, (Npgtu under study), considered the creation of a High Pedestrian Zone (Zap) in the historic nucleus.

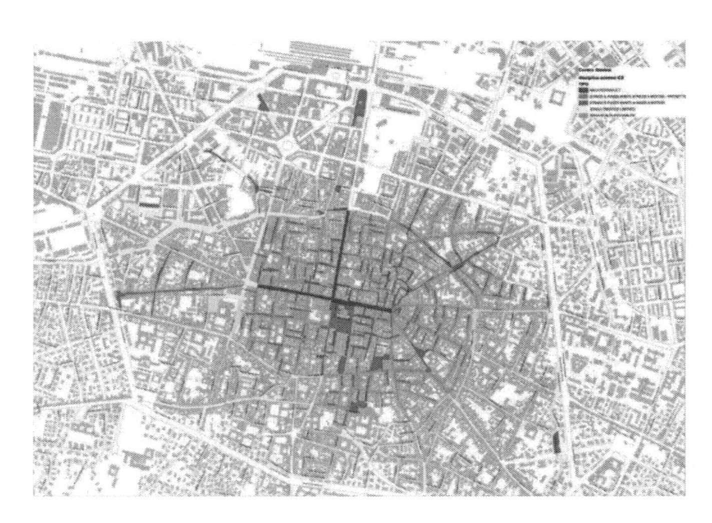

Figure 2. ZAP and ZTL.

Thanks to the establishment of the Zap, the Ztl, the pedestrian areas and the T system remained in force.

In the project area, complying with the European model that has spread over the last few years, access rules have been applied: during the week and the whole day, access to cars and motorcycles was allowed to residents only, public transport on main axles, taxis and operating means at strictly defined timetables.

According to this, in the pedestrian area, seven days a week, the pedestrians and the cyclists were the only categories to be able to circulate freely (Figure 2).

Motor vehicles access (cars and mopeds) was guaranteed only to residents or on strict permits.

Besides allowing continuous cycling and pedestrian paths, these restrictions brought cleaner air, more silence, road safety and, last but not least, the public space was taken away from cars and returned to the citizens.

Both the current PGTU (2006) and the future NPGTU still follows the the Winkler plan guidelines.

3 RESULT OF THE RESEARCH AND CONCLUSIONS

The project we worked on from May-July 2013 mainly concerned the analysis and the processing of data regarding the stalls and the moped's ones.

The analysis was useful in understanding whether there was the possibility, within the new ZAP, to remove the stalls on the road and move them to neighboring areas.

Regarding the mopeds study, the inspections took place in eight days and they were structured in two distinct moments: in the morning we detected the motors in a specific area stall by marking the partial; in the afternoon we re-detect stalls and compare the results to understand the area turnover.

From the municipal center's moped data (16,700 daily motorcycle accesses to the ZTL, 14,000 daily motorcycle accesses in the T, 5,577 moped stalls), we were expecting full stalls and mopeds parked outside the parking spots.

Analyzing the collected data, the total number of mopeds in the first round was 5507, while the second round was 4656, for a total turnover of 2341 mopeds (almost a one on two motorcycles) on a total of 693 stalls/6465 spots.

The situation was much less serious than we expected: there were few red stalls, predominantly in the areas already spotlighted as parking problems (Major Square and court areas) and areas with more parked mopeds in the morning than in the afternoon (due to work or study reasons). In addition, 50% of the roads were full of mopeds.

In the afternoon the overall situation tended to improve. Both data point out that the real problem wasn't the lack of parking spots, but the citizens' bad habits. In many cases the stall located near an attractive pole (ex. Piazza Malpighi) was full or exuberant, while the stalls nearby were empty.

We have used the same method to study the moped's turnover. The data highlighted that the hinge areas and attractive poles were not affected by the residential stall phenomenon, thanks to the movement of the mopeds during the day.

In conclusion, the situation of the stalls was positive and not in crisis as claimed by many, and that at the time the ZAP had entered into full implementation and operation, all the seats on the road could have been relocated to outside areas or within the parking lots in the structure.

The idea of the ZAP highlighted the intention of the Public Administration to "protect" and to enhance the historic center of Bologna through its application in the central core of the ZAP itself, externally limited by a ZTL. Within both ZAP and ZTL, pedestrian areas were particularly sensitive to urban traffic.

The intervention of Bologna Municipality had therefore imposed a strong restriction on the access of motorized vehicles to the historic center through very specific impediments: pedestrians and bicycles could enter only in the pedestrian islands; in the ZAP area the access was allowed to the residents' cars and motorbikes only; and finally, in the ZTL the access was granted to authorized vehicles, all public transport and bicycles.

On this scenario we defined our project guidelines.

The main purpose was to reduce the number of private vehicles entering the historic center, to improve the environmental status, the urban living and the road safety.

The intervention strategy involved the construction of three intermodal hubs, as platforms linked to public transport and to the Metropolitan Rail Service.

The goal was to offer the opportunity to arrive to the city through the orbital road, park the private vehicle and reach the center with public transport.

The intermodal hubs enhance the infrastructure itself and also to the neighborhood where they are located, linked to public transport, bike sharing (present in each hub) and integrating with pre-existing bike lanes and pedestrian paths.

Moreover, they allowed to connect to the historic center with alternative means (rather than cars and motorcycles).

It was necessary to start transformations into the public transport system, improving the fleet of vehicles and, above all, adapting the means to the routes in the historic center with vehicles long less than 18 meters.

Our design hypothesis consists in completing the same objectives of the city by extending the area heavily restricted to motorized traffic.

This idea is simple and inexpensive, thinking of the great infrastructural investments that our country and even Bologna had planned (we do not know how realistically).

By integrating the road system with the SFM, through three intermodal hubs that can collect flows from outside areas to the historic center, it is possible to ensure sustainable travel to the precious historic city.

REFERENCES

Bonfantini B., Evangelisti F., (2009). Bologna. Leggere il nuovo piano urbanistico PSC+RUE+POC. Edisai srl.

Boschi F., (2002). Servizio ferroviario metropolitano e sviluppo urbano, "Metronomie", n. 23. Clueb editore.

Boschi F., Gresceri G., Scannavini R., Trebbi A., (2012). Centro storico. Pedonalità e qualità urbana. Proposta per una Bologna futura. Ascom provincia di Bologna. Confcommercio.

Rovinetti A., (1975). Bologna non deve soffocare: proposte per il riordino del traffico cittadino. Comune di Bologna.

Scavone, A., (1991). Progetto di metropolitana leggera per la città di Bologna. "Parametro", n. 184. Faenza editrice. Pages 20–69.

Winkler, B., (1990). Piano della mobilità per la città di Bologna. "Parametro", n. 177. Faenza editrice. Pages 19–67.

Town and Infrastructure Planning for Safety and Urban Quality – Pezzagno & Tira (Eds)

Livability of Italy's traffic-restricted zones: What do the citizens think?

M. DeRobertis
Università degli Studi di Brescia, Brescia, Italy

ABSTRACT: As early as the 1960's, many Italian cities began to rebel against the intrusion of automobiles in city centers, beginning with bans on cars in piazzas. In the 1970's cities began restricting nonresidents from driving in certain sections of the center, creating what became known as Zone a Traffico Limitato (ZTL) or Traffic-Restricted Zones. Today, at least 300 Italian cities/towns have ZTL. Some cities have documented that ZTL have reduced congestion, improved transit travel times, reduced traffic accidents, and improved pedestrian safety. There has been little or no research on the opinions of the citizens who are the primary beneficiaries.

This paper present the results of a survey of the residents, workers and visitors to the city center and ZTL in Brescia, Italy, conducted in the Spring of 2017 about how they perceive the ZTL. The differences in commute mode, car ownership and car use are also compared.

1 INTRODUCTION

As early as the 1960's, many Italian cities began to rebel against the intrusion of the automobile in the city center, beginning with bans on parking cars in the center of piazzas, then by creating pedestrian-only piazzas and then pedestrian-only streets. Some cities, instead of or in addition to these measures, began to restrict nonresidents from certain sections of their historic city center (centro storico), creating what eventually became known as Zone a Traffico Limitato (ZTL) or Traffic-Restricted Zones. ZTL were formally recognized in Italian law in 1989. ZTL are areas, most often in the city center, in which only authorized vehicles may enter and furthermore, only residents and a few other categories of vehicles can receive authorization (Figure 1). The purpose of a ZTL, according to the Codice della Strada, (CdS,

Figure 1. One of the entrance points to the ZTL in Brescia.

the Italian Highway Code), is to reduce pollution and preserve the artistic, natural, and environmental heritage of city centers.

Today, there are at least 300 cities and towns in Italy with ZTL in their city centers, ranging in size from 1,000 to 2.2 million population. While most of the smaller towns (less than 10,000 population) tend to be tourist/vacation destinations, such as Bellagio with a population of 3,000, many towns with ZTL are not. The vast majority of towns with ZTL do, however, have a verified Centro Storico, (historical center) which is defined by law. Some cities such as Rome have documented that ZTL have reduced congestion, improved transit travel times, reduced traffic accidents, improved pedestrian safety and increased pedestrian and bicyclist mode share. (Roma PUT 2004).

2 RESEARCH APPROACH

The survey questions were developed to address three main areas:

1. Travel behavior and car ownership;
2. Rating the livability of the neighborhoods, both within and outside the ZTL;
3. The advantages and disadvantages of the ZTL from both the resident and visitor perspective.

The surveys were distributed in two main ways: 1) to the students and attendees of meetings and conferences at the University of Brescia and at other public gatherings; and 2) via a link to an online survey form, which was distributed throughout the city center of Brescia. As of the cut-off date for the Living and Walking in Cities conference, 115 responses had been received.

3 DESCRIPTION

3.1 *Innovative aspects*

This survey was unique in several ways. It was the first to ask residents and visitors their opinion of ZTL. It also asked respondents to rate the street where they live by their own definition of livability. It then asked residents to rate specific aspects of livability such as traffic noise, pedestrian safety and how often they greeted their neighbors. Lastly, the residents were asked to identify the advantages as well as the disadvantages of living in a ZTL. Nonresidents such as visitors and people who work in the ZTL were asked similar questions as to whether and how the ZTL makes the city center more attractive to them.

3.2 *Respondents*

Of the responses received at the time of the writing of this paper, 37% came from residents outside the City of Brescia, 37% lived in Brescia but outside the city center, 9% lived in the City Center but not the ZTL and 18% of the respondents lived within the ZTL (Table 1).

Table 1. Residence location of survey respondents.

Residence location	Number	Percent
Live in the ZTL of Brescia.	19	18%
Live in the historic city center of Brescia but not ZTL.	11	9%
Live in Brescia but outside the city center.	42	37%
Live in the province of Brescia but not the city of Brescia.	37	32%
Live outside of the province of Brescia.	6	5%
Total	115	100%

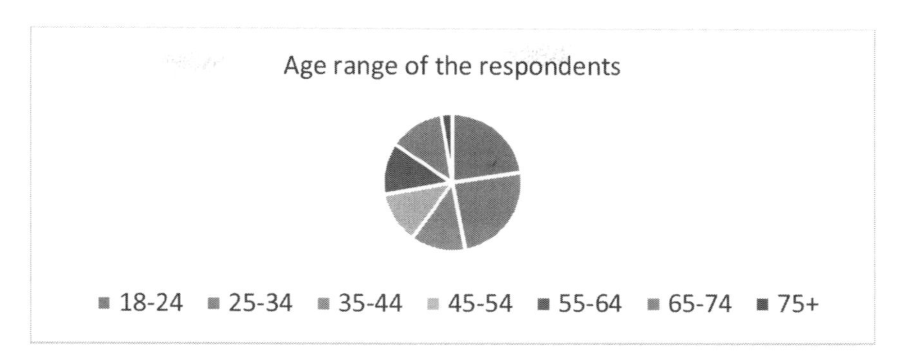

Figure 2. Age distribution of survey respondents.

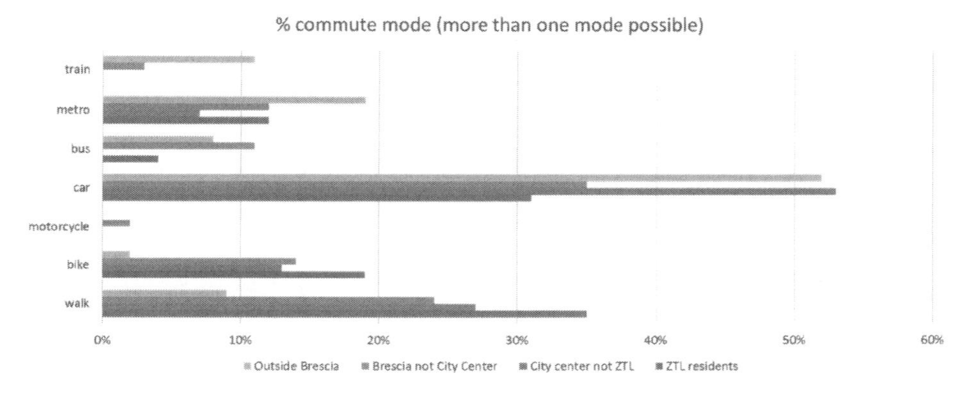

Figure 3. Primary commute mode by location of residence.

With respect to demographics, 57% of the respondents were male and 43% were female; 43% of the respondents were employed, 37% were students and 21% were retired or otherwise not currently employed. The age ranges were well distributed between 18 and 65 with a few over 75, as indicated in Figure 2. Thus, despite the small sample size, a fairly representative range of the population was obtained.

4 RESULTS OF THE RESEARCH

4.1 *Travel behavior*

First the survey asked about travel behavior and car ownership. Respondents were allowed to choose more than one mode, for example bike to the metro station. The responses reveal that those who live in the ZTL are more likely to walk and bike to work/school (35% and 19% respectively), than those who live outside of the city cener and outside of Brescia. ZTL residents have the lowest car modal split, 31%, whereas city center residents and outside of Brescia residents had the highest car modal split, both more than 50% (Figure 3). Interestingly, for travelling to the center, the mode split was more green: 48% of nonresidents said that when they visited the city center for free time or shopping, they use walking, biking or public transit versus 34% who said they use these modes to go to work or school.

Car ownership did not vary much between residence location; for example, 56% of ZTL resident households had two cars compared with 53% and 48% of those who lived outside the city center and outside of Bresica, respectively. Interestingly, the percentge of 3+ car families was similar for those who live in the ZTL and those who live outside of Brescia (25–31%) and it was much higher compared to those who live in Brescia but not the city center (11%). One preliminary conclusion is that the ZTL does not appear to be inhibiting car ownership.

Most respondents regardless of residence location have a private parking parking place in their building or courtyard, 72%, 68% and 87% for ztl residents, Brescia residents and outside of Brescia, respectively.

4.2 *Rating neighborhood livability*

The livability of the neighborhood was assessed by several questions. The first question asked: "Why do you live in your neighborhood"? The most common response of those who live outside the City of Brescia was "family home" or "live with parents". Of those who live in the City of Brescia but outside the historic city center, the most common response was "I like the neighborhood" followed by "Convenient to go to work/school". Of those who live in the city center (including the ZTL), the most common responses were "I like the neighborhood" and "I like being able to walk to destinations".

Next, respondents were asked to rate their neigborhood on a scale of 1 (not very livable) to 5 (very livable). There was a slight difference between those who live in the ZTl and those who live in the center but not the ZTL: 88% vs 82% rated their neighborhood 3 (average) or better.

There was also a difference between those who live in Brescia outside the center vs. those within the center: 60% rated their neighborod 4 or better vs 45% of city center and ZTL residents. Interestingly the only areas which had responses of 1 – not very livable were outside the city center, i.e. Brescia outside the city center and outside the city of Brescia (Figure 4). This illustrates the variation of neighborhoods present in the city as well as in the province.

Respondents were given an opportunity to write in their own words why (or why not) they found their neighborhood livable. Some of the responses received are presented in Table 2. Given that those who live outside the city could be living in a wide range of location types, from another city or in the country, it is hard to draw any firm conclusions, other than from those who specifically stated that they live in the country.

Those who did not live in the historic city center were also asked to rate the livability of the city center. Those who live outside the city center did not find the city center to be as livable as did those who live in the city center (88% vs 72%, respectively, rated the city center 3 or above). Those who live outside the city center also did not find the city center to be as livable as their own neighborhood; 30% rated the city center 4-livable or 5-very livable vs 60% who rated their own neighborhood 4 or 5 (Figure 5).

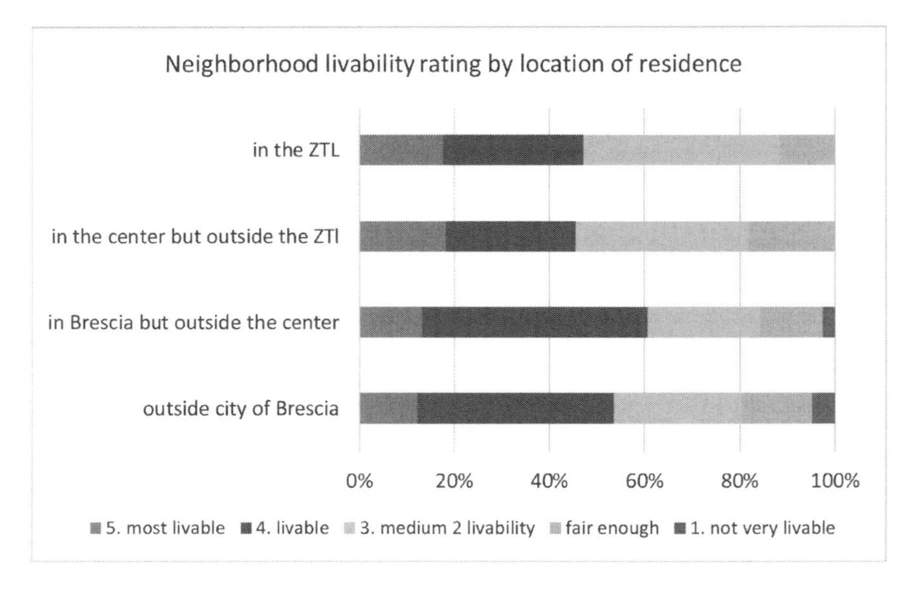

Figure 4. Neighborhood livability rating by location of residence.

Table 2. Why do you consider your neighbourhood livable?

Residence location of survey respondents	Why do you consider your neighborhood livable? (write-in responses)
Live in the ZTL of Brescia	quiet, I always use my bike, clean and lively, social and aesthetic reasons
Live in the historic city center of Brescia but not ZTL	It is very lively, Convenient and quiet, lots of services
Live in Brescia but outside the city center	quiet, lots of stores,
Live in the province of Brescia but not the city of Brescia or Live outside of the province of Brescia	quiet, like greenspace, live in a small town where walking is very safe, live in the country.

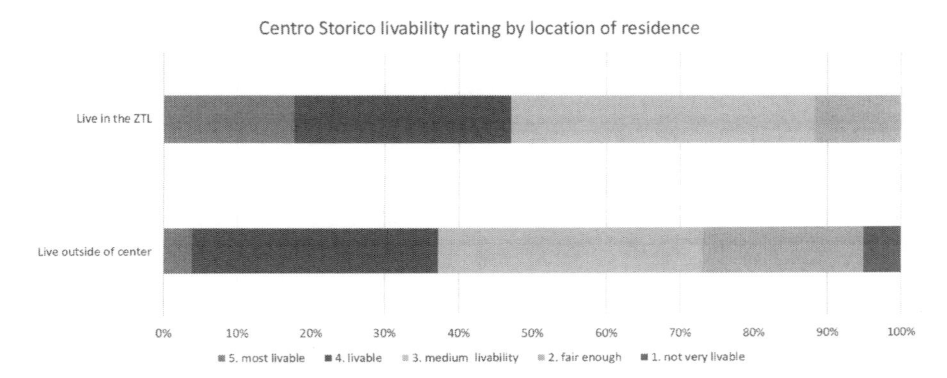

Figure 5. City center livability rating by location of residence.

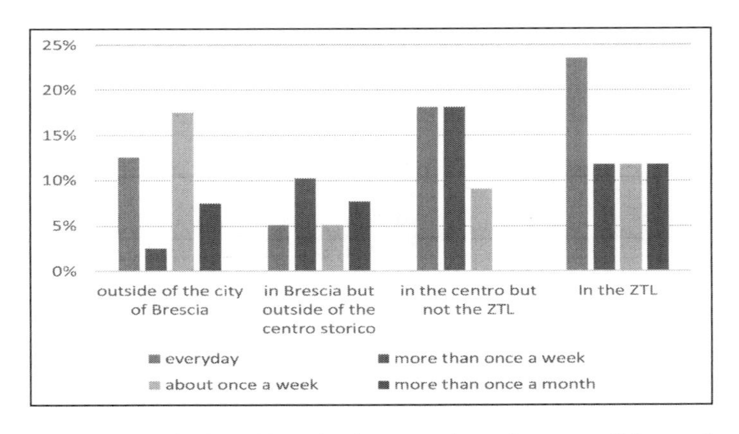

Figure 6. How often are you bothered by noise from people on the street talking out loud?

To assess particular aspects of street life that can make streets less livable, respondents were asked, with respect to the street where they live, how often they were bothered by common urban nuisances. Four types of nuisances were investigated:

- noise from voices of people walking outside (Figure 6)
- noise from public places such as bars, cafes, restaurtants, etc. (Figure 7)
- noise from traffic outside on the street (Figure 8)
- annoyance from auto exhaust fumes (Figure 9)

As shown in Figure 6, residents of the ZTL and the city center are bothered more often by the raised voices of people on the street about one-third more than once a week or everyday

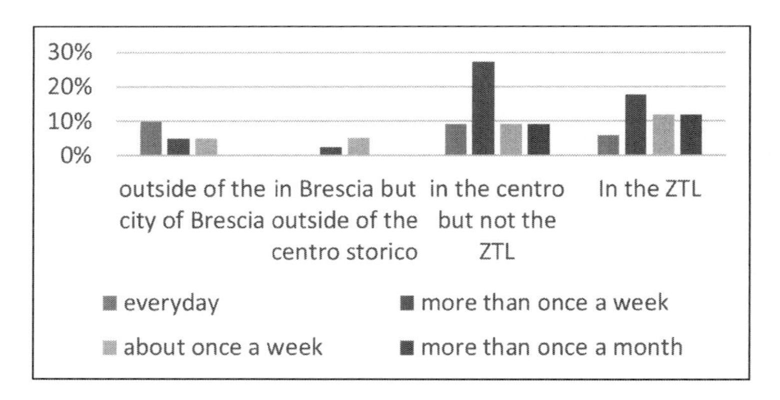

Figure 7. Respondents bothered by noise from the street: Noise from bars and public establishments.

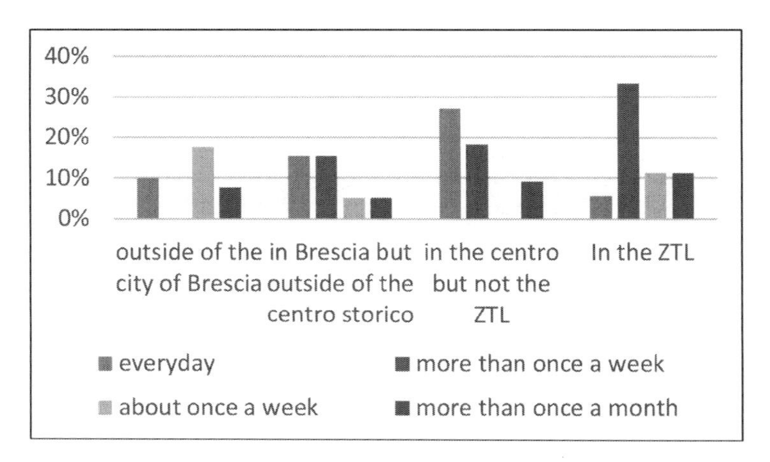

Figure 8. Respondents bothered by noise from the street: Traffic noise.

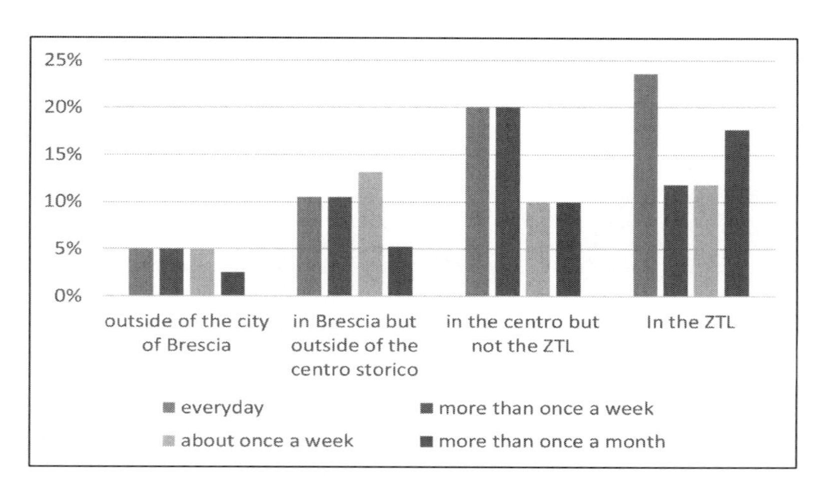

Figure 9. Respondents bothered by traffic fumes.

As shown in Figure 7, residents of the city center and ZTL are also more likely to be bothered by noise from public places like bars, cafes and music venues.

The third category was whether residents were bothered by noise from cars ands traffic. As shown in Figure 8, even residents of outside the centro storico experience annoyance from

Table 3. What else bothers you about your street? Write-in reponses by residence location.

Residence location of survey respondents	What else bothers you about your street?
Live in the ZTL of Brescia	Undisciplined cyclists; drug smuggling; the presence of dangerous subjects; Litter on the ground; purse-snathcing; cars parked illegally; drivers who don't respect STOP signs,
Live in the historic city center of Brescia but not ZTL	Often used as a public toilet; trash cans; presence of homeless people; bicyclists travelliing too fast descending from the hill; the tiny parking lots for the academy.
Live in Brescia but outside the city center	Some neighbors, proximity of the rail yard; wild parking; dog feces; dogs; lack of respect for the speed limit; parked cars; too many traffic lights; speed of cars; sirens; dogs, management of the trash collection.
Live in the province of Brescia but not the city of Brescia	lack of public parking; mothers doubleparking taking kids to school; smells of stables and fertilizing fields; the speed of the cars; at-grade rail crossings; speed of the car; few trees; soccer matches on Sunday; only the dogs barking.
Live outside of the province of Brescia	No write-in responses were received.

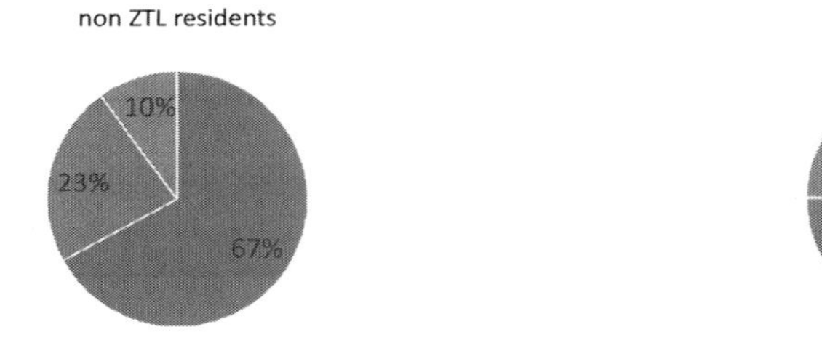

Figure 10. Is the ZTL a problem when you need to visit someone in the ZTL or or have visitors if you live in the ZTL?

traffic. But residents of the ZTL reported beingn less bothered by traffic than those in the city center outside the ZTL.

Lastly, as shown in Figure 9, residents of the centro storico and ZTL are more likely to report being bother by traffic fumes, Interestinly, even 20% of residents outside the city center reported being bothered more than once a week by traffic smells.

In addition, the residents were asked to write in their own words what else bothers them about their street. The write-in responses are presented in Table 3.

4.3 *Disadvantages of the ZTL & modifications to the ZTl*

The last goal of the survey was to determine how the citizens felt about the ZTL. Several questions were asked about whether and how the ZTL was a probem and if they would like to change it. The first question asked if the ZTL was a problem when they wanted to visit someone in who lives in the ZTL or when they have visitors, for those who live in the ZTL. Half the ZTL residents and one-third of the nonresidents considered the ZTL to be a problem or a big problem (Figure 10).

They were also asked if the ZTL was a problem when they needed to give someone a ride who lives in the ZTL or needed a ride from someone outside the ZTL. Residents and nonresidents were similar, in that two thirds did not find the ZTL to be a problem whereas one-third considered the ZTL to be a problem or a big problem (Figure 11).

The respondents were asked who should be permitted to enter the ZTL besides residents. The vast majority of both ZTL residents and nonresidents think that bicyclists, vehicles of disabled persons and public transit vehicles should be permitted within a ZTL. The majority of both residents and non residents did not feel that motorcycles (80%) or uber vehicles (70%) should be permitted. There was less agreement on taxis, vespas/scooter, hotel guests (Figure 12).

The residents of the ZTL were asked how they liked living in the ZTL; 75% said they liked it or really liked it. No one responded that they didn't like it at all Figure 13.

To assess how much they might want to change the conditions of the ZTL, we asked one final question regarding changing the hours or the physical extension of the ZTL. The majority of both residents and nonresidents were content to see the ZTL remain the same or even expand the area included within the ZTL (Figure 14).

Figure 11. Is the ZTL a problem when you need to give someone a ride or need aride?

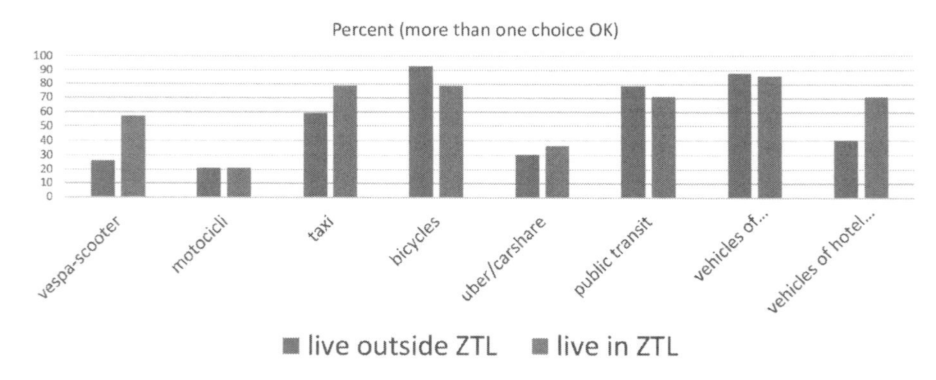

Figure 12. Who should be permitted to enter the ZTL besides residents?

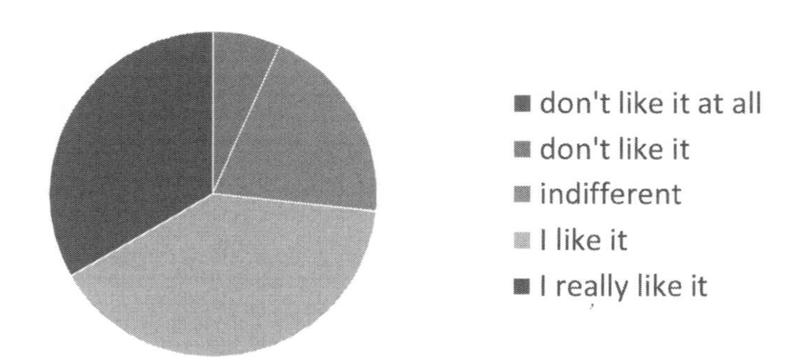

Figure 13. Overall how do you like living in the ZTL?

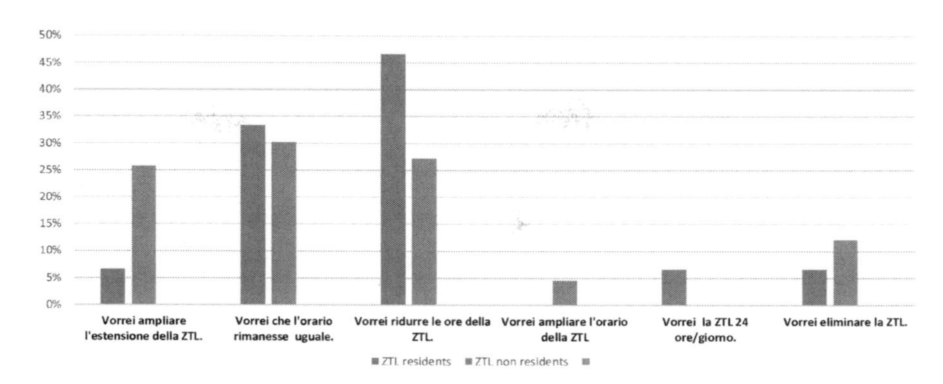

Figure 14. How would you change the ZTL?

5 CONCLUSIONS

It is acknowledged that residents' and visitors' opinions are only one aspect of livability. For example, actual improved traffic safety is more important that perceived traffic safety. For this reason, additional research on actual and traffic collisions within ZTL has been and will be conducted. DeRobertis M. (2016). In addition, the real problems of air and noise pollution are detrimental to the livability urban areas. But it has proved to be difficult to isolate the benefits of reduction of traffic and its reduction in auto exhaust and noise on just the areas encompassed by the ZTL.

It would be very informative to have similar surveys of other ZTL of other medium to large cities throughout Italy. The findings could help city officials in designing the extent and hours of future expansions of ZTL. The future surveys could also help cities both within and outside of Italy decide whether or not to implement ZTL or similar strategies.

REFERENCES

Codice della Strada (1992). 30 aprile 1992. Articolo 3.

Comune di Roma, City of Rome. (2004) Piano Urbano del Traffico/Urban Transportation Plan (PUT).

DeRobertis M. (2017), unpublished data. Italian Cities with ZTL.

DeRobertis, M. and Tira M. (2016) "The most widespread traffic control strategy you've never heard of: Traffic-restricted zones in Italy". ITE Journal, Vol. 86, Issue 12:44–49.

DeRobertis M. and Tira.M. (2016) "Effectiveness of Italian Traffic Limited Zones (ZTL)". Presented at European Conference on Mobility Management (ECOMM). Athens, Greece.

Morici, I. (1994) Le Zone a Traffico Limitato (ZTL). In "Il sistema mobilità: verso una gestione manageriale". RCS Libri & Grandi Opere. pp 214–230.

Town and Infrastructure Planning for Safety and Urban Quality – Pezzagno & Tira (Eds)
© 2018 Taylor & Francis Group, London, ISBN 978-0-8153-8731-2

Principal component analysis and cluster analysis for the assessment of urban mobility in Italy

C. Gargiulo & L. Russo
Università degli Studi di Napoli Federico II, Naples, Italy

ABSTRACT: In Italy, a great number of projects and researches have been promoted in order to increase the use of ecological vehicles and bikes, increase the supply of public transportation and reduce private transport dependence. Therefore, investigating how urban mobility varies across main Italian cities may be of great interest. The aim of this paper is to employ a principal component analysis combined with a cluster analysis to assess urban mobility in Italy, and identify the key characteristics that structure mobility behaviors and mobility administrative strategies.

The results of these analyses shows that the demographic dimension of an urban area and its economic level are positively associated with the supply of public transport and car sharing, and that air pollution depends on the number of vehicles per inhabitnat and the age of the car fleet, but not on either the concentration of eco-vehicles or the implementation of more sustainable mobility policies.

1 INTRODUCTION

Urban areas have been facing extremely important challenges in recent years, especially due to the dramatic consequences of climate change and because of the rapid process of urbanization. Therefore, policy makers, researchers and planners have been working hard to support the implementation of strategies and actions aimed at reducing the carbon footprint of urban areas while promoting economic growth, social inclusion and higher quality of life (Papa et al. 2017; 2014).

In this context, urban mobility represents a crucial factor. Mobility, indeed, has very important impacts on cities, both in terms of environmental sustainability (e.g. air pollution) and citizens' quality of life (e.g. congestion resulting in longer travel time, noise pollution and road safety) (Dameri & Benevolo, 2017). Therefore, urban mobility initiatives have recently attracted significant investments by governments and policy makers all around the world.

Also in Italy, in recent years, innovative/traditional solutions have been implemented/ improved at local level to increase the use of ecological vehicles and bikes, increase the supply of public transportation and reduce private transport dependence in order to foster the transition towards a more sustainable urban mobility (Gargiulo, 2014; Papa et al. 2016). Consequenlty, it is of interest to assess how urban mobility differs across main Italian cities, so to understand if there are significant relationships between the dimensional and economic characteristics of cities and their mobility systems.

2 OBJECTIVE OF THE PAPER

Based on the previous considerations, the aim of this paper is to employ a principal component analysis combined with a cluster analysis to investigate mobility across main Italian cities. This multivariate statistical approach allows us to identify the key characteristics that structure mobility behaviors and mobility administrative strategies within the Italian context and to verify their impacts on both road safety and the reduction of air pollution, which are two of the main goals of smart mobility initiatives. The results of our analyses provided interesting findings, which are very useful for assessing both the effectiveness of smart mobility interventions across main Italian cities and the efforts of citizens in engaging in virtuous behaviors.

3 METHODOLOGICAL APPROACH

The methodological approach used in this study includes four main steps: (1) knowledge acquisition and construction of the conceptual model; (2) selection of the sample and variables to be included in the model; (3) principal component analysis (PCA); (4) cluster analysis; (5) interpretation of the results and validation of the conceptual model.

1. The conceptual model represented in Figure 1 illustrates the theoretical background that we want to test and investigate through a multivariate statistical approach. We want to verify the extent to which both the urban and economic dimension of a city affect the mobility behaviours of its citizens as well as the mobility policies implemented by local policy makers to foster the transition towards a smarter mobility. Furthermore, we want to investigate how both aspects—mobility behaviours and strategies—impact two of the most important smart mobility objectives, i.e. increasing people safety and reducing pollution (Dameri & Benevolo, 2017).
2. In order to test the conceptual model described above, we selected a set of nineteen variables (Table 1). The variables chosen for this study can be grouped into five categories: the first category is called "urban dimension" and includes three variables describing the physical and demographic dimension of an urban area (i.e. *population; total area; population density*); the second group is called "economic dimension" and corresponds to the variable income, which outlines the economic welfare of a city; the third category is called "mobility behaviours" because it includes six variables that describe different mobility habits, and in particular the citizens' car fleet and their inclination to use public transportation (i.e. *car ownership; gas/LPG/methane-powered vehicles; vehicles Euro 0, 1, 3; demand of public transport*); the fourth group is called "mobility strategies" and includes seven variables that provide useful information about both innovative and traditional mobility solutions that local policy makers can implement at urban level (i.e. supply of public transportation; restricted traffic zone; pedestrian zone; bike lanes; toll parking; car sharing; bike sharing); lastly, the fifth group, which is called "smart mobility objectives" includes two variables that allow the assessment of urban mobility across different cities (i.e. PM10 exceedance days; index of accidents).

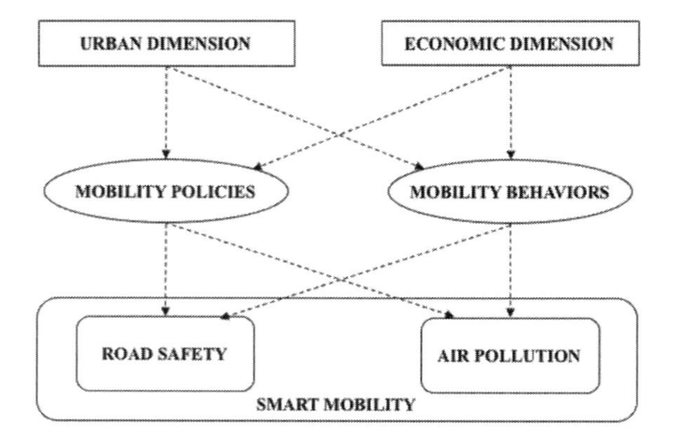

Figure 1. Conceptual model.

Table 1. List of variables included in the statistical model.

Category	Variable	Description	Source (year)
Urban dimension	Population	inhabitants	Osservatorio Mobilità Sostenibile (OMS) (2014)
	Total area	sqm	ISTAT (2014)
	Population density	inhabitants per sqm	OMS (2014)
Economic dimension	Income	euro	MEF (2014)
Mobility behaviors	Car ownership	vehicles/100*inhabitants	OMS (2014)
	gas-powered vehicles/100*inhabitants	vehicles/100*inhabitants	OMS (2014)
	LPG-powered vehicles/100*inhabitants	vehicles/100*inhabitants	OMS (2014)
	methane powered vehicles/100*inhabitants	vehicles/100*inhabitants	OMS (2014)
	vehicles Euro 0, 1 and 3/100*inhabitants	vehicles/100*inhabitants	OMS (2014)
	Public transportation—Demand	Number of passengers/ inhabitants	OMS (2013)
Mobility policies	Public transportation—Supply	Number of seats*km/ inhabitants	OMS (2013)
	Restricted traffic zone	sqm/inhabitants	OMS (2013)
	Pedestrian zone	sqm/inhabitants	OMS (2014)
	Bike lanes	km/10.000 inhabitants	OMS (2014)
	Toll parking	Number of stalls/1.00* inhabitants	OMS (2014)
	Car sharing	users/10.000 inhabitants	OMS (2014)
	Bike sharing	users/10.000 inhabitants	OMS (2014)
Smart mobility objectives	PM10 Exceedance days	number of days	OMS (2014)
	Index of accidents	number of accidents/ inahbitants	ISTAT (2014)

These data were gathered for a sample of forty-one Italian capital cities with a population higher than 100.000 inhabitants[1], which account for about 23% of the total national population.

3 and 4. After collecting and elaborating all the information, we obtained a 41 × 19 matrix that was analysed employing a principal component analysis first and a cluster analysis later. Both multivariate analyses were conducted using the SPAD 55 software. More specifically, multivariate statistical analysis is particularly useful and effective for the study and interpretation of urban phenomenon, which are often complex and multidimensional, because this technique allows us to look into large volumes of data and find significant relationships without a prior hypothesis. In particular, the PCA is an exploratory method that "replaces the p original variables by a smaller number, q, of derived variables, the principal components, which are linear combinations of the original variables" (Jolliffe, 2002). On the other hand, cluster analysis identifies groups of elements with similar characteristics; groups are formed "in such a way that objects in the same group are similar to each other, whereas objects in different groups are as dissimilar as possible" (Kaufman & Rousseeuw, 2009).

1. Venice was excluded from the sample because of its unique geomorphological and physical characteristics, which result in a very specific mobility system that cannot be compared to the others.

2. As part of the autonomus region of Trentiono-Alto Adige, Trento and Bolzano have access to lower car registration taxes and fees. This difference with all other Italian capital cities turns into a much higher car ownership, which would impact the reliability of the statistical results. Therefore, these two cities were not included in the sample.

In the case of our dataset, the PCA provided a straightforward method for investigating urban mobility across main Italian cities, while cluster analysis helped to identify groups of cities with homogeneous mobility patterns.

5. Lastly, the results of both techniques were used to validate the conceptual model described in the first part of this paragraph. Both PCA and cluster analysis, indeed, allowed the identification of the main relationships amongst the nineteen urban-mobility variables, that were compared to those hypothesized in the theoretical model.

4 RESULTS

4.1 *Results of the principal componenet analysis*

The PCA provides three main outputs: (i) the principal components, i.e. the factors that best fit the clouds of points (corresponding to the forty-one Italian cities) in the vector spaces; (ii) the loadings of the nineteen variables on the principal components, which show the contribution of each variable to each factor; (iii) the principal components scores, which can be interpreted as the coordinates of each data point, with respect to the principal component axes.

With respect to the first output, the PCA for the 41×19 matrix of data shows that the first two components dominate, and account for about 47% of the total variance of the dataset, while the first four factors together explain 66% of the total variance of the system (Tab. II). Therefore, components 1 to 4 are described in detail in this paper.

With respect to the second output, the loadings of the nineteen urban-mobility variables on the four principal axes are shown in Table 2, and for the purposes of the following discussion loadings higher than 0.50 (represented in bold in Table 2) are considered as high, loadings between 0.50 and 0.20 are moderate and loadings lower than 0.20 are considered as weak (Kassomenos et al., 2014).

The inverse of the first component (PC1) can be interpreted as a measure of sustainable urban mobility policies implemented in rich and populated urban areas. Specifically,

Table 2. Principal component loadings and variance explanation for the 19 variables included in the model (component loadings higher than ±0.50 are in bold).

Category	Variable	PC1	PC2	PC3	PC4
Urban dimension	Population	**–0.57**	0.26	**–0.68**	–0.14
	Total area	0.07	0.14	**–0.81**	–0.01
	Population density	**–0.72**	0.15	0.03	–0.23
Economic dimension	Income	**–0.77**	–0.41	0.01	–0.03
Mobility behaviors	Car ownership	0.50	–0.23	–0.24	**–0.53**
	gas-powered vehicles	0.30	**–0.84**	–0.31	0.04
	LPG-powered vehicles	0.22	**–0.80**	–0.33	–0.21
	Methane powered vehicles	0.34	**–0.72**	–0.24	0.24
	Euro 0, 1 and 3 vehicles	0.33	**0.53**	–0.09	**–0.52**
	Public transportation—Demand	**–0.82**	–0.03	–0.43	0.19
Mobility policies	Public transportation—Supply	**–0.83**	0.07	–0.30	0.15
	Restricted traffic zone	–0.46	–0.39	0.39	–0.09
	Pedestrian zone	–0.28	–0.32	–0.11	0.34
	Bike lanes	0.22	**–0.80**	–0.17	–0.05
	Toll parking	–0.18	**–0.57**	0.24	0.18
	Car sharing	**–0.75**	–0.14	0.25	–0.09
	Bike sharing	–0.40	–0.20	0.06	–0.30
Smart mobility objectives	PM10 Exceedance days	–0.43	–0.32	0.01	**–0.65**
	Index of accidents	–0.13	**–0.53**	0.47	–0.18
Variance (%)		24.75	21.78	11.84	7.83
Cumulative variance (%)		24.75	46.53	58.37	66.20

PC1 shows high loadings (<–0.50) for two "mobility policies" variables—supply of public transportation and car sharing—as well as for three dimensional measures—income, population density and population—accompanied by PM10 and index of accidents loadings equal to –0.43 and –0.13 respectively. These results reveal three types of relationships. First, they show a strong association between innovative and traditional initiatives promoted by local policy makers for building a more sustainable mobility system and both the economic and demographic size of urban areas: richer, more populated and denser cities accommodate a broader range of public mobility policies. Second, the results highlight a moderate (weak) concordant relationship between sustainable mobility initiatives and the "smart mobility objectives" variables, meaning that the promotion of smart mobility policies is moderately associated with air pollution (and not-significantly associated with the index of accidents). In other words, data do not support the hypothesis that smart mobility initiatives correspond to smarter outcomes, suggesting that other variables (e.g. or citizens' behaviors) may play a more important role.

The second principal component (PC2) is a measure of the citizens' car fleet. Four "mobility behaviors" variables are consistently heavily loaded: gas-powered vehicles (–0.84), LPG-powered vehicles (–0.80), methane-powered vehicles (–0.72) and Euro 0,1,3 vehicles (0.53). However, differently from what expected, the loading of the PM10 variable is moderate and negative (–0.32), revealing that a higher concentration of eco-vehicles does not correspond to a reduction in air pollution, for our sample of cities.

The third component (PC3) is heavily loaded on the total area and population variables (–0.81 and –0.68 respectively). Therefore, PC3 can be interpreted as a measure of the urban size, both in terms of surface and population. The form of this third component shows that the association between "urban dimension" and the index of accidents is moderate (0.47), while the association with the PM10 is not significant (0.01). In other words, we cannot argue that bigger cities correspond to higher air pollution and vice versa.

The fourth principal component (PC4) displays high PM10, car ownership and Euro 0,1 and 3 vehicles loadings (–0.65, –0.53 and –0.52 respectively) and negligible total area, income and population loadings (–0.01, –0.03 and –0.14 respectively), meaning that air pollution is not associated with either the size of a city or its economic strength, but, not surprisingly, it is strongly related to the number of cars per inhabitant and the turnover of the car fleet.

With respect to the third output, the coordinates/scores of the forty-one Italian cities included in the sample on PC1-PC2 and on PC3-PC4 are presented in Fig. 2 and Fig. 3 respectively.

The map in Figure 2 shows several interesting results. In the upper-right quadrant of the map we find small and low-income cities with a weak public transportation system and a very old and polluting car fleet. It is worth noting that many of these cities are concentrated in the southern part of the country. On the other hand, the cities in the bottom-left quadrant

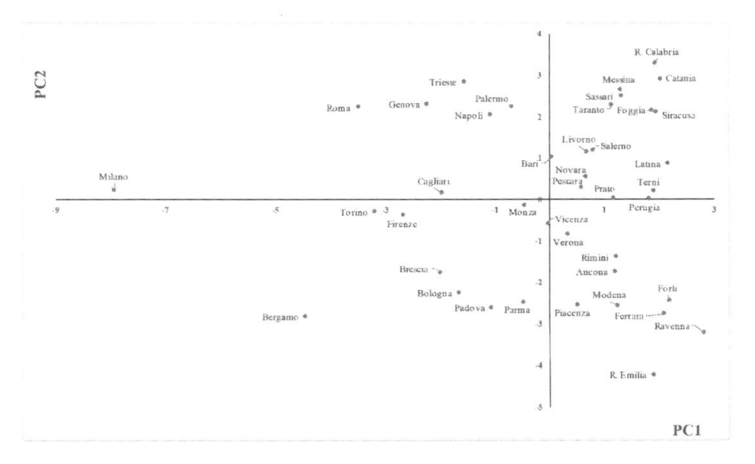

Figure 2. Map of the 41 Italian cities on the first two principal components.

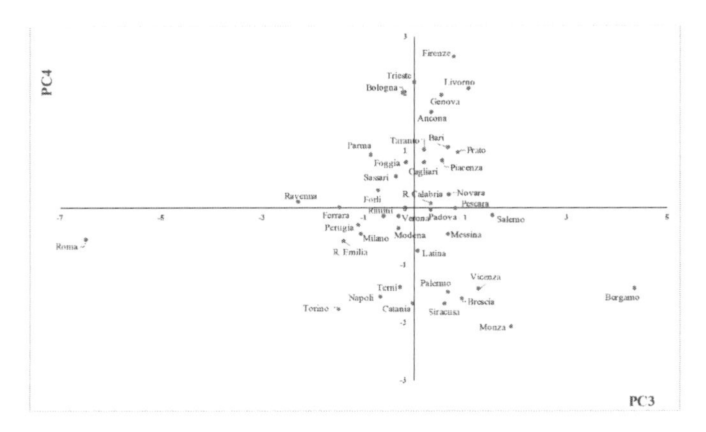

Figure 3. Map of the 41 Italian cities on the third and fourth principal components.

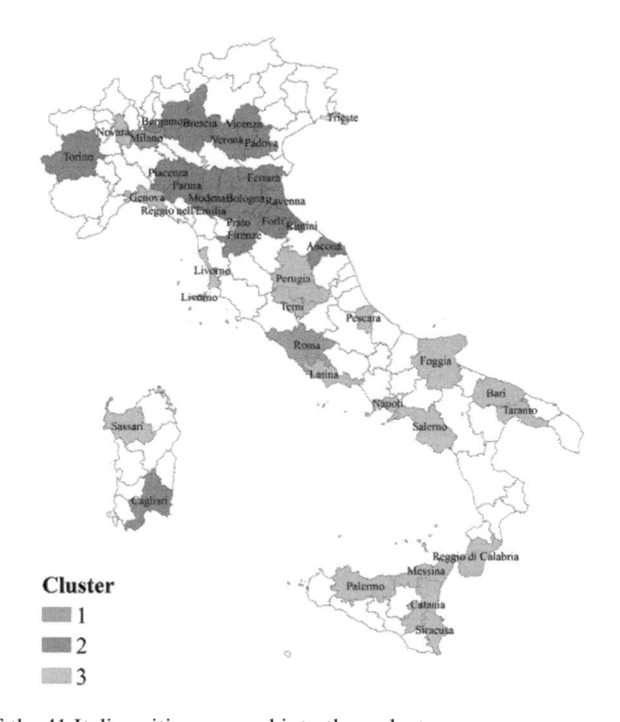

Figure 4. Map of the 41 Italian cities grouped into three clusters.

are characterized by high levels of both innovative and traditional mobility policies as well as by a high concentration of ecological vehicles and bike lanes. Milano stands out as the best-performing city on the first principal axe by far, revealing an excellent supply of public transportation and car-sharing, but its score on PC2 (close to zero) reveals a not equally outstanding performance in terms of car fleet quality.

Additional findings emerge by looking at the distribution of the cities along the third and fourth principal components (Fig. 3). First of all, the cities appear to be concentrated in the central part of the map, which means that their performances on PC3 and PC4 do not significantly differ from each other. Only two cities stand out from the group: Roma and Bergamo. The opposite scores of these two cities on PC3 highlight the dimensional difference between them; but, at the same time, the two urban areas show similar scores on PC4 (i.e. similar levels of air pollution, car ownership and age of the car fleet).

4.2 *Results of the cluster analysis*

The cluster analysis was conducted using the SPAD 55 software, which uses the Ward algorithm when applying the hierarchical classification, and produced a dendrogram of the forty-one Italian capital cities based on the set of nineteen variables considered for this analysis. We cut the clustering tree obtaining three independent clusters.

The first cluster includes only two cities (5% of the sample), the second cluster includes 19 cities (46%) and the third cluster comprises 20 cities (49%). The variables that characterize each cluster are presented in Tables 3, 4 and 5. Moreover, Figure 4 shows the geographical

Table 3. Cluster 1 – characteristic variables.

Characteristic variables	Mean of the class	Mean of the sample	Test values	Probab.
Population	2104590	331747	5.30	0.00***
Public transportation—Supply	10229.100	3167.010	4.33	0.00***
Public transportation—Demand	447.680	122.632	4.29	0.00***
Total area	734.516	219.865	3.38	0.00***
Income	28420	22300	3.14	0.00***

*,**,***Indicates the level of significance at the 10, 5, 1 percent level respectively.

Table 4. Cluster 2 – characteristic variables.

Characteristic variables	Mean of the class	Mean of the sample	Test values	Probab.
Bike lanes	5.129	2.936	4.25	0.00***
gas-powered vehicles/100*inhabitants	7.953	5.685	4.04	0.00***
LPG-powered vehicles/100*inhabitants	4.721	3.633	3.83	0.00***
Index of accidents	0.013	0.011	3.56	0.00***
methan powered vehicles/100*inhabitants	3.096	1.952	3.43	0.00***
Toll parking	1.784	1.297	2.76	0.00***
Income	23580	22300	2.69	0.00***
Pedestrian zone	0.398	0.286	2.57	0.01***
vehicles Euro 0, 1 e 3/100*inhabitants	13.826	15.923	–2.89	0.00***

*,**,***Indicates the level of significance at the 10, 5, 1 percent level respectively.

Table 5. Cluster 3 – characteristic variables.

Characteristic variables	Mean of the class	Mean of the sample	Test values	Probab.
vehicles Euro 0, 1 e 3/100*inhabitants	18.053	15.923	3.09	0.00***
Toll parking	0.868	1.297	–2.43	0.01***
Pedestrian zone	0.182	0.286	–2.49	0.01***
Public transportation—Demand	77.557	122.632	–2.59	0.00***
PM10 Exceedance days	26.421	36.300	–2.88	0.00***
methan powered vehicles/100*inhabitants	1.018	1.952	–2.94	0.00***
Index of accidents	0.008	0.011	–3.24	0.00***
LPG-powered vehicles/100*inhabitants	2.656	3.633	–3.61	0.00***
gas-powered vehicles/100*inhabitants	3.727	5.685	–3.66	0.00***
Bike lanes	0.868	2.936	–3.80	0.00***
Income	20473	22300	–4.04	0.00***

*,**,***Indicates the level of significance at the 10, 5, 1 percent level respectively.

distribution of the three clusters, which represents a useful support for the following interpretation of results.

Cluster 1 includes the two major Italian cities (i.e. Roma and Milano), which stand out from the crowd not simply because of their bigger size (both in terms of population and total area), but also because of their solid economic ground and the higher level of both public transportation supply and demand (Table 3).

If we exclude these two metropolises, the cluster analysis shows that the remaining thirty-nine Italian capital cities can be grouped into two groups, numerically similar, with opposite characteristics. The first group—cluster 2 – is concentrated in the northern part of the country (Fig. 4) and includes those cities with a sustainable urban mobility system, both in terms of car fleet and bike lanes, and favorable economic conditions (Table 4). By contrast, the second group—cluster 3 – comprises those cities with an old car fleet, without bike lanes and pedestrian zones, where the use of public transportation is limited and the average income is low (Table 4).

However, cities included in cluster 3 also show a better air quality and a lower number of accidents per capita; this result substantiates the findings of the PCA previously described, namely that lower air pollution concentrations in Italian cities are not associated with higher concentration of eco-vehicles.

5 VALIDATION OF THE CONCEPTUAL MODEL

Based on the findings described above, we have updated the conceptual model presented at the beginning of this paper. Figure 5 summarizes the main associations highlighted by both PCA and cluster analysis.

Comparing the two conceptual models (Fig. 1 and Fig. 5), two differences emerge: (i) some arrows have switched from dashed to solid; (ii) the others have disappeared. (i) Solid arrows correspond to those relationships that appear to be significant according to the statistical analyses. (ii) All the other relationships were erased because they are not supported by data.

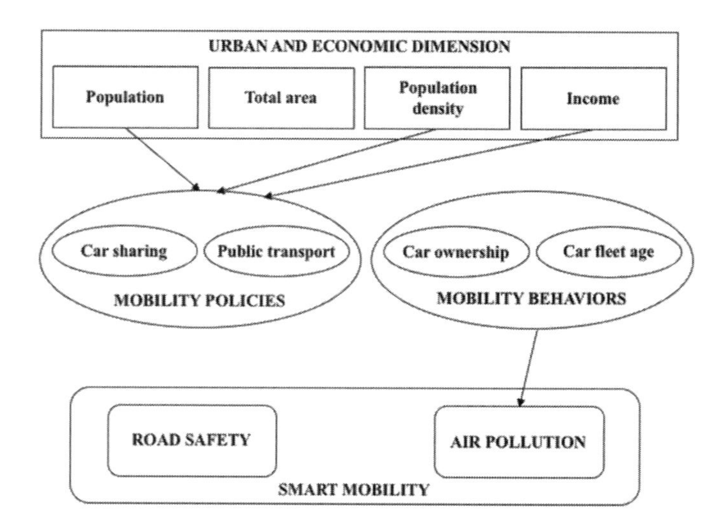

Figure 5. Revised conceptual model.

6 CONCLUSIONS

This work aimed to assess urban mobility across main Italian cities. More specifically, we used a principal component analysis and a cluster analysis to analyze two types of relationships: (i) the relationships between the urban and economic size of a city and both its mobility policies and citizens' mobility behaviors; (2) the relationships between both aspects of urban mobility (i.e. policies and behaviors) and road safety and air pollution, which are two of the most important elements considered for moving towards a more sustainable and smarter mobility system. The results of the two multivariate statistical techniques showed that the demographic dimension of an urban area and its economic level are positively associated with the supply of public transport and car sharing; in other words, more populated and wealthier cities offer a better service in terms of public transportation and car sharing. Furthermore, air pollution depends on the number of vehicles per inhabitant and the age of the car fleet, but not on either the concentration of eco-vehicles or the implementation of more sustainable mobility policies.

The main limitation of this study is that the availability of open access data on urban mobility at a city scale is limited. If more data were available, also at European level, an interesting future development of this work would be to compare urban mobility across main European cities.

REFERENCES

Dameri, R.P., & Benevolo, C. (2017). ICT Intensity in Smart Mobility Initiatives. In Smart City Implementation. Springer International Publishing. pp. 85–108.

Gargiulo, C. (2014). Integrazione trasporti-territorio-strumenti, interventi e best practices verso la Smart City. CLEAN.

Jolliffe, I. (2002). Principal component analysis. John Wiley & Sons.

Kassomenos, P.A., Vardoulakis, S., Chaloulakou, A., Paschalidou, A.K., Grivas, G., Borge, R., & Lumbreras, J. (2014). Study of PM 10 and PM 2.5 levels in three European cities: analysis of intra and inter urban variations. Atmospheric Environment, 87, 153–163.

Kaufman, L., & Rousseeuw, P.J. (2009). Finding groups in data: an introduction to cluster analysis (Vol. 344). John Wiley & Sons.

Papa, R., Gargiulo, C., Battarra, R., Niglio, R., Fabricatti, K., Pappalardo, G., ... & Oliva, J.S. (2016). Città Metropolitane e Smart Governance Iniziative di successo e nodi critici verso la Smart City.

Papa, R., Gargiulo, C., Franco, S., & Russo, L. (2014). Urban smartness Vs urban competitiveness: a comparison of Italian cities rankings. Tema. Journal of Land Use, Mobility and Environment.

Papa, R., Gargiulo, C., Russo, L., & Franco, S. (2017). On The Relationship Between The Promotion Of Environmental Sustainability And The Increase Of Territorial Competitiveness: The Italian Case. International Journal of Sustainable Development and Planning, 12(4), 655–666.

Town and Infrastructure Planning for Safety and Urban Quality – Pezzagno & Tira (Eds)
© 2018 Taylor & Francis Group, London, ISBN 978-0-8153-8731-2

Soft mobility & old town centres. Case study: Genoa

F. Pirlone, I. Spadaro & S. Candia
Università degli Studi di Genova, Genoa, Italy

ABSTRACT: The aim of this paper is to revitalize and renovate historic centres through the identification of new soft mobility solutions for tourists and residents.

Old town centres face several problems that affect the quality of life, many of them are a result of the inability of existing infrastructure to cope with the transportation needs of the population.

The historic centre of Genoa is characterized by a labyrinth of ancient narrow streets, medieval and renaissance palaces. Residents and tourists should move easily in a sustainable way, but it has also some negative aspects: there are many architectonical barriers, security problems and blight zoned. This paper identifies specific practices to improve the quality of life and the attractiveness for tourists and residents of the historic centre of Genoa improving current mobility solutions. Soft mobility represents the best solution for this area, and the research defines or improves pedestrian zones and bike paths.

1 INTRODUCTION

Old town centres face several problems that affect the quality of life, many of them are a result of the inability of existing infrastructure to cope with the transportation needs of the population. "The realization of soft mobility networks offers the opportunity to enjoy a slow landscape, gently moving through the territory where the natural and historic matrix of the place are fused together with the person" (Busi, Pezzagno, 2006).

The historic centre of Genoa is one of the largest in Europe. It is the testament of the glorious past of the city and its maritime hegemony across the Mediterranean sea. The historic centre is characterized by an intricate urban fabric made up of narrow streets (caruggi) unexpectedly spaced out by small squares. These little streets with their tastes, smells and sounds are the real soul of the city. Over the years, Genoese merchants built different beautiful masons, with magnificent frescoes and little courtyard; most of them are now inscribed on the Word Heritage List of the UNESCO. The historic centre has a medieval matrix but it is not difficult to identify different architectural styles that were overlapped throughout the years. It is a real labyrinth of ancient lanes, palaces, medieval walls, gothic loggias, turrets and bell towers.

Genoa's traditions are melded with its historic centre where is it possible to find centenary shops selling craft objects or local specialities prepared according to ancient recipes. The past is the cradle of the present and the future; different ancient masons are now trendy restaurants, bar, conference hall... Tourists and citizens live and experience Genoa old town centre at the same time. The city must provide the best conditions possible to ensure the good functioning of the area including transport respecting tourists and citizens needs.

The carruggi are really narrows, sometimes even only 1,50 meter larger. For this reason almost all the city centre is a pedestrian zone; cars and public transports, excluding the underground, are not allowed to circulate inside, residents and tourists instead can move easily in a sustainable way.

This is a perfect solution for the environment, but it has also some negative aspects. There are many architectonical barriers (damaged pavements, stairs, sloping roads), security problems (narrow streets, insufficient light) and blight zoned (waste management, social problems).

2 OBJECTIVE OF THE PAPER

The aim of this paper is to revitalize and renovate historic centres through the identification of new soft mobility solutions for tourists and residents. "Even though there is not yet a unique definition, we can argue that soft mobility (pedestrian, cycle and other not motorized displacements) is a zero impact mobility trying to be alternative to the cars use" (La Rocca, 2010). There is the need for new solutions capable of getting this heritage accessible and usable for all. "Sustainable mobility provides an alternative paradigm within which to investigate the complexity of cities, and to strengthen the links between land use and transport." (Banister, 2008).

This paper identifies specific practices and new sustainable itineraries to discover: historic small shops and markets (food and wine experience); multi-ethnic everyday life (cross-cultural ways); Genoa's history (cultural itineraries). These good practices are important to improve the quality of life and the attractiveness for tourists and residents of the historic centre of Genoa improving current mobility solutions. In order to prevent that the proposed interventions remain isolated measures, it is required an integrated planning tool that includes a longer-term, global, and organic design. Since the area under study is an important tourist attraction, it is strategically important to draw up a specific Plan for Tourism Development. "This Plan should propose a series of actions that lead to a tourism management which is sustainable and aware, and therefore able to consider the main issues that intersect with the theme of tourism—mobility, transport, waste, water resources, energy... – or, at least, that can collect and analyze, in a sustainable vision, the results reported inside other spatial Plans" (Pirlone, Spadaro, 2017).

3 METHODOLOGICAL APPROACH

Soft mobility represents the best mobility solution for the historic centre of Genoa. The main aspects considered by the authors are: a long-term vision, a clear implementation strategy, an assessment of the current performance, a monitoring process, a transparent approach, a stakeholder involvement action.

Initially, the research analyses the global mobility system with a focus on the accessibility of the historic centre. Particular attention, in the specific case of the historic centre of Genoa, should be paid to the analysis of the architectural barriers that represent a limit in terms of mobility, accessibility and security. Simultaneously, the paper studies best practices for soft mobility to be adapted to the features of the case considered. In the last session, the authors propose new solutions (identification of: means of transport, paths and connections) that are environmentally, economically and socially sustainable. These solutions should be implemented within a spatial plan for territorial management: the Tourism Development Plan.

4 DESCRIPTION

The methodological approach is innovative in several aspects. In particular, the research involves objective parameters and indicators in the selection and definition of soft mobility itineraries.

The main parameters used are (Table 1):

- Accessibility. The presence and the proximity of public transports such as buses, trains, subways...; the morphological characteristics of the urban fabric such as slope or staircase presence;
- Security. It can be determined by the analysis of the lighting and the flows of people;
- Attractiveness. The presence of museums, churches, shops that attract people's gravitation;
- Flows. It is quantifiable measuring the presence of people during day/night, winter/summer…;
- Services. Services are important to make the itinerary sustainable and interesting for users (whether they are resident or tourists).

Table 1. Objective parameters and indicators used by the authors to select and define soft mobility itineraries.

Parameters	Indicators
Accessibility	– proximity to means of transport – reachability – accessibility for disadvantages people – morphological aspects (slope, stairs, damaged pavements)
Security	– lighting – presence/absense of people – waste management – criminality
Attractivines	– museums – historic palaces – churches – typical products – shops
Flows	– day/night – winter/summer
Services	– street forniture (benches, fountains) – toilets – food courts – signage – apps for mobile phones

The analysis considers also one subjective aspect: people perception crossing historic centre of Genoa. This parameter could be quantified through questionnaires or interviews with different samples. The quantification of the above mentioned indicators helps in the selection of the soft mobility paths, and in the identification of the existing good practices suitable for the historic fabric under study.

As regards the methodological approach, the authors initially analyzed the global mobility system and the accessibility of Genoa old town.

Fig. 1 shows the state of the art of the local public transport (bus, subway, railway lines, lifts), the current bike sharing service (Mobike) and the network of roads. The plan reproduced here demonstrates also that the old town centre is inaccessible to car traffic as a restricted area. MoBike is the bike sharing system of the city of Genoa that supports the use of bicycle as a everyday mode of transport. The service includes a number of stations located in close proximity to the main points of interest of the city. It is possible to rent a bike with a membership card or using a cash card.

Data analyzed by the Municipality of Genoa show that people use more this service around the main railway station—Piazza Principe Railway Station—and inside the city centre. The 40% of transfers aren't done to go to work but for short occasional travels (for example to go shopping or for leisure activities).

The research provides a specific section for the identification of parameters and indicators aimed at defining new solutions for soft mobility paths inside the historic centre of Genoa.

Going into detail, an excerpt from the indicator analysis, for the first three parameters, is reported below:

– Accessibility: The plan in Fig. 2 shows the travel times necessary to walk across the city centre. All the main attractions and places can be easily reached within a walking distance of 10 minutes. De Ferrari square is considered the core of the old town centre. This square is equidistant from the two main railway stations and it is well connected with the underground and many different bus lines.

In terms of accessibility, another important indicator that should be considered is the morphological aspect. The slope gradient is one of the main critical aspects that represents a real architectonical obstacle not so much for pedestrian as for cyclists. As reported in

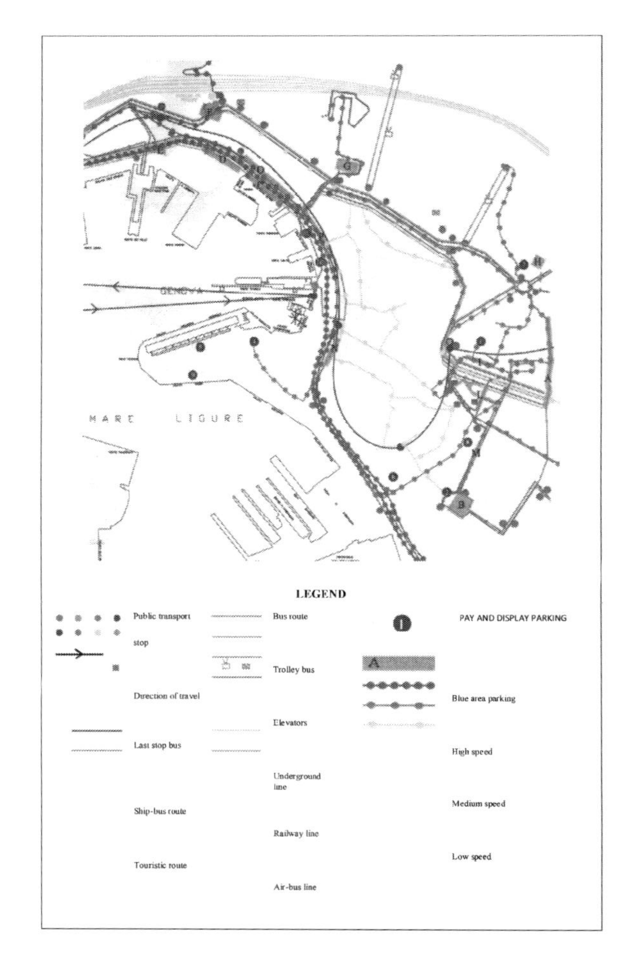

Figure 1. State of the art: Local public transport, bike sharing service, network of roads.

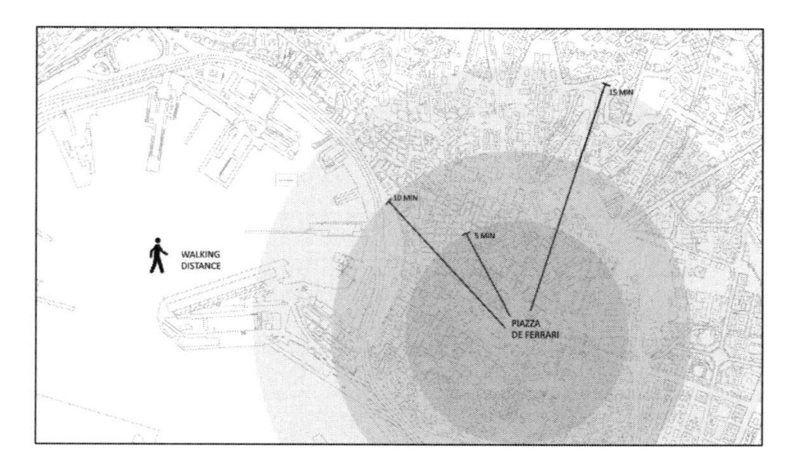

Figure 2. Walking distance inside the historic centre of Genoa.

Figure 3 the average slope inside the old town centre is of 10%. For this reason in Genoa there are many electric bicycles, more than other European cities. Also the first bike sharing system texted in Genoa had different electric bikes. During the first year ware reported some problems related to electrical batteries and some normal bicycles were introduced.

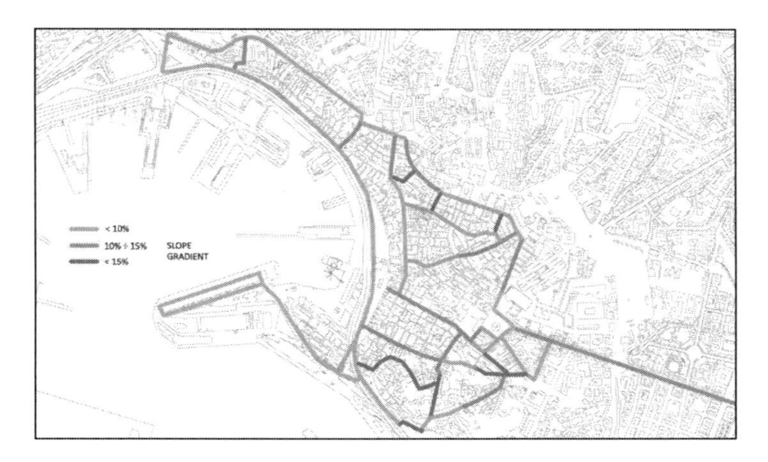

Figure 3. Slope analysis.

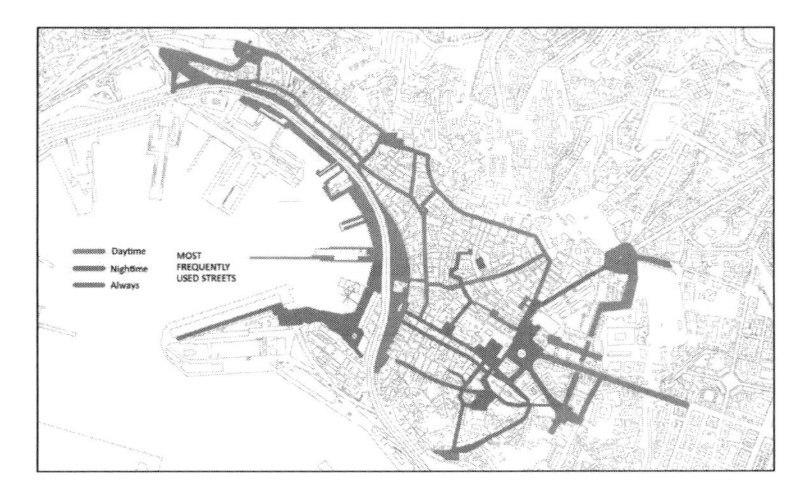

Figure 4. Most frequently used streets during daytime and nigh time.

– Security. In Figure 4 are reported the most frequently used streets during daytime and nigh time. Constant presence of people usually generates feelings of safety.

Anyway during the day the city centre is generally considered as safe place (Fig. 4).

– Attractiveness. To define the most attractive areas, the authors identified the main points of interest according to different categories: cultural attractions—monuments, museums, theatres… – food and wine attractions—restaurants and shops—leisure facilities, offices and services—school, post offices…–. In Fig. 5, is reported the map with the main historic shops that sell traditional products for the food and wine category.

At the same time, good practices regarding soft mobility in old town centres have been analyzed. Cycling remains somewhat marginal in the current policies for the old town centre of Genoa and budgetary allocation reflects this status. Cycling can have many advantages as short-distance mean of travel in urban areas: it is environmentally friendly, provides cost-effective mobility, and offers an opportunity for health and physical fitness by regular exercise. Cycling however isn't the main solution for the old town centre of Genoa because pedestrian zones will remain the favourite practice. In particular, small electric cars, rickshaws, and Segway are good examples. All the itineraries proposed by

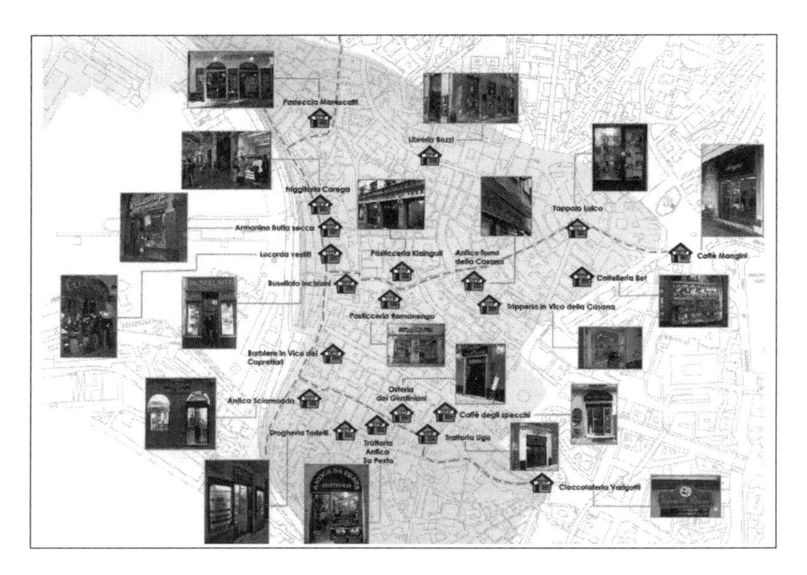

Figure 5. Historic shops inside Genoa city centre.

the research should be interconnected with the public transportation system enhancing intermodality solutions.

5 RESULT OF THE RESEARCH

"The improvement of the pedestrian and cycling mobility, especially in urban areas, is a key action to reduce the huge environmental costs of the motorized transport which is still today the main transport mode for urban and regional travels" (Galderisi A., Ceudech A. 2008). The research suggests a list of objective parameters—with their specific indicators—to identify the best soft mobility itineraries inside city centres. This methodology has been applied to the case study of Genoa. In this case transversal and vertical bike paths are proposed to create a preferential way to cross the historic centre without disturbing pedestrian mobility. These tracks valorise streets that actually are unsafe because isolated. New different pedestrian ways are defined according to people and tourist needs. These ways are related to thematic itineraries: culture, food-wine and historic markets, shopping, cross-cultural ways. In Figures 6–7 are reported the localization of the cycling and pedestrian itineraries identified during the research. Coloured lines could be used to recognize the itineraries and to ensure a good interaction between visitors and places. These lines indicate the soft mobility paths without modifying the historic pavement and interfering with the existing signage (Fig. 8). In parallel with the realization of cycle and pedestrian paths, it is suggested to develop new services such as wifi area, urban street furniture and recycling bins. Electric means of transport (small electric cars, rickshaws, and Segway) are introduced to help people with restricted mobility, improving the accessibility to the city centre. The soft mobility network is also integrated with the local public transport system (metro, buses, trains) creating a unique sustainable mobility offer.

"In support of the solutions presented, the resident population and those directly involved in the tourism sector should be made aware of the advantages of soft mobility through educational programmes" (Sustrans, 2007). These routes contribute also to the ideas of 'functioning city' and 'touristic friendly centre'. As regards to this last concept, all the thematic itineraries and the bikeways designed are important elements towards a Plan for Tourism Development.

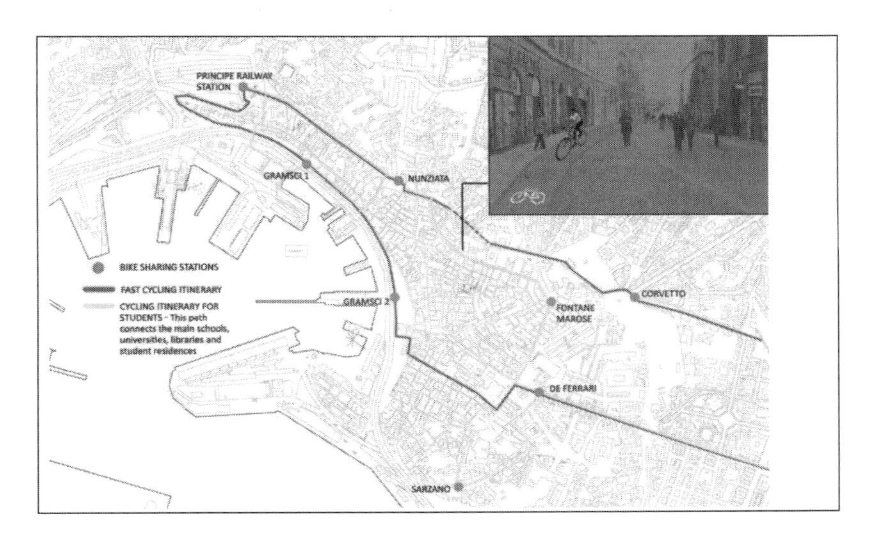

Figure 6. New cycling itineraries.

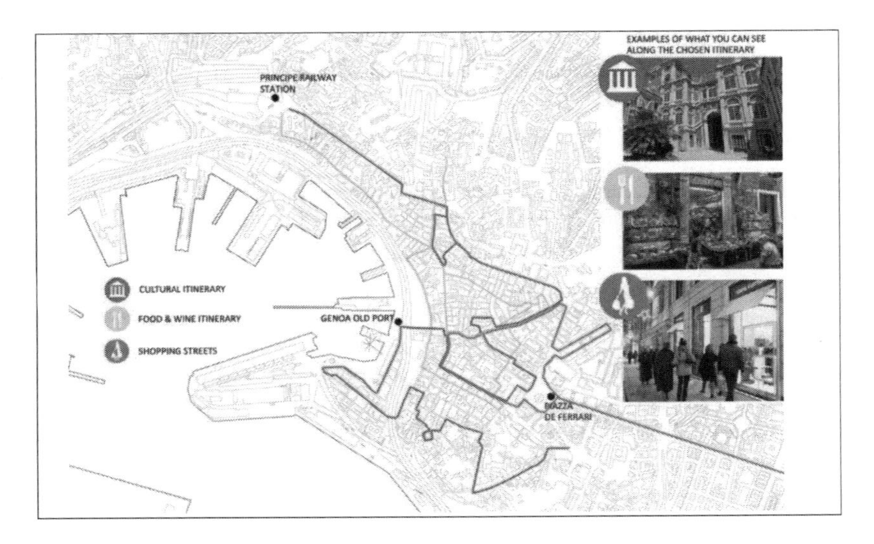

Figure 7. Thematic pedestrian itineraries.

Figure 8. Thematic itineraries: coloured lines in Via Balbi and Via San Lorenzo.

6 CONCLUSIONS

For about 15 years Genoa has been experiencing an industrial transition characterized by the fragmentation of the productive system. The city started a transformation process towards a smart city, which improves quality of life through sustainable economic development based on innovation. Genoa is developing new components of the post-industrial economy as creative industries, technology and tourism. The Municipality of Genoa has recognized tourism as a strategic factor of sustainable development as regards economic, social, cultural and environmental aspects. The current touristic offer is too much fragmented. Moreover there is not a municipal spatial plan that addresses the problems related to tourist mobility. According to this paper, soft mobility is the preferred solution to visit, live and cross the city centre of Genoa. One important aspect that should not to be underestimated is the need for change of mentality/culture in favour of cycling. "Of primary importance to transfer a best practice from a city to another, it's to ensure that the context from which the practice is derived is comparable to the context in which it will be applied. It's necessary to choose the best correction factors to use. One of this factors that mainly influences cycling success is the diffusion of a cycling culture between both citizens. Also the presence or the absence of a strong political commitment could influence cycling diffusion" (Candia, Pirlone, 2015).

Therefore, in light of the current tourist vocation of the city of Genoa and the need for greater attention to soft mobility, it is important to define a urban planning tool that coordinates tourist activities inside the historic centre. The thematic itineraries and the bikeways designed in the paper are important elements that should be considered and included in the Plan for Touristic Development of the city of Genoa.

REFERENCES

Banister, D. (2008). The sustainability mobility paradigm, Elsevier, pp. 73–80.

Busi R., Pezzagno M. (Eds.) (2006). Mobilità dolce e turismo sostenibile, Gangemi editore, pp. 1–382.

Candia S., Pirlone F. (2015). Cycling as best practice for urban renovation. Study case: The city of Genoa, CSE, pp. 79–88.

Ceudech A., Galderisi A. (2008). Soft Mobility and Pedestrian Networks in urban areas, TeMA Journal vol. 2, pp. 21–28.

La Rocca, R.A. (2010). Soft Mobility and Urban Transformation. TeMA, Università degli Studi di Napoli Federico II.

Pirlone F, Spadaro I. (2017). A Sustainable Tourism Action Plan in the Mediterranean coastal areas, International Journal of Sustainable Development and Planning, WIT Press, pp. 995–1005.

Extra-European approaches to town and infrastructure planning

Town and Infrastructure Planning for Safety and Urban Quality – Pezzagno & Tira (Eds)
 ISBN 978-0-8153-8731-2

Challenges of transport node in public space: Itaquera—São Paulo case

Y. Labronici Baiardi
Universidade Presbiteriana Mackenzie, São Paulo, Brazil

F. Anwar
Università degli Studi di Brescia, Brescia, Italy

ABSTRACT: The territory surrounding stations zones is not only a geographical space to acess a transport infrastructure, but also a living, social and creative space. The central issue to be discussed is how to integrate an urban station node into the territory that promotes diversity of uses and urban vitality.

This paper stresses how the infrastructure of urban mobility became a culture of erosion in territory, using the experience of Corinthians-Itaquera subway stations in the city of São Paulo (Brazil).

The objective of recognizing the negative forces that act on territory is to understand them, fight them and turn them into constructive forces.

The analyses are carried out through plans, sketches and related pictures from the site to identify these areas before and after the node-station intervention. The results identify the challenges of urban design for the space appropriation by the community, and the predominance of orthodox urbanism ideology.

1 INTRODUCTION

The connection places among different networks have great importance and transform the *station area* into key challenge of urban dynamics (Ascher, 2010) and very peculiar locations.

They are important nodes in a heterogeneous transportation networks, but also *a place*, both temporarily and permanently inhabited part of the city. There is often dense and diverse assemblage of uses and forms that may, or may not, share the life of the node (Bertolini, 1996).

Therefore, it is faced with *mobility structures*, usually with imposing dimensions and scales and are part of the contemporary urban landscape (Secchi, 2006, p.113). Smets and Shannon (2010) reinforce also the new role of landscape and *contemporary infrastructure*, *dialoguing with architecture, mobility* and city by integrating *territories*, reducing marginalization and segregation as a result and encourage mobility modes (p. 8. stress by author).

The *public space* plays an important role in the city and can contribute to micro accessibility of a station if it has the elements to make it *legible* in the eyes of those who seek to access a node in the metropolitan transport network (Baiardi, 2012). *Space* where occurs an intensification of different people, has accumulation of different activities.

The core of the neighborhoods can be *places* with good combinations of main uses, huge amount of streets, dense mix of buildings with different architectural styles made in different times, high concentrations and flows of people. The surrounding areas of the subway stations are key elements in this wide network of small urban territories.

It's challenging to articulate the *node* and dimensions of *place* in the territory of stations, making them compatible, balanced and beneficial. The areas of stations, especially those with a high passenger demand, can offer advantages to increase mobility, to promote real estate development, urban cohesion, social vitality and increasing the environmental gains

(Conceição, 2015), especially in the public space. However, these benefits are not optimized in the spaces near the stations often enough.

2 THE EROSION OF TERRITORY IN SÃO PAULO, BRAZIL

Robert Moses worked in São Paulo—South America when he was hired in 1949 to prepare the "Public Improvements Plan for São Paulo"[1] by Intern...onal Basic Economic Corporation, IBEC, trading company placed in New York.

Moses brought to São Paulo the model of the 'Highway Research Board', which intended to adapt cities *to accommodate the horizontal peripheral expansion in residential suburbs of motorized middle class*.

This model is adapted to "Avenue Plan"[2], with a wide differentiation from road typology proposed in this plan, explains Anelli (2006, p. 3),

"*His 'urban expressways' would be more appropriate to high volumes of traffic compatible with the complementary policy of road transport. No crossings level and without interference entries and vehicle exits in buildings, expressways Moses constitute a diverse city of Prestes Maia boulevards. For its full efficiency, grid expressways must have independence from the urban fabric through, regardless of whether or not destroys it*"[3].

After the development of "Avenues Plan" – the big plan was reinforced by 'Public Improvement Plan for São Paulo'.

Parallel to this moment, a rampant incentive to the automotive industry took place, which began to be installed in the country and the vertiginous growth of the road infrastructure. And at the 50 s effectively began the abandonment of the railway system as a public policy of collective passenger transport. It was due to lack of public policy for appropriate transport that is modernized and integrated to a multimodal system and the beginning of a new explosion of population growth.

As the result of this policy it encouraged the motorized transport and the construction of hundreds of kilometers of streets, widenings of roads, small sidewalks, bridges and viaducts. Only 78,7 km of subway lines to transport at least 12 million people that live only in São Paulo[4].

In São Paulo, the prolonged implementation of the "Avenues Plan" over the past century and *extreme functionality of transport nodes* built in many territories resulted in *eroded zones*.

The result of the sum of these ideas will be more and more implemented by urban interventions *disconnected with the territory*, extremely functional, leading not only to highways, but any intervention linked to urban form as subway bus stations or transport node (Baiardi, 2012).

3 OBJECTIVE OF THE PAPER

RegardinG the mobility issue, Jacobs (2011, p. 379[5] was in harmony with the relationship between land use and mobility. She challenged in 1961 that: "How can we provide urban

---

1. Translated by author from Portuguese to English—Plano de Melhoramentos Públicos para São Paulo.

2. Plano de Avenidas" or "Avenue Plan", translated by author was structured by a system of radial and perimeter avenues to organize the flows, decongest the central area and enable the endless expansion based on successive road rings. The highways would not be just simple means of transport, would also be urbanization instruments. The spatial structure for main avenues, were in the form of boulevards, sidewalks, trees, finally, a concern with local connections and the artistic aspect of urban composition (Maia, 1930).

3. Translated by author from Portuguese to English and highlighted by author.

4. Avaiable at: http://www.metro.sp.gov.br/metro/numeros-pesquisa/demanda.aspx. Accessed at: 21th ago. 2016.

5. Own Translation from Portuguese to English.

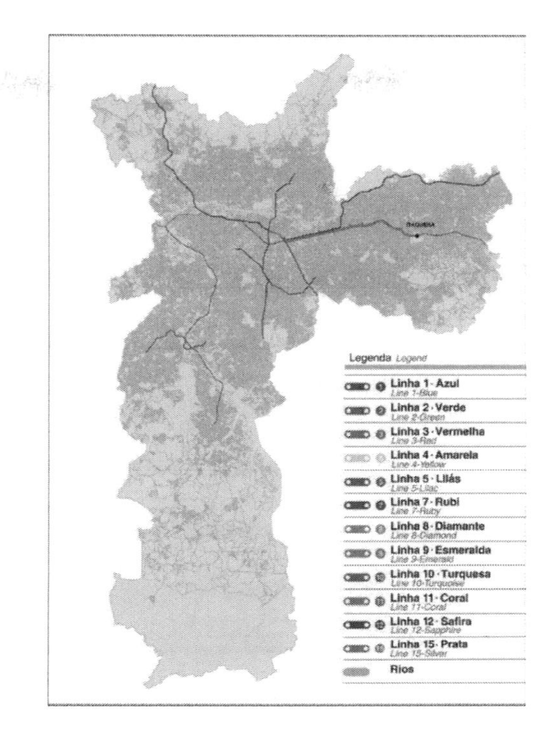

Figure 1. Location of Itaquera subway in the subway network in São Paulo city. Source: Geoportal, Baiardi treatment, 2017.

transport without destroying the related, complex and concentrated use? – that's the question, or by adopting another point of view, how to provide a use of the complex and focused land without destroying the related transport?"

The complexity of these territories, the sidewalks use, the busy street corners, the legibility of urban space brings with it a permanent succession of *eyes* and set movements of a complex uninterrupted urban ballet that results in the appropriation of public space.

The aim of this paper will be on expressing the critical discussion about creating an actual diverse urban scenario between Itaquera station and its surrounding areas.

4 RESEARCH APPROACH

For this paper, one set of graphical analysis was produced to survey the *Urban level* detailed below. It is separated in two main categories: as a 'place' (P) and 'transport node' (N) based on construction Bertolini's (1999), Baiardi's (2012) concepts. The desired analysed area is within 'walkable radius' around 500 m in 10 minutes.

5 DESCRIPTION

The station is located in the east side of the city (approx 18,5 kilometers from Sé station—downtown). It is the fnal station in Line 3 (Red), opened in October 1988 (Fig. 1).

6 URBAN AREA LEVEL: PLACE

a. Infrastructure and the tracing of streets:

The evolution of the maps (Fig. 2) show that the area was completely split until the seventies. The route of the railways were structural elements that stand out in a territory now characterised by a fine grid, without clear hierarchies and the permanence of the void in the study area. In a scheme from 2004, it is possible to observe the changes in the opening of the subway station in 1988, and the transfer of the railway line to the same region of the subway tracks—parallel to a new highway—"Av. Radial Leste". And where there were the train tracks, the "Av. do Contorno" was built. In 2015 there were more changes in the trace of streets motivated by the construction of the soccer stadium.

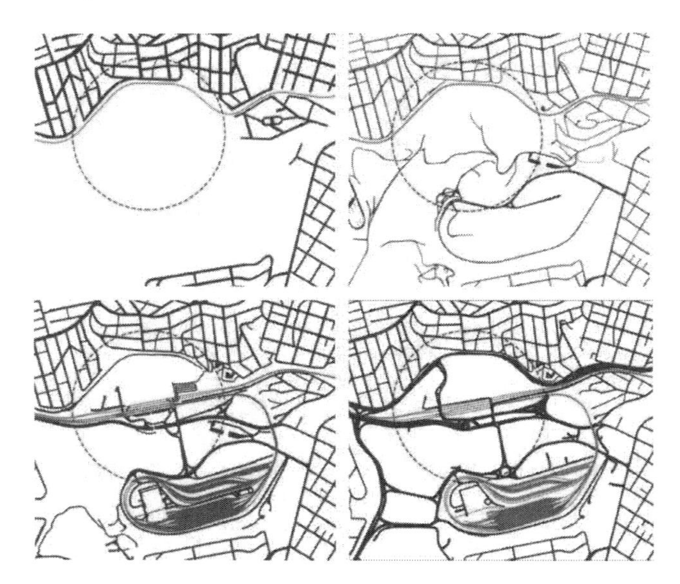

Figure 2. Infrastructure and trace of streets at Itaquera area in 1930, 1974, 2004, and 2015. Source: Sara Brasil, Gegran and MDC maps, Baiardi treatment.

In the 1st image: 1930 – red line was train line, contouring the big void. Nowadays is *Av Contorno*, main access for people to access the subway. Here it is possible to visualize the potential of void of the big public area.

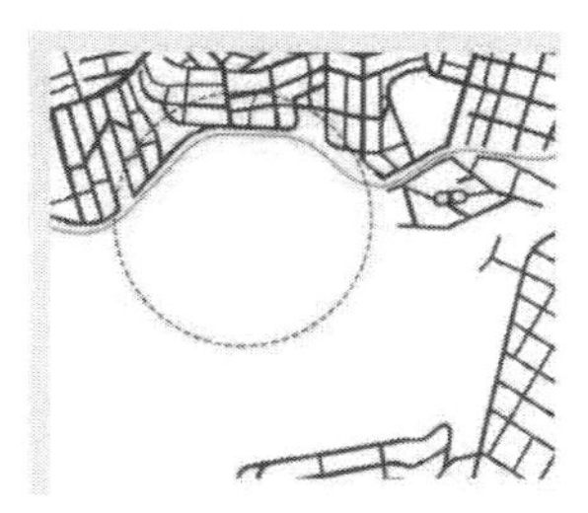

2nd image: 1974 – Start of subway infrastructure in the middle of the empty area, used to have a small river and small mountains surroundings.

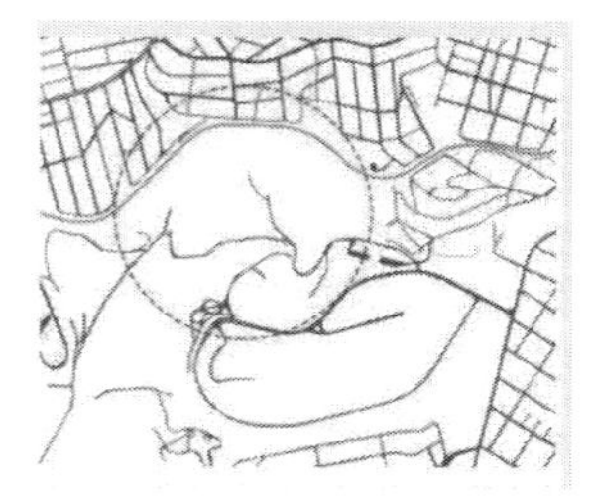

3rd image: 2004 – Main axes is 'Radial Leste' higway and the subways infrastructure was built during 70 s and 80 s, cutting the empty area, covering the river, splitting the neighborhood.

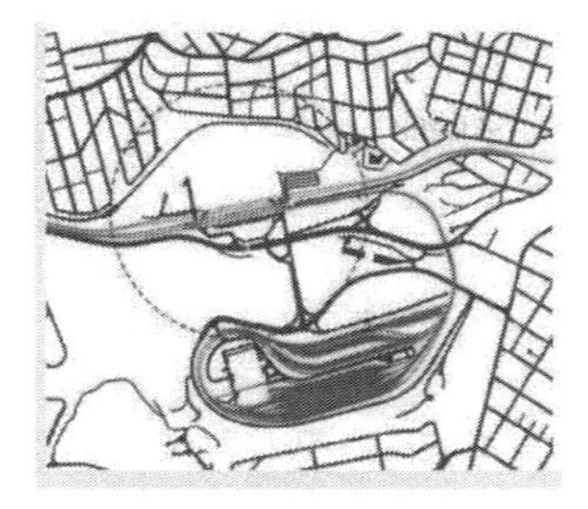

4th image: 2015 – Construction of new streets most of them for car reinforcing the void for people. There are no small passages for people, only big infrastructure cutting from east to west.

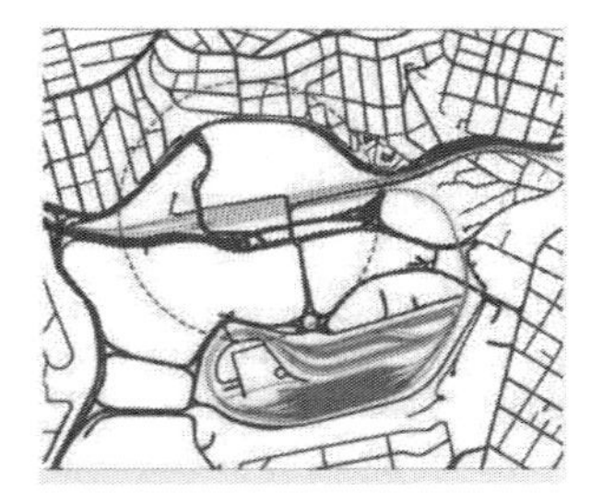

b. Built-up areas

It is observed that the spaces between streets were filled, until 1974, by an intense residential occupancy of low density. The contemporary maps show the arrival of intermodal centrality and an incipient occupation of 'large plots' (Fig. 3).

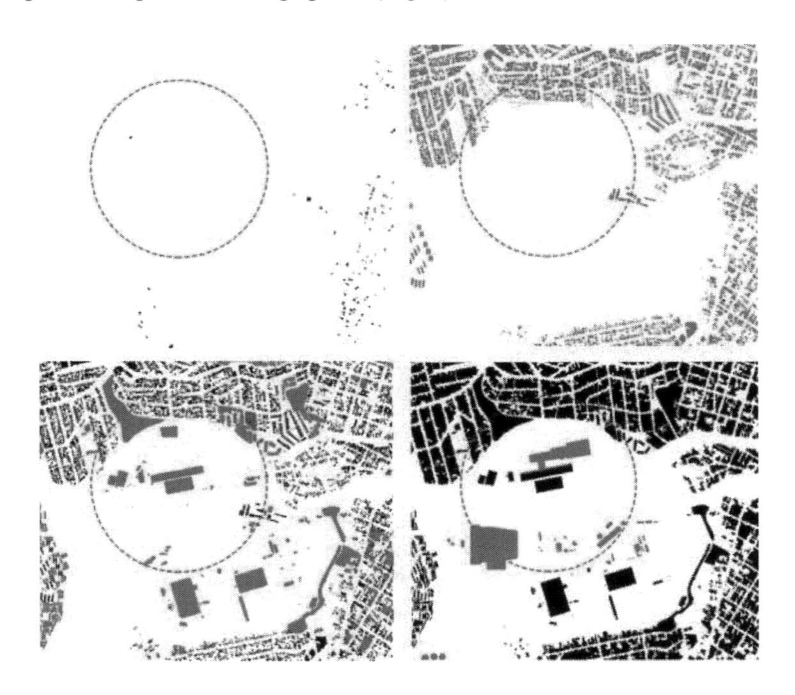

Figure 3. Built-up areas and void at Itaquera area in 1930, 1974, 2004, and 2015.
Source: Sara Brasil, Gegran, MDC maps, Baiardi treatment.

In the 1st image: 1930 – It is clear to see the circle of 500 m. By 1930 used to have the trace of streets (Fig. 1) but few houses.

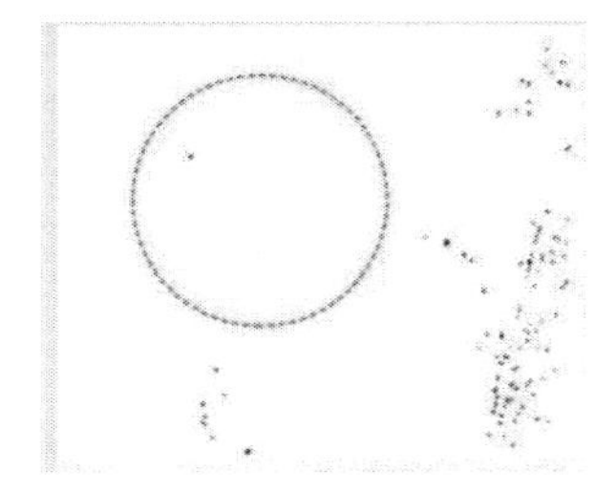

2nd image: 1974 – The area is completely filled with new houses (red color).

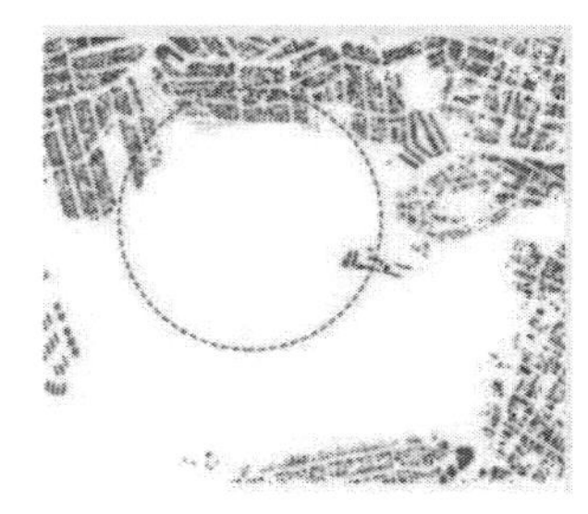

3rd image: 2004 –Most of the new houses after 1974 are slums. It is possible to see also the public buildings as the subway in the middle of circle.

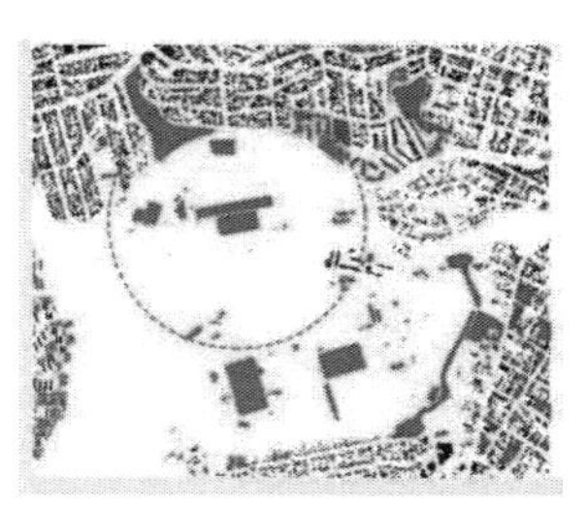

4th image: 2015 – Construction of the stadium and shopping mall.

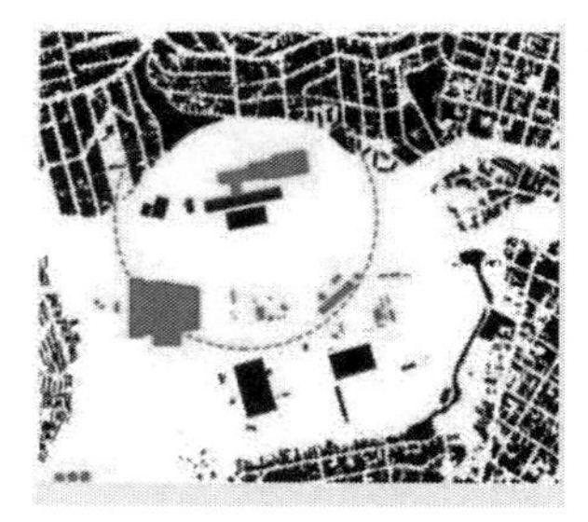

c. *Main activities*

There are several institutional activities in the 'large plot', quite isolated and most of the times with fences. They have predominance of low-density housing at the edges (Fig. 4).

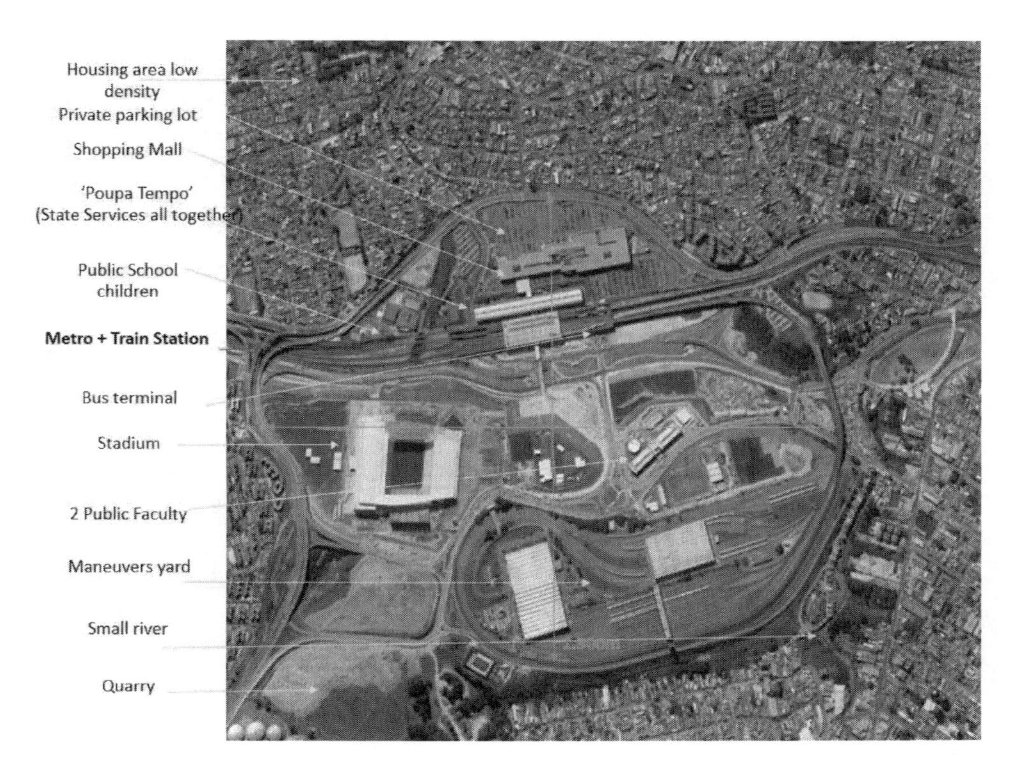

Figure 4. Main activities surrounding at station. Source: Yara Baiardi, based on Google Maps.

7 URBAN AREA LEVEL: NODE

As subway station, it is frequented by 2,493,000 passengers/month (May 2015/Metrô). As a terminal station it has a huge train maintenance yard. It is connected to Line-11 (Coral—Eastern Express) of CPTM, reopened in 2000. There is also a municipal bus terminal, but without bus lanes around the station. The main access road is "Av. Radial Leste" and "Av. do Contorno". It was implemented by the "Metrô Company" that used the expansion of the subway to re-urbanise part of the bordering regions of Line 3. With the implementation of a soccer stadium in 2014, new road interventions and the implementation of a small grid of cycle lanes have been foreseen. Also, there are large areas dedicated for parking private cars and residual areas called 'public spaces' (Figs. 5, 6, 7, 8) in the surroundings in the south entrance of subway station. At north entrance (Figs. 9, 10), you must get inside in a box—shopping mall that is beng expanded or get around it through parking car areas (Figs. 11, 12).

8 RESULTS

After the urban analyses it is evident the absence of station's relationship with territory. The newly inserted buildings (shopping, stadium and schools) are a breakthrough for improving territory vitality to attract new flows, but it does not dialogue with the urban structures of the region because it maintains the implementation of paradigm of the plot closed itself without urban and design strategies. Railways and highways are classic borderline examples (Jacobs, 2011) and it is an urban challenge that must be faced.

Figures 5 and 6. South Entrance to subway and train station. Poor public space, only for fast passage, no connections with the 'eyes' of the city. Source: Baiardi, 2017.

Figures 7 and 8. Radial Leste Avenue in the South part of station area. Highway for cars with poor quality of sidewalk or safety, or cycle lanes. Main access from East to downtown. Source: Baiardi, 2017.

Figures 9 and 10. Av Contorno in the north part of station area, with high possibility of connectivity with urban fabric. But the presence of fences and shopping mall destroying the potentiality of better connections and public uses in the station area. Source: Baiardi, 2017.

Figures 11 and 12. Expansion in a closed box of the shopping mall, with no connections with eyes of the city with surroundings occupied by car parking in area without good public spaces.

9 CONCLUSIONS

Based on the analysis of spatial transformations of the urban insertion of Itaquera station this paper is grounded on the investigation of their functional role in the territory which has become more complex due to intensification of displacement possibilities within a metropolis.

The understanding of this process regarding the modernized of transport system complex addressed as part of Urban Hub integrated in a systematic territory, avoiding the 'large plot' with the predominance of emptiness and many waste areas. This could become the opportunity focus that through the right project solution can guide the spatial transformations and the connections with territory and its surrounding area.

As a work in progress[6], it is still premature to establish all the effects of the spatial transformations surrounding Itaquera subway station. So far, the experience means more than defining conclusions, stress on some challenges to overcome, regarding station zone, its surrounding area and new role in a contemporary territory.

The complexity of the territory, use of sidewalks, the busy corner that brings a permanent succession of eyes arranging in uninterrupted movements in a complex urban ballet that results in confidence in the appropriation of public space as written by Jane Jacobs. It must also be applied in the surrounding of a 'node-place station'.

Corinthians-Itaquera Subway station is not only an attrition with automobiles or highways but an attrition with a complex transport node. Corinthians-Itaquera Subway station is an erosion in public space, in an erosion in the city.

There are no short blocks, there are no crosses on street corners, there is no a complex combination of crossover uses, there is no urban structure that supports the urban diversity of its surroundings. There are no eyes in the 'blind area' in which predominates the 'Great Plague of Monotony'. There is no urban ballet surrounding the station, because people rushed hastily to the entrance of the station.

The objective of recognizing the negative forces that act to the blind territory is to understand them is to try to fight them, or better yet, turn them into constructive forces (Jacobs: 2011, p. 268).

This means that a node of transport in a contemporary city needs to be upgraded, (re)build, (re)discussed and (re)connected to the urban territory, in order to create integrated spaces. So it is not only a transport system in a fragmented practice. It is necessary to (re)discover the territory, re(discover) the place surrounding the station to influence its own growth of diversity and urban vitality.

REFERENCES

Anelli, R. L. S. (2007) Redes de Mobilidade e Urbanismo em São Paulo: das radiais/perimetrais do Plano de Avenidas à malha direcional PUB. Arquitextos, São Paulo, ano 07, n. 082.00, Vitruvius [online]. Available at: http://www.vitruvius.com.br/revistas/read/arquitextos/07.082/259.

Ascher, F 2010 Novos princípios do urbanismo. São Paulo: Romano Guerra.

Baiardi, Y. C. L, Alvim, A. A. B, 2014. Mobilidade urbana e o papel da microacessibilidade às estações de trem. O caso da Estação Santo Amaro, SP. Arquitextos, São Paulo, ano 14, n. 167.07, Vitruvius, [online] Available at: <http://www.vitruvius.com.br/revistas/read/arquitextos/14.167/5185>. [Accessed 27 Jul. 2014].

Bertolini, L. (1999). Spatial development patterns and public transport: the applications of an analytical model in the Netherlands. Planning Practise and Research, 14(2), pp. 199–210.

Bertolini, L. (1996). Nodes and places: Complexities of railway station redevelopment. European Planning Studies, 4(3), pp. 331–346.

6. This paper is part of a Ph.D research still in a progress at Universidade Presbiteriana Mackenzie UPM—School of Architecture and Urbanism (São Paulo, Brazil) in a co-tutelle with Leibniz Universität Hannover. Yara had CNPq scholarship during her stay in Hannover University (2016) and from UPM in 2017.

Conceição, A. L. M, (2015). From city's station to station city: an integrative spatial approach to the (re) development of station areas. PhD. Delft: Technische Universiteit Delf.
Jacobs, J. (2011). Morte e vida de grandes cidades. Rosa, C.S.M. (translation), Cavalheiro, M.E. H (translation review). São Paulo:WMF Martins Fontes – Coleção Cidades, 3° ed.
Maia, F. P, (1930) Plano de Avenidas. São Paulo: Prefeitura de São Paulo.
Smets, M.; Shannon, K. (2010), The landscape of contemporary infrastructure. Rotterdam: NAI Publishers.
Secchi, B. (2006). Primeira Lição do Urbanismo. São Paulo: Perspectiva.

Town and Infrastructure Planning for Safety and Urban Quality – Pezzagno & Tira (Eds)
© 2018 Taylor & Francis Group, London, ISBN 978-0-8153-8731-2

Urban strategy to enhance the safety and health of citizens: A case study in Shanghai

R. De Lotto, V. Gazzola, C. Morelli di Popolo & S. Sturla
Università degli Studi di Pavia, Pavia, Italy

ABSTRACT: Infrastructural system has a strategic and fundamental role for the economic development of a society but it is also one of the sectors that surely produces the greatest pressures on the environment and landscape with reduction of their value and, in the time, of the quality of citizens life.

In context such as megacities in China (such as Shanghai), spatial planning decisions reinforced a strong car oriented approaches. Starting from the results obtained in Shanghai in 2015, the authors present experiences about multilayer infrastructural theme as innovative and efficient urban planning to solve safety and healthy problems.

For the safety, multilayer infrastructural planning approach can ensure a more flexible distribution of displacement opportunities and alternatives to transfer vehicles within the city depending on their destinations.

For the healthy, the use of innovative materials can help to solve the problem of urban pollution the pollution-reducing (photocatalytic materials as titanium dioxide TiO_2).

1 INTRODUCTION

1.1 *Development of urban environment in China about safety and health*

The continuous increasing of urbanization rate[1] and of the cities' dimension that defines megacities in China such as Shanghai, makes Chinese environment and people that move around under a huge pressure.

Social aspects (such as communities' improvement) and infrastructural ones (such as new networks creation and multimodal development) becomes fundamental in Chinese's policy in relation with the expansion.

Regarding to this aspect, Tunney Lee (2016) emeritus professor of MIT, underlines the need of some specific characteristics of infrastructure: the first issue is the increment ability and flexibility, the second is sustainability and low-impact. The first characteristics is related to the fast capacity of development of the city: when the city continuously change, the infrastructure network must be able to expand (in "long-term") and to be flexible in order to adapt to different uses (in "short-term"). For the second issue, it is clear the importance of a glance to environment: the infrastructure plan needs to be advanced, with a strong reduction of carbon emission, and low level of energy use.

There are two "opposite" aspects of Chinese strategy (described in the years Plans[2]): from one point of view the implementation of urban area, with the focus on efficiency of infrastructures, from the other the importance of the environment in polluted Chinese mega cities.

The strong and fast urbanization model, used in the last 30 years, cannot be the correct way because of the enormous use of land resources for cities expansion. In addition public services, quality of urban life, efficiency at the moment must be improved. (Sha et al. 2014).

1. In 2030, 60% of global population will live in a city (Un Habitat).
2. 11th, 12th, 13th Five Year Plan.

2 OBJECTIVE OF THE PAPER

2.1 *New strategies for the protection of Chinese environment*

For the developing of the future China, the main goal is to find an equilibrium between the economic growth (connected to urban growth) and the protection of environment. Chinese government tried to enforce the environmental policy in the last decade.

In 2015, the expectation for the 13th Five Year Plan (2016–2020) by the population are principally connected with the Environmental protection, in which people mostly care about (73,8% of people)[3]. Starting from the interest of people, two of the Major Objectives of current Plan are the improvement of the standards of living and quality of life and the enhancement of environment and ecosystem. In this Plan, the new focus is to diminish the use of land for construction, to reduce the use of water, to reduce the energy consumption and to increase the non-fossil energy from 12% to 15% of primary energy consumption. Other important investments are related with the air quality, strictly connected to the infrastructure system, passing from 76,7% of the year/days of good air to more than 80, with also a reduction in PM 2.5 intensity in cities.

According to this strategy, the academic research tries to focus more on the thematic of environmental protection, health protection, and safer people.

The International Summer School 2015, themed on "Design Against Smog—Thermodynamic methodology for Chinese Architecture" and hosted by the College of Architecture and Urban Planning (CAUP) of Tongji University, has been the opportunity to define ideas, methods and paradigms in response to the environmental crises in contemporary China. Interdisciplinary researches in the field of architecture, planning and design permit to study original solutions in terms of the relationship of air/smog and architecture/urban environment that are subsequently implanted into a specific site, Lujiazui District.

Lujiazui is a central business district (CBD) located in Shanghai, on the eastern part of the Huangpu River in Pudong district and sits opposite to the old financial and business district of the Bund. Since the early 1990s, Lujiazui has been developed specifically as a new financial district of Shanghai and today it covers an area of 6.8 square kilometers (equivalent to Manhattan Downtown financial zone). After more than twenty years developments, there are more than 30 high-rise buildings among which the Oriental Pearl TV Tower (467.4 m, 1995), the Jing Mao Tower (421 m, 1998), the Shanghai World Financial Center (492 m, 2008), the Shanghai Tower (632 m, 2014), important landmarks defining the skyline of the city.

From the semantic point of view, Lujiazui CBD is a gravity center of the whole Shanghai metropolis and it is a symbolic element of the megacity in wich all the critical issues are amplified. Principal problem of Lujiazui CBD is the strong car oriented approach of its planning; it causes traffic pollution the absence of an efficient and well-balanced system of pedestrian and environmental connection. Infrastructural road system produces great pressures on the environment and landscape with reduction of their value and, in the time, of the quality of citizens life. The excessive use of cars represents in fact a double threat for the citizens' health and safety (Figs. 1, 2).

About health issue, Chinese statistic reports show that tailpipe emissions account for up to 30% of PM2.5. So, in urban areas, diseases related to the air pollution, affect tens thousands of people each year, posing serious health threats as respiratory and cardiovascular diseases, cancers, … (WHO, 2005). About safety issue, in China, the deaths by vulnerable user road (pedestrians and cyclists) are over 35% of all road users (Traffic Management Bureau of the Ministry of Public Security of China, 2010).

It is necessary to act in this area with a deep reorganization of utilities and activities related to mobility system that can bring important benefits to traffic management, economic regeneration, environmental and social improvement.

3. http://english.gov.cn/policies/infographics/2015/10/28/content_281475222032525.htm.

Figure 1. Heavy smog shrouds in Lujiazui CBD. Source: Getty Images.

Figure 2. High traffic level in Lujiazui CBD. Source: Flickr site.

3 RESEARCH APPROACH

Research approach on the theme combines together concepts relative of traditional architectural and urban patterns (with their dependence on infrastructural system) and the use of innovative building materials to develop new forms, new aesthetics, new organizational relationship and social experience.

In particular, developed strategy consideres the use of new building materials and technologies/experimentation able to contrast pollution problems applied to multilayer infrastructure and mobility system. The combining of the benefits of these approaches can be considered an innovative and efficient urban planning strategy to solve safety and healthy problems.

3.1 *Multilayer infrastructure*

With its multilevel structure, layering technique permits the organization on overlapping layers and the creation of a complex system where different urban functions exist together. In particular in the case of infrastructural system, layering approach is able to ensure more balanced distribution between vehicular and pedestrian traffic and consequently a widespread quality (ecological and social) of infrastructures in the city.

The idea of stratified and vertical city is not new. Starting from Renaissance, when Leonardo da Vinci presented a project for Milan as ideal multilevel city (Città ideale), to arrive at the beginning of last century, town planners and architects have largely experimented the third dimension of public spaces and its infrastructures. Several visions of multilayer cities, where new transport systems (vehicular, railway, air,..) released from the ground level, starts to catch on. Important examples are *The Cosmopolis of Future* of Moses King (1908), the *Visionary City* of William R. Leight (1908), the *Rue Future* of Eugène Hénard (1910) the *City of the Future* of Harvey W. Corbett (1913), the *Città futura* of Antonio Sant'Elia (1914), Le Corbusier's projects of *Ville Contemporaine* (1922), *Rio de Janeiro* (1929) and *Plan Obus* (1930), the *Vertical city* of Hilbeseimer (1924) and the *Aeroporto urbano* (1931) of the Italian painter Tullio Crali. In the 1960s, as result of the application of high technology to urban multilevel infrastructural systems, other projects are presented: the proposal of George Candilis, Alexis Josic and Shadrach Woods for the *Downtown of Frankfurt* (1962), the *Paris Sous la Seine* of Paul Maymont (1964), the ideas of Archigram group about high-tech cities (*CityInterchange* (1963) and *Plug-in City* (1964)) and Paul Rudolph's project of *Lower Manahattan Expressway* (1970).

At the middle of 1970s the proposals for multilayer cities stopped but they restarted in the 1990s when the problems of urban soil (scarcity and degradation) bring architectural researches towards typological experimentations of new multifunctional buildings: *Millenium Tower* of Norman Foster (1989), *Hyper Building* of Rem Koolhaas (1996) and *KM3* of MVRDV group. Figure 3 shows a few examples of these multilayer infrastructural experiences.

Innovative aspects, costs and technical feasibility

From the analysis of the past experiences about multilayer infrastructures, it is clear that the layering technique is an approach mainly applied for the organization of hard transport systems (vehicular and underground) with little specific attention to pedestrian mobility. In 1970s successfully experiences relative to creation of urban multilevel spaces used only by pedestrians spread in several cities in North America (Robertson, 1994). Skyways and subways represent interesting applications of multilayer approach as solution for the separation of vehicular and pedestrian traffic able to determine good levels of safety, accessibility and comfort in urban dense areas (Hass-Klau, 1990). Respectively these kind of pedestrian passages link buildings (tertiary or commercial maily) with elevated networks or with under-

Figure 3. Multilayer infrastructural experiences.

ground tunnels. The project +15 Walkway developed in Calgary (1970), the Minneapolis Skyway System in Minnesota (1962), the Ville Intérieure in Montreal city (1962) and the PATH in Toronto (1970) are just a few examples (Fig. 3).

Even if there is a great number of projects about urban multilayer pedestrian networks, it is important defining key factors (both positively than negatively) to consider in terms of strengths and weaknesses to understand the real economic and technical feasibility of these projects (Papa, 2008). To synthesize concepts, authors propose a strength and weakness scheme.

Strengths

- Pedestrian safety: creation of specific spaces for exclusive pedestrian use and absence of crossroads with vehicular traffic for the reducing of the deaths degree by vulnerable road users (pedestrians and cyclists).
- Reduction of car use for short displacements: creation of networks with high connections to public transport system and reduction of general infrastructure load and level of air pollution.
- Environmental pedestrian comfort: creation of new pedestrian infrastructures characterized by environmental standards, high quality street furniture and paving.
- Attractiveness of the area: contribution to development and to revitalization of the urban area (especially in commercial areas as in the case of RESO network in Montreal).

Weaknesses

- Building and maintenance cost: in relation to the type of project (elevated or underground) and to the building techniques, costs are greater in comparison to traditional infrastructural systems.
- Visual and landscape impact: in particular skyway projects are often invasive and hard solutions for urban environment.
- Management activity: the absence of public/private partnerships limits the use of pedestrian spaces 24/7, exclusively during the opening of tertiary/commercial activities.

3.2 *Ecological materials*

The use of materials such as photocatalytic materials, in particular when they are applied to infrastructural system, can contribute to enhance air quality and to improve sustainability levels. As a matter of fact, photocatalytic materials represent a strategic way in air quality enhancement, thanks to their property to accelerate natural chemical reactions and promoting a faster decomposition of pollutants.

Titanium dioxide (TiO_2) is the most suitable photocatalytic material, used in constructions, for its property of stability, strong photo-oxidation power and its safety on human body. Ongoing researches show that cement-based materials containing TiO_2 have a good potential in urban pollution control and can reduce both organic pollutants and oxides such as NO, NO_2 and SO_2[4]. An important aspect, for vehicular traffic, is that TiO_2 break down air pollutants at ordinary temperature using ultraviolet light irradiation. The photocatalytic principle is at the base of Titanium dioxide cements properties: under the influence of ultraviolet light irradiation, TiO_2 is activated, subsequently, the pollutants are oxidized due to the presence of the

4. Successfully experiences can be mentioned:
5. Segrate (MI), 2002: cement-base coating on a bituminous road was used to cover the asphalt surface. Monitoring proved a reduction in NO (nitrogen oxides) on this urban road of around 60%.
6. Bergamo, 2006: self-locking blocks were laid with an average values recorded of around 45% of NO reduction;
7. Rome, 2011: photocatalytic cement-based paint was employed in order to renovate a tunnel and it was showed a NOx abatement capacity of over 20%.

photocatalyst and precipitated on the surface of the material. Finally, they can be removed from the surface by the rain or cleaning/washing with water.

In the past few years, research experiences had recognized (at urban—scale) both the benefits than the level of criticality of photocatalytic technology. Figures 4,5,6,7 show a few examples of the material applications. To synthesize concepts, authors propose a strength and weakness scheme.

Strengths

- Environmental quality improvement: reduction of air pollutants, self-cleaning and self-disinfecting properties of materials.
- Versatility of material applications: TiO_2 can be used in several ways. It can be added to the mixture proportion of cement-based materials (photocatalytic cement blocks pavements, photocatalytic pavements combining asphalt and cement mortars) or it can be applied as a surface coating (bituminous pavement sprayed with photocatalytic pollutants/emulsions.
- Abrasion resistance: the addition of TiO_2 does not modify the adhesion between wheels and pavements.
- Economic feasibility: It is very important to underline that to transform the building surface into a photocatalytic surface, it is enough to add around 100 € to the cost of a tradi-

Figure 4. Cittadella Bridge (1996, Alessandria). Source: www.richardmeier.com.

Figure 5. Morandi street (2002, Segrate). Source: Italcementi.

Figure 6. Dives Misericordia (2003, Rome). Source: www.richardmeier.com.

Figure 7. Tunnel Umberto I (2007, Rome). Source: Italcementi.

tional paint and for paving in photocatalytic blocks it costs on average between 10–20% more than traditional paving (Italcementi Group).
- Good efficiency and durability.

Weakness

- Physical properties: it is worth stressed that the addition of TiO2 into concrete material can influence on some properties as a heat of hydration, a workability, a setting time, a chemical shrinkage, a mechanical strength, an abrasion resistance, a fire resistance, a freeze resistance, a water absorption.
- Depending on environment conditions: the best environment conditions for the photocatalytic process are related to light intensities, low relative humidity in the atmosphere, no wind. In different conditions, the efficiency of material is not obtained.
- Limited intervention area: the photocatalytic air purification function is usually restricted to pollutants which are absorbed on the surface of the building materials.

4 RESULT OF THE RESEARCH: DESIGN PROPOSAL

New masterplan of Lujiazui CBD is designed to improve the quality of air and public space using a responsive system of growing networks and materiality to combat smog (Fig. 8). In particular the proposal (called by authors "the Vine" for its shape) is to connect important spatial nodes within the area with a vine-like structure and to superimpose a layer system that serves pedestrians and cyclists (in addition to the current car-oriented traffic pattern) in order to increase their safety and ease of movement (Fig. 9). On the street level, bicycle hubs are placed within a comfortable walking distance of the existing bus stops. These hubs provide bike rentals as well as space for personal bikes, making an easy transition for users of public transportation.

Another layer of modularizable prefabricated fleabane-like structure resolves problem of smog with the application of titanium dioxide (TiO_2). The perforated design increases the surface area to volume ratio, thereby increasing the efficiency of this photocatalytic material (Fig. 10).

Since that the condition of the air pollution is related to the presence of the sun and wind, the design of the new structure also aims to maximize thermal comfort (providing shade

Figure 8. Masterplan proposal.

Figure 9. The "Vine" structure proposal.

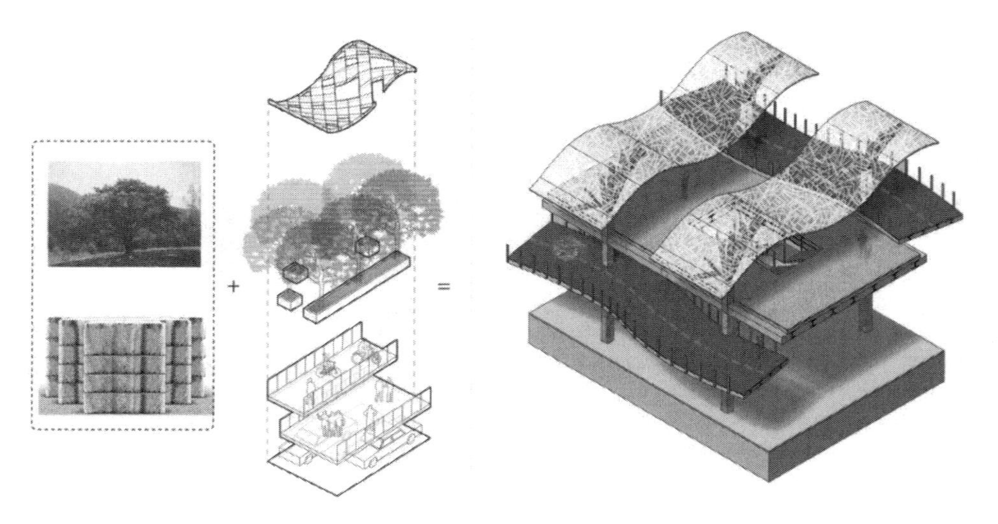

Figure 10. Element combination of the "Vine" pathway.

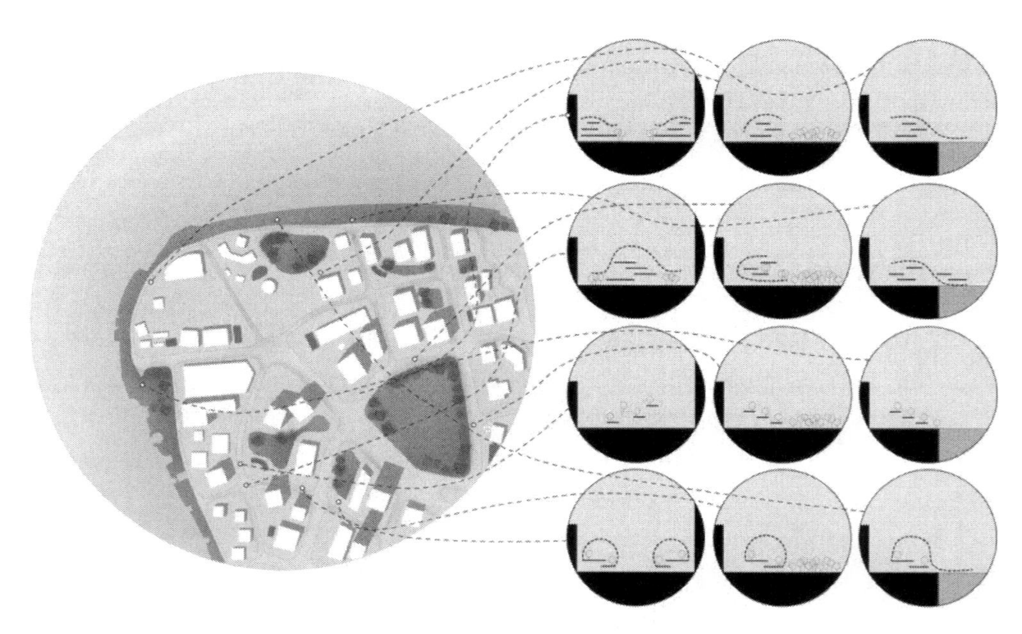

Figure 11. Development of the structure in Lujiazui District.

for people) and natural ventilation of the Lujiazui District to provide locals and tourists an enjoyable environment. With the help of specific softwares, major wind and sun contribution as well as the turbulence in Lujiazui district are identified to define different configurations, adapting to the surrounding environment (Figs. 11, 12, 13). Through air purification, temperature control corresponding to different seasons and airflow movement, the system provides a comfortable environment for the interior and clean air.

Furthermore space for activators such as small kiosks of coffee, food, and small shop vendors along the layers are allowed in order to improve the connectivity throughout the Lujiazui District and enhance the Vine structure. New multi-functional pathways give the citizens possibilities of healthier and safier lifestyle.

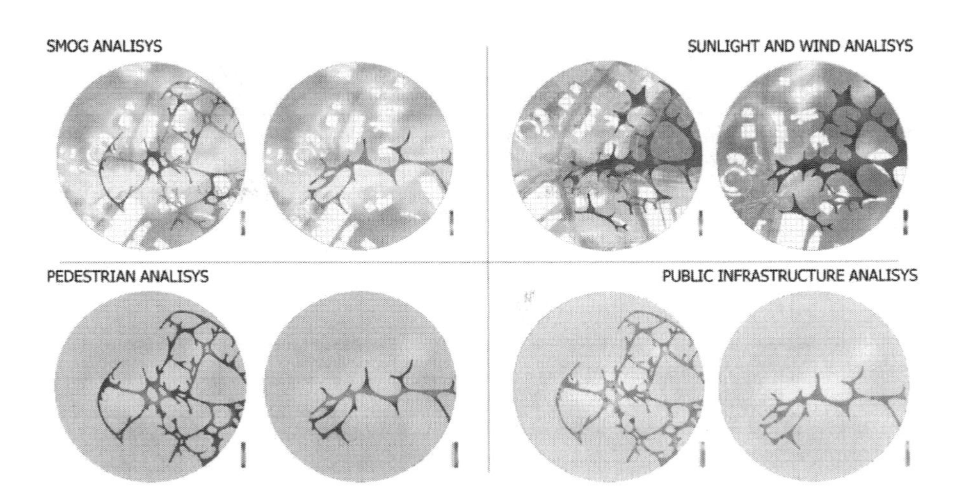

Figure 12. Thermal comfort and natural ventilation studies.

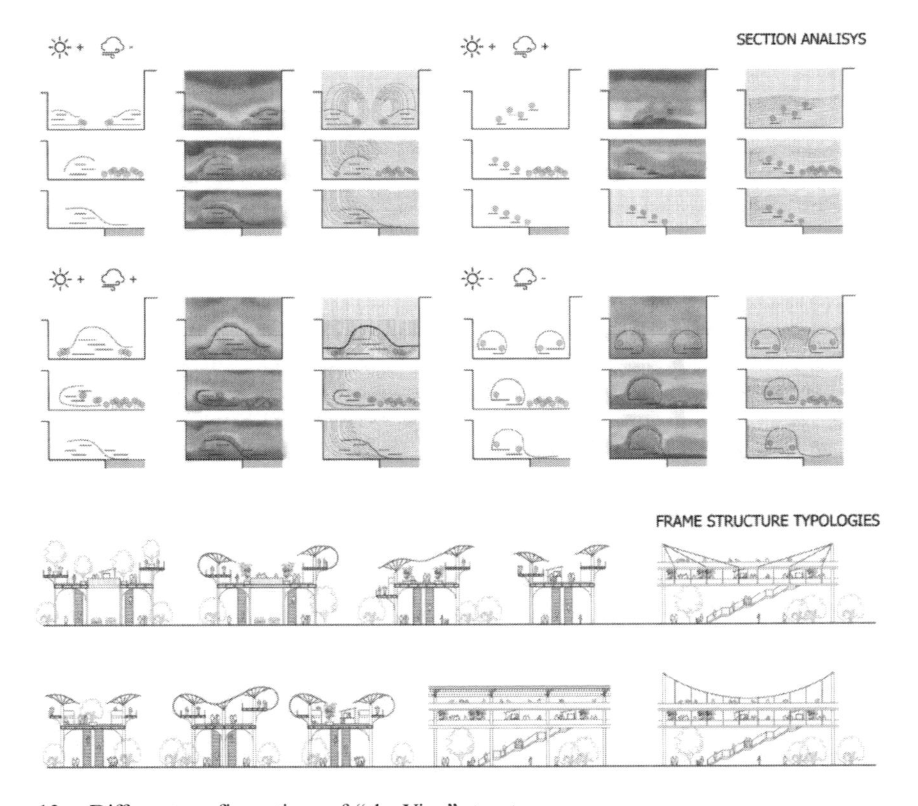

Figure 13. Different configurations of "the Vine" structure.

5 CONCLUSIONS

The proposal can be applicable to other areas in Shanghai or in other similar contexts and change everyone's life style in a real sense: it can be considered as a new paradigm. As Shanghai continues to develop in the future, this network can extend throughout the city to adapt to various needs and conditions. Authors' proposal provides a long term, holistic solution for urban multimodel mobility that is able to grow with the city or metropolis. By implementing this system, the future of Shanghai can look forward to healthy public spaces, greater connectivity, and a blue, smog-free sky. A cost-benefit analysis is needed to better address feasibility and extension of the intervention. Infrastructure can be implemented and flexible, sustainable and with low-impact (Tunney Lee, 2016).

According to the Chinese vision of the future, and analyzing the actual situation, the authors try to balance two important topics of safety and health themes. New ecological way to develop a "safer and more efficient as well as smart, eco-friendly, and interconnected, thereby ensuring they better play their role in supporting and leading economic and social development" (13rd Five Years Plan, p. 78).

ACKNOWLEDGMENTS

Students of Group 6 (The Vine project): Davide Masserini (Politechnico di Milano), Veronica Gazzola (University of Pavia), Alyssa Maristela (The University of Iowa), ZHANG Zhenwei (Beijing University of Civil Engineering and Architecture), WEN Zishen (Southeast University), SUN Tongyue (Tongji University).

Tutors: Minqu Michael DENG, YANG Feng.

REFERENCES

Bacchi M., Venturini L., (2009). Research, Design and Development of a Photocatalytic Asphalt Pavement. Proceeding on 2nd International Conference Environmentally Friendly Roads. ENVIROAD.

Beeldens A., Crispino M., D'Ambrosio G., Guerrini G.L., Vismaral S., (2012). Environmental benefits of innovative photocatalytic cementitious road materials. 10th International Conference on Concrete Pavements: Sustainable Solutions to Global Transportation Needs. Pages 912–923.

Boivin D.J., (1991). Montreal's Underground Network: A Study of the Downtown Pedestrian System. Tunnelling and Underground Space Technology. Vol. 6, N. 1. Pages 83–91.

Boonen E., Beeldens A., (2014). Recent Photocatalytic Applications for Air Purification. Belgium in Coatings, 4. Pages 553–573.

Cattaneo T., (2016). Study on Architecture and Urban Spatial Structure in China's Mega-Cities Suburbs. T. Cattaneo (Editor). Universitas Studiorum, Mantova.

Cassar L., Beeldens A., Pimpinelli N., Guerrini G.L., (2007). Photocatalysis of cementitious materials. In: Baglioni P., Cassar L., (Editors). RILEM International symposium on photocatalysis 'environment and construction materials', (Florence/Italy, 8–9 October 2007), TDP 2007, (Florence/Italy, 8–9 October 2007) Rilem Proc. PRO55. Pages 131–146.

Chen J., Poon C., (2009). Photocatalytic construction and building materials: From fundamentals to applications. In: Building and Environment 44. Pages 1899–1906.

De Lotto R., (2014). Riflessi operativi sugli assetti infrastrutturali e dei sottoservizi nella città flessibile. Urbanistica Informazioni. Vol. 257. Pages 31–33.

De Marco T., Fava G., Guerrini G.L., Manganelli G., Moriconi G., Riderelli L., (2013). Use of photocatalytic products for sustainable construction Development in Third International Conference on Sustainable Construction Materials and Technologies.

De Richter R., Caillol. S., (2011). Fighting global warming: The potential of photocatalysis against CO2, CH4, N2O, CFCs, tropospheric O3, BC and other major contributors to climate change. Journal of Photochemistry and Photobiology C: Photochemistry Reviews, Elsevier, 12. Pages 1–19.

Gazzola V., (2015). Reti infrastrutturali multilayer. Urbanistica Informazioni. Edizioni INU. Vol. 263 s.i.. Pages 9–12.

Hass-Kla C., (1990). The Pedestrian and City Traffic, Belhaven Press, London.

Janus M., Zając K., (2016). Concretes with Photocatalytic Activity. InTechOpen, Pages 141–161.

Künzli N et al., (2000). Public-health impact of outdoor and traffi c-related air pollution: a European assessment. Lancet, 356. Pages 795–801.

Lee T. (2016). Sustainable Neighborhoods in China. Through Inclusiveness, Connection & Environment. In Study on Architecture and Urban Spatial Structure in China's Mega-Cities Suburbs. Edited by T. Cattaneo, Universitas Studiorum, Mantova. Pages 25–47.

Linxue L., Jianjia Z., Zheng T., (2015). Air through the Lens of Thermodynamic Architecture: DESIGN AGAINST SMOG. Tongji University Press. Pages 178–193.

Papa E., (2008). Subway e skyway: infrastrutture sostenibili per la mobilità pedonale?. Laboratorio Territorio e Ambiente – TeMALab. Vol. 3. Pages 49–56.
Rashad A.M., (2015). A synopsis about the effect of nano—titanium dioxide on some properties of cementitious materials – a short guide for civil engineer. Reviews on Advanced Materials Science. Pages 72–88.
Robertson K. A., (1994). Pedestrianization Strategies for Downtown Planners: Skywalks versus Pedestrian Malls. Journal of the American Planning Association, Vol. 59.
Rowe P.G., (2008). Urbanizing China. In: Gil I. (Editor). Shanghai Transforming. Actar Edition, Barcelona. Pages 76–79.
Sassen S., (2008). Disaggregating the Global Economy: Shanghai. In: Gil I. (Editor). Shanghai Transforming. Actar Edition, Barcelona. Pages 80–85.
Sha Y., Wu J., Ji Y., Ting Chan S.L., Qi Lim W., (2014). Shanghai Urbanism at the Medium Scale. Springer.

SITOGRAPHY

National Bureau of Statistics of China: http://www.stats.gov.cn/english/
People's daily, China, English version: http://en.people.cn/index.html
United Nations, Department of Economic and Social Affairs, Population Division (2014). World Urbanization Prospects: The 2014 Revision, CD-ROM Edition. https://esa.un.org/unpd/wup/
World Health Organization (2005) Health effects of transport-related air pollution edited by Michal Krzyzanowski Birgit Kuna-Dibbert and Jürgen Schneider
World Meterological Organization: 'Globally averaged CO2 levels reach 400 parts per million in 2015':
https://public.wmo.int/en/media/press-release/globally-averaged-co2-levels-reach-400-parts-million-2015
11th Five Years Plan (2006–2010):
http://www.gov.cn/english/special/115y_index.htm
12th Five Years Plan (2011–2015):
http://english.gov.cn/12thFiveYearPlan/
13thFive Years Plan: for economic and social development of the People's Republic of China (2016–2020), central compilation & Translational Press. Download at: http://en.ndrc.gov.cn/newsrelease/

Town and Infrastructure Planning for Safety and Urban Quality – Pezzagno & Tira (Eds)
© 2018 Taylor & Francis Group, London, ISBN 978-0-8153-8731-2

Shared mobility service for the current and future challenges in mobility: From policy to implementation

C. Marques Zyngier
Instituto Metodista Izabela Hendrix, Brazil

José Luiz Noronha Cintra & Mário Sérgio Rocha Cintra
Gepro Gestão De Projetos, Brazil

ABSTRACT: This article is an inventory of good manners and what should be done to the future of urban mobility in Brazil. This paper analyzes Brazilian transportation policies and how they affect the population. The objective of this paper is to study the stagnation of urban mobility policies in Brazil, aiming to know the full nature of the process and to identify the causes and roots of this vicious circle. The aim is to identify the major impediments and obstacles to the implementation and maintenance of a consistent and sustainable urban transport plan. Finally, the paper proposes ways to make a short and long term comprehensive and sustainable plan, applied to the conditions and scenario of Brazil, considering all the steps (conceptual, legal, behavioral and physical) needed to its viability. The work also aims to suggest possible strategies, even if gradual, to open paths and create the conditions to overcome these obstacles.

1 INTRODUCTION TO THE REALITY OF BRAZILIAN TRANSPORTATION

Brazil has maintained a policy of transportation and urban mobility based on individual transportation. All transport infrastructure initiatives prioritize the use of the automobile considering it a symbol of progress and development.

In the last decades, several good specific public transportation projects were implemented in different Brazilian cities, always in a timely manner and based on personal initiatives, which should be continued and improved seeking the continuity and integration of solutions.

The predominance of car use in Brazil is mainly due to the road model implemented in the 1950s and enshrined in the quote "governing is to open roads".

This policy and culture led, over the last years, to the degradation and obsolescence of the passenger rail system transportation, throughout the country. Nowadays, we have limited subway systems in some cities, and railroads exclusively for the transport of cargo, mainly iron ore.

In Brazil, most of the infrastructure projects consist of large road constructions which prioritize assisting the circulation of cars through the creation of facilities and increasing the safety of pedestrians and users of other means of transportation.

2 OBJECTIVE

The objective of this project is to study the stagnation of urban mobility policies in Brazil, aiming to know the full nature of the process and to identify the causes and roots of this vicious circle.

The aim is to identify the major impediments and obstacles to the implementation and maintenance of a consistent and sustainable urban transport plan.

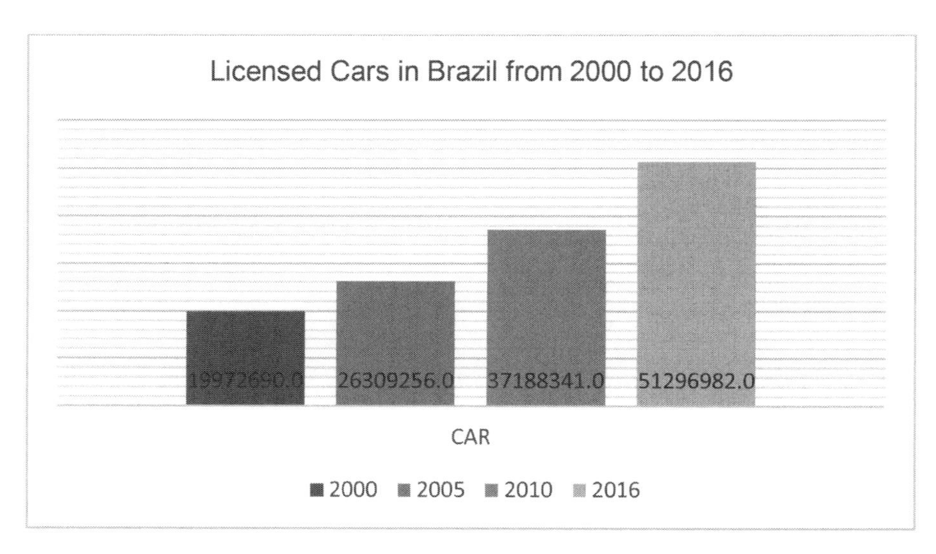

Graphic 1. Licensed cars in Brazil from 2000 to 2016.[1]

Finally, the paper proposes ways to make a short and long term comprehensive and sustainable plan, applied to the conditions and scenario of Brazil, considering all the steps (conceptual, legal, behavioral and physical) needed to its viability. The work also aims to suggest possible strategies, even if gradual, to open paths and create the conditions to overcome these obstacles.

3 THEORETICAL FRAMEWORK

According to Raia (2000), the displacement in cities is analyzed and interpreted in terms of a conceptual scheme which articulates urban mobility, the population masses and their movements, represented by the infrastructure that channels the displacements in space and time; and fluxes, which are macro decisions or constraints that guide the process in space".[2]

The cities had growth beyond expectation, such as Belo Horizonte, Rio de Janeiro, Curitiba and São Paulo, which were designed to be model cities and ended up becoming victims of contemporary urban chaos.

For Lerner (qtd. in Pereira), former mayor of Curitiba, Architect and Urbanist, "Brazil could be a great example in the area of mobility, but it is not because we keep on complicating the problem while we try to implant what is already obsolete in other countries. This movement of replicating what is obsolete makes you buy it as the latest novelty"[3]

Encouraging the varied transport system can be considered the basic prerequisite to the formulation of any long-term mobility policy. Thus, educational campaigns can be held, as well as the gradual implementation of localized solutions that become windows for perceiving the viability and benefits generated.

1. Denatran. Frota de Veiculos Published on Wednesday April of 2017 at 15h29.
2. Raia JR., A.A. 2000 Acessibilidade e Mobilidade na Estimativa de um Índice de Potencial de Viagens utilizando Redes Neurais Artificiais e Sistemas de Informação. São Carlos, São Paulo.
3. Pereira, Neli. "Brasileiro fica encantado com cidades europeias, mas não reproduz soluções aqui', diz lerner".

4 METHODOLOGICAL REFERENCE

Due to the limitation of the present presentation, it was not possible to extend the research to a sample of urban nuclei which would be sufficient to represent all the diversity of demands and topographic, demographic and sociocultural conditions of the cities in the country.

Specific aspects of some cities had been studied, and the current scenario of the city of São Paulo was chosen as a case study which, due to its size, political and economic importance, has a strong repercussion throughout the country of its problems and solutions.

The sequence of events related to the topics of traffic safety and citizen mobility, that have been happening in São Paulo in recent months, show a wide and surprising adherence to almost all the concepts stated in the objective of this research.

The fact that these events are extremely controversial and have wide media coverage allows for a detailed and documented approach to the various aspects involved in the process. And it also highlights the need for great changes we will propose in this paper.

5 CULTURAL AND BEHAVIORAL ASPECTS

The initial difficulty lies in the cultural and behavioral pattern of the population; whose change is indispensable to enable the formulation of any long-term mobility policy. For this to happen, it is essential to increase the interest of the population.

This cultural pattern is characterized by exacerbated individualism imposed by the socioeconomic model of the country and encouraged by the great media vehicles. The basic rule is to take personal advantage in everything, where it is only "good" what offers me the greatest benefit, even at the cost of other's loss.

This state of art takes practically all segments of society to a passive stance, omitting participation during the formulation of mobility or transportation policies, undergoing the influence of the media.

Behavioral change involves carrying out educational campaigns to society, initially promoted by class entities, study and research groups, neighborhood associations or segments of the population, NGOs, etc.

These same entities should lead the movement towards the participation of the society in the definition of priorities and the approval of mobility plans and projects of transport infrastructure.

The main objectives of these campaigns should be:

- Understanding the prevalence of collective interest over individual interest;
- Encouraging the acceptance and practice, by all social classes of the population, of the use of public transportations;
- The dissemination of the concepts which value practices that generate benefits to one's health and to other citizens;
- Awareness of the importance of respect for the rights of others;
- The prioritization of safety and pedestrian rights, especially children and mobility impaired people;
- The minimization of population passivity and "manipulated mass" syndrome;
- Convincing the population to accept restrictions and prohibitions of the indiscriminate use of cars in routine activities in the central areas of the urban centers where other means of transportation are available.

Educational campaigns can be valued by gradually implementing localized mobility improvement solutions which serve as windows for perceiving the viability and benefits generated and, also, encouragement to support these initiatives.

5.1 *Updating the applicable legislation*

In Brazil's current situation, one of the major causes of the stagnation of urban mobility and traffic safety policies is the lack of a sufficient and legal framework. This allows the approval of infrastructure constructions, the modernization of codes and regulations from the point of view of security and well-being of the citizens and ensures the continuity of constructions carried out or started.

In Brazil, there is a specific federal law, sanctioned in 2012, which governs the National Policy on Urban Mobility.[4] This law establishes the principles, guidelines and objectives for urban development and transportation and defines the responsibilities of municipal administrations, as well as the rights of the citizen in relation to this topic.

However, many of its guidelines and determinations are still not operating because of the lack of "how to" regulations.

An example of this situation is that, although the law determines the rights of citizens to participate in the definition and elaboration of Mobility Plans, there is no legal or regulatory mechanism that requires holding public hearings or popular referendums to carry out participation.

A long-term urban mobility plan calls for extensive data research, slow developing projects, and phased implementation projects that usually go beyond the mandates of governments who set out to develop them.

Nowadays, most of the governors'' changes, especially when they are from opposite political currents, generate alteration or abandonment of studies in advanced stages and project interruptions, delays or cancellations. It is necessary society lobby for the creation of a legal basis that offers a minimum guarantee of continuity and permanent solutions proposed and accepted by them.

A sustainable policy presupposes that there are laws that discipline the approval of projects of social interest and prevent the interruption or radical alteration of projects of urban mobility, transport infrastructure and urbanism for political-partisan reasons. Only the mobilization of society can lobby for these laws to be created, even if they represent amendments to the Constitution.

It is imperative that the regulation of these laws be debated within society, establishing the participation of the population in the approval of long-term plans and projects, and conditioning any interruption or cancellation of these plans or projects to the agreement of the society, through popular referendums or similar measures.

5.2 *Case Study – "São Paulo, the curious case of Brazilian retrograde"*

We present below a series of facts and events that have been happening in the city of São Paulo in recent months as a result of the election of a new city mayor.

The previous mayor implemented some changes in July 2015 to increase citizens' safety, such as the decreasing the current city's speed limit of two main avenues: Marginal Tietê and Marginal do Pinheiros (Graphic 2).[5] In these avenues, the number of accidents and deaths by road killings was absurd.

This led to reactions from various levels of society, such as the Brazilian Bar Association, which filed a lawsuit, saying the city administrator could not make such decision without consulting the population. The action demonstrates how the will of a minority with economic power and ease for motor vehicles still have priority over the interest of the people.

---

4. Federal Law for Urban Mobilty nº 12.587.

5. Rodrigues, Artur; Zylberkan, Mariana. Marginais sob Doria têm acidentes em alta e ambulante 'fixo' nas vias. Published on Sunday May of 2017 at 02:00h.

6. CET Companhia de Engenharia de Tráfego. Marginal Segura.

7. Portal R7.COM "Justiça derruba liminar e Doria poderá aumentar velocidade das marginais Tietê e Pinheiros".

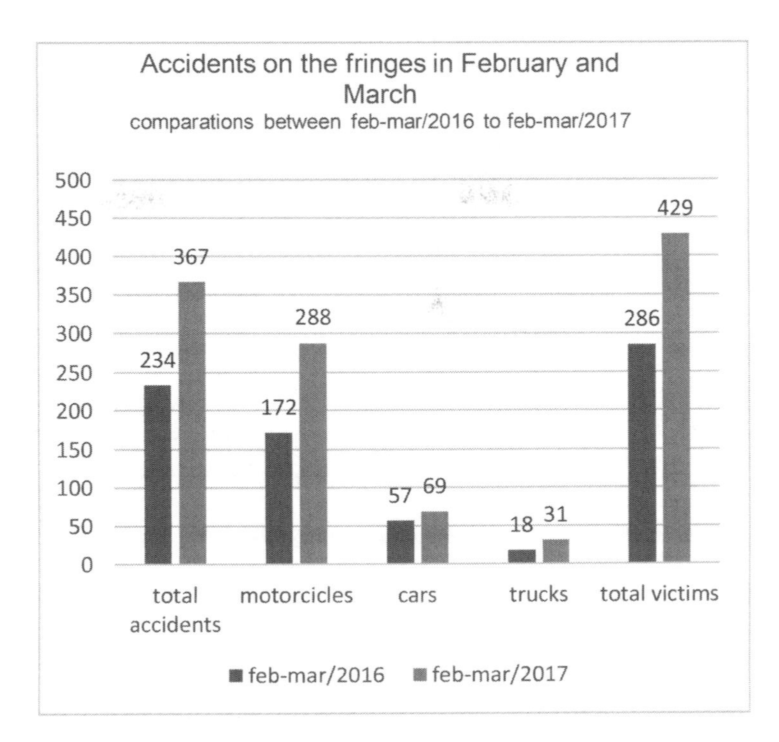

Graphic 2. Accidents on the fringes: Comparisons between February/March 2016 and 2017.

The text also says that the "'Safe Marginal' Program"[6] aims to increase traffic flow and, at the same time, provide security to road users" expecting to put in place "a) educational actions, b) improved road and traffic signs, c) efficient inspection and d) adoption of precautionary measures for prompt service to users".[7]

It is important to take into consideration the community's involvement and carry out detailed studies on the subject which needs to be presented in it's entirety to the population and to provide a better understanding of the issue. From this spectrum, for example, a vote in a plebiscite could be proposed in which the decision of the population could be heard and respected by public managers without intervention of the judiciary from lower levels, many with apparent private or political-partisan motives, with Provisional decisions that paralyze works, impose changes or force them to be undone. These changes remain at the center of a legal battle with successive concessions and appeals of judicial measures.

6 SHORT-TERM PLANS AIMING AT LOCALIZED BENEFITS

For the short-term plans aiming at localized benefits as possible solutions for improvements in the urban mobility area, small size and quick implementation solutions are required, which aim at localized results and can be widely viewed.

For example, it is possible to invest in infrastructure projects focused on the quality of life in districts farthest from the center of cities, implementing on the main street and trunks connecting to it, elements such as bicycle paths for small displacements, walking paths or Race, stimulating the residents to interact more with their neighborhood and neighbors, as well as a landscape project that makes that space be admired.

One can also make that any reform in the sidewalks, is discounted in the form of IPTU (tax on the urban property and territorial). Adding motivators and urban equipment that prioritize the pedestrian and seeking project quality and safety of all who pass through.

A short-term plan may also include improving lighting conditions on city streets and avenues. A well-lit city, with posts that consider the scale of the pedestrian, for example.

The discussion with society and proposition to create a series of restrictions on the indiscriminate use of the car in the central regions of cities, such as the prohibition of parking on the arterial roads, limiting the circulation of cars at certain times, the creation of Automatic toll collection along with the licensing of vehicles, and others.

With the change of awareness about mobility and sustainability, it is possible, for example, to encourage the creation of study and research groups that will lead to the participation of the development of projects that take into account the opinion of the communities involved and can put into practice the concepts Analyzed.

Educational campaigns in public and private education systems should also be discussed. Ghel[8] e Lerner[9] Propose that education for a more sustainable urban mobility and a more human city that should be presented starting at the kindergarten level. Actions for children to become familiar with and understand that the city can bring long-term benefits as they contribute to the formation of conscious citizens.

6.1 *Long term policy for transport and mobility infrastructure*

A long-term policy for transport infrastructure and urban mobility should reflect a set of comprehensive and integrated strategies and solutions taking into account current and future issues of urban mobility. Its complexity requires a slow development of concepts and a sequential, closely linked sequential implementation plan.

The main premises for policies of this type are the creation of infrastructure conditions that make feasible and attractive the collective transportation in arterial trunks and the prioritization of the collective interest of all segments of society and the well-being of the people.

The long-term strategy should also include updating the zoning of urban centers, defining the regions with the greatest population density or large flow of people, where there would be the gradual imposition of restrictions on the use of individual transport.

It is also possible to gradually increase the reduction of street speed limits, which, in addition to reducing pollution, induce a wave of kindness beneficial to all. In this case, the use of spring bars, create specific areas for pedestrians (as in most Italian cities) and limit the speed to limits that do not exceed the absurd, nor are they compromising the traffic flow, calling them *green zones* and deploy *traffic-calming* associated with limiting parking along public roads in order to avoid accidents and reduce traffic jams by people in double-row waiting or attempting to maneuver the car.

The investment in a good urban network that prioritizes walking and lowers car dependency can contribute to the decrease of rates of obesity. These campaigns must be recurrent, and ongoing management to management.

Long-term plans will then be construction planning and infrastructure that require years of study and execution that will target all classes of the population from the periphery to downtown, these works include actions for all.

7 CONCLUSION

We are on the verge of a collapse in mobility and this is a fact, but how could we then make what is shown in the law, indeed, done? The proposal is *popular referendum*. This would be an emergency plan option so that any order, law or addition to the law, implemented not only by the will of the ruler, but also had the participation of the population.

8. Gehl, Jan. Cidade para pessoas. 2. ed. São Paulo: Perspectiva, 2003. p. 3–245.
9. Pereira, Neli. "'Brasileiro fica encantado com cidades europeias, mas não reproduz soluções aqui', diz Lerner".

The sequence of events that are happening in São Paulo at this time and in recent months, related to the issue of traffic safety and citizen mobility, show a wide and surprising adherence to almost all the concepts addressed in the objective of this research.

The fact that these events are extremely controversial and have wide media coverage, allows a detailed and documented approach to the various aspects involved in the process and show the need for major changes that we seek to propose in this article.

Implementing guidelines and thinking about the city from the micro-scale to the macro, and taking into account the participation of the population in these actions is fundamental. Prioritizing the pedestrian should be a prevalence in the evaluation of any discussion or mobility project. By not understanding the flow of people, we will not understand how the city works, politically or at the urban level.

REFERENCES

(2;9) Pereira, Neli. "'Brasileiro fica encantado com cidades europeias, mas não reproduz soluções aqui', diz Lerner". Disponível em: <http://www.bbc.com/portuguese/brasil-39238128> Access on MAR 10 of 2017.

CET Companhia de Engenharia de Tráfego. Marginal Segura. Disponível em: <http://www.cetsp.com.br/consultas/seguranca-e-mobilidade/marginal-segura.aspx> 19 JAN 2017.

Denatran. Frota de Veiculos. Available at: <http://www.denatran.gov.br/frota.htm> Access on APR 15 of 2017.

Gehl, Jan. Cidade para pessoas. 2. ed. São Paulo: Perspectiva, 2003. p. 3–245.

Portal R7.COM "Justiça derruba liminar e Doria poderá aumentar velocidade das marginais Tietê e Pinheiros" Disponível em http://noticias.r7.com/sao-paulo/justica-derruba-liminar-e-doria-podera-aumentar-velocidade-das-marginais-tiete-e-pinheiros-24012017) Acess on APR 28 of 2017.

Different perspective in road safety:
Prevention, infrastructure, sharing

Town and Infrastructure Planning for Safety and Urban Quality – Pezzagno & Tira (Eds)
© 2018 Taylor & Francis Group, London, ISBN 978-0-8153-8731-2

Measuring the hindering effect of intersections on walkability. A practical application in the city of Alghero

Dario Canu & Tanja Congiu
Università degli Studi di Sassari, Sassari, Italy

Giovanna Fancello
University of Paris-Dauphine, Paris, France

ABSTRACT: In this paper a ranking method of urban intersections based on their hindering or facilitating effect on walking is proposed and tested on a sample of street crossings in the city of Alghero. The study investigates the effects of geometric and operational characteristics of urban intersections on pedestrian accessibility. Tthrough a detailed audit conducted on each node we collected a set of micro level data on the impact of intersections on wakability expressed by pedestrian: the comparison between objective and subjective information allowed to identify the most important influencing factors and their relative weight in the measure of walkability. Then we proceeded to classify intersections: the variety and diversity of attributes suggested the employment of a multicriteria ranking evaluative method: more precisely, we adapted the ELECTRE TRI rating method (Roy and Bouyssou, 1993) with the purpose of rating urban opportunities according to their influence on walking.

1 AIM OF THE STUDY

Scholars recognise walkability as an important factor for both the quality of urban space and people's quality of life (Talen, 2002; Frank et al. 2010; Speck, 2012). It is a spatial requisite of the built environment which greatly contributes to its liveability and enables people to more effectively and fully use, and benefit from, urban opportunities. The possibility for the people with different motility (Kauffmann et al. 2004) (ages, gender, residential location, socio-economic status, personal abilities,...) to reach valuable destinations and places "on their own" and by foot is an important capability with respect to their "right to the city".

Improving walkability entails, among other things, improving the spatial qualities of those living environments perceived as physically uncomfortable and emotionally unpleasant, especially for pedestrians. One set of improvements for this purpose regards spatial connectivity: the creation of continuous, "fluid" routes that give access to different destinations and are able to accommodate modes of mobility alternative to cars. Connected, well-maintained pedestrian networks allow people to safely and conveniently perform local activities (related to both urban and agricultural production), reach basic facilities (commercial, offices, schools, businesses, leisure, etc.) and collective spaces (green open spaces, public amenities), thus providing greater opportunities for social engagement and economic transactions.

Among the spatial factors affecting people's choice to walk, in this work we focus on the role of road intersections and their hindering effect on pedestrian accessibility. Intersections are topological interruptions of a pathway which, depending on the geometric and operational characteristics, tangibly hamper the trip experience towards the destination, causing delays, exposing pedestrians to risks of traffic accidents, and generally interrupting the continuity of the walk. For these reasons the represet key impedance factors of walkability.

We propose a method for evaluating the impact intersections exercise on the pedestrian experience. The primary purpose of the method is to assist planners and policy makers by assessing the condition and the effect of intersections on walkability, and by suggesting the priority of interventions and measures of street improvements.

The paper is structured as follows: part 2 presents a review of the state of the art in methods and tools for evaluating intersections effects on walking; part 3 presents the evaluation procedure we used to identify intersection attributes most relevant for the walking and to classifiy nodes with respect to pedestrian accessibility. An application of the method to a sample of road intersections in the city of Alghero is described; part 4 illustrates the procedure for prioritising actions and interventions.

2 METHODS AND TOOLS FOR EVALUATING INTERSECTIONS EFFECTS ON WALKING

In the most widely used measurement methods of urban walkability road crossings enter in the form of spatial connectivity indicators, being node type, presence and multitude, important factors that make urban space more or less conducive to walking (Cervero, Kockelman 1997; Talen, Koschinsky, 2013). According to these measures pedestrian capability to reach urban destinations is associated with high levels of street connectivity generally expressed as number of crossings per square meter or number of legs joining in each node. These measures are based on large spatial units (km^2, block, neighborhood) and neglet analysis of built environment at a finer-grain scale (Berrigan, *et al.* 2010; Stangl, Guinn, 2011). As a result, the incorporation of spatial connectivity indicators into the composite indicators of walkability can lead to unreliable aggregate results, given that they combine detailed data on street segments, with rougher and less detailed description of intersections.

In our proposal, we hold that a more detailed micro-analysis of intersections is essential for the overall proper assessment of urban walkability, since often even some micro-spatial, functional and operational characteristics of crossings may be truly hindering, or conducive, of the practicability, comfort and pleasantness of a walking route to pedestrians.

The so called pedestrian level-of-service measures and assessments—PLOS (Fruin, 1971; TRB 2000) is a field of studies which deals more deeply with the factors of the built environment responsible of the hindering effect of road intersections on the pedestrian accessibility. Researchers broadly employ this method to enrich the set of factors affecting pedestrian behaviour, both along sidewalks and at crossings. The concerns for traffic safety, the comfort requirements and the pedestrian delay are among the key performance indicators of walking impedance of crossings (Landis *et al.* 2005). More recent studies combine the physical and operational variables with the pedestrian's perception of safety and comfort, measured by collecting data on pedestrians through direct observation or by means of surveys. Sisiopiku and Akin (2003) studied the pedestrian perceptions of various signalised and un-signalised intersections and suggested that the distance of the crosswalk to the desired destinations is the most influential factor in making a decision to cross. Basile *et al.* (2010) used the AHP method to assess the safety level of pedestrians at regulated and not regulated crossings: a composite indicator for crossing safety, together with some other specific indicators included in the assessment allowed to rank the crossings on the basis of their spatial characteristics and to suggest the priority of interventions by highlighting the issues that need to be addressed. The absence of pedestrian refuge islands, improper traffic light timing, on-street car parking obstructing the visibility, and accessibility problems due to obstacles along the pedestrian crossing, emerged as the most influential factors for pedestrian safety.

The majority of studies resort to field-calibrated statistical models based on multiple linear, logit, or probit models (Petritsch *et al.* 2005; Muraleetharan *et al.* 2005, Bian, *et al.* 2013; Hubbard, 2009). According to this set of studies, the barrier effect caused by both motorised and non-motorised traffic and the crossing distance represent the two main obstacles perceived by pedestrians at intersections. A careful spatial configuration of the crossing areas, together with their equipment, are thus essential in order to encourage walking.

3 METHODOLOGICAL APPROACH

The study consists of two phases:

1. *field survey* aimed at collecting objective measures of spatial and operational characteristics of crossings and at gathering subjective judgments expressed by pedestrians. Objective and subjective measures were then compared for the purpose of identifying the relative importance of factors for the perceived level of walkability;
2. *classification of intersections* with the ELECTRE TRI rating method (Yu, 1992; Roy and Bouyssou, 1993) based on their conduciveness to walk, with the aim to provide planners and policy makers a decision aiding tool to prioritise improvement actions.

We have tested the procedure in the city of Alghero, a town of about 40,000 inhabitants in Sardinia, Italy. The city has a compact urban fabric along the coastline, with gradually lower density urban expansions in the surrounding rural area. Urban facilities and services are located mainly in the central area closed to the old town, while residential neighbourhoods extend in the surroundings. According to the limited extension of the town the distances between origin and destinations are moderately short, being a spatial condition which facilitates and encourages walking.

3.1 *Field survey*

We conducted a contingent field survey on 180 crossings belonging to 45 intersections. Each intersection was subdivided into individual lane crossings in order to consider all the routing alternatives a pedestrian can walk (Fig. 1).

The intersections were selected evenly distributed over the town to have a heterogeneous, more representative, sample of different conditions for the pedestrians. The importance of intersections, based on their localisation with respect to urban facilities, and on the spatial and operational characteristics of approaching roads (functional classification, number of lanes, crossing distances, etc.) was also considered.

With the support of a group of postgraduate students, we conducted a field survey of all the crossings composing each intersection. Two sets of measures, defined according to the literature, were collected for each crossing. Table 1 summarised the 24 audited attributes concerning the spatial configuration of crosswalks (geometry, dimensions), the obstacles to visibility and the practices of use (count of motorised and non-motorised flows, ...). Table 2 offers a synthetic subjective judgment of the "cross-ability" (4-level Likert scale) based on the safety and comfort features perceived by pedestrians.

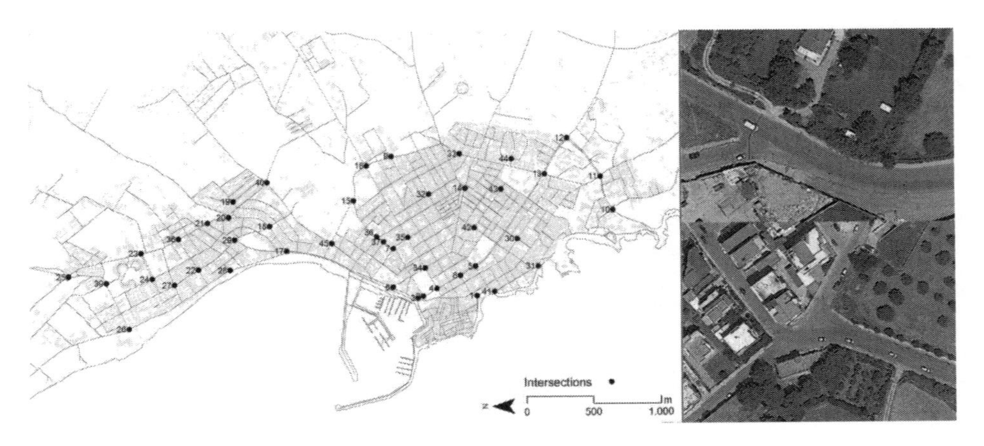

Figure 1. The 45 surveyed intersections in the city of Alghero (left), and detail (right) of the subdivision of each intersection into crossing lanes (for intersection 15 and 40).

Table 1. Attributes of crossings.

	Attributes (Variables)	Scale levels
X1	Carriageway width (C_w)	(continuous)
X2	Sidewalk width (left & right)	(continuous)
X3	Number of vehicular lanes	(continuous)
X4	One-way traffic	1 Yes, 0 No
X5	Bicycle lanes	1 Yes, 0 No
X6	Traffic light	1 Yes, 0 No
X7	Couple of curb cut	1 Yes, 0 No
X8	Elevated sidewalk	1 Yes, 0 No
X9	Crossing island (median)	1 Yes, 0 No
X10	Physical elements directing pedestrian movements (left & right)	1 Yes, 0 No
X11	Sidewalk extensions	2 Yes, both side, 1 Yes, one side, 0 No
X12	Crosswalks condition	0 Absent, 1 Yes, faded, 2 Yes, well-defined
		3 Wide (4–6) or more people await and comfortable passage)
X13	Space at corner (left & right)	2 Medium (2–3 people await and enough passage space)
		1 Limited (1 person await and no passage space)
X14	Position of zebra crossing	2 In-line (less that 10 steps from the street corner), 1 Lateral (more that 10 steps from the street corner), 0 Absent
X20	Bicycle flow rate	(continuous) [frequency in 10 min]
X21	Car flow rate	(continuous) [frequency in 10 min]
X22	Bus and truck flow rate	(continuous) [frequency in 10 min]
X23	Motorcycle and scooter flow rate	(continuous) [frequency in 10 min]
X24	Pedestrian flow rate	(continuous) [frequency in 10 min]
X25	Mean vehicle speed	(continuous) [mean value in km/h]
X26	Waiting time at intersection	(continuous) [mean value in seconds]
X30	Presence of obstacles on the curb	1 Yes (presence of obstacles in street and/or on curb), No (absence of obstacles)
X31	Presence of "inner" obstacles	1 Yes (presence of obstacles further back than 2 m from the curb), No (absence of obstacles)
X32	"Summary level" of obstacles	Presences + Height + Transparency + Permanence

Table 2. Synthetic judgments on pedestrian friendliness of road crossings.

	Synthetic judgment	Description of crossings
4	Very good	Comfortable, with good level of safety; road enough easy to cross (space and time); waiting area wide enough; traffic rather safe
3	Good	medium level of comfort and safety; crossing the road not difficult (space and time); waiting area enough comfortable; traffic averagely safe paying proper attention
2	Sufficient	minimum standards of comfort and safety; road quite difficult to cross (space and time); waiting area limited, traffic dangerous requiring proper attention
1	Insufficient	uncomfortable and dangerous; inconvenient to cross (space and time); waiting area inadequate; traffic very dangerous impossible to pass.

Table 3. Comparison of residuals between predicted and empirically observed values.

	Residuals					Classification		
	Min	1st Qu.	Median	3rd Qu.	Max	Correct	1 cl. Off	2 cl. Off
Mod. A	−1.50	−0.36	−0.05	0.37	1.39	63.89%	36.11%	0.00%
Mod. B	−1.76	−0.39	−0.05	0.37	1.38	61.67%	37.78%	0.56%
Mod. C	−1.54	−0.38	−0.01	0.47	1.66	56.11%	42.78%	1.11%
Mod. D	−1.86	−0.51	−0.11	0.49	2.49	41.67%	57.22%	1.11%

Table 4. Multivariate linear regression—Model B.

	Model B (R-Squared = 0.66)			
	Est.	St. err.	*p*-val.	
Incpt.	0.4754	0.0404	$<2 \times 10^{16}$	***
X1	−0.3533	0.1393	0.0121	*
X6	0.2083	0.0353	2.1×10^{8}	***
X7	0.0800	0.0266	0.0030	**
X9	0.1189	0.0371	0.0016	**
X11	0.2055	0.0698	0.0037	**
X12	0.1340	0.0405	0.0011	**
X14	0.1471	0.0400	0.0003	***
X20	−0.1739	0.0675	0.0108	*
X21	−0.1394	0.0515	0.0075	**
X24	0.4382	0.0782	8.3×10^{8}	***

Significance codes: 0 '***' 0.001 '**' 0.01 '*' 0.05 '.' 0.1 ' ' 1.

Both sets of measures were collected twice, once in the morning and once in the evening, to capture possible variability in flows, traffic and environmental conditions at the intersections.

3.2 *Factors relevant for pedestrian accessibility*

After processing data we implemented several multiple linear regressions (Table 3) with the purpose of exploring correlations between the perceived evaluations of pedestrian quality of crossing (dependent variable) and the attributes of crossings (independent variables).

The comparison of statistical models showed their relative performance with the Model B resulting the most representative (R-squared = 0.66, based on 10 variables and 99.5% of probability to classify crossing at most one class off).[1]

The model B (Table 4) is based on the variables with p-value <0.01 in the model A: (X1) carriageway width; (X6) traffic light; (X7) couple of curb cut, (X9) crossing island; (X11) sidewalk extensions; (X12) crosswalks condition; (X14) position of crosswalks; (X20) bicycles flow rate, (X21) car flow rate and (X24) pedestrian flow rate. The sign of coefficients of each variable confirms the intuitive idea that attributes X1, X20 and X21 have a negative influence on the safety and comfort of crossing, while the other statistically significant variables influence it positively, considering that they consist of pedestrian-friendly equipment such as medians, sidewalk adjustments, control systems, etc.

Given its relative greater simplicity, we use the factors from model B for the next stage in our evaluation procedure.

1. For more detailed discussion of this regression analysis and the comparisons among the three models see Blecic et al. (2016).

3.3 *Classification of intersections*

The second step of the procedure, was to classify intersections according to their conducive or hindering effect on pedestrian mobility, with the final objective being to identify critical factors of intersection design and operation which limit walking, and thus setting the stage for the subsequent prioritisation of intersections for street improvements. For this purpose we employed ELECTRE TRI outranking rating/classification procedure which is widely used in urban and transportation planning and decision aiding (Fancello *et al.* 2014; Sousa *et al.* 2017).

Methodologically, among the multiple criteria evaluation rating methods, ELECTRE TRI is endowed with properties useful for the purpose of the present study: (i) it allows for the sorting of street intersections by priority based to their performance on the criteria; (ii) the criteria aggregation is flexible and permits the elicitation of the relative weights, clusters of coalitions (majority rule) and possible veto powers; (iii) it allows a prudential non-compensatory aggregation with limited loss of information during the consecutive stages of evaluation; (iv) it is a procedure that resembles individual models of judgment; and (v) the outcomes are reasonably simple to interpret and communicate.

Operationally for the evaluation of the 180 road crossings in our sample, we defined three classes for the levels of walkability: C_1—"Conducive to walk", C_2—"Supports walking", C_3—"Obstacle to walk" (Table 5).

According to the results of the previous section, each crossing within an intersection was evaluated on six criteria (h_1, h_2, …, h_6). Table 6 provides details on the six criteria, their relative weights and the conditions of inclusion in the three classes.[2]

The ELECTRE TRI model included 5 crossings in the class C_1, 104 in C_2 and 71 in class C_3.

We then test the concordance of the rating model with the perceived judgments of the crossing quality expressed by pedestrians (Table 7).

Table 5. Performance classes of crossing pedestrian quality.

C_1	C_2	C_3
"Very good to walk"	"Supports walking"	"Obstacle to walk"
Very safe and comfortable crossing, Narrow street width or traffic flow regulated by crossing facilities.	Crossing safe and comfortable enough. Medium street width and some crossing facilities.	Crossing not safe nor comfortable wide street width and crossing facilities lacking.

Table 6. ELECTRE TRI model.

W	Criteria	C_3 "Obstacle to Walk"	C_2 "Supports Walk"	C_1 "Very Good to Walk"
44.33%	X1 + X6 + X9 + X11	3NNY, 3NYN, 2NNN, 3NNN	3YNY, 3YSN, 2YNN, 2NNY, 2NYN, 1NNN, 3YNN	1YNY, 1YSN, 2YNY, 2YYN, 1NNY, 1YNN
14.06%	X12 + X14	00	12, 22, 21	11
4.00%	X7	No	Yes	Yes
8.70%	X20	>18	18 ≤ x < 10.5	≤10.5
6.97%	X21	>549	549 ≤ x < 285	≤285
21.93%	X24	>27	27 ≤ x < 60.75	≤60.75

2. In Table 6 we defined two groups of superindicator (Brüggemann, and Patil, 2011) represented by a sequence of four values, one for each variable. For example, the alternative 3YNY indicates an intersection with carriageway width ≥9 m (3), with traffic light (Y), no crossing island (N), with sidewalk extension (Y).

Table 7. Confusion matrix of comparison between predicted and observed rating of crossings.

		Predicted			
		C_3	C_2	C_1	Sensitivity
Observed	C_3	28	5	0	85%
	C_2	40	74	0	65%
	C_1	3	25	5	15%
	Precision	39%	71%	100%	59%

The results show that the predicted crossings in C_3 and C_2 are most of the time classified in concordance with the pedestrian's judgment classes, while alternatives in C_1 frequently "support walk", underestimating the effective quality of the crossing to promote walking. With regards to the capacity of the model to avoid wrong classifications, we obtain good results for the classes C_1 and C_2, but lower precision for the class C_3 (39%). This means that the model adopts a precautionary principle by classifying the alternative in the lowest class.

More generally the comparison reveals a good correspondence between the rating model and the preferences revealed by pedestrian.

4 PRIORITIZING INTERSECTIONS

The final objective of the study is to outline an operational method to orient policy makers in recognition of the needs, and in the prioritisation of interventions to make urban areas more pedestrian-friendly.

With respect to this objective, the above-described classification alone is not sufficient to establish an order of priority among possible interventions of improvement. We thus employ a second classification procedure that considers the urban quality of the context in which the intersections are located, and its potentials as a walkable space (Forsyth, 2015). To capture this urban quality, we resort to the "capability-wise walkability score" (CAWS) evaluation method (Blecic *et al.* 2015) which is an analytical measure of the mutual relations between the urban organisation and the people's attitudes in space. From an operational point of view, CAWS was designed to bring together both the opportunity sets distributed in space, as well as the characteristics of urban environment which affect walking. According to these assumptions, a walkability score is assigned to each point in space with a resulting capability-wise walkability map.

Figure 2 shows the CAWS walkability map obtained for the town of Alghero with intersections and the limit of the rural-urban fringe. Three classes of urban walkability are defined based on walkability scores: CAWS1 for score <20; $20 < \text{CAWS2} < 40$ and CAWS3 for values >40.

Therefore, we considered the following two criteria to establish an order of priority among crossings:

- spatial and operational performance class, based on our rating classification C1, C2, C3;
- walkability of the site: each crossing is assigned to one of the three classes (CAWS1, CAWS2 or CAWS3) based on the CAW score of the location.

Considering the preferences of a hypothetical decision maker, and assigning a numeric value to each crossing according to the classes and equivalent to their scale order (Borda rule with CAWS1 = 1, CAWS2 = 2 or CAWS3 = 3) we finally arrive at a priority rating.

For example, if we assume that for a given decision maker the intersection performance has the priority (weight 60%) over site's walkability (40%), the resulting rating (with the percentage of crossings included in each class) would be that shown in Table 8.

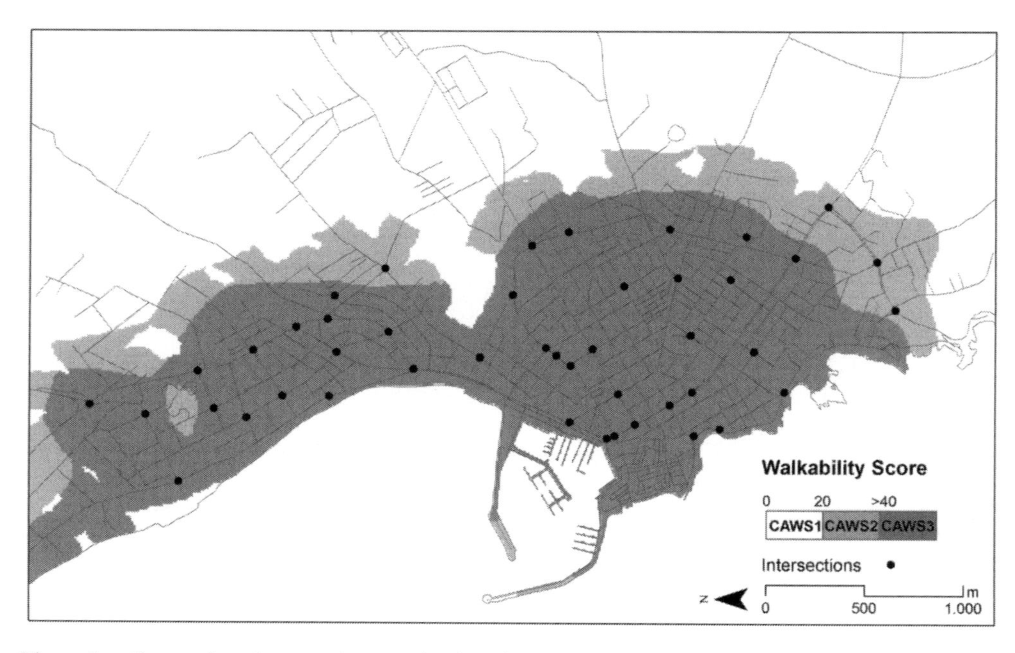

Figure 2. Comparison between intersection location and CAWS walkability score.

Table 8. Crossings classification according to criteria of priority (total crossings 180).

Performance class	Site walkability class (Order of priority)		
	CAWS1	CAWS2	CAWS3
C_1 conducive	0.0% (9)	8.9% (8)	29.6% (6)
C_2 support	0.0% (7)	1.1% (5)	59.4% (3)
C_3 obstacle	0.0% (4)	0.0% (2)	3.9% (1)

5 CONCLUSIONS

Most of the research on urban walkability reported in literature is generally focused on spatial features on uninterrupted sidewalks. Much less attention has been devoted to intersections, although they are often the predominant factor of the quality of walk and of pedestrians' perception of safety, comfort and pleasantness of a walking route. In this paper we presented an application of an evaluation model, based on ELECTRE TRI, for rating of urban intersections according to their effect in limiting or facilitating walking, and a subsequent procedure for prioritising interventions of improvement. The approach we propose provides meaningful information on the spatial conditions of intersections related to features which mainly influence walking. Therefore, we believe it represents a useful tool for planners and decision makers to implement policies and interventions aimed at improving the liveability of the city.

REFERENCES

Basile O., Persia L., Usami D.S., (2010). A methodology to assess pedestrian crossing safety. Eur. Transp. Res. Rev., 2. 129–137.

Berrigan D., Pickle L.Z., Dill J., (2010). Associations between street connectivity and active transportation. International Journal Health Geogr., 9. 20.

Bian Y., Jian L., Zhao L., (2013). Method to determine pedestrians level of service for unsignalized intersections. Appl. Mech. Mater., 253–255. 1936–1943.
Blečić I., Canu D., Cecchini A., Congiu T., Fancello G., (2016). Factors of Perceived Walkability: A Pilot Empirical Study. Lect. Notes Comput. Sci., 9789. 125–137.
Blečić I., Cecchini A., Congiu T., Fancello G., Trunfio G.A., (2015). Evaluating walkability: A capability-wise planning and design support system. Int. J. Geogr. Inf. Sci. 29. 1350–1374.
Brüggemann R., Patil G.P., (2011). Ranking and Prioritization for Multi-Indicator Systems; Springer: New York, NY, USA.
Cervero R., Kockelman K., (1997). Travel demand and the 3Ds: Density, diversity, and design. Transp. Res. D, 2. 199–219.
Fancello G., Carta M., Fadda P.A., (2014). Decision support system based on Electre III for safety analysis in a suburban road network. Transp. Res. Procedia, 3, 175–184.
Forsyth A., (2015). What is a Walkable Place? The Walkability Debate in Urban Design. Urban Design. Int., 20. 274–292.
Frank L.D., Sallis J.F.; Saelens B.E., Leary L., Cain, K.; Conway, T.L., Hess P.M., (2010). The development of a walkability index: Application to the neighborhood quality of life study. Br. J. Sports Med. 44. 924–933.
Fruin J.J., (1971). Pedestrian Planning and Design; Metropolitan Associations of Urban Designers and Environmental Planners: New York, NY, USA.
Hubbard S.M.L., Bullock D.M., Mannering F.L., (2009). Right Turns on Green and Pedestrian Level of Service. Statistical Assessment. J. Transp. Eng. 2009, 135. 153–159.
Kaufmann V., Bergman M. M., & Joye D., (2004). Motility: mobility as capital. International Journal of Urban and Regional Research, 28(4). 745–756.
Landis B.W., Petritsch T.A., McLeod P.S., Huang H.F., Guttenplan M., (2005). Video Simulation of Pedestrian Crossings at Signalized Intersections. Transp. Res. Rec., 1920. 49–56.
Muraleetharan T., Adachi T., Hagiwara T., Kagaya S., (2005). Method to determine pedestrian level-of-service for crosswalks at urban intersections. J. East. Asia Soc. Transp. Stud., 6. 127–136.
Petritsch T.A., Landis B.W., Mcleod H.F., Huang H.F., Challa H.S., Guttenplan M., (2005) Level of Service Model for Signalized Intersections for Pedestrians. Transp. Res. Rec., 1939. 55–62.
Roy B., Bouyssou D., (1993). Aide Multicritère à la Décision: Méthodes et Cas. Economica. Paris, France.
Sisiopiku V.P., Akin D., (2003). Pedestrian behaviors at and perceptions towards various pedestrian facilities: An examination based on observation and survey data. Transp. Res. F, 6. 249–274.
Sousa N., Coutinho-Rodrigues J., Natividade-Jesus E., (2017). Sidewalk Infrastructure Assessment Using a Multicriteria Methodology for Maintenance Planning. J. Infrastruct. Syst., 23. 05017002-1-05017002-9.
Speck J., (2012). Walkable City: How Downtown Can Save America, One Step at a Time; Macmillan: New York, NY, USA.
Stangl P., Guinn J.M., (2011). Neighborhood design, connectivity assessment and obstruction. Urban Design Int., 16. 285–296.
Talen E., (2002). Pedestrian Access as a Measure of Urban Quality. Plan. Pract. Res., 17. 257–278.
Talen E., Koschinsky J., (2013). The Walkable Neighborhood: A Literature Review. International Journal of Sustainable Land Use and Urban Planning, 1(1). 42–63.
Transportation Research Board, (2000). Highway Capacity Manual; TRB National Research Council: Washington DC., USA.
Yu W., (1992). ELECTRE TRI: Aspects Méthodologiques et manuel D'utilisation; Document du LAMSADE N.74; Université Paris-Dauphine. Paris, France.

Town and Infrastructure Planning for Safety and Urban Quality – Pezzagno & Tira (Eds)
© 2018 Taylor & Francis Group, London, ISBN 978-0-8153-8731-2

Some infrastructural elements for cyclists' injury prevention

G. Maternini & A. Guga
Università degli Studi di Brescia, Brescia

ABSTRACT: The increase of the number of accidents is partly due to inadequacy of cycling infrastructures. The aim of this study is to deepen by what means a safe and high-quality bike path can be realized. Different discomforts for the users can been noted, especially regarding bike paths that usually are not enough visible or separated from the traffic lane.

A new concrete kerb has been designed to separate the cycle path from the vehicular passage, reducing the risk of injury. Its shape and size have been investigated in order to find the correct solution that provides a better protection: the Italian regulation has no indication about the shape, dimensions and material but it gives just indications about the width. To fill this gap, an innovative "illuminating concrete" will be used with the aim to increase the visibility of the track.

1 INTRODUCTION

In terms of infrastructure, there are some very clear principles underlying the design of measures in the best cycling cities. However, there is no single physical 'model' that is either clearly optimal or directly transferable to each case.

In attempting to promote a cycle-friendly infrastructure, three planning levels may be distinguished: network, connecting (route) and provision. The main requirements of cyclists in relation to network and route are related with attractiveness, safety and comfort. Therefore, the functional requirement for a successful cycling infrastructure is that traveling from point A to point B should be convenient and quick, and the journey should be a safe and pleasant experience.

There are good grounds for taking safety and security as a starting point in bicycle traffic planning, to encourage more people to cycle. First of all, safety and a sense of security are not the same thing. A sense of security is the cyclist's subjective perception of what cycling in traffic feels like, whereas safety is the objective registration of accidents.

Since segregation between motor traffic and cars can make cyclists feel more secure, cities with the highest cycling levels generally afford cycling good physical protection or effective spatial separation from motor traffic, unless traffic speeds and volumes are low. Separating cycles & motor traffic—options can be various (i.e. stepped cycle tracks, vertical features impossible to overrun, painted lines, offside car parking, trees and street furniture), thus in this study we propose an innovative raised kerb to be used for protected bicycle lanes in urban areas.

It is known that the incidence of traffic accidents is an important criterion when assessing the main requirement of safety, in fact concentrations of cycle accidents involving injured victims are always cause for an investigation into the link between accidents and design.

Looking into the number of incidents involving cyclists (Table 1) in the Lombardy region (Italy) since 2010 to 2015, it can be noted how the number of accidents, of injured and of deaths have increased, showing the urgency of increasing cyclist safety.

We extracted data in which the cyclist was alone in the circumstance of the incident in order to understand how the incorrect design of the kerb can affect the dynamics of cyclist accidents.

Table 1. Number of incidents, injured and deaths involving cyclists in Lombardy region from 2010 to 2015.

Lombardy region			
Years	Accidents	Injured	Deaths
2010	4045	4277	44
2011	4428	4673	64
2012	4738	5037	66
2013	4615	4889	49
2014	4547	4821	56
2015	4308	4547	50
% Increasing since 2010	+6,5%	+6,3%	+13,64%

Source: Eupolis Lombardia.

Table 2. Number of incidents, injured and deaths involving cyclists in Lombardy region from 2010 to 2015 caused by "kerb-related" dynamics.

		Fall off the bicycle	Left the bike lane	Hit an object	Total "kerb-related"	Total in Lombardia
2010	Accidents	152	263	101	516	4045
	Injured	156	261	102	519	4277
	Deaths	2	4	1	7	44
2011	Accidents	184	314	135	633	4428
	Injured	185	316	135	636	4673
	Deaths	2	4	0	6	64
2012	Accidents	170	393	113	676	4738
	Injured	172	397	114	683	5037
	Deaths	2	4	0	6	66
2013	Accidents	200	269	109	578	4615
	Injured	198	263	108	569	4889
	Deaths	4	9	3	16	49
2014	Accidents	200	269	109	578	4547
	Injured	198	263	108	569	4821
	Deaths	4	9	3	16	56
2015	Accidents	162	223	83	468	4308
	Injured	159	224	84	467	4547
	Deaths	3	2	0	5	50

Source: Eupolis Lombardia.

The causes reported in traffic police's minutes regarding accidents with only a bicycle involved are:

- fall off the bicycle;
- sudden braking;
- leave bicycle lane;
- pedestrian running over;
- impact with an obstruction;
- impact with parked vehicle.

Among these causes, we investigated "fall off the bicycle", "leave bicycle lane" and "impact with an obstruction", in which a proper kerb design can reduce risk that accident happens. Those factors will be called in this paper "kerb-related" (Table 2).

As can be seen in the Table 3, the number of wounded cyclist and the number of accidents, which can be affected by an inadequate design of the kerb, are respectively above 12% of the

Table 3. Percentage of total number of incidents, injured and deaths involving cyclists in Lombardy region from 2010 to 2015 caused by "kerb-related" dynamics, on the total in Lombardy.

	From 2010 to 2015		
	Accidents	Injuries	Deaths
Total Lombardy region	26681	28244	329
"Kerb-related"	3449	3443	56
% Kerb-related/total	12,9	12,2	17,0

Source: Eupolis Lombardia.

Table 4. Number of incidents, injured and deaths involving cyclists in Italy in 2015 caused by "kerb-related" dynamics.

Cyclists in 2015	In Italy	% kerb-related/total in Lombardia	Kerb-related in Italy
Accidents	17437	12,9%	2249
Deaths	252	17%	43
Injured	16827	12,2%	2053

Source: ACI-ISTAT, Table 2.24.

Table 5. Cost per year caused by kerb-related accidents, deaths and injured in Italy.

Kerb-related in Italy, 2015		Average cost	Total cost per year
Accidents	2249	€ 10.986,00	€ 24.711.611,78
Deaths	43	€ 1.503.990,00	€ 64.430.931,60
Injured	2053	€ 42.219,00	€ 86.671.131,79
			€ 175.813.675,16

Source: Ministry of Transport and Infrastructures, "Studio di valutazione dei Costi Sociali dell'incidentalità stradale", 2013.

total of accidents and injured cyclist in Lombardy from 2010 to 2015. The percentage of the deaths is even higher, precisely 17% of the total number in Lombardy, with the involvement of one cyclist alone in the circumstances of the accident.

Considering the data of Italy in 2015, we can point out the number of kerb-related accidents, deaths and injured, assuming the same percentage of Lombardia.

Knowing the average cost for each accident, death and injured, the total cost caused by kerb-related dynamics per year for the community can be calculated.

These data point out the influence that may have to act on the design of the kerb in reducing the number of cyclist accidents.

2 STATE OF THE ART OF THE ITALIAN REGULATION

In Italy, the standards that provide guidance on raised kerbs are D.P.R. 16/12/1992 n. 495, "Regolamento di esecuzione e di attuazione del nuovo codice della strada", and D.M. n. 557 30/11/1999 "Regolamento per la definizione delle caratteristiche tecniche della piste ciclabili". The D.M. 557/1999, to Chapter I, Item 4 introduces the following types of cycle paths, realised within built-up areas or connecting bordering city centres:

- *Protected bike lane*: in 2nd paragraph, item 6, chapter II, are identified by virtue of the physical separation of their location from those of motor vehicles and pedestrians, through a "appropriate physically insurmountable longitudinal divider".
- *Buffered bike lane*: they can be obtained from the carriageway or from the sidewalk, with the difference that in the first case they must be one-way path in the same direction of the vehicular lane to which they are adjacent and separated by means of longitudinal pavement marking or physical barriers. In the second case, they may be one-way or two-way if their width does not affect the pedestrian movement.

For these two types, bike lane has to be separated from motor traffic by means of different elements.

Regarding the *buffered bike lane*, D.P.R. 495/1992 item 178, paragraph (2) lays down that "reserved lanes, where only certain categories of vehicles are permitted to transit, may be physically delimited by the lane lines as provided for in item 140, paragraph (6) and (7), or by embossed elements. In this case, the raised elements replace the yellow lane lines". These items, according to paragraphs (3) and (4), must be made of plastic or yellow rubber, they must be between 15 and 30 cm in width and 5 cm and 15 cm in height, with a consistency and a profile such that it can be overcome if necessary.

Concerning *protected bike lanes*, D.M. 557/1999 imposes in item 7 paragraph (4) that "the width of the insurmountable divider, separating the cycle lane from the roadway intended for motor vehicles, has not be less than 0.50 m wide". In D.M. 557/1999, and neither in other standards, other characteristics are not laid down, such as height, cross section or surface finish.

Therefore, these are the current regulatory limits that must be respected in the design of the separating element, which is the subject of this study.

3 STATE OF THE ART OF THE REGULATION IN THE SOME EUROPEAN COUNTRIES

In France, in the "*Recommandations pour les aménagements cyclabes*"[1] manual, it is considered that separating kerb may be used for the creation of separated cycling ways where there is not enough room for a margin to be created. Such elements may have angular or semicircular sections and double kerbs can be used with in-between blocks or asphalt pavements. Moreover, it is recommend the height of the kerbs to be used on cycling and road side, 5–7 cm and 10–12 cm respectively.

In England, the "*London Cycle Network*"[2] manual suggest to the use kerbs for the separation of cycle lanes from the roadway with a height ranging from 5 to 10 cm, in order to avoid impacts with bicycle pedals. The manual suggests the realization of separate cycle lanes at an intermediate level between sidewalk and the vehicular carriageway, using for the separation from traffic on vehicular lane, kerbs with a height between 7.5 and 10 cm, although it is at grade with pedestrian way. The aim of creating cycling tracks at a higher level than the vehicular roadway is to give a greater sense of security. Regarding the type of kerb to use, it is suggested those with a rectangular section or with a 45° sloped profile on the side of the bicycle lane in order to avoid the collision with the cyclists pedals and to facilitate the drainage of water.

In Denmark, the Cycling Embassy of Denmark in "*Collection of Cycle Concepts 2012*" point out cycle track design and maintenance should be of such a standard that cyclists never choose to ride on the carriageway instead. The manual provides the following diagram indicating the degree of separation appropriate to the speed and volume of motorized traffic.[3]

---

1. Certu, Reccommandations pour les aménagements cyclables, France, 2008.
2. Director of Environmental Services, London Cycle Network Design Manual, London, 1988, p. 44.
3. Andersen T., et al., Collection of Cycle Concepts, Cycling Embassy of Denmark, 2012, p. 52–53.

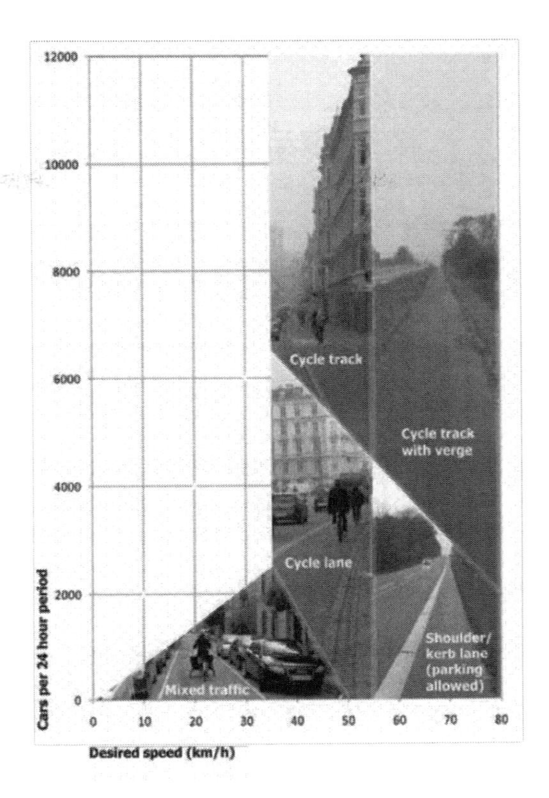

Figure 1. Cycling solutions in relation to motor traffic volume and speed.
(Source: Collection of Cycle Concepts, Cycling Embassy of Denmark, 2012).

Separation to a bicycle lane is suggested at greater than 4,000 vehicles per day and posted speeds of more than 35 km/hr. Cycle tracks are suggested at speeds greater than 40 km/hr and high vehicle volumes, and on lower-volume routes with speeds greater than 55 km/hr.

Regarding the kerb, the manual suggest the height between the separated bike lane and the roadway to be in the range of 7 to 12 cm. The advantages of this choice include a perfect drainage—operation, drivers of motorized vehicles will be prevented from parking on the bicycle path and will drive in the built-up areas at a lower speed. In addition, there is a range ranging from 5 to 9 cm as separation between the sidewalk and the bicycle path, in order to disincentive the use of the pedestrian path by the cyclist and to make the pedestrian conscious when getting out of the sidewalk.

3.1 *Dutch context*

In the Netherlands, the CROW[4] manuals that represent a benchmark for bicycle mobility are "ASCC Reccomendations for traffic provisions in built-up areas" and "Design manual for bicycle traffic".

The figure below depicts facility selection guidelines from the "*Design Manual for bicycle traffic*", based on motor vehicle volumes and speeds. Separation is recommended via bicycle lanes beyond volumes of 5,000, where speeds are greater than 40 km/hr. Bicycle lanes are also recommended on low-volume cycling routes with speeds less than 50 km/hr. For speeds greater than 50 km/hr, separation using cycle tracks is recommended.[5]

4. CROW: The national Information and Technology Centre for Transport and Infrastructure.
5. CROW, Design manual for bicycle traffic, Rik de Groot, Herwijnen, 2007, p. 119, 123.

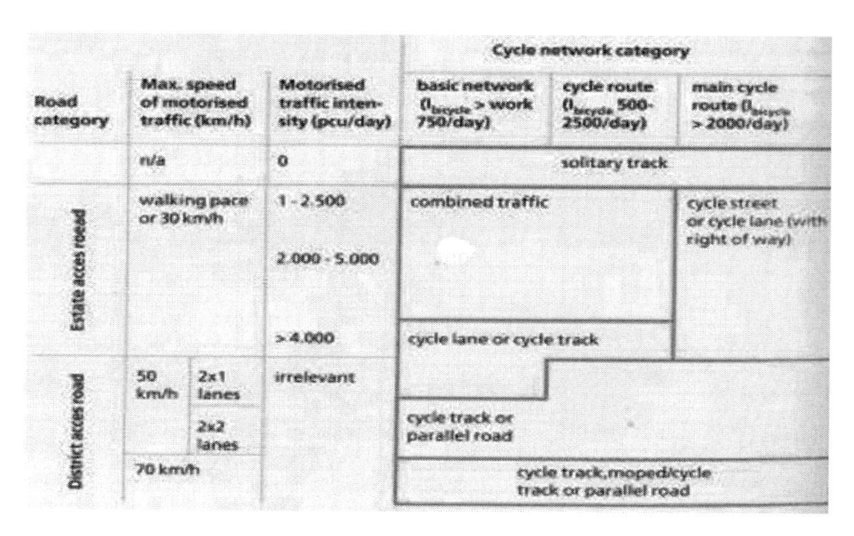

Figure 2. Cycling solutions in relation to motor traffic volume and speed in Netherlands. (Source: Design manual for bicycle traffic, CROW, 2007).

Table 6. Physical partition between cycle and vehicular traffic.

Cycle track and/or carriageway with paving stone or clinker type brick paving	
	Carriageway — Cycle track
Two concrete kerb blocks with paving-stones or bricks laid between them	h_1 h_2
Two concrete kerb stones placed back-to-back	h_1 h_2
Convex-profile kerb stone	h_1 h_2
Concave kerbing profile	h_1 h_2
Extended asphalt pavement	
Asphalt kerb	b h_1
Wide concrete kerb blocks or plates	b h_1
Dimensioning	$b = 0.40$ to 0.50 m (minimum 0,35 m) $h_1 \leq 0{,}10$–$0{,}12$ m $h_2 = 0{,}05$ $(0{,}07)$ m; if it is $\geq 0{,}07$ m, choose a profile that will not be hit by cycle pedals

Source: "ASCC recommendations for traffic provisions in built-up areas", CROW.

The CROW Manual suggests different solutions depending on whether the cycle track pavement is made up with. It shows the profiles and heights on the side of carriageway and on bicycle way side, but not the width. The manual specifies that separating kerb should have a higher difference in altitude than the carriageway and smaller on the side of the cycle lane. The aim is to prevent motorized vehicles from driving along this kerb and to make cyclists circulation more comfortable, in fact in the presence of a too high kerb cyclists tend to move far from the margin. In addition, the shape of the separation kerbs must be capable to avoid contact between the pedal of the velocipede and the kerb itself.

In Table 4, the first four typologies are used when the pavement of the bicycle lane is composed of self-locking screeds, different then that of the carriageway. Instead, the other typologies are used when the road surface of the cycle way is the same as the vehicular lane and therefore the bicycle flooring is in bituminous conglomerate.

4 NEW KERB PROPOSALS

Concrete separating elements proposed in this paper, whose application is limited to built-up areas, on one hand comply the applicable regulatory requirements, while on the other they seek to meet the needs of cyclists comfort and safety. These kerbs with lowered profiles on the side of the bicycle track prevent any accidental bump with the bicycle pedal.

In the design phase of the new infrastructural element, based on the fact that around 90% of cycling incidents happen in the built-up areas, attention has been focused on this context. In particular, it has been decided to analyse the infrastructure type that offers greater protection to the cyclist, i.e. the separeted cycle lane with a longitudinal raised kerb physically insourmontable with a width of not less than 50 cm, as the D.M. 557/1999 requires.

The kerb is made by concrete and it is designed with a length of one meter, as it is optimal for handling and transporting, and laying the elements.

4.1 *Geometry*

Carriageway-side kerb' profile was studied, taking into account the following features:

- discourage the parking on the bicycle way by motorized vehicles;
- redirection of wandering motorized vehicles
- possibility of the pedestrians to go over;
- encourage a proper runoff of rain water.

This profile has to redirect motorized vehicles at low impact and low speed, as required by D.M. 557/1999 and by the Ministry of Infrastructure and Transport, which stated that the kerb' height must be such as to prevent overtaking of motor vehicles and suggests a minimum size of at least 15 cm.[6]

This height will not increase because we want to make the structure permeable by the pedestrian component.

Bicycle path side profile has to:

- avoid contact with the pedals of bicycle;
- not avoid lateral displacement of the tires of the bicycle.

With regard to the first aim, in other European countries, there is a range of heights ranging from 5 to 7.5 cm, only in Denmark it rises to 12 cm.

In addition, to avoid that the contact between the kerb and the tires of the velocipede, causing a cyclist's fall by overturning, it will be used a curvilinear profile that would start at level of the pavement of the bicycle track, up to the top of the kerb.

6. Circular 6573, 29 October 2013, Ministry of Infrastructure, Italy.

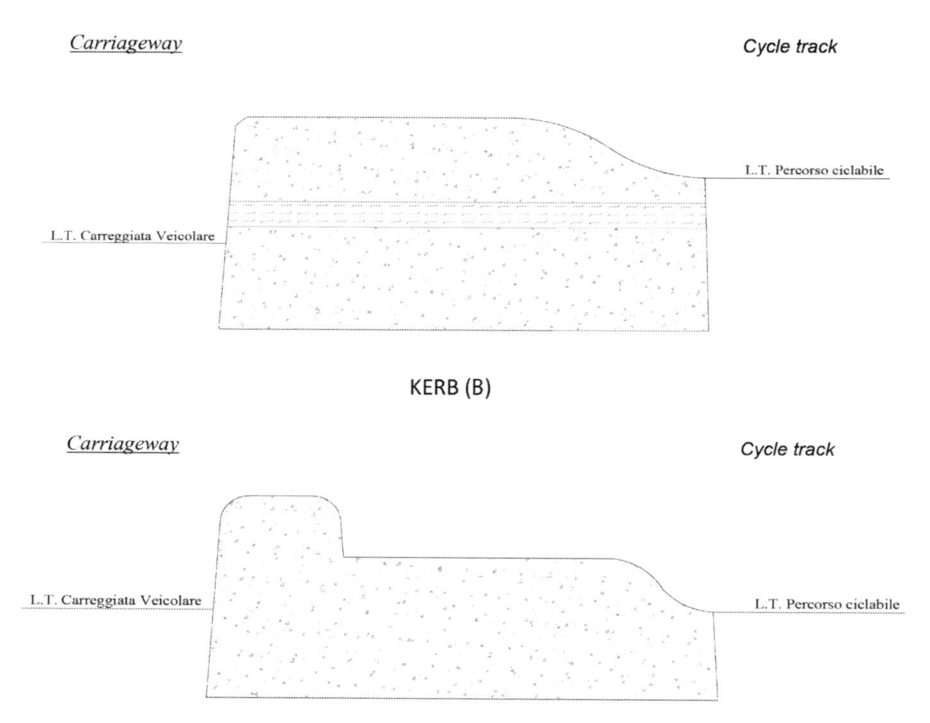

Figure 3. Kerb solutions proposed.

Two variants have been designed; the first one provides for a bicycle path at an 8 cm height higher than that of the carriageway, while the second is proposed for a non-draining cycling track at altitude with the carriageway.

4.2 *Kerb type (A)*

This first variant (Fig. 3) has a nearly vertical carriage-side profile, differently the bicycle-track-side profile has a concave curve that starts at level of the drainage floor, followed by an opposite curve connected with the flat top. The two curves have a total width of 20 cm. This allows to have a flat top part of 27.9 cm, similar to a step size, so pedestrian' overhanging can be easy.

Furthermore, the shape of this profile, while not obstructing the lateral movement of the tires of the velocipede, entails the cyclist a vertical acceleration in an attempt to exceed the difference in height, which could cause him to notice the erroneous trajectory, without a loss of stability. If the urban water collection system is outside the bicycle track and before the roadway, the kerb cooperates with the drainage floor (Fig. 3). The canal in the kerb, measuring 9 cm in width for 3 cm in height, is placed at a height of 2 cm from the road surface with the aim of limiting the entrance of debris inside the canal.

4.3 *Kerb type (B)*

In order to enable the construction of a low-cost infrastructure, a variant has been conceived in the absence of drainage flooring and cycle paths at altitude with the carriageway. This kerb has a 13 cm carved side elevation.

To allow outflow of rain waters, kerbs must be at a distance of above 3 cm one from the other (Fig. 6).

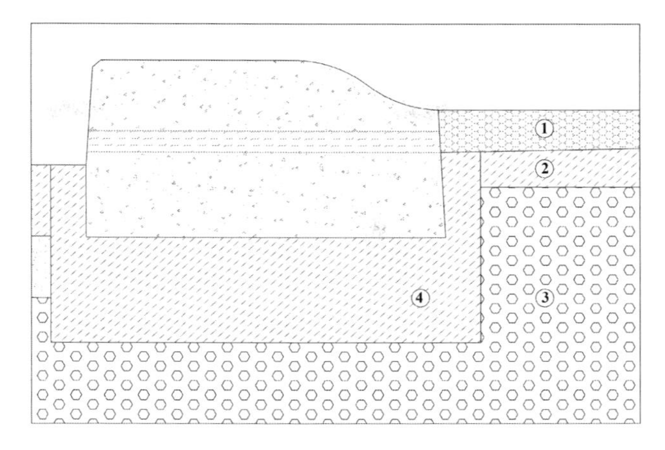

Figure 4. Kerb A, section detail. 1) Drainage flooring; 2) Waterproofing substrate with inclination necessary to drain the rain-water towards the inner channel of the kerb; 3) Kerb foundation.

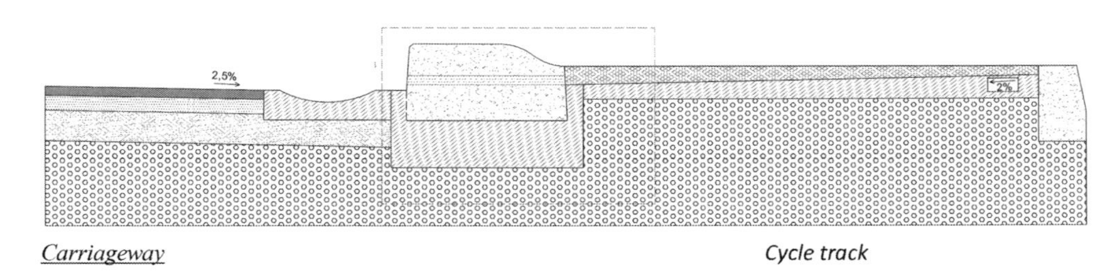

Figure 5. Kerb A in place.

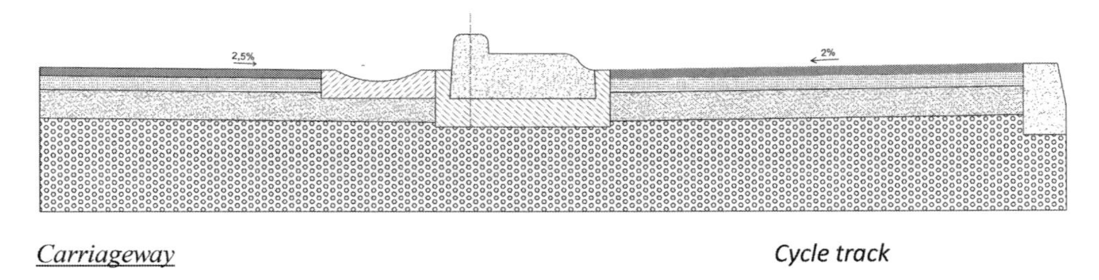

Figure 6. Kerb B in place.

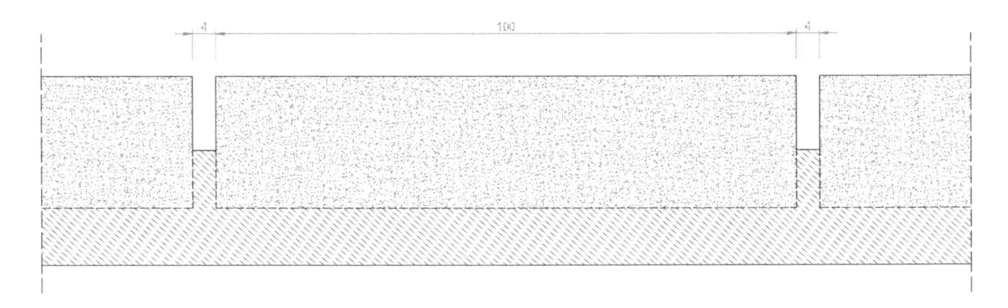

Figure 7. Kerb B, longitudinal section.

Figure 8. Kerb A prototype.

Figure 9. Kerb A prototype.

Table 7. Technical characteristics of the kerb type.

Kerb Type A	
Length	100 cm
Width	50 cm
Cross section area	1180 cm^2
Weight	230 Kg
Material	C 25/30
Surface finishes	Upper part with improved friction

Figure 10. Kerb B prototype.

Table 8. Technical characteristics of the kerb type B.

Kerb Type B	
Length	100 cm
Width	50 cm
Cross section area	849 cm^2
Weight	187 Kg
Material	C 25/30
Surface finishes	Upper part with improved friction

For both kerbs, the central flat part is designed to maximize the safety level for pedestrians and cyclists, in fact it can be seen how the non-slip surface made in negative by means of an embossed sheet placed at the bottom of the mould.

4.4 *Brightness*

Another challenge is that of lighting the cycle ways sufficiently for the purposes of navigation and personal security—so that they are not merely part-time facilities.

For an increased visibility of the kerb during daytime and night-time hours and in low visibility conditions (adverse weather conditions, foggy areas, etc.), it is recommended to use a white cement with a surface finish characterized by refractive elements or varnishing with white paint, added with glass microspheres, generally used for horizontal marking. Glass microspheres (premixed and/or spun-off) are transparent glass spherical particles that, by exploiting the principle of retro reflection of incident rays of a vehicle's headlamps towards the driver, allow and/or increase visual perception. Such materials may be premixed during the production of the paints or added to liquid products just before their application. Alternatively, microspheres can be sprayed onto any liquid-applied product.

4.5 *Prototype*

Prototypes have been realized by Calubini S.r.l., in Montichiari (BS), Italy.

5 CONCLUSIONS

Both curbs allow to reach high safety standards and higher comfort level for cycle paths. More investigation has to be done about how to increase visibility and about the material. Furthermore, the experimentation in a real case could be important in order to verify their potentiality and merchantability, in terms of implementation, handling and installation, to identify any improvement.

REFERENCES

Andersen T., et al., "Collection of Cycle Concepts", Cycling Embassy of Denmark, Denmark, 2012.
Certu, "Recommandations pour les aménagements cyclables", France, 2008.
CROW, "ASVV Recommendations for traffic provisions in built-up areas", The Netherlands, Oland, 1988.
CROW, "Design manual for bicycle traffic", The Netherlands, Oland, 2007.
Director of Environmental Services, "London Cycle Network Design Manual", London, 1998.
D.M. n. 557 30/11/1999, "Regolamento per la definizione delle caratteristiche tecniche della piste ciclabili".
D.P.R. n. 495 16/12/1992, "Regolamento di esecuzione e di attuazione del nuovo codice della strada".
Maternini G., et al., "Mobilità ciclistica. Metodi, politiche e tecniche.", in Tecniche per la sicurezza in ambito urbano, Volume XVI, EGAF Edizioni S.r.l., 2012.
Maternini G., et al., "Mobilità ciclistica", EGAF Edizioni S.r.l., 2016.
Ministry of Transport, Circular 18982, 27 February 2008.
Ministry of Transport, Circular 25807, 19 March 2008.
Ministry of Infrastructure and Transport, Circular 6573, 29 October 2013.

Author index